Sustainable Microbial Technologies for Valorization of Agro-Industrial Wastes

This book provides an overview of the different aspects of microbial bioconversion methodologies for valorization of underutilized wastes of varied nature. It covers microbiological/biotechnological aspects, environmental concerns, bioprocess development, scale-up aspects, challenges, and opportunities in microbial valorization at commercial scale. It explains sustainable microbiological processes for bioconversion and valorization of the wastes for production of various products of commercial interests, including biofuels, bioenergy, and other platform chemicals.

The book

- presents potential biotechnological topics and strategies for the valuation of agricultural waste materials;
- provides technical concepts on the production of various commercially significant bioproducts;
- introduces various microbial bioprocesses to sustainably valorize various potential wastes as renewable feedstocks for production of biofuels and biochemicals;
- explores the relevant scale-up opportunities, commercialization aspects, and critical technological advances; and
- explains concepts and recent trends in life cycle analyses in waste valorization.

It is aimed at researchers and graduate students in bioengineering, biochemical engineering, microbial technology/microbiology, environmental engineering, and biotechnology.

Novel Biotechnological Applications for Waste to Value Conversion

Series Description:

Solid waste and its sustainable management is considered as one of the major global issue due to industrialization and economic growth. Effective solid waste management (SWM) is a major challenge in the areas with high population density, and despite significant development in social, economic and environmental areas, SWM systems is still increasing the environmental pollution day by day. Thus, there is an urgent need to attend to this issue for green & sustainable environment. Therefore, the proposed book series is a sustainable attempt to cover waste management and their conversion into value added products.

Series Editors:
Neha Srivastava
IIT BHU Varanasi, Uttar Pradesh, India
Manish Srivastava
IIT BHU Varanasi, Uttar Pradesh, India

Utilization of Waste Biomass in Energy, Environment and Catalysis
Dan Bahadur Pal and Pardeep Singh

Nanobiotechnology for Safe Bioactive Nanobiomaterials
Poushpi Dwivedi, Shahid S. Narvi, Ravi Prakash Tewari and Dhanesh Tiwary

Sustainable Microbial Technologies for Valorization of Agro-Industrial Wastes
Jitendra Kumar Saini, Surender Singh and Lata Nain

Enzymes in Valorization of Waste: Enzymatic Pre-treatment of Waste for Development of Enzyme based Biorefinery (Vol I)
Pradeep Verma

Enzymes in Valorization of Waste: Enzymatic Hydrolysis of Waste for Development of Value-added Products (Vol II)
Pradeep Verma

Enzymes in Valorization of Waste: Next-Gen Technological Advances for Sustainable Development of Enzyme based Biorefinery (Vol III)
Pradeep Verma

Enzymes in Valorization of Waste, Three Volume Set
Pradeep Verma

Biotechnological Approaches in Waste Management
Rangabhashiyam S, Ponnusami V, Pardeep Singh

For more information about this series, please visit: www.routledge.com/Novel-Biotechnological-Applications-for-Waste-to-Value-Conversion/book-series/NVAWVC

Sustainable Microbial Technologies for Valorization of Agro-Industrial Wastes

Edited by
Jitendra Kumar Saini,
Surender Singh and
Lata Nain

CRC Press is an imprint of the
Taylor & Francis Group, an informa business

First edition published 2023
by CRC Press
6000 Broken Sound Parkway NW, Suite 300, Boca Raton, FL 33487–2742

and by CRC Press
4 Park Square, Milton Park, Abingdon, Oxon, OX14 4RN

CRC Press is an imprint of Taylor & Francis Group, LLC

ISBN: 978-1-032-04269-5 (hbk)
ISBN: 978-1-032-04274-9 (pbk)
ISBN: 978-1-003-19124-7 (ebk)

DOI: 10.1201/9781003191247

Typeset in Times
by Apex CoVantage, LLC

Dedication

To My Late Mother.

—Jitendra Kumar Saini

To All Corona Warriors.

—Surender Singh

To All My Dear Students.

—Lata Nain

Contents

About the Editors

Dr. Jitendra Kumar Saini received his BSc (industrial microbiology) and MSc (microbiology) from Gurukula Kangri University, Haridwar. He obtained his PhD in microbiology from Gobind Ballabh Pant University of Agriculture and Technology, Pantnagar, in 2010, after which he worked as a postdoctoral associate at GADVASU, Ludhiana, in a World Bank–funded NAIP project on rumen microbiology. Later, he joined DBT-IOC Centre for Advanced Bioenergy Research, Indian Oil Corporation Ltd., Research and Development Centre, Faridabad, as a scientific officer, where he led a team on enzyme development for advanced biofuels. He joined the Department of Microbiology, Central University of Haryana, Mahendergarh, in 2016 as an assistant professor and teaches industrial microbiology and food and dairy microbiology to PG students. His current research focuses on enzyme and microbial technologies for sustainable development of energy and environment. Dr. Saini is a recipient of an early career research grant from the Science and Engineering Research Board, Department of Science and Technology, Government of India, and a twinning grant from the Department of Biotechnology, Government of India. He has supervised one research scholar and twenty-one MSc dissertation students, has cosupervised one postdoc in the past, and is currently supervising three research scholars. He has filed one US patent, is an author of more than forty five articles, has edited two international books, and is an active reviewer for many reputed journals in biofuel and bioenergy research. He is the review editor for journal *Frontiers in Energy Research.* Dr. Saini conducted a one-week Global Initiative of Academic Networks (GIAN) course on Integrated Lignocellulosic Biorefineries for Sustainable Development as a course coordinator. Recently, he organized the International Conference AMI-2019 as an organizing secretary. Dr. Saini is a life member of the Association of Microbiologists of India (AMI) and the Asian Federation of Biotechnology (AFOB).

Prof. Surender Singh received his BSc (hons.) in agriculture from Chaudhary Charan Singh Haryana Agricultural University, Hisar (India). He obtained his MSc and PhD in microbiology from the Indian Agricultural Research Institute (IARI), New Delhi, in 2005 and 2009 respectively before joining as a scientist (ARS) in IARI, New Delhi. He joined the Central University of Haryana as an associate professor in 2018. His current research focuses on organic matter recycling, including bioethanol production from lignocellulosic material. He was awarded the prestigious Endeavour Research Fellowship (2011) by the Government of Australia for his academic excellence to carry out six months' research in the University of South Australia, Adelaide. He was also awarded the Young Scientist award in 2015 by the Association of Microbiologists of India (AMI), Young Scientist Award (2015–2016), and associateship by the National Academy of Agricultural Sciences (NAAS), New Delhi, and Haryana Yuva Vigyan Ratan (2018) by the Department of Science and Technology, Government of Haryana. Dr. Singh has been involved in eight research projects funded extramurally by ICAR, DBT, DST, and MoEF (Government of India). He is currently supervising four doctoral scholars and has supervised fourteen PG dissertations in the past. He has authored more than eighty research articles, fifteen book chapters, and is an active reviewer for many reputed journals in biofuel and bioenergy research. He is also an editorial board member of the *Electronic Journal of Biotechnology*. Dr. Singh is a life member of the Association of Microbiologists of India (AMI) and Indian Science Congress.

Dr. Lata Nain has a research and teaching experience of thirty years and received her MSc and PhD in microbiology from Gobind Ballabh Pant University of Agriculture and Technology, Pantnagar, in 1981 and 1985 respectively. She started her career as a scientist (ARS) at Central Potato Research Institute Shimla (India) and then moved to her present institute, IARI, New Delhi, in 1988. Dr. Lata also worked in the Department of Biology, University of Waterloo, Ontario, Canada, for her post-doctorate. Dr. Lata handled various sponsored projects funded by external agencies. Dr. Lata has made pioneering contributions in the area of agrowaste management and utilization of biomass for bioenergy using microbial resources and enzymes. She has developed and commercialized a promising consortium of hyperlignocellulolytic fungi for rapid bioconversion of agro-residues to produce high-quality compost. She has also developed formulations of bacteria and cyanobacteria as synergistic bioinoculants for rice-wheat cropping system. Dr. Lata has successfully transferred the technologies developed at IARI to entrepreneurs, leading to increased revenue generation and coverage of land under microbial inoculants. She has more than 275 research papers in journals of repute along with one hundred other publications, including book chapters, monographs, and popular articles. She is a member of several national-level scientific committees and is on the editorial board of many peer-reviewed journals. She has guided six MSc and ten PhD students. She has been a recipient of the prestigious IARI Best Teacher Award (2004), is a fellow of the National Academy of Agricultural Sciences (FNAAS) and the Academy of Microbiological Sciences (FAMSc), and has received the Rangaswamy award by the Association of Microbiologists of India (AMI).

Contributors

A. K. Lavanya
ICAR-Central Research Institute for Jute and Allied Fibers, Barrackpore, Kolkata, India

Abdeshahian, P.
Engineering School of Lorena, University of São Paulo, Lorena, Brazil

Akhilesh Kumar Singh
Mahatma Gandhi Central University, Motihari, East Champaran, Bihar, India

Akshaya Gupte
Natubhai V. Patel College of Pure and Applied Sciences, Vallabh Vidyanagar, Gujarat, India

Ana Angélica Feregrino-Pérez
Universidad Autónoma de Querétaro, Campus Amazcala, Querétaro, México

Anamika Sharma
Agricultural Research Organization, Volcani Center, Rishon LeZion, Israel

Antunes, F. A. F.
Engineering School of Lorena, University of São Paulo, Lorena, Brazil

Anuj Kumar Chandel
Engineering School of Lorena (EEL), University of São Paulo, Lorena, Brazil.

Arindam Kuila
Banasthali Vidyapith, Rajasthan, India

Arruda, G. L.
Engineering School of Lorena, University of São Paulo, Lorena, Brazil

Asish K. Binodh
Tamil Nadu Agricultural University, Coimbatore, Tamil Nadu, India

Ayekpam Bimolini Devi
Mizoram University, Aizawl, Mizoram, India

Balbino, T. R.
Engineering School of Lorena, University of São Paulo, Lorena, Brazil

Barbosa, F. G.
Engineering School of Lorena, University of São Paulo, Lorena, Brazil

Bijan Majumdar
ICAR-Central Research Institute for Jute and Allied Fibers, Barrackpore, Kolkata, India

Castro-Alonso, M. J.
Engineering School of Lorena, University of São Paulo, Lorena, Brazil

Claudia Gutiérrez-Antonio
Universidad Autónoma de Querétaro, Campus Amazcala, Querétaro, México

Cruz-Santos, M. M.
Engineering School of Lorena, University of São Paulo, Lorena, Brazil

Cunha, M. L. S.
Engineering School of Lorena, University of São Paulo, Lorena, Brazil

da Silva, S. S.
Engineering School of Lorena, University of São Paulo, Lorena, Brazil

Dhanya M. S.
Central University of Punjab, Bathinda, India

Faizal Bux
Durban University of Technology, Durban, South Africa

Gouranga Kar
ICAR-Central Research Institute for Jute and Allied Fibers, Barrackpore, Kolkata, India

Gretty K. Villena
Universidad Nacional Agraria La Molina. Av. La Molina s/n Lima 12, Perú

Harinder Singh Oberoi
ICAR-Indian Institute of Horticultural Research, Bangalore, India

Ilanit Samolski
Universidad Nacional Agraria La Molina. Av. La Molina s/n Lima 12, Perú

Ingle, A. P.
Engineering School of Lorena, University of São Paulo, Lorena, Brazil

Juan Fernando García-Trejo
Universidad Autónoma de Querétaro, Campus Amazcala, Querétaro, México

K. Tamreihao
ICAR-Research Complex for NEH Region, Manipur Centre, Imphal, India

Kiruthika Thangavelu
Tamil Nadu Agricultural University, Coimbatore, Tamil Nadu, India

Kokila, V.
ICAR-Indian Agricultural Research Institute, New Delhi, India

Lata Nain
ICAR-Indian Agricultural Research Institute, New Delhi, India

Laura G. Covinich
IMAM, UNaM, CONICET, FCEQYN, Programa de Celulosa y Papel (PROCYP), Misiones, Argentina, Félix de Azara 1552, Posadas, Argentina

Laxmi Sharma
ICAR-Central Research Institute for Jute and Allied Fibers, Barrackpore, Kolkata, India
Mahendra Engineering College, Namakkal, Tamil Nadu, India

María Cristina Area
IMAM, UNaM, CONICET, FCEQYN, Programa de Celulosa y Papel (PROCYP), Misiones, Argentina, Félix de Azara 1552, Posadas, Argentina

Meghna Diarsa
Natubhai V. Patel College of Pure and Applied Sciences, Vallabh Vidyanagar, Gujarat, India

Melo, Y. C. S.
Engineering School of Lorena, University of São Paulo, Lorena, Brazil

Mier-Alba, E.
Engineering School of Lorena, University of São Paulo, Lorena, Brazil

Monika Chaudhary
Banasthali Vidyapith, Rajasthan, India

Muñoz, S. S.
Engineering School of Lorena, University of São Paulo, Lorena, Brazil

Naganandhini Srinivasan
Tamil Nadu Agricultural University, Coimbatore, Tamil Nadu, India, 641 003

Nirmal Renuka
Durban University of Technology, Durban, South Africa

Pangambam Langamba
ICAR-Research Complex for NEH Region, Manipur Centre, Imphal, India

Pintubala Kshetri
ICAR Research Complex for NEH Region, Manipur Centre, Imphal, India

Prado, C. A.
Engineering School of Lorena, University of São Paulo, Lorena, Brazil

Pratik Satya
ICAR-Central Research Institute for Jute and Allied Fibers, Barrackpore, Kolkata, India

Praveen Jain
Government Chandulal Chandrakar Arts and Science PG College Patan, Durg, Chhattisgarh, India

Radha Prasanna
ICAR-Indian Agricultural Research Institute, New Delhi, India

Rameshwar Tiwari
Ulsan National Institute of Science and Technology (UNIST), Ulsan, Republic of Korea

Ranjitha K.
ICAR-Indian Institute of Horticultural Research, Bangalore, India

Reyes-Guzman, R.
Engineering School of Lorena, University of São Paulo, Lorena, Brazil

Ribeaux, D. R.
Engineering School of Lorena, University of São Paulo, Lorena, Brazil

Rocha, T. M.
Engineering School of Lorena, University of São Paulo, Lorena, Brazil

Ruiz, E. D.
Engineering School of Lorena, University of São Paulo, Lorena, Brazil

Sachitra Kumar Ratha
CSIR-National Botanical Research Institute, Lucknow, India

Sajan Kurien
Karunya Institute of Technology and Sciences, Coimbatore, Tamil Nadu, India

Santos, J. C.
Engineering School of Lorena, University of São Paulo, Lorena, Brazil

Sashi Sonkar
Bankim Sardar College, Tangrakhali, South 24 Parganas, West Bengal, India

Sergio Iván Martínez-Guido
Universidad Autónoma de Querétaro, Campus Amazcala, Querétaro, México

Sivakumar Uthandi
Tamil Nadu Agricultural University, Coimbatore, Tamil Nadu, India

Srinjoy Ghosh
ICAR-Central Research Institute for Jute and Allied Fibers, Barrackpore, Kolkata, India

Subhra Saikat Roy
ICAR-Research Complex for NEH Region, Manipur Centre, Imphal, India

Sugitha Thankappan
Karunya Institute of Technology and Sciences, Coimbatore, Tamil Nadu, India

Suman Roy
ICAR-Central Research Institute for Jute and Allied Fibers, Barrackpore, Kolkata, India

Sunanda Joshi
Banasthali Vidyapith, Rajasthan, India

Surender Singh
Central University of Haryana, Mahendergarh, India

Susheel Kumar Sharma
ICAR-Research Complex for NEH Region, Manipur Centre, Imphal, India

Thangjam Surchandra Singh
ICAR-Research Complex for NEH Region, Manipur Centre, Imphal, India

Valeria Caltzontzin-Rabell
Universidad Autónoma de Querétaro, Campus Amazcala, Querétaro, México

Varsha Upadhayay
Banasthali Vidyapith, Rajasthan, India

Vijay Rakesh Reddy
ICAR-Indian Institute of Horticultural Research, Bangalore, India

Virthie Bhola
Durban University of Technology, Durban, South Africa

Preface

Increased urbanization due to population explosion has led to an exponential rise in industrial production. A lot of waste is being generated by such industries, which needs to be managed in a sustainable manner. Various agro-industrial wastes contain many nutrients and minerals in good quantities, which could be harnessed as renewable and sustainable sources for the production of many value-added products. Microorganisms, their enzymes, and their metabolic reactions play a crucial role in such bioconversions. Microbial technologies are being utilized in almost all important industrial sectors, be it energy, petroleum, environment, health, nutrition, or agriculture. Considering the vast potential of microorganisms and microbial technologies in sustainable management and valorization of abundantly available varied types of agro-industrial wastes, this edited book, *Sustainable Microbial Technologies for Valorization of Agro-Industrial Wastes*, was conceptualized.

The broad areas considered in this book are microbial bioconversions of various agro-industrial wastes into biofuels, bioenergy, biochemicals, and other useful products, which are envisaged to help solve the global problems of waste generation and management and energy crisis. This edited book is targeted to provide an overview of the different aspects of microbial bioconversion methodologies for the valorization of agro-industrial wastes of varied nature, including the microbiological and biotechnological aspects, environmental concerns, bioprocess development, and scale-up aspects, as well as the challenges and opportunities in the microbial valorization of such wastes at a large scale. The book has been divided into sixteen chapters which cover major aspects of agro-industrial waste valorization. **Chapters 1** and **2** provide an insight on lignocellulosic waste availability and their prospects in sustainable production of fuels, biochemicals, and other value-added products. **Chapter 3** covers applications of bioprospecting and genomics tools for cost-effective utilization of agro-industrial wastes. **Chapter 4** discusses metagenomics tools for prospecting glycosyl hydrolases. **Chapter 5** highlights lignin as a source of value-added chemicals, whereas **Chapter 6** includes information on bioconversion of food industry wastes nutraceuticals. Sustainable production of polyhydroxyalkanoate from food wastes has been covered in **Chapter 7**. The biotechnological potential of feruloyl esterase enzyme has been highlighted in **Chapter 8**. Next, in **Chapter 9**, biobutanol production has been covered. **Chapters 10** and **11** cover the bioethanol production potential of jute-mesta and cyanobacterial biomass, respectively, whereas poultry feather waste valorization has been included in **Chapter 12**. Productions of levulinic acid and organic fertilizer from lignocelluloses and Parthenium weed have been respectively targeted in **Chapters 13** and **Chapter 14**, respectively**. Chapter 15** discusses the utilization of waste oil for production of value-added products. Towards the end, **Chapter 16** covers the concept life cycle analyses in econoenvironmental analysis of waste valorization for value-added products.

This book is designed to keep in view the needs of students, teachers, researchers, and entrepreneurs. The targeted audiences of this book include science and technology researchers, professionals, environmentalists, microbiologists, biotechnologists, researchers, engineers, and scientists working in the area of sustainable microbial technologies for waste valorization, biorefinery, and circular bioeconomy. Bioeconomy analysts and policy-makers will also benefit from this book. As nothing in nature is considered perfect, the same is also true with this book. Therefore, we take full responsibility for any shortcomings and fully look forward to any critical observations, suggestions, and recommendations of our highly valued readers.

We are thankful to our family, friends, students, teachers, and mentors, who acted as a source of inspiration to us. This book project could not have been initiated without the support of the CRC press team and the series editors, who bestowed their full support to us and believed in our capabilities. We were lucky enough to convince renowned international researchers from different parts of the globe to contribute chapters in this book, which tremendously improved its quality, enriched it with global perspectives, and gave it an international flavor.

We dedicate this work of ours to the frontline workers who lost their lives while serving the people affected by the COVID-19 pandemic.

India, October 2021

Jitendra Kumar Saini
Surender Singh
Lata Nain

Foreword

The economic and social development of any country is governed by the sustainable use of accessible resources. Sustainability is intimately tied to the utilization of renewable resources to meet the demand for energy and chemicals. Recently, the concept of circular economy, which tackles these challenges by utilizing the resources and producing valuable products, has been envisaged to solve this global problem of waste valorization. Agricultural activities generate a lot of secondary wastes comprising a lignocellulosic biomass which remains underutilized due to the lack of feasible technologies to produce any value-added product. The editors have selected sixteen topics to provide an overview of the valorization of underutilized wastes of varied nature and their conversion to biofuel, chemicals, and value-added products. Bio-based feedstocks may present a greener route with the breakdown of their lignocellulosic biomass to sugars, their oligomers, and other useful chemicals employing pretreatment and a cocktail of hydrolytic enzymes. The monomers may be utilized for the production of ethanol, butanol, and polyhydroxyalkanoate. Similarly, feather waste may be hydrolyzed to produce amino acid–rich bioproducts. The search of new enzymes, e.g., feruloyl esterase and its application in hemicellulose processing, is also covered in this book. Lignin-derived levulinic acid, ferulic acid, and vanillin are the most promising platform chemicals, owing to their convertibility into various industrial products having diverse applications. One of the chapters also gives an insight into the management of parthenium weed and its conversion to biomanure. A biorefinery approach decreases the dependence on fossil fuels for a sustainable production of fuels and chemicals. New technologies, however, can only be deployed commercially if they lead to cost-efficient production. Therefore, life cycle analyses provide an appropriate methodology during the process development to decide future competitiveness and long-term sustainability of the process. The inclusion of the last chapter on concepts and recent trends in life cycle analyses in waste valorization provides an insight to budding researchers in the development of an economically efficient process.

The efforts of the editors are commendable, and the book would be very useful for the researchers involved in developing bio-based technologies and will encourage sustainable use of renewable resources. Overall, this edited volume, *Sustainable Microbial Technologies for Valorization of Agro-Industrial Wastes*, is a comprehensive effort concerning research perspectives on the global problem of waste valorization in an eco-friendly manner. The chapters have been contributed by the experts of the area across the globe and offer clear and concise information on technologies to use agricultural and food wastes. This book will serve as an invaluable reference for the scientific community involved in the management of agro-industrial wastes for the production of second- and third-generation biofuel and biochemicals.

Dr. Anil Kumar Saxena, ***FNAAS, FIAMS, FIMS***
Director
ICAR-National Bureau of Agriculturally Important Microorganisms (NBAIM)
Kusmaur, Mau, Uttar Pradesh 275103, INDIA

1 Lignocellulosic Waste Availability for Microbial Production of Fuels, Biochemicals, and Products

Antunes, F. A. F., Ingle, A. P., Abdeshahian, P., Ribeaux, D. R., Prado, C. A., Muñoz, S. S., Barbosa, F. G., Balbino, T. R., Castro-Alonso, M. J., Reyes-Guzman, R., Arruda, G. L., Cruz-Santos, M. M., Mier-Alba, E., Rocha, T. M., Ruiz, E. D., Melo, Y. C. S., Cunha, M. L. S., Santos, J. C., and da Silva, S. S.

Department of Biotechnology, Engineering School of Lorena, University of São Paulo, Lorena, Brazil

CONTENTS

DOI: 10.1201/9781003191247-1

1.1 INTRODUCTION

The primary use of petroleum-based raw material has played and yet plays a crucial role in the world's economy. Indeed, plenty of products can be derived from fossil resources, such as petrochemicals, fertilizers, and industrial inputs. Most importantly, it fulfills the necessity for the energy supply in the transportation sector (Tye et al., 2016). Nonetheless, petroleum has a constrained reservoir that is foreseen by several econometric methods to reduce significantly for the next decades (Shafiee and Topal, 2009). Additionally, it also carries inherent burdens that could compromise ecosystems through extensive emissions of greenhouse gases (GHGs) and particulate matter (Patel et al., 2016). Thereafter, the setup of a new perspective aimed at the utilization of renewable sources has emerged as a potential and low-cost alternative for its replacement.

Accordingly, among several forecasted materials, lignocellulosic wastes (LCWs) comprises high bioavailability, biocompatibility, and less GHG emissions (Farrell et al., 2006). At this glance, prior to understanding the suitability of LCWs in fermentative processes, it is important that they are classified into nonwood and wood categories (Tye et al., 2016). The latter is often referred to as softwood and hardwood lignocellulose, whereas the former is generally categorized as agro- and forest-residues. This distinctive classification relies on their differences in chemical composition and physical properties (Zhu and Pan, 2010). Above all, LCWs are primarily composed of cellulose fibrils folded within an amorphous matrix of hemicelluloses and cross-linked lignin to form a semicrystalline complex heteropolymer. The respective fraction of each component may considerably vary according to the species and climate conditions during plant growth (Koutsianitis et al., 2015).

In contrast, a few constraints are found within microbial metabolism as well as difficulties in the material processing operations which hinder their wide implementation in biotechnological industries. Within this context, the polyphenolic structure of lignin may pose a biochemical barrier to microbial enzymes and mitigate the digestion of fermentable sugars (Den et al., 2018). Although a few microorganisms have been reported to exhibit the ability to degrade lignin through a variety of metabolic pathways, it is known by its recalcitrance, which is addressed to the enzymatic inhibitory effect imposed by its chemical structure, especially in the respiratory chain metabolism. Nonetheless, recent advances in lignin-selective depolymerization as well as metabolically engineered microorganisms may scavenge a path towards the potential utilization of lignin in the biosynthesis of several market-attractive products (Liu et al., 2019).

Bearing those trends in mind, the establishment of the so-called pretreatments (PTs) and enzymatic hydrolysis took place in order to facilitate the solubilization of cellulose and hemicellulose, depolymerization, and removal of lignin, followed by microbial fermentation (Den et al., 2018). A multitude of PTs have been studied and developed, including chemical, physical, physicochemical, and biological treatments and combined (Rezania et al., 2020); however, only a few are environmentally, economically attractive. It is noteworthy that their outcome differs depending on the method and technique of choice. Still, there is no such "best option," but rather what is suitable to the targeted approach. In the very last, there are differential means by which PTs can be evaluated; processing variables and parameters are directly related to the efficiency and profitability of any process of operation, and hence, it could provide guidance to achieve robustness in PT technology. This multivariate system may indicate whether the process is viable or inappropriate, depending on which constraints are used. Thereby, *in silico* analyses, such as life cycle assessment (LCA) and techno-economic assessment (TEA), are extensively deployed in order to address limitations and econo-feasibility of LCWs conversion technology (Patel et al., 2016).

Forwardly, through the stepwise combination of PT, enzymatic hydrolysis, and fermentation operations, LCW conversion into value-added biomolecules has taken utmost steps, approaching reliable results that may be in potential congruence with an industrial commercialization figure. Numerous strategies may be conducted to facilitate LCWs implementation, and hence, it is paramount to underpin remarkable questions aimed at the enhancement of their usage and conversion efficiency. In accordance, enzymatic hydrolysis is continuously improved (Luft et al., 2019), fermentation configuration and modes have been deeply explored (Zabed et al., 2017, Sánchez-Muñoz et al., 2020), and yet a wide revenue is still lying forth. Thus, this chapter focuses on giving an ample description over LCW availability and history regarding forestry, agricultural, and wood wastes for microbial production of fuels and biochemicals. Henceforth elucidating current approaches, advancements, and trends that concur to the consolidation of a well-designed, modern biorefinery concept.

1.2 WOOD

1.2.1 Characteristics and Types

Wood represents a basic tissue of plants which acts in the conduction of nutrients in plants while strengthening the plant body. Wood also functions in the synthesis of biological chemicals and the storage of chemicals obtained from photosynthesis (Tsoumis, 2020, Wiedenhoeft, 2013). If a transverse section of a tree in the trunk is provided, generally different parts are observed which are placed from outside to inside, including bark, cambium, wood, and pith. Bark is composed of an outer bark and an inner bark. The outer bark is a dead part of the bark which protects the inner bark and restricts high evaporation and water loss of trees. The inner bark contains living cells which carry nutrients obtained from photosynthesis in leaves to other parts of the plant. The cambium is a thin layer between the inner bark and the wood which can be observed with a microscope. The cambium is composed of living cells which grow and increase the diameter of the trunk. The growth of cambium cells leads to the formation of cells of wood and bark layers (Miller, 1999; Woodford, 2017).

Wood is the major constituent part of the trunk composition. Wood consists of two parts, namely, the outer layer of sapwood and the inner part of heartwood. Sapwood thickness may range from 4 to 6 cm in radial thickness. Generally, trees with more growth process have higher thickness of sapwood. The sapwood includes living cells and dead cells. The cells of the sapwood form tubes for the translocation of chemicals. The sapwood functions in the conduction of water (sap) and nutrients from roots to leaves. Furthermore, sapwood carries out the long-term storage of photosynthate (biochemicals synthesized by photosynthesis process), such as starch and lipids (Miller, 1999; Wiedenhoeft, 2013). The active cells of sapwood also produce chemicals which are deposited in dead cells and thus could form the heartwood (Wiedenhoeft, 2010). The heartwood constitutes the biggest part of the wood construction in ageing trees. The heartwood forms the harder and darker part of wood. The heartwood is composed of dead cells which do not have any role either in storage of food or transfer of water (Miller, 1999). The heartwood acts in the storage of chemicals produced by ageing sapwood cells, including those placed next to the heartwood boundary. These chemicals are totally called extractives since they can be extracted by solvents. The extractives contain different chemicals, such as fats, resin, gum starch, phenolics, and waxes (Miller, 1999; Wiedenhoeft, 2010). Extractives are responsible for the dark color of heartwood. Moreover, extractives make heartwood more resistant to decay and water (Wiedenhoeft, 2013). Cambium cells are responsible for the growth of wood. These cells are divided to produce a large number of new cells at a determined time, thus growing wood tissue. These cell groups produced in wood during a distinct time interval are called growth increments or growth rings, which are shown as concentric circular bands in wood (Wiedenhoeft, 2013). In temperate regions, growth rings are prominently observed where the cross section of the trunk is studied. The wood formed in the early growth season is called earlywood, displaying as a light-color growth ring, while the wood formed in the latter period of the seasonal growth is called latewood, shown as a dark-color growth ring. An annual growth ring includes an earlywood and an outer latewood. The age of the tree can be relatively calculated by counting the growth rings at stump areas (Wiedenhoeft, 2010). Normally, sapwoods have a low number of growth rings, which are less than 1 cm in thickness (Miller, 1999). The term *hardwood* refers to the wood obtained from angiosperms (flowering trees with fruits and pods surrounding seeds), such as ash, beech, birch, mahogany, maple, oak, teak, and walnut. The term *softwood* refers to the wood obtained from gymnosperms (trees containing uncovered seeds by fruit or ovary with needle-like leaves), such as cedar, cypress, fir, pine, spruce, and redwood (Woodford, 2017; Wiedenhoeft, 2010).

Wood cells are mainly composed of lumen (the space inside the cell) and a cell wall that forms around the lumen. The axial system of softwoods is mainly composed of long cells, called tracheids, which act similar to tubes for the conduction of nutrients and water. Furthermore, tracheids can provide mechanical strength for wood tissue. On the other hand, in the transverse view of softwoods, there are brick-shaped cells which are known as ray parenchyma cells. The major functions of these cells are the synthesis, storage, and translocation of biochemicals and, to a lesser extent, water conduction. The cell types of hardwoods include wood cells in the axial part and horizontal direction. The axial system of hardwood contains vessel elements, which carry out water conduction in wood tissue; fibers, which provide mechanical support of wood structure; and axial parenchyma cells. The horizontal system is composed of ray parenchyma cells. The wood cells in the horizontal parts form an organized structure known as rays, which have been elongated in radial direction from pith to bark. Rays simply can be observed in edge-grained or quarter-sawn view (Miller, 1999; Wiedenhoeft, 2010). The biochemical structure of wood in a dry form includes three main components, namely, cellulose, hemicellulose, and lignin, with a low amount of extraneous substances (5–10% wood weight). In this regard, cellulose forms the major fraction of wood, constituting about 50% wood composition. Lignin represents a brittle compositional material, constituting 23% to 33% of softwood weight and 16% to 25% hardwood structure (Miller, 1999; *Wiedenhoeft*, 2013). The properties of wood are related to various characters, such as color, odor, moisture content, strength, and durability. In this regard, woods obtained from different trees may display varied colors, ranging from white to black. Tannin, resin, and pigments in cells make specific colors in wood. The woods obtained from softwoods, such as deodar and pine, reveal a white color. Aspen is a hardwood tree which has white wood. Walnut trees have dark-brown wood. Teak trees are characterized by a golden-yellow color. Odor is a typical characteristic of woods which could be related to substances such as essential oils, resins, and tannins found in wood composition. For instance, teakwood has an aromatic odor, and the wood of pines gives a resin odor (CivilSeek, 2021; ArchiLine Wooden Houses, 2021). The moisture content of wood is the amount of moisture in the wood cells either as free water in the lumen of the wood cell or as moisture bound in the cell wall. Wood absorbs water under damp conditions and swells physically, while it releases moisture into the environment where dry air and increased temperature occur. Durability represents the characteristic of wood for resistance to physical factors, such as tension, compression, bending, and shear. (CivilSeek, 2021; Woodford, 2017; ArchiLine Wooden Houses, 2021).

1.2.2 History of Use Applied to Production of Fuels, Biochemicals, and Products

The history of wood as fuel is rooted since the first appearance of fire. Since ancient times, trees have provided material for several types of tools, for construction, and to make fire, which allowed man adaptation to relatively cold climates and, consequently, to the conquest of the world (Jones and Sandberg, 2020). Thus, throughout ancient history until the contemporary era, wood has been part of every civilization in the world. Thus, wood fuel has the potential to contribute to a cleaner environment based on the use of biomass as a renewable source and the CO_2-neutral supply of energy (Singh et al., 2021a). Moreover, the specific applications will depend on different factors related to the type of wood (Table 1.1)

Currently, wood can be used for bioenergy applications, such as electricity, the production of heat, and biofuels. Gas from wood can be obtained by heating the biomass in the absence of oxygen (pyrolysis) to break it down until solid charcoal fuel remains. Subsequently, a vaporization process allows the conversion of solid wood into potentially flammable gases that are captured to be used in internal combustion engines or power boilers (Situmorang et al., 2020). Liquid fuel from wood follows the same process at the beginning, with the difference that the gases are condensed. Thus, the mixture (carbon monoxide, hydrogen, carbon dioxide, and methane) called synthesis gas (syngas) can be converted to liquid fuels, such as ethanol or diesel or value-added chemicals (Singh et al., 2021b). Another alternative involves the use of

TABLE 1.1
Factors Involved in Wood Applications

Source	Pine, oak, cedar, cherry, spruce, redwood, eucalyptus, walnut, acacia, maple, mahogany, aspen, bamboo, ebony, Poplar, beech, yew, teak, balsa	(Dillen et al., 2016)
Types	Softwood (fast-growing, light in weight and density, very poor resistance to fire, less expensive; used for interior moldings, construction framing, the manufacturing of windows, construction framing, and sheet wood, among others)	Ramage et al., 2020
	Hardwood (slow-growing, heavy in weight and density, less expensive, good resistance to fire; applied in building, utility structures, furniture, musical instruments, flooring, and boatbuilding)	(Woodard and Milner, 2016)
Available Shapes	These are derived from logs, firewood, blocks, stovewood, bolts, charcoal, sheets, sawdust, shavings, and residues of other wood processing industries.	(Bartlett et al., 2019)

enzymes to transform cellulose and hemicellulose to fermentable sugars which are used by microorganisms to synthesize ethanol. By contrast, fuelwood is the dominant source of energy in many developing countries, comprising 60–80% of total wood consumption, and can represent between 50% and 90% of all household energy (FAO, 2010, FAO, 2016; Singh et al., 2021b). In general, the main sectors include the residential and commercial for activities such as cooking, baking, food processing, lime and brick burning, among others.

On the other hand, extraction companies, paper manufacturers, and wood biorefineries produce a wide variety of chemical products from wood biorefineries. In this sense, two general categories can be identified. The first group belongs to products obtained from the chemical processing of wood. It is the case for cellulose, which is recovered from chemical processing after complete removal of the other components (hemicelluloses, lignin, and extractives) and can be used in drug delivery, wound dressing, tissue engineering, varnishes, inks, photographic films, magnetic tapes, cellophane, and many others (Stafford et al., 2020). Lignin is another by-product obtained from the chemical modification of wood and is applied in adhesives, solvents, foam materials, soil conditioners, rubber, ceramics, pharmaceuticals, vanillin, and other products (Ragauskas et al., 2014). The second group involves the extractives and their derivatives. Pine resins and tannins are the typical wood extractives of practical value. Resins are produced by epithelial cells of living trees and act as both a mechanical and chemical barrier to avoid the entry of organisms. Among the resins, those from pine stand out in the production of soap, paper, food, ink, pesticides, spices, and new fuels (Kelkar et al., 2006; Lai et al., 2020). Whereas, tannins are produced by most higher plants and are mainly made up of glycosides that belong to the polyphenol family. These natural compounds protect plants against pathogenic bacteria, fungi, and insects (Pizzi, 2019). Although the traditional application is associated with the preservation of leather skins, these compounds are used as part of animal feed, antioxidants, clarifying agents to produce wine and beer, natural dyes, and adhesives, among others (Pagliaro et al., 2021). Thus, the use of wood takes on a special role in the context of the bioeconomy due to products and formulations with high added value with a more ecological approach and reduced carbon footprint (Temmes and Peck, 2020).

1.3 AGRICULTURAL WASTES

1.3.1 Characteristics and Types

Agricultural waste is liquid or solid materials produced during agricultural activities, in direct consumption of primary products, or in industrial processes, including residues of cultivation, processing, and consumption of grains, fruits, vegetables, poultry, and dairy products (Obi et al., 2016; Onu and Mbohwa, 2021). The generation of agricultural waste follows a long way, starting with production in the field (planting, harvesting, and transportation), through processing in industries (generation of waste and effluents), and finishing in the retail and marketing sector, where food wastes (after cooking and disposal) are generated, reaching a total of 30% of all agricultural productivity worldwide (Urbaniec and Bakker 2015).

Residues from agricultural production are divided into field residues and process residues. Field residues are residues remaining in the postharvest planted area, such as leaves, stems, seedpods, roots, among other things, and commonly left in the planted area to increase irrigation efficiency and prevent soil erosion. On the other hand, the process residues are those generated after the processing of the vegetable step. Such as seeds, bark, bagasse, molasses, bran, leaves, stems, straws, and spares, the majority of which are used as fertilizers in soil improvement, feed formulation for animals, and marketed as by-products for other industries and processes (Asim et al., 2015; Sadh et al., 2018).

Agricultural residues can also receive a second classification, based on the percentage of cellulose: in agricultural lignocellulosic residues (by-products of rice, corn, sugarcane, and fruit bagasse, coffee husk, coconut fiber, and vegetable bran) and nonlignocellulosic agricultural waste (eggshell, dairy waste, and those with cellulose amounts below 10%) (Girelli et al., 2020).

With a growing production worldwide, agricultural waste is mostly underutilized and untreated and is still disposed of with burning and deposition in inappropriate landfills, mainly because they still have economic values lower than the expenses acquired with the collection, transportation, and processing (Barcelos et al., 2020). However, the vast majority of agricultural residues are sources of nutrients and available industrial components, such as fermentable sugars, vitamins, fibers, phenolic compounds, and others, which they ensure as a renewable source of agricultural by-products (Zuin et al., 2018).

The composition of agricultural waste is directly related to the type of agricultural crop and can be available in liquid, mud, or solid form (Obi et al., 2016). The composition of agricultural waste is directly related to the type of agricultural crop and can be available in liquid, mud, or solid form (Obi et al., 2016). With a varied chemical composition, they present different proportions of proteins (grain bran), sugars (cellulosic and hemicellulosic fractions), minerals, and bioactive compounds (phenolic, antioxidant, and aromatic compounds). These compounds can be used as raw material in microbial fermentation, in nutritional supplementation, or in the production of dietary supplements (fiber, vitamins, and minerals) (Satari and Karimi, 2018).

1.3.2 Historical Aspects of Production of Fuels, Biochemical, and Products from Agricultural Wastes

Agricultural residues have been considered as important cellulosic storage since they do not compete with food availability and are largely accessible worldwide, representing sources of fiber, plant nutrients, and energy (Bergmann et al., 2018; Ali et al., 2019). The main global agricultural residue production comes from corn and sugarcane, produced in America, and wheat and rice, in Asia (Feng-Wu et al., 2019).

Energy and biochemical production is of the main sectors in which the replacement of fossil resources is necessary since the global demand is ascending, along with fossil fuel price instability and the energy sector being the primary polluter, responsible for GHG emissions (Bergmann et al., 2018; Clauser et al., 2021). Therefore, the demand for bioproducts from alternative sources grew in the last decade (Ali et al., 2019; Kapoor et al., 2020). Besides, agricultural residues have been more space-energetic after the first crises of petrol in 1975 (Santos et al., 2020). In 1980, forest residues were material to generating electricity in most of the world, and in 1980, the demand for energy from biomass and wastes was 749 Mt, increasing to 1,031 Mt in 2000 and 1,436 Mt during the last twenty years (Bergmann et al., 2018; Leontopoulos and Arabatzis, 2021). In 1990, flex-fuel cars in Brazil, and in 1991 in the USA, were the biggest producers of ethanol from corn (Manna et al., 2018; MNRE, 2021). During 2000–2005, ethanol from biomass had more space in the world's energetic system (Ali et al., 2019; Kapoor et al., 2020). However, the traditional measures of waste utilization are still the most common, such as in livestock feed, organic fertilizers, or burning straw in fields, which lowers the utilization efficiency (Coutinho et al., 2011; Feng-Wu et al., 2019). Brazil has been producing agricultural biomass in 2009/2010 of 714.31 Mt, generating 20% of this quantity of agricultural residue, with a perspective of growing more than 20% in ten years of this production (Bergmann et al., 2018). During this period, bigger companies such as BASF and Brasken started the production of biopolymers from biomass, such as polyethylene derived from ethanol produced from biomass feedstock (Coutinho et al., 2011).

1.4 FORESTRY WASTES

1.4.1 Characteristics and Types

Forest wastes are commonly represented by wood bark, hardwood, softwood, sawdust, and a fraction of remnants resulting from tree felling, including stumps,

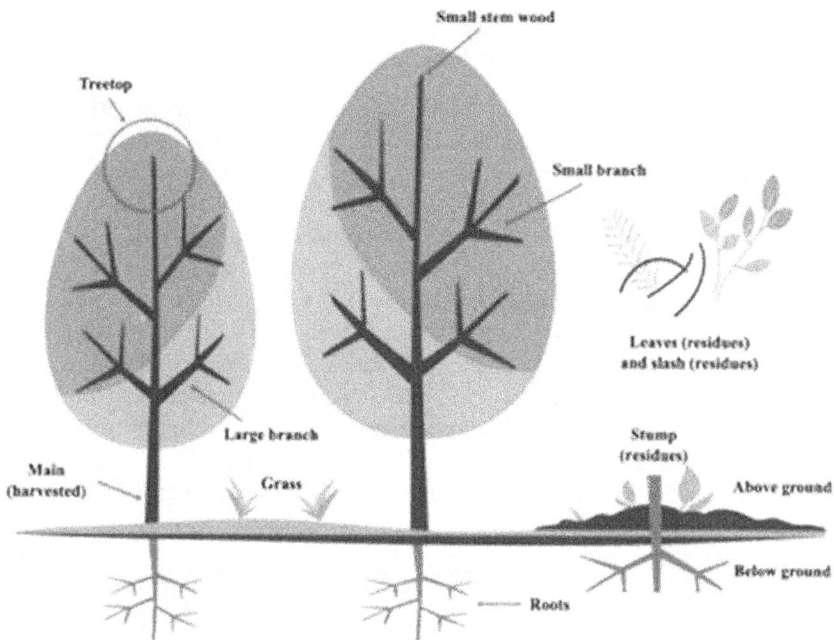

FIGURE 1.1 Schematic of forest wastes of the tree.

Source: Based on Searle and Malins (2013).

leaves, small branches, and small stems at the top of the tree (Figure 1.1) (Hu et al., 2008; Searle and Malins, 2013). Besides, the vegetation that covers the soils is also considered a forest residue, for example, coastal Bermuda grass (Wang et al., 2010).

Forest residues are considered sources of carbon. According to the ultimate analysis, their carbon content is approximately 50% (Cao et al., 2011; Lin and Ku, 2012; Ngo et al., 2014). This type of waste is abundant. Its production potential is found in the forest area (Bilgen and Sarıkaya, 2016). Cao et al. (2017) reported that the carbon availability in a global scenario from forest residues is equivalent to 393.4 petagrams of carbon per year. Although it is known that forest biomass is rich in carbon, more studies are necessary related to its chemical structure, morphology, composition, and thermal properties (Moriana et al., 2015). Some studies related to forest waste report that it is also named lignocellulosic products, due to their structure composed of lignin (18–33%), cellulose (23–42%), and hemicellulose (16–25%) (Shafiei et al., 2010; Moriana et al., 2015). The variation of the lignocellulosic content in forest biomass differs depending on the type of forest residue (remembering that it goes from the leaves to the bark of the trees). Deepening the studies of forest biomass will help exploit its value as a waste, contributing to the technological development of renewable resources.

1.4.2 History of Use Applied to Production of Fuels, Biochemicals, and Products

The use of forest biomass has increased in the last decades. Forest resources include all parts of trees, such as trunk, bark, branch, leaves, and roots (Delucis et al., 2017; Stafford et al., 2020). As shown in Figure 1.2, these residues are used in the bioenergy sector, in the production of biofuels for the generation of heat and electricity, liquid biofuels for transportation, among other bioproducts (Palgan et al., 2016; Kumar et al., 2020).

Over the years, there has been an increase in the use of these forest residues in processes involved in the production of products with high added value. In the 1980s, after the oil crisis in the 1970s, there was a substantial increase in the price of oil. During that period, approximately 1.5×10^{11} tons of plant material were available in the world, less than 1.0% of which was made up of forest resources and of those 30–50% were turned into waste. In 1981, the United States generated approximately 220 million tons of wood and bark residues, such as paper pulp and wood sawdust, annually in the forest products industry (Janshekar and Fiechter, 1982; Ghose, 1978). From these residues, the ability of microorganisms such as *Clostridia* to produce acetate, lactate, acetone, butanol, etc. from cellulosic residues, such as sawdust, straw, bark, was already observed (Grossbard et al., 1979; Janshekar and Fiechter, 1982). In that period, organic ethanol production was already technically feasible; however, its economic viability was not yet defined. Thus, the ethanol obtained was not commercially available (Janshekar and Fiechter, 1982).

Since the eighties and nineties, there were limitations in the processes of obtaining ethanol from biomass due to the pretreatments for sugar release. Studies have already used acid hydrolysis to produce fermentable sugars, and this would later be metabolized in ethanol by microorganisms. In the work realized by Ogundipe and Lu (1989), sawdust was treated with hydrogen peroxide, followed by acid hydrolysis. This pretreatment provided better yield in the concentration of sugars to be fermented by *Saccharomyces uvarum* and *Pachysolen tannophilus*. Forest residues have the potential to contribute approximately an additional equivalent amount of ethanol.

In the 1990s, fuels from plant biomass were responsible for supplying 13% of the world's energy demand. The main biomasses consisted of forest and agrarian waste (Engstrom, 1999). It is observed that the demand for renewable products has increased in the last decade, so research has been increasing to improve the efficiency of bioprocesses using forestry biomass. For example, Stenberg's work in 1997 tested the pretreatment of *Pinus sylvestris* by impregnating with SO_2 and steam (Stenberg et al., 1998).

For the beginning of the twenty-first century, there still is a constant search to establish bioenergy and biofuel industries around the world, including research to adjust these industries to developing countries, to get an economical and environmentally sustainable energy supply (Junginger et al., 2001; Mier-Alba et al., 2020). In Europe, between the years 2009 and 2014, there was an increase in the use of forest residues applied in biorefineries. Since the main waste used was the sawmill, and as shown in Figure 1.3, this raw material increased between these years worldwide (Kumar et al., 2020).

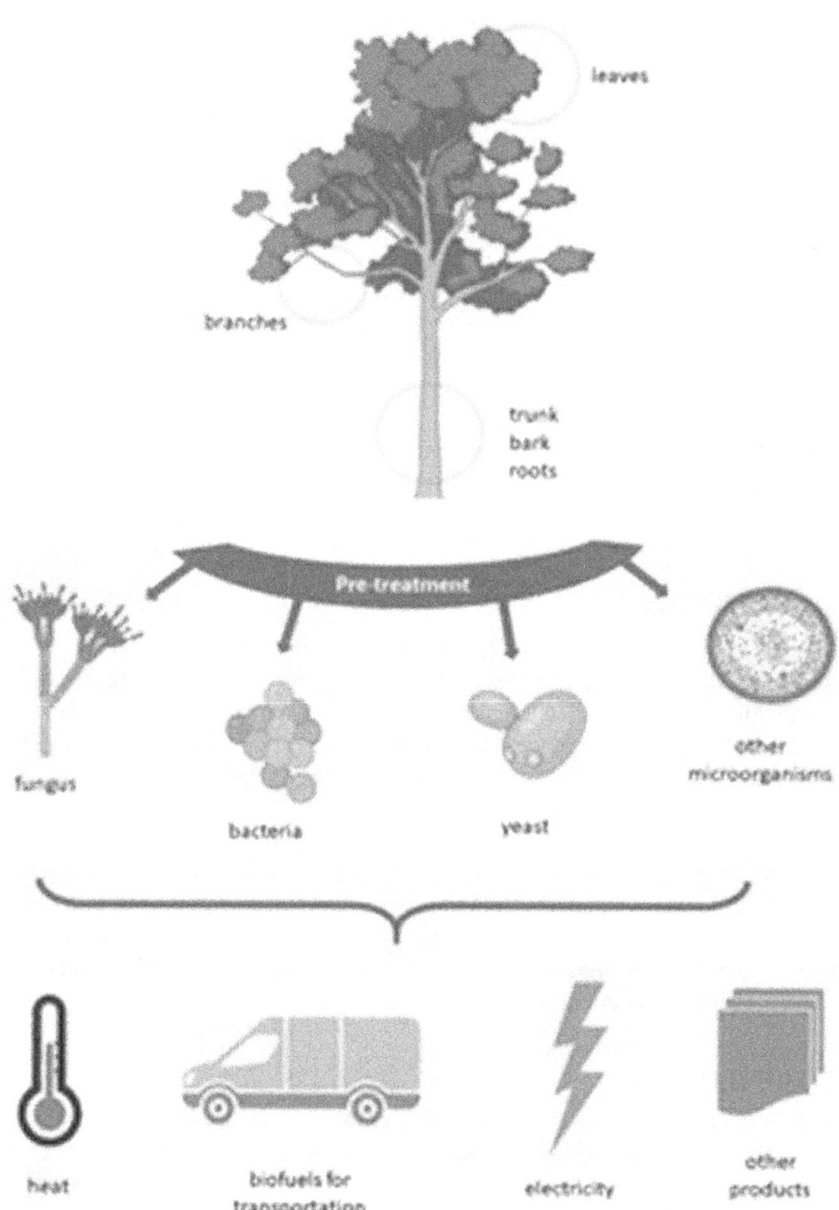

FIGURE 1.2 Diagram of biotechnological services and products produced from forestry residues.

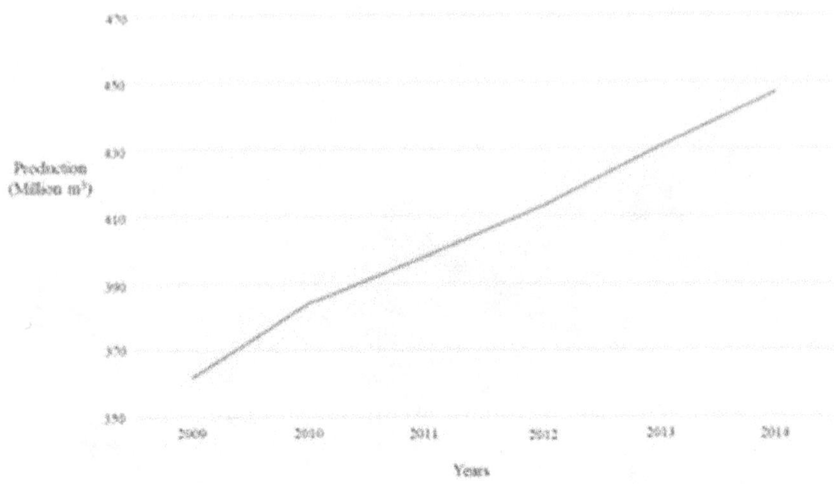

FIGURE 1.3 Worldwide sawmill production between 2009 and 2014.

Source: Based on Sawmill database.

1.5 CONCLUSION

The chemical structure of lignocellulosic biomasses has been covered as well as the types which are derived from different industrial fields, overviewing the opportunities to address economical solutions to related issues of each area. Furthermore, it is undoubtedly true that lignocellulosic wastes have a major historical background which has provoked several former practices to change. Their abundance has driven many governments, especially those whose economic trade-offs are empowered by croplands and agriculture, to seek alternatives to aid the hurdles of the excessive use of fossil fuels and the consequent global warming. A shift in policy-making from benefiting petrochemicals to biomass-based products poses quite a challenge. Therefore, it is of paramount importance to emphasize the role of lignocellulosic materials towards a sustainable era, and to move forward, the acknowledgment of the former policies is highly important to establish future frameworks.

REFERENCES

Ali, M., Saleem, M., Khan, Z., and Watson, I. A. (2019). The use of crop residues for biofuel production. *Biomass, Biopolymer-Based Materials, and Bioenergy*: 369–395.

ArchiLine Wooden Houses. 2021. Wood main physical properties. https://ownwoodenhouse.com/index.pl?act=NEWSSHOW&id=2011062001.

Asim, N., Emdadi, Z., Mohammad, M., Yarmo, M. A., and Sopian, K. (2015). Agricultural solid wastes for green desiccant applications: An overview of research achievements, opportunities and perspectives. *Journal of Cleaner Production* 91: 26–35. https://doi.org/10.1016/j.jclepro.2014.12.015.

Barcelos, M. C., Ramos, C. L., Kuddus, M., Rodriguez-Couto, S., Srivastava, N., Ramteke, P. W., Mishra P. K., and Molina, G. (2020). Enzymatic potential for the valorization of agro-industrial by-products. *Biotechnology Letters* 42(10): 1799–1827. https://doi.org/10.1007/s10529-020-02957-3.

Bartlett, A. I., Hadden, R. M., and Bisby, L. A. (2019). A review of factors affecting the burning behaviour of wood for application to tall timber construction. *Fire Technology* 55(1): 1–49. https://doi.org/10.1007/s10694-018-0787-y.

Bergmann, J. C., Trichez, D., Sallet, L. P., e Silva, F. C. D. P., and Almeida, J. R. (2018). Technological advancements in 1G ethanol production and recovery of by-products based on the biorefinery concept. In *Advances in Sugarcane Biorefinery*, 73–95. Elsevier.

Bilgen, S., and Sarıkaya, İ. (2016). Utilization of forestry and agricultural wastes. *Energy Sources, Part A: Recovery, Utilization, and Environmental Effects* 38(23): 3484–3490.

Cao, J. P., Xia, X. B., Zhang, S. Y., Zhao, X. Y., Sato, K., Ogawa, Y., Wei, X. Y., and Takarada, T. (2011). Preparation and characterization of bio-oils from internally circulating fluidized-bed pyrolyses of municipal, livestock, and wood waste. *Bioresource Technology* 102: 2009–2015.

Cao, L., Zhang, C., Chen, H., Tsang, D. C., Luo, G., Zhang, S., and Chen, J. (2017). Hydrothermal liquefaction of agricultural and forestry wastes: State-of-the-art review and future prospects. *Bioresource Technology* 245: 1184–1193.

CivilSeek. (2021). Top 6 properties of wood and timber used in construction. https://civilseek.com/properties-of-wood-timber Accessed March 17, 2021.

Clauser, N. M., Felissia, F. E., Area, M. C., & Vallejos, M. E. (2021). A framework for the design and analysis of integrated multi-product biorefineries from agricultural and forestry wastes. *Renewable and Sustainable Energy Reviews* 139: 1106–1187.

Coutinho, P., et al. (2011). Technological roadmap in renewable raw materials: A base for the construction of policies and strategies in Brazil. *Quimica Nova* 34: 910–916.

Delucis, R. D. A., Santos, P. S. B. D., Beltrame, R., and Gatto, D. A. (2017). Chemical and fuel properties of forestry wastes from pine plantations. *Revista Árvore* 41(5).

Den, W., Sharma, V. K., Lee, M., et al. (2018). Lignocellulosic biomass transformations via greener oxidative pretreatment processes: Access to energy and value-added chemicals. *Frontiers in Chemistry* 6: 141.

Dillen, J. R., Dillén, S., and Hamza, M. F. (2016). Pulp and paper: Wood sources. In *Saleem Hashmi. Reference Module in Materials Science and Materials Engineering*, 1–6. Elsevier. https://doi.org/10.1016/B978-0-12-803581-8.09802-7.

FAO. (2010). *Global Forest Resources Assessment 2010*. FAO.

FAO. (2016). *FAOSTAT DATA. Forest Production and Trade*. FAO.

Farrell, A. E., Plevin, R. J., Turner, B. T., et al. (2006). Ethanol can contribute to energy and environmental goals. *Science* (5760) (January): 506–508. https://science.sciencemag.org/content/311/5760/506/tab-pdf.

Feng-Wu, B., et al. (2019). Fuel ethanol production from lignocellulosic biomass. *Comprehensive Biotechnology* 3: 49–65.

Ghose, T. K. (1978). Bioconversion of cellulosic substances into energy chemicals and microbial protein. Conference: Symposium on bioconversion of cellulosic substances into energy, chemicals and protein, New Delhi, India, 7–23 Feb 1977, India: N. p., 1978. Web.

Girelli, A. M., Astolfi, M. L., and Scuto, F. R. (2020). Agro-industrial wastes as potential carriers for enzyme immobilization: A review. *Chemosphere* 244: 125368. https://doi.org/10.1016/j.chemosphere.2019.125368.

Grossbard, E. (1979). Straw decay and its effect on disposal and utilization. In *Symposium on Straw Decay (1979: Hatfield Polytechnic); Workshop on Assessment Techniques (1979: Hatfield Polytechnic)*. J. Wiley.

Hu, G., Heitmann, J. A., and Rojas, O. J. (2008). Feedstock pretreatment strategies for producing ethanol from wood, bark, and forest residues. *BioResources* 3(1): 270–294.

Janshekar, H., and Fiechter, A. (1982). Engenharia bioquímica. Em *New Trends in Chemistry*, 97–126. Springer.

Jones, D., and Sandberg, D. (2020). A review of wood modification globally—updated findings from COST FP1407. *Interdisciplinary Perspectives on the Built Environment*: 1. https://doi.org/10.37947/ipbe.2020.vol1.1.

Junginger, M., Faaij, A., Van den Broek, R., Koopmans, A., and Hulscher, W. (2001). Fuel supply strategies for large-scale bio-energy projects in developing countries. Electricity generation from agricultural and forest residues in Northeastern Thailand. *Biomass and Bioenergy* 21(4): 259–275.

Kapoor, R., Ghosh, P., Kumar, M., Sengupta, S., Gupta, A., Kumar, S. S., . . . Pant, D. (2020). Valorization of agricultural waste for biogas based circular economy in India: A research outlook. *Bioresource Technology* 304: 123036.

Kelkar, V. M., Geils, B. W., Becker, D. R., Overby, S. T., and Neary, D. G. (2006). How to recover more value from small pine trees: Essential oils and resins. *Biomass and Bioenergy* 30(4), 316–320. https://doi.org/10.1016/j.biombioe.2005.07.009.

Koutsianitis, D., Mitani, C., Giagli, K., et al. (2015). Properties of ultrasound extracted bicomponent lignocellulose thin films. *Ultrasonics Sonochemistry* 23: 148–155.

Kumar, A., Adamopoulos, S., Jones, D., and Amiandamhen, S. O. (2020). Forest biomass availability and utilization potential in Sweden: A review. *Waste and Biomass Valorization*: 1–16.

Lai, M., Zhang, L., Lei, L., Liu, S., Jia, T., and Yi, M. (2020). Inheritance of resin yield and main resin components in Pinus elliottii Engelm at three locations in southern China. *Industrial Crops and Products* 144: 112065. https://doi.org/10.1016/j.indcrop.2019.112065.

Leontopoulos, S. V., and Arabatzis, G. (2021). The contribution of energy crops to biomass production. *Low Carbon Energy Technologies in Sustainable Energy Systems*: 47–113.

Lin, T. Y., and Kuo, C. P. (2012). Study products yield of bagasse and sawdust via slow pyrolysis and iron-catalyze. *The Journal of Analytical and Applied Pyrolysis* 96: 203–209.

Liu, Z. H., Le, R. K., Kosa, M., et al. (2019). Identifying and creating pathways to improve biological lignin valorization. *Renewable and Sustainable Energy Reviews* 105: 349–362.

Luft, L., Confortin, T. C., Todero, I., et al. (2019). Ultrasound technology applied to enhance enzymatic hydrolysis of brewer's spent grain and its potential for production of fermentable sugars. *Waste and Biomass Valorization* 10 (8) (February): 2157–2164.

Manna, M. C., et al. (2018). Bio-waste management in subtropical soils of India: Future challenges and opportunities in agriculture. *Advances in Agronomy* 152: 87–148.

Mier-Alba, E., Sánchez-Muñoz, S., Barbosa, F. G., Garlapati, V. K., Balagurusamy, N., da Silva, S. S., . . . Chandel, A. K. (2020). Comparative analysis of biogas with renewable fuels and energy: Physicochemical properties and carbon footprints. In *Biogas Production*, 125–143. Springer.

Miller, R. B. (1999). Structure of wood. In *Wood Handbook: Wood as an Engineering Material*, 2.1–2.4. USDA Forest Service, Forest Products Laboratory. General technical report FPL; GTR-113.

MNRE (Ministry of New, Renewable Energy Resources). (2021). *Govt of India*. www.mnre.gov.in/biomassrsources.

Moriana, R., Vilaplana, F., and Ek, M. (2015). Forest residues as renewable resources for bio-based polymeric materials and bioenergy: Chemical composition, structure and thermal properties. *Cellulose* 22(5): 3409–3423.

Ngo, T. A., Kim, J., and Kim, S. S. (2014). Characteristics of palm bark pyrolysis experiment oriented by central composite rotatable design. *Energy* 66: 7–12.

Obi, F. O., Ugwuishiwu, B. O., and Nwakaire, J. N. (2016). Agricultural waste concept, generation, utilization and management. *Nigerian Journal of Technology* 35(4): 957–964. https://doi.org/http://dx.doi.org/10.4314/njt.v35i4.34.

Ogundipe, A., and Lu, J. Y. (1989). Ethanol production from dilute acid hydrolyzate of sawdust treated with hydrogen peroxide. *Biomass* 20(3–4): 291–299.

Onu, P., and Mbohwa, C. (2021). Waste management and the prospect of biodegradable wastes from agricultural processes. *Agricultural Waste Diversity and Sustainability Issues*: 1–20. https://doi.org/10.1016/b978-0-323-85402-3.00006-1.

Pagliaro, M., Albanese, L., Scurria, A., Zabini, F., Meneguzzo, F., and Ciriminna, R. (2021). Tannin: A new insight into a key product for the bioeconomy in forest regions. *Preprints*: 2021020266. https://doi.org/10.20944/preprints202102.0266.v1.

Palgan, Y. V., and McCormick, K. (2016). Biorefineries in Sweden: Perspectives on the opportunities, challenges and future. *Biofuels, Bioproducts and Biorefining* 10(5): 523–533.

Patel, M., Zhang, X., and Kumar, A. (2016). Techno-economic and life cycle assessment on lignocellulosic biomass thermochemical conversion technologies: A review. *Renewable and Sustainable Energy Reviews* 53: 1486–1499.

Pizzi, A. (2019). Tannins: Prospectives and actual industrial applications. *Biomolecules* 9: 344. https://doi.org/10.3390/biom9080344.

Ragauskas, A. J., Beckham, G. T., Biddy, M. J., Chandra, R., Chen, F., Davis, M. F., Davison, B. H., Dixon, R. A., Gilna, P., Keller, M., Langan, P., Naskar, A. K., Saddler, J. N., Tschaplinski, T. J., Tuskan, G. A., and Wyman C. E. 2014. Lignin valorization: Improving lignin processing in the biorefinery. *Science* 344(6185). http://dx.doi.org/10.1126/science.1246843.

Ramage, M. H., Burridge, H., Busse-Wicher, M., Fereday, G., Reynolds, T., Shah, D. U., Rezania, S., Oryani, B., Cho, J., et al. (2020). Different pretreatment technologies of lignocellulosic biomass for bioethanol production: An overview. *Energy* 199: 117457.

Sadh, P. K., Duhan, S., and Duhan, J. S. (2018). Agro-industrial wastes and their utilization using solid state fermentation: A review. *Bioresources and Bioprocessing* 5(1): 1–15. https://doi.org/10.1186/s40643-017-0187-z.

Sánchez-Muñoz, S., Mariano-Silva, G., Leite, M. O., Mura, F. B., Verma, M. L., da Silva, S. S., and Chandel, A. K. (2020). Production of fungal and bacterial pigments and their applications. In *Biotechnological Production of Bioactive Compounds*, 327–361. Elsevier.

Santos, C. M., et al. (2020). Produção de Bioetanol a partir dos Residuos da Indústria de Papel. *Vir. Química* 12(4).

Satari, B., and Karimi, K. (2018). Citrus processing wastes: Environmental impacts, recent advances, and future perspectives in total valorization. *Resources, Conservation and Recycling* 129: 153–167. https://doi.org/10.1016/j.resconrec.2017.10.032.

Searle, S., and Malins, C. (2013). *Availability of Cellulosic Residues and Wastes in the EU*. International Council on Clean Transportation.

Shafiee, S., and Topal, E. (2009). When will fossil fuel reserves be diminished? *Energy Policy* 37 (1) (September): 181–189.

Shafiei, M., Karimi, K., and Taherzadeh, M. J. (2010). Pretreatment of spruce and oak by N-methylmorpholine-N-oxide (NMMO) for efficient conversion of their cellulose to ethanol. *Bioresource Technology* 101(13): 4914–4918.

Singh, D., Zerriffi, H., Bailis, R., and LeMay, V. (2021b). Forest, farms and fuelwood: Measuring changes in fuelwood collection and consumption behavior from a clean cooking intervention. *Energy for Sustainable Development* 61: 196–205. https://doi.org/10.1016/j.esd.2021.02.002.

Singh, M., Babanna, S. K., Kumar, D., Dwivedi, R. P., Dev, I., Kumar, A., Tewari, R. K., Chaturvedi, O. P., and Dagar, J. C. (2021a). Valuation of fuelwood from agroforestry systems: A methodological perspective. *Agroforestry Systems*: 1–17. https://doi.org/10.1007/s10457-020-00580-9.

Situmorang, Y. A., Zhao, Z., Yoshida, A., Abudula, A., and Guan, G. (2020). Small-scale biomass gasification systems for power generation (< 200 kW class): A review. *Renewable and Sustainable Energy Reviews* 117: 109486.

Stafford, W., De Lange, W., Nahman, A., Chunilall, V., Lekha, P., Andrew, J., . . . Trotter, D. (2020). Forestry biorefineries. *Renewable Energy* 154: 461–475.

Stenberg, K., Tengborg, C., Galbe, M., and Zacchi, G. (1998). Optimisation of steam pretreatment of SO2-impregnated mixed softwoods for ethanol production. *Journal of Chemical Technology & Biotechnology: International Research in Process, Environmental AND Clean Technology* 71(4): 299–308.

Temmes, A., and Peck, P. (2020). Do forest biorefineries fit with working principles of a circular bioeconomy? A case of Finnish and Swedish initiatives. *Forest Policy and Economics* 110: 101896. https://doi.org/10.1016/j.forpol.2019.03.013.

Tsoumis, G. T. (2020). Wood. *Encyclopedia Britannica.* www.britannica.com/science/wood-plant-tissue. Accessed March 17, 2021.

Tye, Y. Y., Lee, K. T., Abdullah, W. N. W., et al. (2016). The world availability of non-wood lignocellulosic biomass for the production of cellulosic ethanol and potential pretreatments for the enhancement of enzymatic saccharification. *Renewable and Sustainable Energy Reviews* 60: 155–172.

Urbaniec, K., and Bakker, R. R. (2015). Biomass residues as raw material for dark hydrogen fermentation—A review. *International Journal of Hydrogen Energy* 40(9): 3648–3658. https://doi.org/10.1016/j.ijhydene.2015.01.073.

Wang, Z., Keshwani, D. R., Redding, A. P., and Cheng, J. J. (2010). Sodium hydroxide pretreatment and enzymatic hydrolysis of coastal Bermuda grass. *Bioresource Technology* 101(10): 3583–3585.

Wiedenhoeft, A. C. (2010). Structure and function of wood. In *Wood Handbook, Wood as an Engineering Material*, ed. R. S. Ross, 3–1, 3–18. Centennial ed. General Technical Report FPL; GTR-190. U.S. Dept. of Agriculture, Forest Service, Forest Products Laboratory.

Wiedenhoeft, A. C. (2013). Structure and function of wood. In *Handbook of Wood Chemistry and Wood Composites*, ed. R. M. Rowell, 9–32. CRC Press, Taylor & Francis Group.

Woodard, A. C., and Milner, H. R. (2016). Sustainability of timber and wood in construction. In *Sustainability of Construction Materials*, 129–157. Woodhead Publishing. https://doi.org/10.1016/B978-0-08-100370-1.00007-X.

Woodford, C. (2017). Wood. www.explainthatstuff.com/wood.html. Accessed March 17, 2021.

Zabed, H., Sahu, J. N., Suely, A., et al. (2017). Bioethanol production from renewable sources: Current perspectives and technological progress. *Renewable and Sustainable Energy Reviews* 71: 475–501.

Zhu, J. Y., and Pan, X. J. (2010). Woody biomass pretreatment for cellulosic ethanol production: Technology and energy consumption evaluation. *Bioresource Technology* (13) (December): 4992–5002.

Zuin, V. G., Segatto, M. L., and Ramin, L. Z. (2018). Plants as resources for organic molecules: Facing the green and sustainable future today. *Current Opinion in Green and Sustainable Chemistry* 9: 1–7. https://doi.org/10.1016/j.cogsc.2017.10.001.

2 Recent Advancements and Prospects of Using Lignocellulosic Wastes for Microbial Production of Fuels, Biochemicals, and Products

Antunes, F. A. F., Ingle, A. P., Abdeshahian, P., Ribeaux, D. R., Prado, C. A., Muñoz, S. S., Barbosa, F. G., Balbino, T. R., Castro-Alonso, M. J., Reyes-Guzman, R., Arruda, G. L., Cruz-Santos, M. M., Mier-Alba, E., Rocha, T. M., Ruiz, E. D., Melo, Y. C. S., Cunha, M. L. S., Santos, J. C., and da Silva, S. S.

Department of Biotechnology, Engineering School of Lorena, University of São Paulo, Lorena, Brazil

CONTENTS

2.1 INTRODUCTION

Lignocellulosic wastes are natural and renewable resource of biomass. These are ubiquitous and abundantly present on the planet Earth (Ingle et al. 2019). Lignocellulosic wastes are an inexpensive, cellulose-based biomass, and the important sources of lignocellulosic wastes mainly include agricultural residues, energy crops, forest

DOI: 10.1201/9781003191247-2

residues, cellulosic wastes, etc. It is estimated that the annual global production of lignocellulosic wastes is about 181.5 billion tons, and the estimated cost of this waste is between $24 and $121 per ton, depending on the crop, yield, region, and method of analysis. Among these, the USA alone produces about 1.3 billion tons of lignocellulosic wastes annually to generate bioenergy (Haq et al. 2021).

As discussed above, a considerable amount of such lignocellulosic wastes is being generated all over the world through agricultural and forest practices and industries associated with these sectors (Ingle et al. 2020). The management of a huge quantity of waste in a sustainable manner is very difficult, and hence, it becomes a major concern for the whole world. However, as a routine management strategy, such lignocellulosic wastes are usually disposed of by burning in both developed and developing countries (Anwar et al. 2014). The burning of waste raises the issue related to air pollution, which directly and indirectly affects the environment and human health. Therefore, utilization of generated lignocellulosic wastes in an appropriate and sustainable manner is the need of the hour.

Recently, lignocellulosic wastes have gained increasing research interests and special attention because of their rich composition and renewable nature. As a result, most countries have recently focused on the valorization of such lignocellulosic wastes and their components into value-added products, including biofuels, value-added fine biochemicals, etc., through microbial fermentation by setting up second-generation (2nd) biorefineries (Nanda et al. 2014).

Usually, microbial systems present in different habitats naturally produce a wide range of bioactive compounds that can be commonly used as fuels, drugs, and other important chemicals (Chubukov et al. 2016; Ahmed et al. 2018). It has been well-known for many years that microbes can play a promising role in the production of biofuel through the biosynthesis of enzymes that act on a variety of lignocellulosic feedstocks (Ceballos et al. 2005). Similarly, microbial systems have the potential to utilize different lignocellulosic wastes and convert them into various high-value products (Jiang et al. 2019). The conversion of lignocellulosic biomass into different valuable products is usually performed through the depolymerization of polysaccharides present in these wastes by the action of specific enzymes produced by these microbes. However, microbial strain development and improvement through genetic engineering and optimization of fermentation parameters are extremely necessary to achieve a higher production yield of such products and also to make the process eco-friendly and economically viable (Adegboye et al. 2021).

Considering the importance of lignocellulosic wastes for smooth functioning of modern biorefineries and the production of different biofuels, biochemicals, and other high-value products under the theme "best from waste," in the present we are focusing on current prospects regarding the availability of such lignocellulosic wastes, technical advancements, and reminiscent challenges to prompt critical insights for the provision of more efficient processes in a modern biorefinery scenario.

2.2 PERSPECTIVES FOR WOOD APPLICATION

The use of wood as an energy source dates back from the mid-1800s. This material continues to be an important feedstock around the world, especially in developing countries, for daily routines (IEA 2020). Actually, the major producers of forest products (from wood) are India (wood fuel—Figure 2.1), United States (industrial roundwood, pulp of paper, wood pellets, and other agglomerates), Brazil (wood charcoal), and China (sawn wood, wood-based panels, recovered paper, paper, and paperboard) (FAO 2019).

The importance of wood ranges from the highest industrial levels to the smallest villages. For example, in 2019, about 2.3% of total US annual energy consumption was from wood and wood by-products. Another example, in 2018, the European Union demand for wood pellets (an alternative product for charcoal, firewood, oil and gas in heating, cooking, boiler, and power plants) amounted to approximately 23 million metric tons (Eurostat 2019: Thrän et al. 2019; Kittler et al. 2020). Indeed, demand for forest products will continue growing according to population increase. The most recent FAO projections estimate that by 2030, global consumption of industrial roundwood will rise by 60% over current levels to around 2,400 million m^3. Substantial rises are also likely in the consumption of paper and paperboard products (FAO 2015).

Nevertheless, several external factors impact this important market. For example, one year after the SAR-CoV-2 (COVID-19) pandemic, our world has turned upside down. The pandemic goes far beyond the disease itself. Today in early 2021, it has been shown that the world faces a series of challenges and makes evident the fragility in sectors of importance for the development of each country (WHO 2020, FAO 2020). The planned projections have had to undergo necessary modifications and adjustments to cope with an uncertain and constantly changing future. However, the demand for renewable and sustainable energy will grow worldwide to meet the Paris Agreement and the global push towards reducing anthropogenic GHG emissions (IEA 2021; UNCTAD 2019). Moreover, the planet continues to struggle with deforestation due to fires caused by droughts or that are man-made (FAO-UN 2021).

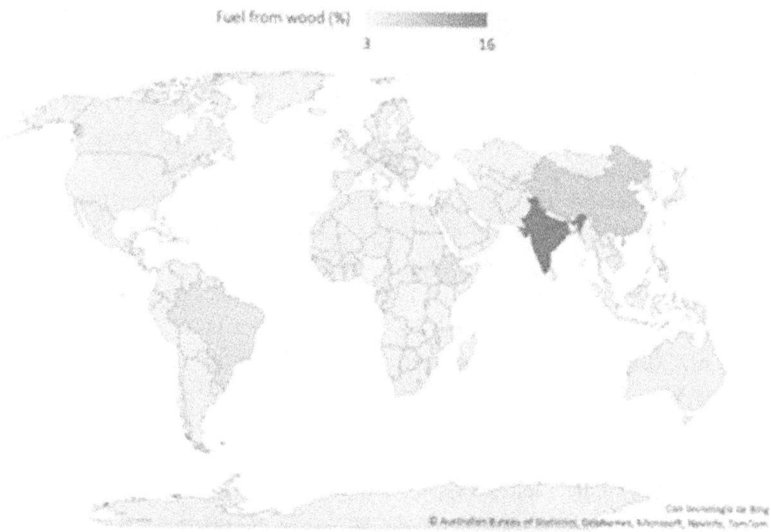

FIGURE 2.1 Major producers of fuel from wood around the world (from FAO 2019).

Events in California in the United States and the Amazon in Brazil are examples of fires that occurred in the past years that presented a loss of 2.5 million hectares in August 2019 alone, according to the National Institute for Space Research of Brazil (INPE 2019).

The wood market is facing widespread competition in global wood-forest-product trade, and the trading and competitive environment may become more complex in the future (Long et al. 2019). For example, in the absence of friction between global markets, the law of one price states that the price for a specific wood assortment will be the same (Chudy & Hagler 2020). However, in certain countries or regions, there is illegal logging with low market prices, which causes losses to an industry that seeks a sustainable way to meet demands within the market. The quality and characteristics of the product (wood) are also bottlenecks in this biological process. The process demands time for the growth and development of the trees because of the varieties and nutritional requirements for their production. In some cases, this situation is commercially negative for the economic development of this market (Nepal et al. 2019; Kuuluvainen et al. 2021; Chudy et al. 2020).

Faced with the challenges previously foreseen and those specific to each place or each region, the markets begin to return to supply and demand after the 2020 situation. They are the new niches of high added-value products that present opportunities of commercial interest and strengthen the circular economy, which seeks to adapt to times of change and generate ideal conditions for markets (OECD 2021; NewBio 2014).

2.3 PERSPECTIVES FOR AGRICULTURAL WASTE APPLICATION

Large quantities of lignocellulosic agricultural wastes are produced throughout the world, approximately 147.2 million metric tons per year (Sarkar et al. 2021). The type of these agriculture residues varies depending on the crop grown in each region. The high availability and variety of these materials allow for generating different products, and their use as renewable, low-cost raw material helps raise the production of several compounds, such as biofuels and biochemicals. For instance, bioethanol production is estimated to increase 49 billion liters per year using the total crop residues and wasted crops added to lignocellulosic biomass rather than using just the biomass (Saini et al. 2015).

In 2019, as per FAO statistics, the annual global production of sugarcane, corn, wheat, and rice was around 1949.3, 1148.4, 765.7, and 755.4 million tons, respectively. China and India were the main world producers of rice and wheat. In general, about 1.0–1.5 kg of rice straw can be produced from the harvest of 1.0 kg of rice grain (Abraham et al. 2016), and the yield of wheat straw is around 7.3 tons.ha^{-1} (Dai et al. 2016). Regarding the crop of corn, the United States of America is the largest producing country in the world (FAO STAT 2019). Approximately 48% of corn straw can be produced concerning the total dry mass of the crop, being that leaves and stems have the largest amount of total dry mass of the straw (Shinners & Binversie 2007). The yield of corn straw is about 4.7 tons.ha^{-1} (Sokhansanj et al. 2002). Considering the culture of sugarcane, Brazil stands out as the leading world producer (FAO STAT 2019). Approximately 0.3 metric tons of cane bagasse can be obtained for every 1.0 metric ton of sugarcane (Bezerra & Ragauskas 2016). Based on the forecast that the global demand for food will double in the future fifty years (Andrade et al. 2021), the generation of these residuals derived from agriculture also tends to rise.

Amid a large amount of agricultural waste, the choice of the type of lignocellulosic material to be used must be evaluated, as this influences the price of the final bioproduct. A case study in India showed that the costs of wheat straw were about 19% higher than of rice straw (Duncan et al. 2020). The price of rice straw is from \$20.5 to 65.4 dry.t^{-1}, varying according to the distance that the material needs to be transported (Diep et al. 2015). So it is needed to evaluate the costs of the raw material, the production procedure, and the price of the final bioproduct to find out if the process is economically rentable.

Agricultural wastes have shown promising results as feedstocks for microbial production of fuels, biochemicals, and products on lab and pilot scale and in an environment-friendly and sustainable manner compared to other biomass sources.

For example, agricultural waste can be applied in energy production in noneffective ways, such as thermal and electric energy, while it could be used as raw material for biorefineries, due to its large availability (Forster-Carneiro 2013; Bergmann et al. 2018). Among the 1.3 billion dry tons of lignocellulose material produced annually in the USA, agricultural residues contribute to 933 million tons per year during 2017 (De Bhowmick et al. 2018; Manna et al. 2018). One value-added application for corncob and corn straw has been 2G ethanol production, which is growing in the USA (Ali et al. 2019). In a ten-year period, from 2012 to 2022, it was estimated that the production of cellulose ethanol grew from 0.25 to 16 billion gallons (Feng-Wu et al. 2019; Clauser et al. 2021).

As an agriculture-based country, India has 1.79 Mkm^2 of agricultural land, producing around 500 MT of agricultural waste per year during 2020 (Kapoor et al. 2020). They are commonly burned by farmers, though these residues may also be applied as raw material in waste valorization processes, obtaining biofuels, energy, and chemicals (Coutinho et al. 2011; Bergmann 2018). An established process that utilizes agricultural residue in India is biomethanation, which is the first option of waste valorization for a great percentage of the generated waste (Manna et al. 2018; MNRE 2021).

The estimation of agro-industrial waste generated from agriculture, food processing, and alcohol industries in China, Brazil, India, and the US reaches 3,500 million tons/year, with agricultural waste accounting for 1,100 million tons/year in China, Europe, and India and for 240 million tons/year in the USA (Kapoor et al. 2020; Clauser et al. 2021). This number shows that the agricultural residues will have an important contribution to the energetic development in the world, such as in 2G ethanol (Santos et al. 2020; Leontopoulos & Arabatzis 2021).

However, the utilization of agricultural wastes still confronts some challenges for the industrialization of the bioprocesses, as shown in Figure 2.2.

The main challenges involve the (1) collection, transportation, and handling of agricultural wastes; (2) reduction of costs in pretreatment technologies; (3) exploration and the application of efficient thermophilic and thermotolerant microorganisms for bioprocesses on an industrial scale; (4) reduction in the cost of cellulolytic

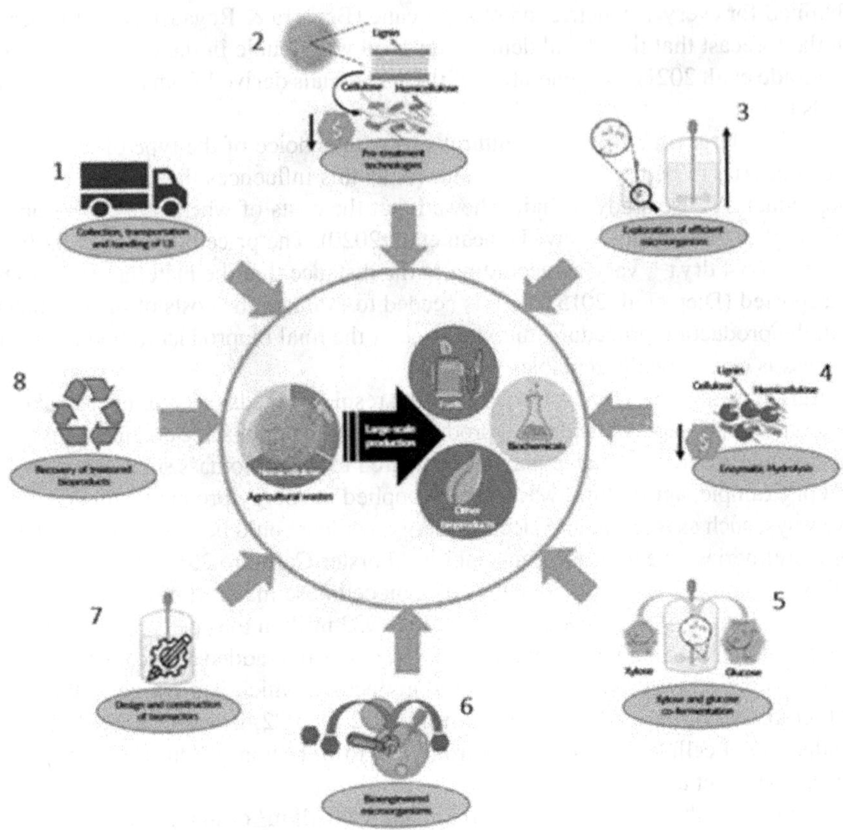

FIGURE 2.2 Some challenges for bioprocesses industrialization.

enzymes for bioconversion of agricultural wastes; (5) advancement of xylose and glucose cofermentation by microorganisms; (6) development and application of recombinant/metabolically bioengineered microbial strains, to obtain higher yields in large-scale production of biofuels, biochemicals, and other bioproducts; (7) design and build of bioreactors with acceptable heat release during the process, to reduce the negative impacts caused in microbial metabolism by the accumulation of heat; and finally, (8) recovery and utilization of nutrients and other potentially treasured bioproducts (e.g., lignin) derived from microbial production of fuels, chemicals, and other products from agricultural biomass (Sarsaiya et al. 2019; Clauser et al. 2021).

Advancements in addressing the challenges mentioned above will lead to the exclusion of current technology and bottlenecks of bioprocesses for the efficient conversion of agricultural wastes. It is important to highlight that the huge availability of feedstocks from agronomy and the remarkable efforts being carried by researchers to make bioprocesses more cost-effective indicate that large-scale production of biofuels, biochemicals, and other bioproducts will be achieved forthcoming (Saini et al. 2015; Sarsaiya et al. 2019; Clauser et al. 2021).

2.4 PERSPECTIVES FOR FORESTRY WASTES APPLICATION

Over the period of the past few years, processes that aim to harness lignocellulosic residues from forests have been gaining more and more attention, mainly because they work in the design of strategies for the development of biorefineries (Stafford et al. 2020). However, for such biomass to be valued, efforts are needed to enable waste management (Braghiroli & Passarini 2020) so that a portion is maintained to fulfill its biological role of providing nutrients for the growth of surrounding vegetation and trees (Gregg et al. 2020), and another part is their use as feedstock for biorefineries (Han et al. 2018).

The residues obtained from forest operations have the potential to be used for the generation of bioproducts according to their composition and availability (Millati et al. 2019). The recovery of such waste must take place through the development of efficient methods of extracting biomass from forests, and its subsequent recovery can contribute to mitigating the costs of forest restoration, to reduce the risk of forest fire and reduce atmospheric contamination by burning (Han et al. 2018). The processing of forestry wastes is expected to obtain benefits: correct disposal of waste, reduction of greenhouse gas emissions due to increased energy autonomy in detriment to fossil fuels, generation of sustainable bioproducts and biofuels, creation of jobs, development of rural economies, among others (Lynd 2017; Clauser et al. 2021a, 2021b).

The overall production of hardwood and softwood (the two major forest residues) in the year 2013 and 2016 was found to be 58 million m^3 and 166 million m^3, respectively (Millati et al. 2019). Therefore, such forest wastes can be used as promising feedstock in biorefineries as they are available in large amounts. The development and consolidation of techniques for the use of these wastes can move billionaire markets of bioplastics, biolubricants, biosolvents, biosurfactants, and biopharmaceuticals (Gregg et al. 2020), in addition to the ethanol market, since forest biomass is an alternative and cheap substrate for the production of this biofuel (Buzała et al. 2017).

Some countries find factors favorable to the use of forest residues, such as Finland, Brazil, Sweden, Canada, and the United States, as they have large annual production, quality sectors for research and development of such feedstocks, in addition to industrial infrastructure with potential for establishment (Hämäläinen et al. 2011). In the meantime, it was reported in Sweden that there is a generation of about 8TWh energy from such materials (Aryapratama & Janssen 2017). Forecasts indicated that lignocellulosic waste from forest should significantly increase the production of biofuels and bioproducts by 2050, strengthening biotechnological engagement policies and opening economic opportunities for the forestry sector (Gregg et al. 2020).

However, despite the benefits that forest waste offers, there are many technological and economic challenges to overcome. Forest biomass has heterogeneity in its properties, derived from the climatic conditions in which it originated, maturity, soil type, and growth conditions (De Menezes et al. 2009). In addition, structural components of biomass, such as cellulose, must be broken down into simpler sugars before they are converted into energy. This conversion is hampered by the fact that the plants have complex structural and chemical mechanisms that resist decomposition by microorganisms and factors, known as recalcitrance, and it demands the application of pretreatments, which represents a large part of the cost of converting biomass into energy (Zoghlami and Paës 2019).

Moreover, in the economic field, previous works evaluated the installation of a transportable biomass conversion system (forest-to-product) and highlighted that the cost depends mainly on the raw material costs for the installation of the biomass conversion system, the types of products to be processed, the scale of the installation, the location of the installation, and the frequency with which the transportable installation is relocated (Han et al. 2018).

In addition to the economic challenges, it is necessary to implement appropriate government policies targeting biorefineries, with a forward-looking perspective to encourage and support research for new technologies in the utilization and recovery of forest waste for the development of innovative bioproducts (Lynd 2017). Previous studies suggest that government agencies should focus on the specific challenges of each state and region in the use of biomass to design effective policies and encourage competition between renewable energy platforms (Nicholls et al. 2018). Developed countries like the United States have implemented federal and state policies, organized into three different approaches: (1) incentive policies, which aim to reduce the cost of capital and operating costs of projects; (2) regulatory policies that require production and consumption standards; and (3) information policies that include feasibility studies and information dissemination (Figure 2.3) (Nicholls et al. 2018), resulting in a significant influence on the development of forest-based bioenergy (Ebers et al. 2016).

Indeed, it should be noted that government policies, combined with the development of technologies for the utilization of forest biomass, can corroborate the consolidation of biorefineries for the utilization of this waste, resulting in the production of bioproducts of interest and the strengthening of an economy based on sustainable processes around the globe.

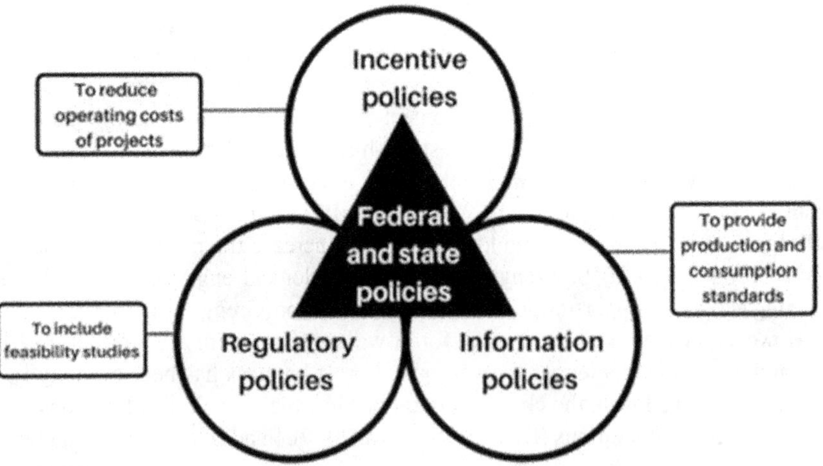

FIGURE 2.3 Use of biomass to design effective policies and encourage competition between renewable energy platforms.

The "biorefinery" concept became stronger, and forestry and other residues are being used as substrate for the biotechnological (or chemical) production of fiber-polymer composites, microbiological cellulose, bioenergy, biofuels, biopigments, among others. For example, in the biotechnological route, bacteria from genera *Gluconacetobacter, Sarcina*, or *Agrobacterium* can be used to produce a particular type of cellulose with multiple applications (Carreira et al. 2011), or fungi and bacteria with human health role, as *Monascus* and *Serratia*, for biopigment production (Sánchez-Muñoz et al. 2020). However, all the available or possible high-value products are derived in services and other products that could attend to diverse areas in society, from health, energy, supplies, nanotechnology, and further (Junginger et al. 2001; Cambero et al. 2015; Väisänen et al. 2016).

Even though the high cost and energy demand from pretreatments for forestry residue biomass still represents a challenge, researchers are continuously working to improve existing pretreatment and are seeking new possibilities. That is how in the second decade of the twenty-first century appeared new options of pretreatment for lignocellulosic materials, such as hydrodynamic cavitation, the application of which could vary in different economic sectors as paper production, biotechnological products, and even in food industry (Hilares et al. 2020; Tsalagkas et al. 2021).

Nowadays, with panorama in 2020–2021 affected by the COVID-19 (SARS-CoV-2) pandemic, global investments have been directed towards pandemic research and public programs to lead with its consequences in different sectors, especially in basic services, such as food and health. However, it brings two possible future scenarios: difficulties for the economy with the effect on supply chains, fossil fuels, and local transport, and, fortunately, a possible focus on green energy and products, sustainability, and an innovation/creativity boost in all sectors, which would lead to improvement of forestry and other waste use, treatment, and application for a sustainable economy, as mentioned above (Rowan & Galanakis 2020).

2.5 CONCLUSION

As one can encounter throughout the chapter, differential types of biomasses share structural commonalities as well as distinguishable aspects which may lead to distinct types of processing conduction. Nevertheless, the likelihood of lignocellulosic utilization is highly country-dependent due to variations in climate, soil, and natural resources. Pretreatments are useful techniques to exploit the full potential of lignocellulose by facilitating enzymatic and microbial accessibility to fermentable sugars. Yet drawbacks involving microbial-efficient biomass fermentation are ubiquitous; thus, the development of new processing design and engineered microbes to aid low productivity is of global interest. Concomitantly, policies towards sustainable practices must follow in order to enable funding and biotechnological investments. In this sense, by gaining a better understanding of microbial fermentative routes, pretreatment design, and downstream processing, one may prompt the development of new technologies which could facilitate and pressure policy-makers to set more ecological goals congruent with the common good.

REFERENCES

Abraham, A., Mathew, A.K., Sindhu, R., Pandey, A., & Binod, P. (2016). Potential of rice straw for bio-refining: An overview. *Bioresource Technology*, 215, 29–36.

Adegboye, M.F., Ojuederie, O.B., Talia, P.M., & Babalola, O.O. (2021). Bioprospecting of microbial strains for biofuel production: Metabolic engineering, applications, and challenges. *Biotechnology for Biofuels*, 14, 5. https://doi.org/10.1186/s13068-020-01853-2.

Ahmed, A.A.Q., Babalola, O.O., & McKay, T. (2018). Cellulase- and xylanase-producing bacterial isolates with the ability to saccharify wheat straw and their potential use in the production of pharmaceuticals and chemicals from lignocellulosic materials. *Waste Biomass Valorization*, 9(5), 765–775.

Ali, M., Saleem, M., Khan, Z., & Watson, I.A. (2019). The use of crop residues for biofuel production. *Biomass, Biopolymer-Based Materials, and Bioenergy*, 369–395.

Andrade, M.C., Silva, C.D.O.G., de Souza Moreira, L.R., & Ferreira Filho, E.X. (2021). Crop residues: Applications of lignocellulosic biomass in the context of a biorefinery. *Frontiers in Energy*, 1–22.

Anwar, Z., Gulfraz, M., & Irshad, M. (2014). Agro-industrial lignocellulosic biomass a key to unlock the future bio-energy: A brief review. *Journal of Radiation Research and Applied Sciences*, 7, 163–173.

Aryapratama, R., & Janssen, M. (2017). Prospective life cycle assessment of bio-based adipic acid production from forest residues. *Journal of Cleaner Production*, 164, 434–443.

Bergmann, J.C., Trichez, D., Sallet, L.P., e Silva, F.C.D.P., & Almeida, J.R. (2018). Technological advancements in 1G ethanol production and recovery of by-products based on the biorefinery concept. In *Advances in Sugarcane Biorefinery* (pp. 73–95). Elsevier.

Bezerra, T.L., & Ragauskas, A.J. (2016). A review of sugarcane bagasse for second-generation bioethanol and biopower production. *Biofuels, Bioproducts and Biorefining*, 10(5), 634–647.

Braghiroli, F.L., & Passarini, L. (2020). Valorization of biomass residues from forest operations and wood manufacturing presents a wide range of sustainable and innovative possibilities. *Current Forestry Reports*, 6(2), 172–183.

Buzała, K. P., Kalinowska, H., Małachowska, E., & Przybysz, P. (2017). The utility of selected kraft hardwood and softwood pulps for fuel ethanol production. *Industrial Crops and Products*, 108, 824–830.

Cambero, C., Sowlati, T., Marinescu, M., & Röser, D. (2015). Strategic optimization of forest residues to bioenergy and biofuel supply chain. *International Journal of Energy Research*, 39(4), 439–452.

Carreira, P., Mendes, J.A., Trovatti, E., Serafim, L.S., Freire, C.S., Silvestre, A.J., & Neto, C.P. (2011). Utilization of residues from agro-forest industries in the production of high value bacterial cellulose. *Bioresource Technology*, 102(15), 7354–7360.

Ceballos, R., Chan, M., Batchenkova, N., Duffing-Romero, A., Nelson, A., & Man, S. (2005). Bioethanol: Feedstock alternatives, pretreatments, lignin chemistry, and the potential for green value-added lignin co-products. *International Journal of Environmental Analytical Chemistry*, 2(5). https://doi.org/10.4172/2380-2391.1000164.

Chubukov, V., Mukhopadhyay, A., Petzold, C.J., Keasling, J.D., & Martín, H.G. (2016). Synthetic and systems biology for microbial production of commodity chemicals. *NPJ Systems Biology and Applications*, 2, 16009.

Chudy, R.P., & Hagler, R.W. (2020). Dynamics of global roundwood prices-Cointegration analysis. *Forest Policy and Economics*, 115, 102155. https://doi.org/10.1016/j.forpol.2020.102155.

Clauser, N.M., Felissia, F.E., Area, M.C., & Vallejos, M.E. (2021a). A framework for the design and analysis of integrated multi-product biorefineries from agricultural and forestry wastes. *Renewable and Sustainable Energy Reviews*, 139, 110687.

Clauser, N.M., González, G., Mendieta, C.M., Kruyeniski, J., Area, M.C., & Vallejo, M.E. (2021b). Biomass waste as sustainable raw material for energy and fuels. *Sustainability*, 13(794). https://doi.org/10.3390/su13020794.

Coutinho, P., et al. (2011). Technological roadmap in renewable raw materials: A base for the construction of policies and strategies in Brazil. *Quimica Nova*, 34, 910–916.

Dai, J., Bean, B., Brown, B., et al. (2016). Harvest index and straw yield of five classes of wheat. *Biomass and Bioenergy*, 85, 223–227.

De Bhowmick, G., Sarmah, A.K., & Sen, R. (2018). Lignocellulosic biorefinery as a model for sustainable development of biofuels and value added products. *Bioresource Technology*, 247, 1144–1154.

De Menezes, A.J., Siqueira, G., Curvelo, A.A., & Dufresne, A. (2009). Extrusion and characterization of functionalized cellulose whiskers reinforced polyethylene nanocomposites. *Polymer*, 50(19), 4552–4563.

Diep, N.Q., Sakanishi, K., Nakagoshi, N., Fujimoto, S., & Minowa, T. (2015). Potential for rice straw ethanol production in the Mekong Delta, Vietnam. *Renewable Energy*, 74, 456–463.

Duncan, A.J., Samaddar, A., & Blümmel, M. (2020). Rice and wheat straw fodder trading in India: Possible lessons for rice and wheat improvement. *Field Crops Research*, 246, 107680.

Ebers, A., Malmsheimer, R.W., Volk, T.A., & Newman, D.H. (2016). Inventory and classification of United States federal and state forest biomass electricity and heat policies. *Biomass and Bioenergy*, 84, 67–75.

Eurostat. (2019). Data on EU pellet imports from the U.S. in 2016–2019. https://ec.europa.eu/eurostat.

FAO. (2015). Towards sustainable forestry: Wood products: Growing demand, growing productivity. www.fao.org/3/y3557e/y3557e10.htm#:~:text=The%20most%20recent%20FAO%20projections,2%20400%20million%20m3.&text=The%20past%20decade%20has%20also,much%20fuller%20use%20of%20timber.

FAO. (2019). Forest product statistics: Forest product consumption and production. www.fao.org/forestry/statistics/80938@180723/en/#:~:text=Major%20producers%20of%20forest%20products&text=Industrial%20roundwood%3A%20United%20States%20of%20America%20(19%25)%3B%20Russian%20Federation,Republic%20of%20Congo%20(5%25).

FAO. (2020). Impactos de la COVID-19 en las cadenas de valor de la madera y la respuesta del sector forestal: Resultados de una encuesta mundial realizada en 2020. www.fao.org/3/cb1987es/CB1987ES.pdf.

FAO-UN. (2021). Building climate-resilient dryland forest and agrosilvopastoral production systems. www.fao.org/3/cb3803en/cb3803en.pdf.

Feng-Wu, B., et al. (2019). Fuel ethanol production from lignocellulosic biomass. *Comprehensive Biotechnology*, 3, 49–65.

Food and A. O. of the United Nations (FAO STAT). (2019). Crop production in 2019. data retrieved from FAO STAT. www.fao.org/faostat/en/#data/QC, Filters: Regions Total, Production Quantity, Rice, Sugar cane, Wheat, Maize, and 2019 (accessed April 12, 2021).

Forster-Carneiro, T., et al. (2013). Biorefinery study of availability of agriculture residues and wastes for integrated biorefineries in Brazil. *Resources, Conservation and Recycling*, 78–88.

Gregg, J.S., Jürgens, J., Happel, M.K., Strøm-Andersen, N., Tanner, A.N., Bolwig, S., & Klitkou, A. (2020). Valorization of bio-residuals in the food and forestry sectors in support of a circular bioeconomy: A review. *Journal of Cleaner Production*, 122093.

Hämäläinen, S., Näyhä, A., & Pesonen, H.L. (2011). Forest biorefineries—A business opportunity for the Finnish forest cluster. *Journal of Cleaner Production*, 19(16), 1884–1891.

Han, H.S., Jacobson, A., Bilek, E.T., & Sessions, J. (2018). Waste to Wisdom: Utilizing forest residues for the production of bioenergy and biobased products. *Applied Engineering in Agriculture*, 34(1), 5–10.

Haq, I.U., Qaisar, K., Nawaz, A., Akram, F., Mukhtar, H., Zohu, X., Xu, Y., Mumtaz, M.W., Rashid, U., Ghani, W.A.W.A.K., et al. (2021). Advances in valorization of lignocellulosic biomass towards energy generation. *Catalysts*, 11, 309. https://doi.org/10.3390/catal11030309.

Hilares, R.T., Dionízio, R.M., Muñoz, S.S., Prado, C.A., de Sousa Júnior, R., da Silva, S.S., & Santos, J.C. (2020). Hydrodynamic cavitation-assisted continuous pre-treatment of sugarcane bagasse for ethanol production: Effects of geometric parameters of the cavitation device. *Ultrasonics Sonochemistry*, 63, 104931.

IEA. (2020). Biomass explained: Wood and wood waste. www.eia.gov/energyexplained/biomass/wood-and-wood-waste.php.

IEA. (2021). International energy agency. Energy and climate leaders from around the world pledge clean energy action at the IEA-COP26 net zero summit. www.iea.org/news/energy-and-climate-leaders-from-around-the-world-pledge-clean-energy-action-at-the-iea-cop26-net-zero-summit.

Ingle, A.P., Chandel, A.K., Antunes, F.A.F., Rai, M., & da Silva, S.S. (2019). New trends in application of nanotechnology for the pretreatment of lignocellulosic biomass. *Biofuels Bioproducts and Biorefining*, 13, 776–788.

Ingle, A.P., Chandel, A.K., & da Silva, S.S. (2020). Biorefining of lignocellulose into valuable products. In *Lignocellulosic Biorefining Technologies* (Eds. Ingle, A.P., Chandel, A.K., da Silva, S.S.). Wiley Blackwell.

INPE. (2019). Programa queimadas. https://queimadas.dgi.inpe.br//queimadas/aq1km/.

Jiang, Y., Wu, R., Zhou, J., He, A., Xu, J., Xin, F., Zhang, W., Ma, J., Jiang, M., & Dong, W. (2019). Recent advances of biofuels and biochemicals production from sustainable resources using co-cultivation systems. *Biotechnology for Biofuels*, 12, 155. https://doi.org/10.1186/s13068-019-1495-7.

Junginger, M., Faaij, A., Van den Broek, R., Koopmans, A., & Hulscher, W. (2001). Fuel supply strategies for large-scale bio-energy projects in developing countries. Electricity generation from agricultural and forest residues in Northeastern Thailand. *Biomass and Bioenergy*, 21(4), 259–275.

Kapoor, R., et al. (2020). Valorization of agricultural waste for biogas based circular economy in India: A research outlook. *Bioresource Technology*, 304, 1230–1236.

Kittler, B., Stupak, I., & Smith, C. T. (2020). Assessing the wood sourcing practices of the US industrial wood pellet industry supplying European energy demand. *Energy, Sustainability and Society*, 10, 1–17.

Kuuluvainen, J., Korhonen, J., Wang, L., & Toppinen, T. (2021). Wood market cartel in finland 1997–2004: Analyzing price effects using the indicator approach. *Forest Policy and Economics*, 124, 102380. https://doi.org/10.1016/j.forpol.2020.102380.

Leontopoulos, S.V., & Arabatzis, G. (2021). The contribution of energy crops to biomass production. *Low Carbon Energy Technologies in Sustainable Energy Systems*, 47–113.

Long, T., Pan, H., Dong, C., Qin, T., & Ma, P. (2019). Exploring the competitive evolution of global wood forest product trade based on complex network analysis. *Physca A*, 525, 1224–1232. https://doi.org/10.1016/j.physa.2019.04.187.

Lynd, L.R. (2017). The grand challenge of cellulosic biofuels. *Nature Biotechnology*, 35(10), 912–915.

Manna, M. C., et al. (2018). Bio-waste management in subtropical soils of India: Future challenges and opportunities in agriculture. *Advances in Agronomy*, 152, 87–148.

Millati, R., Cahyono, R.B., Ariyanto, T., Azzahrani, I.N., Putri, R.U., & Taherzadeh, M.J. (2019). Agricultural, industrial, municipal, and forest wastes: An Overview. *Sustainable Resource Recovery and Zero Waste Approaches*, 1–22.

MNRE (Ministry of New, Renewable Energy Resources). (2021). Govt of India. www.mnre.gov.in/biomassrsources.

Nanda, S., Mohammad, J., Reddy, S.N., Kozinski, J.A., & Dalai, A.K. (2014). Pathways of Lignocellulosic biomass conversion to renewable fuels. *Biomass Conversion and Biorefinery*, 4, 157–191.

Nepal, P., Korhonen, J., Prestemon, J.P., & Cubbage, F.W. (2019). Projecting global planted forest area developments and the associated impacts on global forest product markets. *Journal of Environmental Management*, 240, 421–430. https://doi.org/10.1016/j.jenvman.2019.03.126.

NewBio. (2014). Market opportunity for lignocellulosic biomass. https://farm-energy.extension.org/wp-content/uploads/2019/04/Biomass-Market-Opportunity_Final-2014_0.pdf.

Nicholls, D.L., Halbrook, J.M., Benedum, M.E., Han, H.S., Lowell, E.C., Becker, D.R., & Barbour, R.J. (2018). Socioeconomic constraints to biomass removal from forest lands for fire risk reduction in the western US. *Forests*, 9(5), 264.

OECD. (2021). Economic Outlook, interim report. Strengthening the recovery: The need for speed. www.oecd.org/economic-outlook/.

Rowan, N.J., & Galanakis, C.M. (2020). Unlocking challenges and opportunities presented by COVID-19 pandemic for cross-cutting disruption in agri-food and green deal innovations: Quo Vadis? *Science of the Total Environment*, 141362.

Saini, J.K., Saini, R., & Tewari, L. (2015). Lignocellulosic agriculture wastes as biomass feedstocks for second-generation bioethanol production: Concepts and recent developments. *3 Biotech*, 5(4), 337–353.

Sánchez-Muñoz, S., Mariano-Silva, G., Leite, M.O., Mura, F.B., Verma, M.L., da Silva, S.S., & Chandel, A.K. (2020). Production of fungal and bacterial pigments and their applications. In *Biotechnological Production of Bioactive Compounds* (pp. 327–361). Elsevier.

Santos, C.M., et al. (2020). Produção de Bioetanol a partir dos Residuos da Indústria de Papel. *Vir. Química*, 12(4).

Sarkar, N., Ganguly, A., & Chatterjee, P.K. (2021). Valorization of food waste. In *Handbook of Advanced Approaches Towards Pollution Prevention and Control* (pp. 157–172). Elsevier.

Sarsaiya, S., Jain, A., Awasthi, S.K., Duan, Y., Awasthi, M.K., & Shi, J. (2019). Microbial dynamics for lignocellulosic waste bioconversion and its importance with modern circular economy, challenges and future perspectives. *Bioresource Technology*, 291, 121905.

Shinners, K.J., & Binversie, B.N. (2007). Fractional yield and moisture of corn stover biomass produced in the Northern US Corn Belt. *Biomass and Bioenergy*, 31(8), 576–584.

Sokhansanj, S., Turhollow, A., Cushman, J., & Cundiff, J. (2002). Engineering aspects of collecting corn stover for bioenergy. *Biomass and Bioenergy*, 23(5), 347–355.

Stafford, W., De Lange, W., Nahman, A., Chunilall, V., Lekha, P., Andrew, J., & Trotter, D. (2020). Forestry biorefineries. *Renewable Energy*, 154, 461–475.

Thrän, D., Schaubach, K., Peetz, D., Junginger, M., Mai-Moulin, T., Schipfer, F., . . . Lamers, P. (2019). The dynamics of the global wood pellet markets and trade—key regions, developments and impact factors. *Biofuels, Bioproducts and Biorefining*, 13(2), 267–280.

Tsalagkas, D., Börcsök, Z., Pásztory, Z., Gogate, P., & Csóka, L. (2021). Assessment of thepapermaking potential of processed Miscanthus× giganteus stalks using alkaline pre-treatment and hydrodynamic cavitation for delignification. *Ultrasonics Sonochemistry*, 72, 105462.

UNCTAD. (2019). Commodity dependence, climate change and the Paris agreement. https://unctad.org/meeting/commodities-and-development-report-2019-commodity-dependence-climate-change-and-paris.

Väisänen, T., Haapala, A., Lappalainen, R., & Tomppo, L. (2016). Utilization of agricultural and forest industry waste and residues in natural fiber-polymer composites: A review. *Waste Management*, 54, 62–73.

WHO. (2020). International day of epidemic preparedness. www.who.int/news-room/events/detail/2020/12/27/default-calendar/international-day-of-epidemic-preparedness.

Zoghlami, A., & Paës, G. (2019). Lignocellulosic biomass: Understanding recalcitrance and predicting hydrolysis. *Frontiers in Chemistry*, 7, 874.

3 Bioprospecting and Genomic-Based Biotechnology for Economic Use of Agro-Industrial Wastes

Gretty K. Villena[1] *and Ilanit Samolski*[1]
[1] Laboratorio de Micología y Biotecnología "Marcel Gutiérrez-Correa." Universidad Nacional Agraria La Molina. Av. La Molina s/n Lima 12. Perú.

CONTENTS

3.1 INTRODUCTION

In recent years, with the advent of the bioeconomy, paradigms have changed to prioritize the use of renewable resources of biological origin and their conversion through biotechnology in value-added products with demand in different sectors, such as energy, food industry, materials and manufacturing, feeds, and others.

The main raw materials used include agricultural and forestry residues, herbaceous and woody crops, oleaginous residues, crops and starchy residues, and algal biomass. Chemically, these residues and biomass are sources of lignocellulose, hemicellulose, triglycerides and phospholipids, starch, alginates, fucoidans, and other polymeric compounds.

Lignocellulosic feedstock, which is composed mainly of cellulose (38–50%), hemicellulose (23–32%), and lignin (10–25%), is one of the most abundant in the Earth, with an annual production of about 181.5 billion tons (Paul and Dutta 2018). The advantages of the use of lignocellulose are related to inexhaustibility, abundance,

DOI: 10.1201/9781003191247-3

degradability, better cleaning performance, and availability of key chemical elements (C,H,O) (Chen 2015), but its complex chemical architecture and recalcitrance, as well as the lack of efficient green technology, makes difficult its use on a large scale. The raw materials derived from starchy and oilseed crops are more technologically affordable, but their great disadvantage is that they compete with agricultural lands, risking food security.

According to bioeconomy, conventional and integrated biorefineries offer facilities for biomass conversion to biofuels, new materials and compounds, and value-added products through biotechnological approaches.

Cutting-edge innovations in biotechnology rely on a deeper understanding of living systems through the application of genomic technologies (omics) and molecular tools that include metabolic engineering, gene editing, and directed evolution, among others, to design cell factories capable of carrying out chemical transformations and producing conventional and new bio-based products.

Biorefineries are the ideal production models for the biotechnological conversion of biomass into high value-added products and biomaterials that can gradually replace the oil refineries and petrochemical industry, but for this goal, knowledge-based innovation is pivotal.

3.2 BIOPROSPECTING FOR NEW ENZYMES AND MICROBIAL CATALYSTS TO CONVERT FEEDSTOCK BIOMASS

Feedstock biomass provides carbon building blocks for production of a wide type of products, including biofuels, chemicals, biomaterials, and power. But because of its polymeric and crystalline structures and the recalcitrance of its chemical bonds, biomass deconstruction has usually required the use of physical and chemical treatments. However, to achieve environmental sustainability, it is necessary to replace the use of these treatments with efficient biocatalysts.

Thus, biotechnological conversion of biomass into value-added products is strongly dependent on enzymatic technology, bioprospection for new and more efficient enzymes, as well as the formulation of new enzyme cocktails according to the chemical composition of the biomass residues (Contesini et al. 2021). Also, alternative strategies include the searching of substrate-disrupting accessory proteins or domains to optimize substrate accessibility and improve enzymatic effectiveness (Eijsink et al. 2008).

Hydrolases that commonly include ligninases, cellulases, pectinases, hemicellulases, monooxygenases, amylases, lipases, and proteases are the most demanded catalysts for biomass breakdown. Figure 3.1 shows the enzymes required for the conversion of biomass as a function of the starting raw material.

Among them, Carbohydrate-Active EnZymes (CAZymes) include all enzymes for the modification, deconstruction, or biosynthesis of carbohydrates and have been classified into glycoside hydrolases, glycosyltransferases (GTs), polysaccharide lyases (PL), carbohydrate esterases (CE), and enzymes with auxiliary activities (AA) (Lombard et al. 2014). At the date (updated to July 2022), about 300 families of catalytic and ancillary modules (families associated with catalytic modules) have been deposited in this database. It includes approximately 100,000 nonredundant entries, corresponding to 23,783 genomes (bacteria, virus, archaea, eukaryota) (www.cazy.org).

Although a wide variety of native and improved enzymes is available, new microbial species and enzyme-codifying genes are continually being isolated from the environment. Precisely because of their ecology and adaptability based on their metabolic versatility, microorganisms represent an inexhaustible source of biocatalysts. However, culturability of microorganisms is very low, only 0.1% for prokaryotes, and for fungi, just ~5% have been described (Mueller and Schmit 2007; Alain and Querellou 2009). Despite this, classical approaches based on culturing methods have contributed enormously with the availability of a wide range of enzymes for different applications. In addition, the deepest understanding of ecological, nutritional, and physiological aspects has allowed to increase the recovery of unculturable microorganisms through improvement in the criteria for: a) media optimization, adequate use of minimal or oligotrophic media, or media enrichment according to specific groups of microorganisms; b) culturing techniques that include the dilution approach, co-culturing, single-cell cultivation, among others; c) optimization of cultivation conditions to simulate the natural environment; and d) dependence on hosts for certain nutrients or signaling (Prakash et al. 2013; Ahrendt et al. 2018). Also, different gelling agents are proposed for solid and semisolid media formulation to cover different pH and temperatures as well as different viscosity and solubility (Das et al. 2015).

Evolutionarily, fungi had developed a great capability for secreting degradative and oxidizing enzymes (Zeiner et al. 2016; Lange et al. 2019). Almost 90% of the global enzyme market corresponds to microbial enzymes (Sharma and Upadhyay 2020), and 50% of industrial and technical enzymes (hydrolytic depolymerases, proteases) come from representative fungal genera including *Trichoderma*, *Aspergillus*, *Rhizopus*, *Mucor*, *Trametes*, and yeasts (McKelvey and Murphy 2017) as well as from heterologous production.

Also, bioprospecting is being expanded from soil to sea and extreme environments, including microbiomes of plants and animals.

On the other hand, culture-independent methods, properly *metagenomics*, have allowed the detection of many genomes and enzyme genes by solving the limitations associated with the complexity of microbial communities and extreme environments (Sysoev et al. 2021) and high metabolic specialization that restrict the cultivation of microorganisms. Metagenomic tools are based in direct isolation of environmental DNA and analysis through two strategies: a) sequence-driven and b) function-driven screening through cloning or cell-free systems expression (Kirubakaran 2020; Robinson et al. 2021). In the last decade, hundreds of enzymes, including cellulases/hemicellulases, chitinases, esterases, lipases, oxidoreductases, proteases, and others, have been identified and characterized (Berini et al. 2017), and some of them patented (Prayogo et al. 2020). Additionally, molecular handling techniques have provided conditions for metatranscriptomics and metaproteomics. Moreover, synthetic biology tools have promoted the birth of synthetic metagenomics (Guazzaroni et al. 2015; van der Helm et al. 2018).

A complementary cost- and time-effective approach, in *silico* bioprospecting, allows the integration and mining of genomic data to obtain new enzyme candidates with better performance (Kamble et al. 2019). Some *in silico* techniques include a) multiple sequence alignment, b) sequence-related consensus sequence design, c) molecular docking, and d) molecular dynamics (Fülöp and Ecker 2020; Zhang et al. 2020).

Microbial and enzyme bioprospecting and complementary bioinformatics tools are being applied towards the improvement of catalytic capabilities and variety of enzymes for biomass conversion:

- *Cellulases and hemicellulases.* Enzymatic deconstruction of cellulose to simple sugars requires the synergistic action of endoglucanases, exoglucanases, and β-glucosidases. Cellulolytic enzymes are produced by fungi and bacteria. Extracellular cellulases are typical for fungi and aerobic bacteria, while anaerobic bacteria like *Clostridium* possess an enzymatic complex called cellulosome.

Since the enzymatic cellulose hydrolysis is still considered as limited and expensive for biomass treatment, the search and screening for cellulolytic microorganisms and genes is a continuous task. Different approaches to recover them from different environments have been used. Marine microbes with cellulolytic capabilities from water, sediments, including bacteria, fungi, and yeast have been extensively reported, as well as cold-active microorganisms and antarctic fungi which display cellulolytic and xylanolytic activity (Trivedi et al. 2016; Danilovich et al. 2018; Duarte et al. 2018; Varrella et al. 2021). Thermal hot springs and other exotic locations attractive for isolation of extremophiles, including archaea (*Pyrococcus*, *Sulfolobus*, and *Thermotoga*), were explored (Acharya and Chaudhary 2012; Thankappan et al. 2018). Bacterial consortia, including Actinobacteria, Proteobacteria, Bacilli, Sphingobacteria, and Flavobacteria, were isolated from a pristine lake in Greece, showing potential for growth in lignin hydrolysates derived from pretreated agricultural wastes (Georgiadou et al. 2021). Insect microbiomes also represent a source of cellulolytic bacteria and fungi, since some of these invertebrates process lignocellulose in their digestive tracts through its microbiota (Alves et al. 2019).

Importantly, rumen microorganisms, including anaerobic bacteria, fungi, archaea, among others, secrete cellulases, hemicellulases, and ligninases, with a synergistic and particular mode of action (Liang et al. 2020).

Natural substrates with cellulose composition also allow the isolation of strains with several glycosyl hydrolases capabilities (Sahoo et al. 2020). Soil is still a source of new microorganisms. For example, potentially unclassified proteobacteria with beta-glucosidases, xylanases, and others were isolated and sequenced by a culture-dependent metagenomic approach from soil litter. At the same time, by stable isotope probing (SIP) combined with metagenomics on topsoil litter amended with extracellular polymeric substances (EPS), it was possible to detect glycosyl hydrolases (GH) between other enzymes (Costa et al. 2020).

The breakthrough in the identification of new cellulase strains and genes has been made by metagenomics in different environments. A large survey of seventy-one hot spring metagenomes analyzed microbial diversity and CAZyme composition, including cellulases and hemicellulases (Reichart et al. 2021). A total of 4,083 putative genes with functional domains involved in the catalysis of carbohydrate degradation was detected in four thermal water reservoirs of 55 to 98°C in India. Many of them corresponded to novel biocatalysts associated with hydrolysis of cellulose, hemicellulose, lignin, and pectin (Kaushal et al. 2018).

Non-described activities with hemicellulose specificity belonging to *Crenarchaeota* have been reported for mud/water pools samples with extreme temperature and pH conditions (Strazzulli et al. 2020).

Mining and heterologous expression of cellulases genes from natural lignocellulosic substrates as decaying wood was achieved and resulted in the identification of at least 4,000 glycoside hydrolase homologues from an anaerobic bacterial community with a successful expression of at least two novel enzymes with cellulolytic activity in the presence of ionic liquids (Li et al. 2011). Similarly, a culture-dependent metagenomic approach of a bacterial consortium obtained from soil and dry straw leftover found that many of the most abundant species carried sequences related to hemicellulose and cellulose deconstruction (Weiss et al. 2021).

Another prospecting results referred to the analysis of cow and buffalo ruminal metagenomes revealed a dependence between bacterial community founded with the feed type in these animals and a great diversity of CAZy enzymes, including cellulases and hemicellulases (Bohra et al. 2019). Also, by combining meta-omics and enzymology approaches, a novel Bacteroidetes family ("Candidatus MH11") composed of cultivated strains predominant in ruminants has been proposed (Naas et al. 2018).

Interestingly, complementary approaches and functional metagenomics allowing the identification of new cellulases in insect guts have revealed the presence of endogenous cellulases. The expression of genes of two endoglucanases and one β-glycosidase was identified in the termite *Reticulitermes speratus* (Ahn and Kim 2021). Previously, evidence for an endogenous beta-1,4-endoglucanase additionally to microbial cellobiase in the gut of the beetle *Oryctes rhinoceros* could explain its capability to degrade cellulose (Shelomi et al. 2019).

Through in silico screening of a huge number of rumen microbiota enzymes, a xylanase active in highly alkaline and temperate conditions was discovered (Ariaeenejad et al. 2019).

- *Ligninases.* Lignolytic enzymes have a key role in decomposing lignin in nature. Lignin has a complex structure and low degradability, and a synergic action of laccases, lignin peroxidases, manganese peroxidase, and versatile peroxidase is required for its degradation. Complementary enzymes (feruloyl esterase, aryl-alcohol oxidase, quinone reductases, lipases, catechol 2, 3-dioxygenase) and mediators could improve their performance. Microbial ligninases from bacteria actinomycetes, α- proteobacteria, and γ-proteobacteria and fungi (mainly ascomycetes and basidiomycetes) are commonly reported (Kumar and Chandra 2020).

Basidiomycetes have gained importance as primary sources of lignin-transforming enzymes. White rot fungi, which includes at least 10,000 species, possess the enzymatic pool to degrade cellulose, hemicelluloses, and lignin. Some genera like

Phanaerochaete, *Trametes*, and *Pleurotus* produce decomposition of lignin and wood decay through the activity of lignin-modifying enzymes (LMEs). The set of LMEs includes phenoloxidase, laccase, and three peroxidases (high oxidation potential class II peroxidases), lignin peroxidase (LiP, EC 1.11.1.14), manganese peroxidase (MnP, EC 1.11.1.13), and a versatile peroxidase (VP, 1.11.1.16). A wide array of isoforms could be produced by many different species and conditions (Peralta et al. 2017). Laccase and/or peroxidase activities have also been reported for *Bjerkandera*, *Ceriporiopsis*, *Irpex*, *Panus*, *Phellinus*, *Pleurotus*, *Polyporus*, *Stereum*, and *Trametes*, and interestingly, a synergistic model, called "microbial sink," demonstrated that a treatment with both the *Pleurotus eryngii* secretome and an aromatic catabolic bacteria increases the lignin depolymerization degree by avoiding repolymerization of monomeric species (Salvachúa et al. 2016). A more restricted enzyme system was reported for the edible mushroom *Schizophyllum commune*, which produces laccases and complementary enzymes, but not lignin peroxidases (LiP), manganese peroxidases (MnP), and versatile peroxidases (VP). Nevertheless, because of its capacity to produce cellulases, hemicellulases, and other enzymes, it registered more than 1,000 patents (applied and granted) in the field of lignocellulosic biomass (Tovar-Herrera et al. 2018).

Filamentous fungi like Ascomycetes have a very limited capacity to degrade lignin. Although some genera, including *Penicillium*, *Fusarium*, *Alternaria*, *Xylaria*, and few others cause soft rot decay, some of them are laccase producers (Guimarães et al. 2017).

Recently, a termite-symbiont fungi, *Termitomyces*, was analyzed through omics techniques, and a set of oxidizing enzymes (manganese peroxidase, dye decolorization peroxidase, an unspecific peroxygenase, laccases, and aryl-alcohol oxidases) was identified. This enzymatic array and a Fenton reaction contribute to biomass degradation (Schalk et al. 2021).

Bioprospecting, metagenomics, and genomics have provided more information on the enzymatic capabilities of archaea and prokaryotes.

Most representative groups recognized as bacterial models used for lignin degradation include Proteobacteria, Actinobacteria, and Firmicutes species. Typical lignin-modifying enzymes include DyP-type peroxidases (dye-decolorizing peroxidases) and laccases (Scully et al. 2013). Genomes mining from predicted proteomes performed by BLASTP identified DyPS homologues (de Gonzalo et al. 2016).

The use of culture-based methods and different screening methodologies had enabled the identification of ligninolytic bacteria, such as *Pandoraea*, *Pseudomonas*, *Klebsiella*, *Bacillus*, *Ochrobactrum*, and *Rhodococcus*, among others (Gonçalves et al. 2020).

A similar lignin-degrading consortium (LigMet) composed by Proteobacteria, Actinobacteria, and Firmicutes members, including the Alcaligenaceae and Micrococcaceae families, was obtained from a sugarcane plantation soil together with several enzymes linked to lignin-degrading metabolic pathways (Moraes et al. 2018).

Some Archaea can degrade lignocellulose at high temperature, and laccase enzymes have been identified in Halobacteriales and Thermoproteales (Cragg et al. 2015).

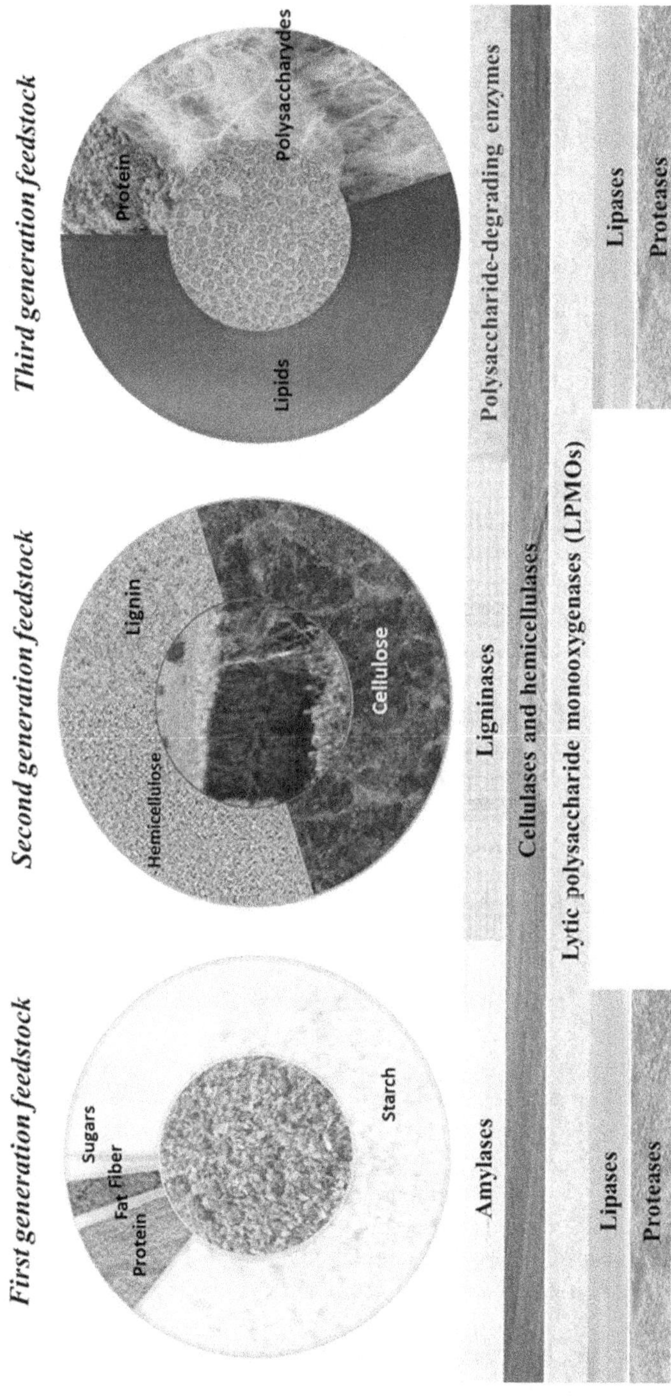

FIGURE 3.1 Overview of the main enzymatic groups required for deconstruction of feedstock related on the basis of its chemical composition.

Metagenomics and sequencing techniques in different locations have added more insights for novel microorganisms and enzymes. The gut metagenome of a beetle (*A. glabripennis*) includes candidate lignin-degrading genes of laccases.

- *Polysaccharide lyases or auxiliary enzymes* (LMPOs). Lytic polysaccharide monooxygenases (LPMOs) catalyze the oxidative deconstruction of cellulose, chitin, starch, and hemicellulose through oxidative reactions through a mechanism involving a copper ion, molecular oxygen, and an electron donor. LPMOs are classified into auxiliary activity (AA) families 9, 10, 11, 13, 14, 15, and 16. Bacterial LMPOs bacterial AA10 family degrade chitin or cellulose, while AA9 corresponds to fungi. AA11 cleaves chitin chains, and AA13 cleaves starch, and A14 are xylan-active (Bomble et al. 2017; Kojima et al. 2020).

AA16 was recently proposed as a putative LPMO family which is encountered previously in fungi and oomycetes through bioinformatic analysis. Experimental comparison of *Aspergillus* secretomes and a proteomic analysis revealed a new auxiliary enzyme from *A. aculeatus* with particular characteristics that referred to substrate specificity, regioselectivity of oxidation, and contribution to cellulose degradation (Filiatrault-Chastel et al. 2019).

As usual, microbiome exploration has contributed to discovering new LMPOs enzymes. The genome analysis of *Teredinibacter turnerae*, a bacterial symbiont of shipworm marine shipworm *Lyrodus pedicellatus*, enables the identification of a cellulose-active AA10 LPMO gene (Fowler et al. 2019).

Similarly, a single gene encoding for a copper-dependent lytic polysaccharide monooxygenase was described for *Photorhabdus luminescens*, a nematode-symbiont bacterium (Munzone et al. 2020).

A metatranscriptome analysis from microbial communities involved in the decomposition of raw straw in conjunction with metagenomic data reveals polysaccharide monooxygenase genes in that thermophilic community (Simmons et al. 2014).

A bioinformatic model for genomes, metagenomes, and transcriptomes mining, based on a multiple-protein alignment, allows for the classification of LPMOs along phylogenetic groups, facilitating the identification of new AA families (Voshol et al. 2017).

- *Amylases*. Amylases are starch-hydrolyzing catalysts including endo and exo-acting enzymes. α-Amylases belong to the GH-13 family of glycosyl hydrolases and cleave α-(1,4)-D-glycosidic bonds in starch. Amylases are the most commercially important enzymes since they are extensively used in many industrial processes and cover at least 30% shares of the total enzyme market. Despite it, continuous improvement of wild amylases is being carried out to confer desirable properties (Paul et al. 2021).

Amylases have bacterial, archaeal, and eukaryotic origin. Bacterial amylases are the most diverse group, showing a wide range of optimal temperature and pH. Archaeal α-amylases are generally thermostable and with acidic optimal pH (Mehta and

Satyanarayana 2016). On the other hand, fungal α-amylases are preferred for starch hydrolysis applications in the food industry since they are generally recognized as safe. *Aspergillus* is the most important genus for the production of amylases, mainly *A. niger*, *A. awamori*, and *A. oryzae*.

Even with the wide distribution of bacterial and fungal amylases, unexplored environments are gaining attention for the search of new amylases.

For example, a novel α-amylase AmyZ1 was selected from a marine sediment bacteria *Pontibacillus* sp. and then cloned for its characterization (Fang et al. 2019).

A functional metagenomics approach for a halite microbiome of the Atacama Desert found a vast repertory of carbohydrate-active enzymes where amylases showed high abundance. From 162 putative glycosyl hydrolases genes, 21% corresponded to amylases (GH13), mainly archaeal amylases. Three novel GH13 subfamilies have been proposed (Gómez-Silva et al. 2019).

The rumen environment harbors several microorganisms and genes that include those for carbohydrate-active enzymes. A metagenome survey in goats to evaluate dietary rumen degradable starch (RDS) resulted in an abundance of amylase genes in animals with increased RDS (Shen et al. 2020).

- *Other hydrolases*. Proteases and lipases play an important role in waste processing mainly from food processing industries. Proteases include proteinases, peptidases, amidases, and microbial proteases, representing 40% of the entire global enzymes market (Naveed et al. 2021). Nevertheless, bioprospecting of new strains and enzymes continues. Recently, cold-tolerant bacteria with serine-type protease activity were isolated from Western Himalayas (Salwan et al. 2020).

Lipases catalyze the hydrolysis of triacylglycerols to di- and monoacylglycerols, fatty acids, and glycerol, which are extensively used for biodiesel production from crude oil and fats (Uçkun Kiran et al. 2014). Lipases are one of the most prevalent new enzymes in soil metagenomes (Lee and Lee 2013). Also, extremophilic lipases identified by metagenomics are being molecularly characterized (Verma et al. 2021).

In perspective, advances in metagenomics approaches will continue to reveal new enzyme genes with different capabilities and specificity from different origins. However, the potential of these new genes requires successful strategies for their heterologous expression and large-scale production. In the same way, the new strains with enzymatic capabilities will require genetic improvement to make them competitively efficient for biomass conversion.

3.3 GENOMIC AND MOLECULAR STRATEGIES FOR OPTIMIZING BIOMASS CONVERSION IN A BIOREFINERY MODEL

Biorefinery, defined as the "sustainable processing of biomass into a spectrum of marketable products and energy" (Cherubini et al. 2010), represents a new productive paradigm designed to achieve an economic, environmental, and social sustainability driving a circular bioeconomy. According to their feedstocks and products, biorefineries are classified in first, second, or third generation. The main characteristics of the types of biorefineries are shown in Table 3.1. First-generation feedstocks refer to amylaceous or sugar-rich food crops (corn, wheat, sugarcane). Second-generation feedstocks correspond to lignocellulosic materials and include energy crops and wastes, whereas algae biomass is considered as third-generation feedstocks (Azapagic 2014). Although the largest production of biofuel comes from first-generation ethanol, it could deal with food security and use of land resources. To ensure maximum use of biomass, an integrated approach should be considered for obtaining not only biofuel, as the main product, but also other valuable products, through a cascade processing of biological residues that can improve economics.

Biorefinery platforms include thermochemical, mechanical, and biological conversion. Biological conversion includes mainly anaerobic digestion, enzymatic hydrolysis, and fermentation for bioenergy products like biogas, biodiesel, and bioethanol (Ubando et al. 2020). For a better biomass valorization, genomics-based biotechnology could contribute to bioeconomy *innovativeness*. That means to transfer basic scientific knowledge into a successful industrial manufacturing and agriculture and other production areas by improving quantity and quality of products. Newly conceived knowledge-based bioeconomy will arise by coupling genomics, modern molecular techniques, and synthetic biology together with engineering strategies and bioinformatics (Aguilar et al. 2018; de Lorenzo et al. 2018; Lokko et al. 2018).

The biotechnological tools, including directed evolution, metabolic engineering, gene editing, and synthetic biology that can rapidly promote the development of

TABLE 3.1
General Classification and Typical Products of Biorefineries

Biorefinery Type	Main Feedstocks	Main Products
First Generation	Starch crops residues, sugar-rich crops residues	Starch-derived biogas Biofuels: ethanol sweeteners, high-fructose corn syrups
Second Generation	Agricultural residues, wood waste	Biogas: methane, Biofuels: ethanol, butanol Lignin derivatives: coumaric acid, ferulic acid, vanillin, furfural, phenol resins, etc. Glucose fermentation products: organic acids (levulinic, lactic, succinic, etc.) Hemicellulose fermentation products: xylitol, mannitol, sorbitol, etc. Biomaterials: nanofibers, cellulose, nanocrystals
Third Generation	Algal biomass (mainly microalgae)	Biohydrogen Biofuels: ethanol, oil biofuels Alkanes Fatty acids Proteins Pigments

knowledge-based biorefineries, in addition to biological engineering, have played a key role in the optimization of fermentations. Figure 3.2 shows the impact of biotechnological tools for moving towards integrated biorefineries.

Biomass conversion in biorefineries relies on physical, chemical, and enzymatic treatments to produce fermentable sugars. Interestingly, a life cycle assessment approach (LCA) analysis on second-generation biorefineries suggests that further integration of enzyme production into the biorefinery is expected to result in a reduction of the total impact, aiming for a scenario in which total environmental burdens can be reduced (Bello et al. 2018). Enzymes represent about 15% to 20% of the total cost of biofuel production (Singh et al. 2019) so that the enhancement of enzyme production or its performance is highly required.

Metabolic engineering makes possible bio-based chemical production by designing organisms for production of commodity chemicals, specialty chemicals, and biomaterial (primary products of biorefineries), but also high-value, high-purity, renewable sugar-based chemicals (secondary products of biorefineries) (Erickson and Winters 2012). Synthetic biology and systems biology and genome editing are strongly supporting advances in metabolic engineering. Conventional and advanced metabolic engineering based on genome edition has revolutionized strain improvement tools. CRISPR-Cas application to bacteria for microbial biofuel production (bioethanol, bioethanol), as well as other hydrocarbons, has been extensively revised by Shanmugam et al. (2020). Engineered yeasts for biofuel production, including ethanol and long-chain fatty acid–derived molecules, as well as a variety of oleochemicals and others, are being achieved (Stovicek et al. 2017; Rebello et al. 2018; Ma et al. 2020; Liu et al. 2021; Zhang et al. 2021). In addition, combinatorial, transcriptional, and posttranscriptional engineering, as well as adaptive laboratory evolution, consolidate yeast as advanced microbial cell factories (Guirimand et al. 2021). Also, for filamentous fungi, CRISPR-Cas technology could be applied for gene silencing, directed mutagenesis, improvement of secondary metabolites, gene overexpression,

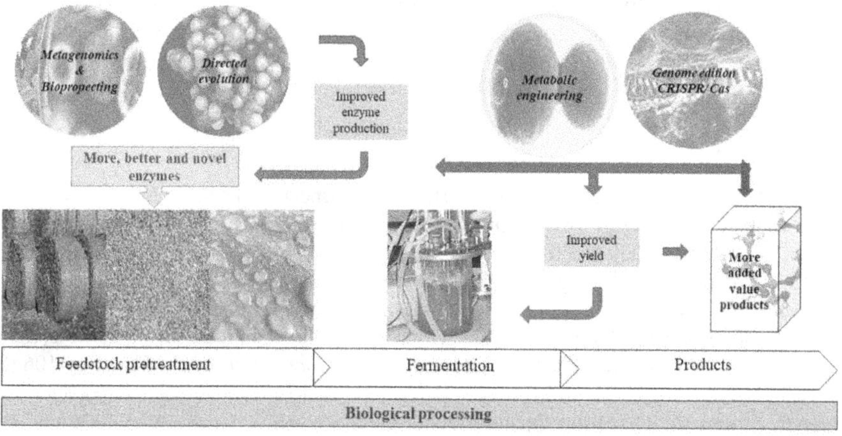

FIGURE 3.2 Expected positive impacts of a cutting-edge biotechnology on biological processing in biorefineries.

and gene regulation to engineering metabolic pathways. For example, to obtain next-generation cellulases for biofuel production (Mondal et al. 2021). Microalgae have also revealed an enormous potential as bio-cell factories for production of key chemicals, recombinant proteins, enzymes, lipids, and alcohol, although synthetic gene network, genome-scale reconstruction, and CRISPR-Cas platforms are still in progress (Jagadevan et al. 2018; Tanwar and Kumar 2020).

Current advances in protein engineering for rational and *de novo* design of enzymes as well as directed evolution (useful when lack of structural information leads to difficulties in the use of a rational design) are contributing to the enzyme improvement in bacteria and fungi for industrial applications (Porter et al. 2016; Asemoloye et al. 2021; Dinmukhamed et al. 2021). Also, enzymatic cocktails for biomass breakdown could be rationally optimized by combining protein engineering and directed evolution (Contreras et al. 2020).

Complementarily, the arising bioinformatic tools and new algorithms, named "computational enzyme engineering," simplified experimental testing by performing a rational in silico screening of enzyme variants (Chowdhury and Maranas 2020).

Furthermore, synthetic biology and bioinformatics, together with fusion protein engineering applied to the obtention of chimeric enzymes, could be a cost-efficient process to break down biomass (Jadaun et al. 2020).

3.4 CONCLUSION AND REMARKS

Lignocellulosic biomass from agricultural residues represents the most abundant source of renewable resources and, together with several feedstock wastes, offers a unique opportunity for conversion into biofuel materials and various value-added products in a biorefinery model. The environmental sustainability of biorefineries demands the progressive replacement of chemical-to-biological conversions as well as a cascade processing of biological residues to better economics.

To ensure bioeconomy *innovation*, knowledge-based utilization of microorganisms and enzymes, through metagenomics and functional genomics and cutting-edge tools as metabolic engineering, directed evolution, and synthetic biology, together with bioinformatics, should be considered for successful integrated biorefineries.

REFERENCES

Acharya, Somen, and Anita Chaudhary. 2012. "Bioprospecting Thermophiles for Cellulase Production: A Review". *Brazilian Journal of Microbiology* 43 (3): 844–856. doi:10.1590/s1517-83822012000300001.

Aguilar, Alfredo, Roland Wohlgemuth, and Tomasz Twardowski. 2018. "Perspectives on Bio-economy". *New Biotechnology* 40: 181–184. doi:10.1016/j.nbt.2017.06.012.

Ahn, Hee-Hoon, and Tae-Jong Kim. 2021. "Three Endogenous Cellulases from Termite, *Reticulitermes speratus* KMT001". *Archives of Insect Biochemistry and Physiology* 106 (3): e21766. doi:10.1002/arch.21766.

Ahrendt, Steven R., C. Alisha Quandt, Doina Ciobanu, Alicia Clum, Asaf Salamov, Bill Andreopoulos, and Jan-Fang Cheng et al. 2018. "Leveraging Single-Cell Genomics to Expand the Fungal Tree of Life". *Nature Microbiology* 3 (12): 1417–1428. doi:10.1038/s41564-018-0261-0.

Alain, Karine, and Joël Querellou. 2009. "Cultivating the Uncultured: Limits, Advances and Future Challenges". *Extremophiles* 13 (4): 583–594. doi:10.1007/s00792-009-0261-3.

Alves, Sérgio L., Caroline Müller, Charline Bonatto, Thamarys Scapini, Aline F. Camargo, Gislaine Fongaro, and Helen Treichel. 2019. "Bioprospection of Enzymes and Microorganisms in Insects to Improve Second-Generation Ethanol Production". *Industrial Biotechnology* 15 (6): 336–349. doi:10.1089/ind.2019.0019.

Ariaeenejad, Shohreh, Morteza Maleki, Elnaz Hosseini, Kaveh Kavousi, Ali A. Moosavi-Movahedi, and Ghasem Hosseini Salekdeh. 2019. "Mining of Camel Rumen Metagenome to Identify Novel Alkali-Thermostable Xylanase Capable of Enhancing the Recalcitrant Lignocellulosic Biomass Conversion". *Bioresource Technology* 281: 343–350. doi:10.1016/j.biortech.2019.02.059.

Asemoloye, Michael Dare, Mario Andrea Marchisio, Vijai Kumar Gupta, and Lorenzo Pecoraro. 2021. "Genome-Based Engineering of Ligninolytic Enzymes in Fungi". *Microbial Cell Factories* 20 (1): 20. doi:10.1186/s12934-021-01510-9.

Azapagic, Adisa. 2014. "Sustainability Considerations for Integrated Biorefineries". *Trends in Biotechnology* 32 (1): 1–4. doi:10.1016/j.tibtech.2013.10.009.

Bello, Sara, Carmen Ríos, Gumersindo Feijoo, and María Teresa Moreira. 2018. "Comparative Evaluation of Lignocellulosic Biorefinery Scenarios under a Life-Cycle Assessment Approach". *Biofuels, Bioproducts and Biorefining* 12 (6): 1047–1064. doi:10.1002/bbb.1921.

Berini, Francesca, Carmine Casciello, Giorgia Letizia Marcone, and Flavia Marinelli. 2017. "Metagenomics: Novel Enzymes from Non-Culturable Microbes". *FEMS Microbiology Letters* 364 (21). doi:10.1093/femsle/fnx211.

Bohra, Varsha, Nishant A. Dafale, and Hemant J. Purohit. 2019. "Understanding the Alteration in Rumen Microbiome and CAZymes Profile with Diet and Host Through Comparative Metagenomic Approach". *Archives of Microbiology* 201 (10): 1385–1397. doi:10.1007/s00203-019-01706-z.

Bomble, Yannick J., Chien-Yuan Lin, Antonella Amore, Hui Wei, Evert K. Holwerda, Peter N. Ciesielski, Bryon S. Donohoe, Stephen R. Decker, Lee R. Lynd, and Michael E. Himmel. 2017. "Lignocellulose Deconstruction in the Biosphere". *Current Opinion in Chemical Biology* 41: 61–70. doi:10.1016/j.cbpa.2017.10.013.

Chen, Hongzhang. 2015. *Lignocellulose Biorefinery Engineering: Principles and Applications*. Woodhead Publishing. doi:10.1016/C2014-0-02702-5.

Cherubini, Francesco, Michael Mandl, Connie Philips, et al. 2010. "IEA Bioenergy Task 42 on Biorefineries: Co-production of Fuels, Chemicals, Power and Materials from Biomass: IEA Bioenergy Task 42—Countries Report". *IEA Bioenergy*: 37.

Chowdhury, Ratul, and Costas D. Maranas. 2020. "From Directed Evolution to Computational Enzyme Engineering—A Review". *AIChE Journal* 66 (3): e16847. doi:10.1002/aic.16847.

Cragg, Simon M., Gregg T. Beckham, Neil C. Bruce, Timothy D. H. Bugg, Daniel L. Distel, Paul Dupree, and Amaia Green Etxabe, et al. 2015. "Lignocellulose Degradation Mechanisms Across The Tree of Life". *Current Opinion in Chemical Biology* 29: 108–119. doi:10.1016/j.cbpa.2015.10.018.

Contesini, Fabiano Jares, Rasmus John Normand Frandsen, and André Damasio. 2021. "Editorial: CAZymes in Biorefinery: From Genes to Application". *Frontiers in Bioengineering and Biotechnology* 9: 622817. doi:10.3389/fbioe.2021.622817.

Contreras, Francisca, Subrata Pramanik, Aleksandra M. Rozhkova, Ivan N. Zorov, Olga Korotkova, Arkady P. Sinitsyn, Ulrich Schwaneberg, and Mehdi D. Davari. 2020. "Engineering Robust Cellulases for Tailored Lignocellulosic Degradation Cocktails". *International Journal of Molecular Sciences* 21 (5): 1589. doi:10.3390/ijms21051589.

Costa, Ohana Y. A., Mattias de Hollander, Agata Pijl, Binbin Liu, and Eiko E. Kuramae. 2020. "Cultivation-Independent and Cultivation-Dependent Metagenomes Reveal Genetic and

Enzymatic Potential of Microbial Community Involved in the Degradation of a Complex Microbial Polymer". *Microbiome* 8 (1): 76. doi:10.1186/s40168-020-00836-7.

Danilovich, Mariana Elizabeth, Leandro Arturo Sánchez, Federico Acosta, and Osvaldo Daniel Delgado. 2018. "Antarctic Bioprospecting: In Pursuit of Microorganisms Producing New Antimicrobials and Enzymes". *Polar Biology* 41 (7): 1417–1433. doi:10.1007/s00300-018-2295-4.

Das, Nabajit, Naveen Tripathi, Srijoni Basu, Chandra Bose, Susmit Maitra, and Sukant Khurana. 2015. "Progress in the Development of Gelling Agents for Improved Culturability of Microorganisms". *Frontiers in Microbiology* 6: 698. doi:10.3389/fmicb.2015.00698.

de Gonzalo, Gonzalo, Dana I. Colpa, Mohamed H. M. Habib, and Marco W. Fraaije. 2016. "Bacterial Enzymes Involved in Lignin Degradation". *Journal of Biotechnology* 236: 110–119. doi:10.1016/j.jbiotec.2016.08.011.

de Lorenzo, Víctor, and Markus Schmidt. 2018. "Biological Standards for the Knowledge-Based Bioeconomy: What Is at Stake". *New Biotechnology* 40: 170–180. doi:10.1016/j.nbt.2017.05.001.

Dinmukhamed, Tanatarov, Ziyang Huang, Yanfeng Liu, Xueqin Lv, Jianghua Li, Guocheng Du, and Long Liu. 2021. "Current Advances in Design and Engineering Strategies of Industrial Enzymes". *Systems Microbiology and Biomanufacturing* 1 (1): 15–23. doi:10.1007/s43393-020-00005-9.

Duarte, Alysson Wagner Fernandes, Juliana Aparecida dos Santos, Marina Vitti Vianna, Juliana Maíra Freitas Vieira, Vitor Hugo Mallagutti, Fabio José Inforsato, and Lia Costa Pinto Wentzel, et al. 2018. "Cold-Adapted Enzymes Produced By Fungi from Terrestrial and Marine Antarctic Environments". *Critical Reviews In Biotechnology* 38 (4): 600–619. doi:10.1080/07388551.2017.1379468.

Eijsink, Vincent G. H., Gustav Vaaje-Kolstad, Kjell M. Vårum, and Svein J. Horn. 2008. "Towards New Enzymes for Biofuels: Lessons From Chitinase Research". *Trends in Biotechnology* 26 (5): 228–235. doi:10.1016/j.tibtech.2008.02.004.

Erickson, Brent, Janet E. Nelson, and Paul Winters. 2012. "Perspective on Opportunities in Industrial Biotechnology in Renewable Chemicals". *Biotechnology Journal* 7 (2): 176–185. doi:10.1002/biot.201100069.

Fang, Wei, Saisai Xue, Pengjun Deng, Xuecheng Zhang, Xiaotang Wang, Yazhong Xiao, and Zemin Fang. 2019. "Amyz1: A Novel A-Amylase from Marine Bacterium *Pontibacillus* sp. ZY with High Activity toward Raw Starches". *Biotechnology for Biofuels* 12 (1): 95. doi:10.1186/s13068-019-1432-9.

Filiatrault-Chastel, Camille, David Navarro, Mireille Haon, Sacha Grisel, Isabelle Herpoël-Gimbert, Didier Chevret, and Mathieu Fanuel, et al. 2019. "AA16, a New Lytic Polysaccharide Monooxygenase Family Identified in Fungal Secretomes". *Biotechnology for Biofuels* 12 (1): 55. doi:10.1186/s13068-019-1394-y.

Fowler, Claire A., Federico Sabbadin, Luisa Ciano, Glyn R. Hemsworth, Luisa Elias, Neil Bruce, Simon McQueen-Mason, Gideon J. Davies, and Paul H. Walton. 2019. "Discovery, Activity and Characterisation of an AA10 Lytic Polysaccharide Oxygenase from the Shipworm Symbiont *Teredinibacter turnerae*". *Biotechnology for Biofuels* 12 (1): 232. doi:10.1186/s13068-019-1573-x.

Fülöp, László, and János Ecker. 2020. "An Overview of Biomass Conversion: Exploring New Opportunities". *PeerJ* 8: e9586. doi:10.7717/peerj.9586.

Georgiadou, Daphne N., Pavlos Avramidis, Efstathia Ioannou, and Dimitris G. Hatzinikolaou. 2021. "Microbial Bioprospecting for Lignocellulose Degradation at a Unique Greek Environment". *Heliyon* 7 (6): e07122. doi:10.1016/j.heliyon.2021.e07122.

Gómez-Silva, Benito, Claudia Vilo-Muñoz, Alexandra Galetović, Qunfeng Dong, Hugo G. Castelán-Sánchez, Yordanis Pérez-Llano, and María del Rayo Sánchez-Carbente, et al. 2019. "Metagenomics of Atacama Lithobiontic Extremophile Life Unveils Highlights on Fungal Communities, Biogeochemical Cycles and Carbohydrate-Active Enzymes". *Microorganisms* 7 (12): 619. doi:10.3390/microorganisms7120619.

Gonçalves, Carolyne Caetano, Thiago Bruce, Caio de Oliveira Gorgulho Silva, Edivaldo Ximenes Ferreira Fillho, Eliane Ferreira Noronha, Magnus Carlquist, and Nádia Skorupa Parachin. 2020. "Bioprospecting Microbial Diversity for Lignin Valorization: Dry and Wet Screening Methods". *Frontiers in Microbiology* 11: 1081. doi:10.3389/fmicb.2020.01081.

Guazzaroni, María-Eugenia, Rafael Silva-Rocha, and Richard John Ward. 2015. "Synthetic Biology Approaches to Improve Biocatalyst Identification in Metagenomic Library Screening". *Microbial Biotechnology* 8 (1): 52–64. doi:10.1111/1751-7915.12146.

Guimarães, L. R. C., A. L. Woiciechowski, S. G. Karp, J. D. Coral, A. Zandoná Filho, and C. R. Soccol. 2017. "Laccases". In *Current Developments in Biotechnology and Bioengineering: Production, Isolation and Purification of Industrial Products*, eds. A. Pandey, S. Negi, and C. R. Soccol, 199–216. Elsevier. doi:10.1016/B978-0-444-63662-1.00009-9.

Guirimand, Gregory, Natalja Kulagina, Nicolas Papon, Tomohisa Hasunuma, and Vincent Courdavault. 2021. "Innovative Tools and Strategies for Optimizing Yeast Cell Factories". *Trends in Biotechnology* 39 (5): 488–504. doi:10.1016/j.tibtech.2020.08.010.

Jadaun, Jyoti Singh, Lokesh Kumar Narnoliya, Archana Srivastava, Sudhir P. Singh. 2020. "Chimeric Enzyme Designing for the Synthesis of Multifunctional Biocatalysts". In *Biomass, Biofuels, Biochemicals*, eds. S. P. Singh, A. P. R. R. Singhania, and C. L. Zhi Li, 119–143. Elsevier. doi:10.1016/B978-0-12-819820-9.00008-9.

Jagadevan, Sheeja, Avik Banerjee, Chiranjib Banerjee, Chandan Guria, Rameshwar Tiwari, Mehak Baweja, and Pratyoosh Shukla. 2018. "Recent Developments in Synthetic Biology and Metabolic Engineering in Microalgae Towards Biofuel Production". *Biotechnology for Biofuels* 11 (1): 185. doi:10.1186/s13068-018-1181-1.

Kamble, Asmita, Sumana Srinivasan, and Harinder Singh. 2019. "*In-Silico* Bioprospecting: Finding Better Enzymes". *Molecular Biotechnology* 61 (1): 53–59. doi:10.1007/s12033-018-0132-1.

Kaushal, Girija, Jitendra Kumar, Rajender S. Sangwan, and Sudhir P. Singh. 2018. "Metagenomic Analysis of Geothermal Water Reservoir Sites Exploring Carbohydrate-Related Thermozymes". *International Journal of Biological Macromolecules* 119: 882–895. doi:10.1016/j.ijbiomac.2018.07.196.

Kojima, Yuka, Anikó Várnai, Vincent G. H. Eijsink, and Makoto Yoshida. 2020. "The Role of Lytic Polysaccharide Monooxygenases in Wood Rotting Basidiomycetes". *Trends in Glycoscience and Glycotechnology* 32 (188): E135–E143. doi:10.4052/tigg.2020.7e.

Kirubakaran, Rangasamy, K. N. ArulJothi, Sundaravadivel Revathi, Nowsheen Shameem, and Javid A. Parray. 2020. "Emerging Priorities for Microbial Metagenome Research". *Bioresource Technology Reports* 11: 100485. doi:10.1016/j.biteb.2020.100485.

Kumar, Adarsh, and Ram Chandra. 2020. "Ligninolytic Enzymes and Its Mechanisms for Degradation of Lignocellulosic Waste in Environment". *Heliyon* 6 (2): e03170. doi:10.1016/j.heliyon.2020.e03170.

Lange, Lene, Bo Pilgaard, Florian-Alexander Herbst, Peter Kamp Busk, Frank Gleason, and Anders Gorm Pedersen. 2019. "Origin of Fungal Biomass Degrading Enzymes: Evolution, Diversity and Function of Enzymes of Early Lineage Fungi". *Fungal Biology Reviews* 33 (1): 82–97. doi:10.1016/j.fbr.2018.09.001.

Lee, Myung Hwan, and Seon-Woo Lee. 2013. "Bioprospecting Potential of the Soil Metagenome: Novel Enzymes and Bioactivities". *Genomics & Informatics* 11 (3): 114. doi:10.5808/gi.2013.11.3.114.

Li, Luen-Luen, Safiyh Taghavi, Sean M. McCorkle, Yian-Biao Zhang, Michael G. Blewitt, Roman Brunecky, and William S. Adney, et al. 2011. "Bioprospecting Metagenomics of Decaying Wood: Mining for New Glycoside Hydrolases". *Biotechnology for Biofuels* 4 (1): 23. doi:10.1186/1754-6834-4-23.

Liang, Jinsong, Mohammad Nabi, Panyue Zhang, Guangming Zhang, Yajing Cai, Qingyan Wang, Zeyan Zhou, and Yiran Ding. 2020. "Promising Biological Conversion of Lignocellulosic Biomass to Renewable Energy with Rumen Microorganisms: A Comprehensive

Review". *Renewable and Sustainable Energy Reviews* 134: 110335. doi:10.1016/j.rser.2020.110335.

Liu, Zihe, Hamideh Moradi, Shuobo Shi, and Farshad Darvishi. 2021. "Yeasts as Microbial Cell Factories for Sustainable Production of Biofuels". *Renewable and Sustainable Energy Reviews* 143: 110907. doi:10.1016/j.rser.2021.110907.

Lokko, Yvonne, Marc Heijde, Karl Schebesta, Philippe Scholtès, Marc Van Montagu, and Mauro Giacca. 2018. "Biotechnology and the Bioeconomy—Towards Inclusive and Sustainable Industrial Development". *New Biotechnology* 40: 5–10. doi:10.1016/j.nbt.2017.06.005.

Lombard, Vincent, Hemalatha Golaconda Ramulu, Elodie Drula, Pedro M. Coutinho, and Bernard Henrissat. 2014. "The Carbohydrate-Active Enzymes Database (CAZy) in 2013". *Nucleic Acids Research* 42 (D1): D490–D495. doi:10.1093/nar/gkt1178.

Ma, Jingbo, Yang Gu, Monireh Marsafari, and Peng Xu. 2020. "Synthetic Biology, Systems Biology, and Metabolic Engineering of *Yarrowia lipolytica* Toward a Sustainable Biorefinery Platform". *Journal of Industrial Microbiology and Biotechnology* 47 (9–10): 845–862. doi:10.1007/s10295-020-02290-8.

McKelvey, Shauna M., and Richard A. Murphy. 2017. "Biotechnological Use of Fungal Enzymes". In *Fungi: Biology and Applications*, ed. K. Kavanagh, 201–225. John Wiley & Sons, Inc. doi:10.1002/9781119374312.ch8.

Mehta, Deepika, and Tulasi Satyanarayana. 2016. "Bacterial and Archaeal A-Amylases: Diversity and Amelioration of the Desirable Characteristics for Industrial Applications". *Frontiers in Microbiology* 7: 1129. doi:10.3389/fmicb.2016.01129.

Mondal, Subhadeep, Suman Kumar Halder, and Keshab Chandra Mondal. 2021. "Tailoring in Fungi for Next Generation Cellulase Production with Special Reference to CRISPR/CAS System". *Systems Microbiology and Biomanufacturing*: 1–17. doi:10.1007/s43393-021-00045-9.

Moraes, Eduardo C., Thabata M. Alvarez, Gabriela F. Persinoti, Geizecler Tomazetto, Livia B. Brenelli, Douglas A. A. Paixão, and Gabriela C. Ematsu, et al. 2018. "Lignolytic-Consortium Omics Analyses Reveal Novel Genomes and Pathways Involved in Lignin Modification and Valorization". *Biotechnology for Biofuels* 11 (1): 75. doi:10.1186/s13068-018-1073-4.

Mueller, Gregory M., and John Paul Schmit. 2007. "Fungal Biodiversity: What Do We Know? What Can We Predict?" *Biodiversity and Conservation* 16 (1): 1–5. doi:10.1007/s10531-006-9117-7.

Munzone, Alessia, Bilal El Kerdi, Mathieu Fanuel, Hélène Rogniaux, David Ropartz, Marius Réglier, Antoine Royant, A. Jalila Simaan, and Christophe Decroos. 2020. "Characterization of a Bacterial Copper-Dependent Lytic Polysaccharide Monooxygenase with an Unusual Second Coordination Sphere". *The FEBS Journal* 287 (15): 3298–3314. doi:10.1111/febs.15203.

Naas, A. E., L. M. Solden, A. D. Norbeck, H. Brewer, L. H. Hagen, I. M. Heggenes, and A. C. McHardy, et al. 2018. "'*Candidatus* Paraporphyromonas Polyenzymogenes' Encodes Multi-Modular Cellulases Linked to the Type IX Secretion System". *Microbiome* 6 (1): 44. doi:10.1186/s40168-018-0421-8.

Naveed, Muhammad, Fareeha Nadeem, Tahir Mehmood, Muhammad Bilal, Zahid Anwar, and Fazeeha Amjad. 2021. "Protease—A Versatile and Ecofriendly Biocatalyst with Multi-Industrial Applications: An Updated Review". *Catalysis Letters* 151 (2): 307–323. doi:10.1007/s10562-020-03316-7.

Paul, Jai Shankar, Nisha Gupta, Esmil Beliya, Shubhra Tiwari, and Shailesh Kumar Jadhav. 2021. "Aspects and Recent Trends in Microbial A-Amylase: A Review". *Applied Biochemistry and Biotechnology* 193 (8): 2649–2698. doi:10.1007/s12010-021-03546-4.

Paul, Subhash, and Animesh Dutta. 2018. "Challenges and Opportunities of Lignocellulosic Biomass for Anaerobic Digestion". *Resources, Conservation And Recycling* 130: 164–174. doi:10.1016/j.resconrec.2017.12.005.

Peralta, Rosane Marina, Bruna Polacchine da Silva, Rúbia Carvalho Gomes Côrrea, Camila Gabriel Kato, Flávio Augusto Vicente Seixas, and Adelar Bracht. 2017. "Enzymes from Basidiomycetes—Peculiar and Efficient Tools for Biotechnology". In *Biotechnology of Microbial Enzymes*, ed. G. Brahmachari, 119–149. Academic Press. doi:10.1016/B978-0-12-803725-6.00005-4.

Porter, Joanne L., Rukhairul A. Rusli, and David L. Ollis. 2016. "Directed Evolution of Enzymes for Industrial Biocatalysis". *Chembiochem* 17 (3): 197–203. doi:10.1002/cbic.201500280.

Prakash, Om, Yogesh Shouche, Kamlesh Jangid, and Joel E. Kostka. 2013. "Microbial Cultivation and the Role of Microbial Resource Centers in the Omics Era". *Applied Microbiology and Biotechnology* 97 (1): 51–62. doi:10.1007/s00253-012-4533-y.

Prayogo, Fitra Adi, Anto Budiharjo, Hermin Pancasakti Kusumaningrum, Wijanarka Wijanarka, Agung Suprihadi, and Nurhayati Nurhayati. 2020. "Metagenomic Applications in Exploration and Development of Novel Enzymes from Nature: A Review". *Journal of Genetic Engineering and Biotechnology* 18 (1): 39. doi:10.1186/s43141-020-00043-9.

Rebello, Sharrel, Amith Abraham, Aravind Madhavan, Raveendran Sindhu, Parameswaran Binod, Arun K. Babu, Embalil Mathachan Aneesh, and Ashok Pandey. 2018. "Non-Conventional Yeast Cell Factories for Sustainable Bioprocesses". *FEMS Microbiology Letters* 355 (21). doi:10.1093/femsle/fny222.

Reichart, Nicholas J., Robert M. Bowers, Tanja Woyke, and Roland Hatzenpichler. 2021. "High Potential for Biomass-Degrading Enzymes Revealed by Hot Spring Metagenomics". *Frontiers in Microbiology* 12: 668238. doi:10.3389/fmicb.2021.668238.

Robinson, Serina L., Jörn Piel, and Shinichi Sunagawa. 2021. "A Roadmap for Metagenomic Enzyme Discovery". *Natural Product Reports*. doi:10.1039/d1np00006c.

Sahoo, Kalpana, Rajesh Kumar Sahoo, Mahendra Gaur, and Enketeswara Subudhi. 2020. "Cellulolytic Thermophilic Microorganisms in White Biotechnology: A Review". *Folia Microbiologica* 65 (1): 25–43. doi:10.1007/s12223-019-00710-6.

Salvachúa, Davinia, Rui Katahira, Nicholas S. Cleveland, Payal Khanna, Michael G. Resch, Brenna A. Black, and Samuel O. Purvine, et al. 2016. "Lignin Depolymerization by Fungal Secretomes and a Microbial Sink". *Green Chemistry* 18 (22): 6046–6062. doi:10.1039/c6gc01531j.

Salwan, Richa, Vivek Sharma, Ramesh Chand Kasana, and Arvind Gulati. 2020. "Bioprospecting Psychrotrophic Bacteria for Serine-Type Proteases from the Cold Areas of Western Himalayas". *Current Microbiology* 77 (5): 795–806. doi:10.1007/s00284-020-01876-w.

Schalk, Felix, Cene Gostinčar, Nina B. Kreuzenbeck, Benjamin H. Conlon, Elisabeth Sommerwerk, Patrick Rabe, and Immo Burkhardt, et al. 2021. "The Termite Fungal Cultivar *Termitomyces* Combines Diverse Enzymes and Oxidative Reactions for Plant Biomass Conversion". *Mbio* 12 (3): e03551–20. doi:10.1128/mbio.03551-20.

Scully, Erin D., Scott M. Geib, Kelli Hoover, Ming Tien, Susannah G. Tringe, Kerrie W. Barry, Tijana Glavina del Rio, Mansi Chovatia, Joshua R. Herr, and John E. Carlson. 2013. "Metagenomic Profiling Reveals Lignocellulose Degrading System in a Microbial Community Associated with a Wood-Feeding Beetle". *PloS One* 8 (9): e73827. doi:10.1371/journal.pone.0073827.

Shanmugam, Sabarathinam, Huu-Hao Ngo, and Yi-Rui Wu. 2020. "Advanced CRISPR/Cas-Based Genome Editing Tools for Microbial Biofuels Production: A Review". *Renewable Energy* 149: 1107–1119. doi:10.1016/j.renene.2019.10.107.

Sharma, Himanshu, and Santosh Kumar Upadhyay. 2020. "Enzymes and Their Production Strategies". In *Biomass, Biofuels, Biochemicals: Advances in Enzyme Catalysis and Technologies*, eds. S. P. Singh, A. Pandey, R. R. Singhania, C. Larroche, Z. Li, 31–48. Elsevier. doi:10.1016/B978-0-12-819820-9.00003-X.

Shelomi, Matan, Shih-Shun Lin, and Li-Yu Liu. 2019. "Transcriptome and Microbiome of Coconut Rhinoceros Beetle (*Oryctes rhinoceros*) Larvae". *BMC Genomics* 20 (1): 957. doi:10.1186/s12864-019-6352-3.

Shen, Jing, Lixin Zheng, Xiaodong Chen, Xiaoying Han, Yangchun Cao, and Junhu Yao. 2020. "Metagenomic Analyses of Microbial and Carbohydrate-Active Enzymes in the Rumen of Dairy Goats Fed Different Rumen Degradable Starch". *Frontiers in Microbiology* 11: 1003. doi:10.3389/fmicb.2020.01003.

Simmons, Christopher W., Amitha P. Reddy, Patrik D'haeseleer, Jane Khudyakov, Konstantinos Billis, Amrita Pati, Blake A. Simmons, Steven W. Singer, Michael P. Thelen, and Jean S. VanderGheynst. 2014. "Metatranscriptomic Analysis of Lignocellulolytic Microbial Communities Involved in High-Solids Decomposition of Rice Straw". *Biotechnology for Biofuels* 7 (1): 495. doi:10.1186/s13068-014-0180-0.

Singh, Anusuiya, Rosa M. Rodríguez Jasso, Karla D. Gonzalez-Gloria, Miriam Rosales, Ruth Belmares Cerda, Cristóbal N. Aguilar, Reeta Rani Singhania, and Héctor A. Ruiz. 2019. "The Enzyme Biorefinery Platform for Advanced Biofuels Production". *Bioresource Technology Reports* 7: 100257. doi:10.1016/j.biteb.2019.100257.

Stovicek, Vratislav, Carina Holkenbrink, and Irina Borodina. 2017. "CRISPR/Cas System for Yeast Genome Engineering: Advances and Applications". *FEMS Yeast Research* 17 (5). doi:10.1093/femsyr/fox030.

Strazzulli, Andrea, Beatrice Cobucci-Ponzano, Roberta Iacono, Rosa Giglio, Luisa Maurelli, Nicola Curci, and Corinna Schiano-di-Cola, et al. 2020. "Discovery of Hyperstable Carbohydrate-Active Enzymes Through Metagenomics of Extreme Environments". *The FEBS Journal* 287 (6): 1116–1137. doi:10.1111/febs.15080.

Sysoev, Maksim, Stefan W. Grötzinger, Dominik Renn, Jörg Eppinger, Magnus Rueping, and Ram Karan. 2021. "Bioprospecting of Novel Extremozymes from Prokaryotes—The Advent of Culture-Independent Methods". *Frontiers In Microbiology* 12: 630013 doi:10.3389/fmicb.2021.630013.

Tanwar, Amita, and Shashi Kumar. 2020. "Genome Editing of Algal Species by CRISPR Cas9 for Biofuels". In *Genome Engineering via CRISPR-Cas9 System*, eds. V. Singh and P. K. Dhar, 163–176. Academic Press. doi:10.1016/B978-0-12-818140-9.00015-5.

Thankappan, Sugitha, Sujatha Kandasamy, Beslin Joshi, Ksenia N. Sorokina, Oxana P. Taran, and Sivakumar Uthandi. 2018. "Bioprospecting Thermophilic Glycosyl Hydrolases, from Hot Springs of Himachal Pradesh, for Biomass Valorization". *AMB Express* 8 (1): 168. doi:10.1186/s13568-018-0690-4.

Tovar-Herrera, Omar Eduardo, Adriana Mayrel Martha-Paz, Yordanis Pérez-LLano, Elisabet Aranda, Juan Enrique Tacoronte-Morales, María Teresa Pedroso-Cabrera, Katiushka Arévalo-Niño, Jorge Luis Folch-Mallol, and Ramón Alberto Batista-García. 2018. "*Schizophyllum* Commune: An Unexploited Source for Lignocellulose Degrading Enzymes". *Microbiologyopen* 7 (3): e00637. doi:10.1002/mbo3.637.

Trivedi, N., C. R. K. Reddy, and A. M. Lali. 2016. "Marine Microbes as a Potential Source of Cellulolytic Enzymes". *Advances in Food and Nutrition Research* 79: 27–41. doi:10.1016/bs.afnr.2016.07.002.

Ubando, Aristotle T., Charles B. Felix, and Wei-Hsin Chen. 2020. "Biorefineries in Circular Bioeconomy: A Comprehensive Review". *Bioresource Technology* 299: 122585. doi:10.1016/j.biortech.2019.122585.

Uçkun Kiran, Esra, Antoine P. Trzcinski, Wun Jern Ng, and Yu Liu. 2014. "Enzyme Production from Food Wastes Using a Biorefinery Concept". *Waste And Biomass Valorization* 5 (6): 903–917. doi:10.1007/s12649-014-9311-x.

van der Helm, Eric, Hans J. Genee, and Morten O. A. Sommer. 2018. "The Evolving Interface Between Synthetic Biology and Functional Metagenomics". *Nature Chemical Biology* 14 (8): 752–759. doi:10.1038/s41589-018-0100-x.

Varrella, Stefano, Giulio Barone, Michael Tangherlini, Eugenio Rastelli, Antonio Dell'Anno, and Cinzia Corinaldesi. 2021. "Diversity, Ecological Role and Biotechnological Potential of Antarctic Marine Fungi". *Journal of Fungi* 7 (5): 391. doi:10.3390/jof7050391.

Verma, Swati, Gautam Kumar Meghwanshi, and Rajender Kumar. 2021. "Current Perspectives for Microbial Lipases from Extremophiles and Metagenomics". *Biochimie* 182: 23–36. doi:10.1016/j.biochi.2020.12.027.

Voshol, Gerben P., Erik Vijgenboom, and Peter J. Punt. 2017. "The Discovery of Novel LPMO Families with a New Hidden Markov Model". *BMC Research Notes* 10 (1): 105. doi:10.1186/s13104-017-2429-8.

Weiss, Bruno, Anna Carolina Oliveira Souza, Milena Tavares Lima Constancio, Danillo Oliveira Alvarenga, Victor S. Pylro, Lucia M. Carareto Alves, and Alessandro M. Varani. 2021. "Unraveling a Lignocellulose-Decomposing Bacterial Consortium from Soil Associated with Dry Sugarcane Straw by Genomic-Centered Metagenomics". *Microorganisms* 9 (5): 995. doi:10.3390/microorganisms9050995.

Zeiner, Carolyn A., Samuel O. Purvine, Erika M. Zink, Ljiljana Paša-Tolić, Dominique L. Chaput, Sajeet Haridas, and Si Wu, et al. 2016. "Comparative Analysis of Secretome Profiles of Manganese(II)-Oxidizing Ascomycete Fungi". *PloS One* 11 (7): e0157844. doi:10.1371/journal.pone.0157844.

Zhang, Yi, Alberta N. A. Aryee, and Benjamin K. Simpson. 2020. "Current Role of *in silico* Approaches for Food Enzymes". *Current Opinion in Food Science* 31: 63–70. doi:10.1016/j.cofs.2019.11.003.

Zhang, Yiming, Jens Nielsen, and Zihe Liu. 2021. "Yeast Based Biorefineries for Oleochemical Production". *Current Opinion in Biotechnology* 67: 26–34. doi:10.1016/j.copbio.2020.11.009. doi:10.1016/j.copbio.2020.11.009.

4 Functional Metagenomics for Bioprospecting Novel Glycosyl Hydrolases in 2G Biofuel Production from Lignocellulosics

Sugitha Thankappan[1], Sajan Kurien[1], Surender Singh[2], and Asish K. Binodh[3]

[1]School of Agriculture and Biosciences, Karunya Institute of Technology and Sciences, Coimbatore, Tamil Nadu, India-641114

[2]Department of Microbiology, Central University of Haryana, Mahendergarh, India-123031

[3]Tamil Nadu Agricultural University, Coimbatore, Tamil Nadu, India-641003

CONTENTS

DOI: 10.1201/9781003191247-4

4.1 INTRODUCTION

The daunting panorama of fossil fuel depletion and energy production has stimulated researchers across the globe to search for cost-economic alternate sources of renewable energy. Many countries have formulated fuel efficiency strategies to curtail the prevailing energy demand. In India and China, fossil fuel consumption has escalated by 5% and 2.9%, respectively, during the year 2018–2019, which shows the increasing trend of oil consumption (Energy, 2018, 2019; Report, 2020). It is evident from the statistics of the past ten years that there is an increasing need for alternative renewable fuels, like biofuel (Pabbathi et al., 2021). The global biofuel has registered a 3% increase, whereas India accounts for approximately 10% increase in energy production from renewable sources against 2018 (BP Statistical Review of World Energy, 2018, 2019). However, the bioenergy demand is expected to surge up to 11% by 2040 (Agency, 2017). In the quest for alternative energy reserves, Indian forest biomass and lignocellulosic crop residues have captured the interests of scientists and policy-makers. Lignocellulosic biomass (LCB) is a reliable source for sugars that yield green fuels and platform chemicals upon fermentation (Sorensen, Lübeck, Lübeck, & Ahring, 2013).

Plant fixes atmospheric CO_2 and incorporates half of the carbon in structural polysaccharides and lignin (lignocellulose). Lignocellulose comprises three macromolecules: cellulose (40–50%), hemicellulose (25–30%), and lignin (15–25%) (Gray, Zhao, & Emptage, 2006). Lignin is more recalcitrant for degradation due to the presence of aromatic rings. Though considerable approaches are available for LCB pretreatment and its destruction, biocatalysts, including microbial and enzymatic pretreatment and hydrolysis processes, are not much explored at commercial scale. There is ample scope to improve the biological LCB valorization using novel, efficient, and engineered enzymes (Gowen & Fong, 2010). The comprehensive breakdown of LCB into sugar monomers is mediated by different enzymes belonging to the glycosyl hydrolases (GHs) family. Microbiome in a lignocellulose-degrading habitat secretes mixture of GHs, which aid in thorough degradation of LC (Bayer, Chanzy, Lamed, & Shoham, 1998; Ljungdahl, 2008). Apart from the GHs of culturables, the search for robust, novel, thermostable, and efficient enzyme systems for bioconversion of biomass into biofuel from natural ecological niches seems to be more worthy. One of the approaches to unravel novel GHs for biomass valorization is metagenomics. In this chapter, emphasis is given to throw light on the metagenomic-derived GHs targeting biofuel production.

4.2 LIGNOCELLULOSICS: A HIDDEN TREASURE OF GREEN CHEMICALS AND BIOFUELS

Bioethanol production from LCB entails four important steps, i) pretreatment of raw biomass, ii) saccharification, iii) fermentation process, iv) product recovery, to extract the chemicals/bioethanol produced. The pretreatment process allows the enzymatic treatment of biomass by disrupting the bonds between recalcitrant lignin and cellulose/hemicellulose. Pre-treatment techniques include the physical/mechanical, physicochemical, chemical, and biological. However, the appropriate choice of pre-treatment is based on the composition and nature of LCB (hard- or softwood). The second step is the saccharification, during which the pretreated biomass rich in cellulose is hydrolyzed to sugars, mainly glucose, by cellulase. Microbes such as bacteria, fungi, archaea, and actinomycetes are efficient cellulase producers, mostly explored for biotechnological and bioenergy applications. During fermentation, yeast belonging to the genera *Saccharomyces* or *Kluyveromyces* converts the hydrolyzed sugars to different products, mainly ethanol. During the final step, bioethanol from fermentation broth is separated by distillation or pervaporation to obtain high-purity ethanol.

4.3 STRATEGIES OF LIGNOCELLULOSICS VALORIZATION

A separate hydrolysis and fermentation (SHF) process is the conventional approach of bioethanol production, wherein processes like saccharification and fermentation are performed individually in two pots (Salehi Jouzani & Taherzadeh, 2015). The major limitations of SHF are the feedback inhibition by cellulase due to the sugars liberated by the substrate during hydrolysis (Goshadrou, 2013). On the contrary, simultaneous saccharification and fermentation (SSF) are efficient in overcoming substrate-induced feedback inhibition by the immediate conversion of hydrolyzed sugars to ethanol, thus increasing enzyme efficiency (Moshi et al., 2014). However, the major drawbacks of the SSF process are high hydrolysis and fermentation temperatures (45–60°C), which require thermostable biocatalysts and thermotolerant yeasts (Bušić et al., 2018). Nonisothermal simultaneous saccharification and fermentation (NSSF) that involves partial enzymatic hydrolysis at optimum temperatures alleviates the issues of SSF (Goshadrou, 2013; Wu & Lee, 1998). Besides, membranes are used for downstream processes and to obtain a concentrated sugar slurry from the saccharified hydrolysate. In single-pot process of hydrolysis and co-fermentation (SSCF), xylose (part of saccharified hydrolysate) is efficiently utilized by employing glucose and xylose utilizing wild-type or engineered microbial strains (S. J. Kim et al., 2007; J. Zhang & Lynd, 2010). Since, xylose is a major monomer next to glucose in the hydrolysate, for efficient xylose utilization, consolidated bioprocessing (CBP) has been proposed as a successful alternative.

Consolidated bioprocessing (CBP) evades the demerits of conventional biofuel production from LCBs. CBP integrates the processes like enzyme production, hydrolysis, and fermentation in a single step by employing a single microorganism or a consortium. Thus, CBP eliminates the dependency on exogenous application of hydrolytic enzymes, decreases the cellulase feedback inhibition by hydrolyzed sugars, and also increases the process efficiency (Lynd et al., 2002). Further, CBP reduces the operations involved, thus reducing considerable process cost (Lynd, Weimer, Zyl, & Pretorius, 2002). More recently, researchers are working on process parameters for production of biomass directly from raw biomass without any pretreatment (Salehi Jouzani & Taherzadeh, 2015). Henceforth, CBP requires microbes/consortium producing robust and stable enzyme systems tolerant to high temperatures. The nature of biomass intended also plays an integral role in the efficiency of CBP.

4.4 GLYCOSYL HYDROLASES (GHs) FOR LCB VALORIZATION

The microfibrillar nature and high degree of self-association of cellulosic biomass make it more recalcitrant to absolute valorization. Cellulose degradation by microbial cellulases and cellulosomes constitutes a predominant C trade from fixed C sinks to atmospheric carbon dioxide (CO_2) pools. In general, microbes possess a cellulose-degrading machinery to fulfill their nutrient requirements in the environment. Cellulase enzymes belonging to different GHs (EC 3.2.1) valorize cellulose completely by cleaving the glycosidic bonds linking the two sugar monomer molecules and also between a sugar monomer and other side chains, like acetyl ferulic acid moieties. According to the nomenclature of IUBMB (International Union of Biochemistry and Molecular Biology), GHs were classified according to molecular mechanism and substrate specificities, more precisely amino acid homology (B. Henrissat, 1991), structural and functional relationships, and mechanistic flow of information (Lombard et al., 2013). The CAZy database has been accessed in public domain to retrieve GH families since 1999. Three different GHs, namely, endoglucanase (EC 3.2.1.4), nonreducing or reducing end-acting cellobiohydrolases (EC 3.2.1.91 or 3.2.1.176), and β-glucosidases (EC 3.2.1.21), are required for complete hydrolysis of cellulose.

The mechanisms of glycosyl hydrolase catalysis of two amino acid residues are classified into two categories: a) inverting mechanism and b) retaining mechanism (McCarter & Stephen Withers, 1994). In the former, β to α anomeric transition occurs through a single displacement, while anomeric carbon remains in the same position due to double displacement in the case of the retaining mechanism (Thuan & Sohng, 2013). However, the position of the proton donor in both mechanisms remains unchanged. In case of an inverting mechanism, a water molecule is inserted within the sugar moiety and catalytic base mechanism, whereas in the retaining mechanism, such interactions are not observed (Bernard Henrissat & Davies, 1997). However, epoxides (Sobala et al., 2020) and oxocarbenium ion–like intermediate states are observed in both mechanisms. Seven GH families exhibit an inverting mechanism of catalysis (6, 8, 9, 45, 48, 74, and 124), and eleven families act through the retaining mechanism (1–3, 5, 7, 12, 30, 39, 44, 51, and 116) (Pabbathi et al., 2021).

The structural analysis of protein represents the similarity between the members of different GHs. As an example, the endoglucanase belonging to GH-9 from *Nasutitermes takasagoensis* (a termite) encompasses three conserved catalytic sites, such as two aspartate and one glutamate moieties. Both aspartic acid residues deprotonate an H_2O molecule (FAOSTAT, 2021) and create a nucleophile which attacks the carbon at anomeric position, followed by breaking up the glycosidic bonds and an inversion at its anomeric position. Likewise, glutamic acid donates protons to the sessile O_2 in the glycosidic linkage (Linton, 2020). Endoglucanases of family GH-5 possess a conserved pair of glutamic acid residues, such as E314 and E179, in the active site of enzyme Cel-1 (Dadheech et al., 2018). However, many families of GHs lack a catalytic proton donor/acceptor and/or nucleophile (Dennis et al., 2006; Hidaka et al., 2006). Hence, alternate catalytic mechanisms, like proton-transferring network, substrate-assisted catalysis, noncarboxylate residues, and exogenous base/nucleophile, were proposed (Vuong & Wilson, 2010).

4.4.1 Endoglucanases/Carboxymethyl Cellulases (CMCase)

The random reaction of CMCases on the cellulose chain results in the insertion of new ends for subsequent hydrolysis by exoglucanases or cellobiohydrolases. Cellobiohydrolase binds to the amorphous regions and cleaves the cellobiose units. Thus, the cellulose hydrolysis is attenuated by the synergistic action of endoglucanases and exoglucanases (Dashtban, Maki, Leung, Mao, & Qin, 2010). Endoglucanases are grouped in the GH families 5–9, 12, 44, 45, 51, 61, 72, 74, and 102 in Cazypedia.

4.4.2 Cellobiohydrolases/Filter Paperases (FPases)

The enzyme cellobiohydrolase, or β- 1,4-exoglucanases, acts either on the reducing or nonreducing ends of cellulose microfibrils. Exoglucanases facilitate the conversion of cellulose into either glucose (glucanohydrolase) or cellobiase (cellobiohydrolases) as the product (Tiwari, Nain, Labrou, & Shukla, 2018). The cellulose chain end links to the catalytic site of exoglucanase, which is tunnel-like, and converts it into short oligomeric chains. In Cazypedia, cellobiohydrolases are representatives of GH5/6/7/9/48/72.

4.4.3 β-Glucosidases

The β-glucosidase (β-D-glucoside gluco-hydrolase) enzyme mediates glycosyl moiety transfer among nucleophilic oxygens. This enzyme hydrolyzes cellobiose, which is one of the rate-limiting steps defining the rate of reaction (Ketudat Cairns & Esen, 2010). Glucose is the end product of glucosidase action, and increased glucose levels inhibit the enzyme by feedback regulation.

4.5 CARBOHYDRATE-BINDING MODULES

Carbohydrate-binding modules (CBMs) or cellulose-binding domains (CBDs) are generally a polypeptide chain next to some GHs and fold independently. CBMs are predominantly associated with cellulose binding, which directs enzymes towards catalytic fragments. More specifically, CBM holds the catalytic fragment at its vicinity while binding to the carbohydrate moieties. Thus, CBM facilitates substrate binding, wherein the tyrosine and tryptophan in ligand-binding sites play an integral role (Abbott & Boraston, 2012; Armenta, Moreno-Mendieta, Sánchez-Cuapio, Sánchez, & Rodríguez-Sanoja, 2017). Till this date, CAzy databases encompass 283,460 modules categorized into eighty-eight families, whereas 1,750 modules are yet to be classified.

CBMs are classified into three types according to their ligand-binding site topography: CBM-A (families 1, 2, 3, 5, 63, 64, and 79) is hydrophobic, with a planar surface, and binds to structural polysaccharides like chitin and cellulose. The substrate binding in CBM-A, B, and C types owe to similar aromatic amino acids. However, type-B topographical anatomy is unique and distinct. Typically, CBM-B enables substrate interaction with two distinct sites on the same protein. A variable loop site (VLS) is located at the protein's tail end, while a concave face site (CFS) is at the concave end. Although CBM-B can bind to different sugars, both hexoses and pentoses, it lacks affinity to cellulose. CBM-B families include 4, 16, 22, 31, 48, 58, 61, 75, and 80. CBM-C differs in its topological anatomy from the other two (A and B) with a pocket-like substrate-binding site which allows the interaction of sugars, including monomers and oligomers. The major examples of CBM-C families are 13, 14, 32, 42, 62, 66, and 71 (Armenta et al., 2017).

4.6 MINING GHS USING CULTURE-INDEPENDENT APPROACH—THE METAGENOME

The traditional cultivable approaches could only explore $< 1\%$ of the microbes existing in any niche (Batista-García et al., 2016). For complete representations of harsh environments such as hot springs and salt lakes, culture-independent approaches have been developed.

4.6.1 Novel Paradigm of Metagenome

The term *metagenomics* denotes the genomic analysis of microbial assemblages and was first proposed by Handelsman and his coworkers (1998). The metagenomics technique combines molecular biology and genetics, which involves extraction, identification, and characterization of genetic information contained within the niche. In the metagenomics approach, sequencing or homologous expression of metagenomics DNA and mRNA is employed by directly cloning into a vector and a host. Some of the metagenomic applications include diversity of unculturable microbes, mining new genome, discovery of novel biocatalysts, small molecules, microbial assemblages structure, and elucidating new biosynthetic pathways. Further, the impact of climate changes on microbial structure requires conservation of metagenomic information. Metagenomics collaborates with other omics-based approaches, like metatranscriptomics, metaproteomics, and/or metabolomics, to provide a throughput profile of unculturable microbial communities. This facilitates the quest for bioprospecting unique proteins/enzymes, novel genes, and small molecules of biotechnological importance, including GHs (Batista-García et al., 2016).

4.6.2 Culturable versus Unculturable Approaches of GHs Mining

The constitutive soil heterogeneity far exceeds other eukaryotic systems, as each one gram of soil harbors approximately ten billion microbes comprising 500,000 diverse species (Rosselló-Mora & Amann, 2001). The total population of prokaryotes on this Earth is estimated to be approximately 4–6 × 10^{30} (Whitman, Coleman, & Wiebe, 1998). With this huge microbial diversity, conventional culturable methods have their own limitations in unraveling the whole microbial world, owing to the vast physiochemical and functional divergence. The "great plate anomaly," explained by Staley and Konopka (Staley & Konopka, 1985), states that 95.0–99.9% of total microbial populations are difficult to culture through traditional plating and serial dilution methods (Alain & Querellou, 2009). Therefore, metagenomics has been envisioning our understanding of untraversed genetic bioresources. Since Schmidt et al. (Schmidt, DeLong, & Pace, 1991) constructed the first metagenomic library, significant progress in the discovery of new biocatalysts has been reported. With the advent of advanced sequencing techniques like NGS and reliable substrates, metagenomics enables researchers to unravel novel GHs from extreme environments (Lubieniechi, Peranantham, & Levin, 2013; Xing, Zhang, & Huang, 2012). Few merits of mining GHs through metagenomics over culturable approaches are detailed in further subsections. The schematic representations of culture-dependent versus culture-independent techniques are given in Figure 4.1.

a) ***Isolation versus Genomic Extraction***
Traditional isolation procedures are not feasible to achieve the holistic representation of a microbial community for screening specific GHs due to the massive divergence in the metabolic, physiological, and phenotypic requirements. However, the genetic information from a single or diverse environmental sample may lead to ubiquitous and unbiased selection of key functional information encoded in the genome (Ufarté, Potocki-Veronese, & Laville, 2015)
b) ***"One Organism to One Gene" versus "Metabiome to Metagenomics"***
In classical techniques, single-gene or gene cascades encoding hydrolytic enzymes were isolated using polymerase chain reaction and cloned subsequently into a suitable host/vector system, after phenotypic selection or activity-based plate screening for the targets. On the contrast, metagenome approaches target complete metabiome from any environment and are directly cloned for selecting GHs gene. It is more successful, as the niche information derived provides insights on the target gene variance (Lepage et al., 2013; Milshteyn, Schneider, & Brady, 2014).
c) ***"Activity-Based Screening" versus "Sequence-Based Screening"***
Gene sequences from the environment can be validated by natural evolution through sequence-based screening. However, functional information in a heterologous host may be lacking in an activity-based screening due to subthreshold expression level (Adrio & Demain, 2014; Lorenz, Liebeton, Niehaus, & Eck, 2002).

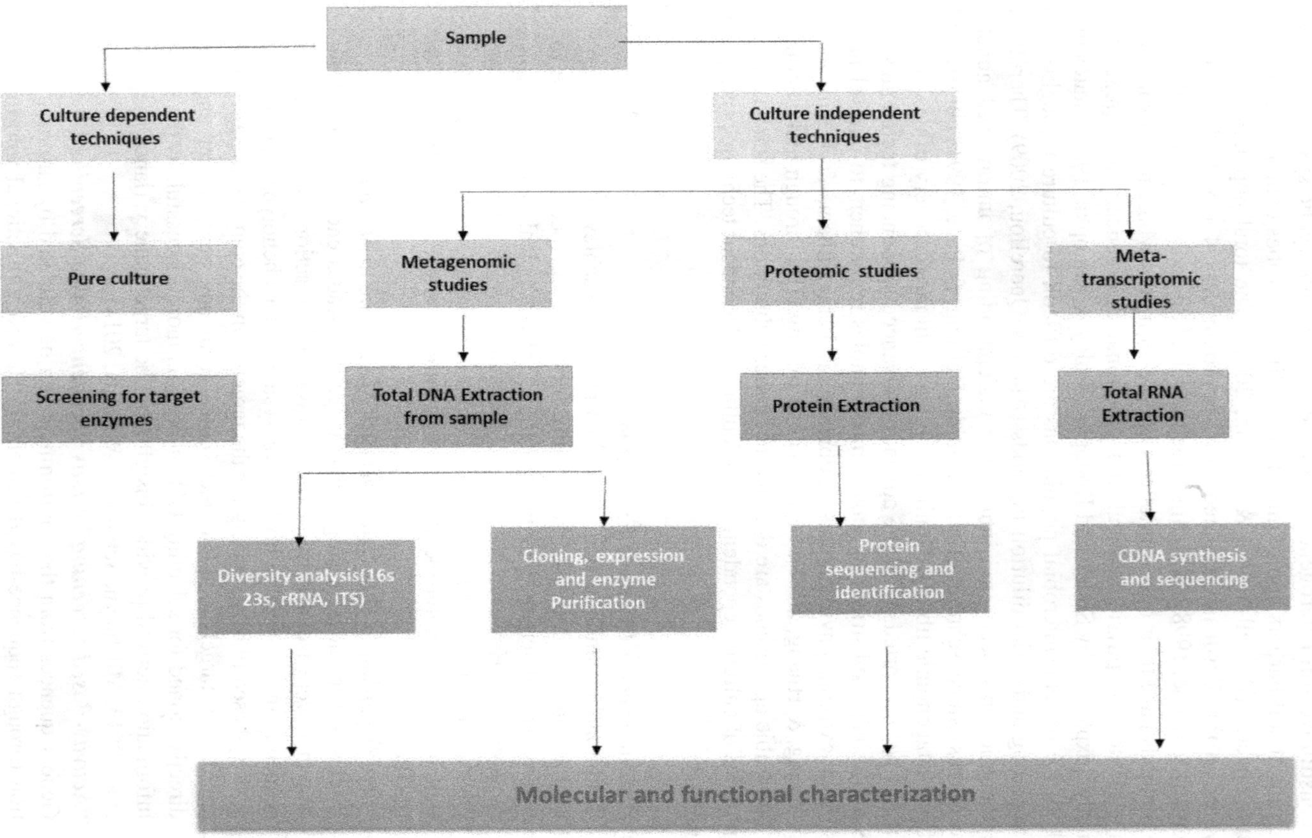

FIGURE 4.1 Various culturable and nonculturable techniques (adapted from Batista-García et al., 2016; License no. 5164081157475).

4.7 FUNCTIONAL AND STRUCTURAL METAGENOMICS FOR UNRAVELING GHS

Metagenomics reveals the taxonomic structure of the entire microbiome (structural) as well as the genes associated with the functional microbiome. Structural metagenomics identifies major microorganisms and their interactions with the environment, evolution, and biogeochemical cycles. Functional metagenomics deals with the genomic diversity in any environmental sample for discovering new genes and metabolic pathways that encode new biomolecules/functional enzymes (Simon & Daniel, 2011). Many researchers employ functional metagenomics in identifying new protein families, especially CAZymes for LCB deconstruction, such as cellulases, esterases, lipases, etherases, and xylanases (Thabata M. Alvarez et al., 2013; Coughlan, Cotter, Hill, & Alvarez-Ordóñez, 2015; Couto et al., 2010). The integral steps involved in functional metagenomics are represented in Figure 4.2.

To understand a given microbiome, two most common methods are generally used: a) amplicon-based metagenomics and b) shotgun metagenomics. The first method, amplicon-based metagenomics, is based on 16S ribosomal RNA, ITS, and 18S regions for characterization of bacteria, fungi, and eukaryote, respectively. The 16S/18S rDNA region contains conserved and nonconserved sequences. Prokaryotic16S rRNA sequencing targets the hypervariable regions (V_1 to V_9) for the determination of bacterial identities. The sequences showing more than 97% similarity are clustered into an operational taxonomic unit (OTU), which is often regarded as a taxon (Morgan & Huttenhower, 2012). Nevertheless, two organisms showing similar 16S rRNA sequences are regarded as a single species, though they may be representative of two different species. Therefore, amplicon-based metagenomics cannot be used to differentiate closely related coliform species: *Shigella flexneri* and *E. coli* (Hilton et al., 2016). The fungal taxonomic analysis depends on large (LSU, 28S) and small (SSU, 18S) and 5.8S subunits of the rDNA region. Fungal diversity is analyzed using ITS regions, whereas 18S rRNA sequences are more specific for fungal taxonomies (Bromberg, Fricke, Brinkman, Simon, & Mongodin, 2015). The 16S/18S rRNA metagenomics gives insights on precise communities among microbiomes as it does not justify the whole microbiome in the habitat (Quince, Walker, Simpson, Loman, & Segata, 2017).

To overcome the bias on amplicon-based sequencing, an untargeted shotgun sequencing method has been designed to analyze all the microbes existing in a niche. Sample collection and processing followed by sequencing, filtering the reads for quality based on Phred scores, reads assembly, binning contigs, and data analysis, are the main steps in amplicon-based sequencing (Ghosh, Mehta, & Khan, 2019; Quince et al., 2017). The relative abundance of species in the given habitat has to be identified before a shotgun study which is based upon 16S/18S rRNA metagenomics analysis. In order to identify the rare taxa, soil samples require more sequencing data, sequencing depth, due to more species abundance when compared to a gut sample (Ghosh, Mehta, & Khan, 2019). The shotgun metagenomic analysis of an endoglucanase-enriched bacterial consortia leads to the reconstruction of six complete genomes. The novel species identified were *Bacillus thermozeamaize*, *Geobacillus thermoglucosidasius*, and *Caldibacillus debilis*. The CAZyme analysis

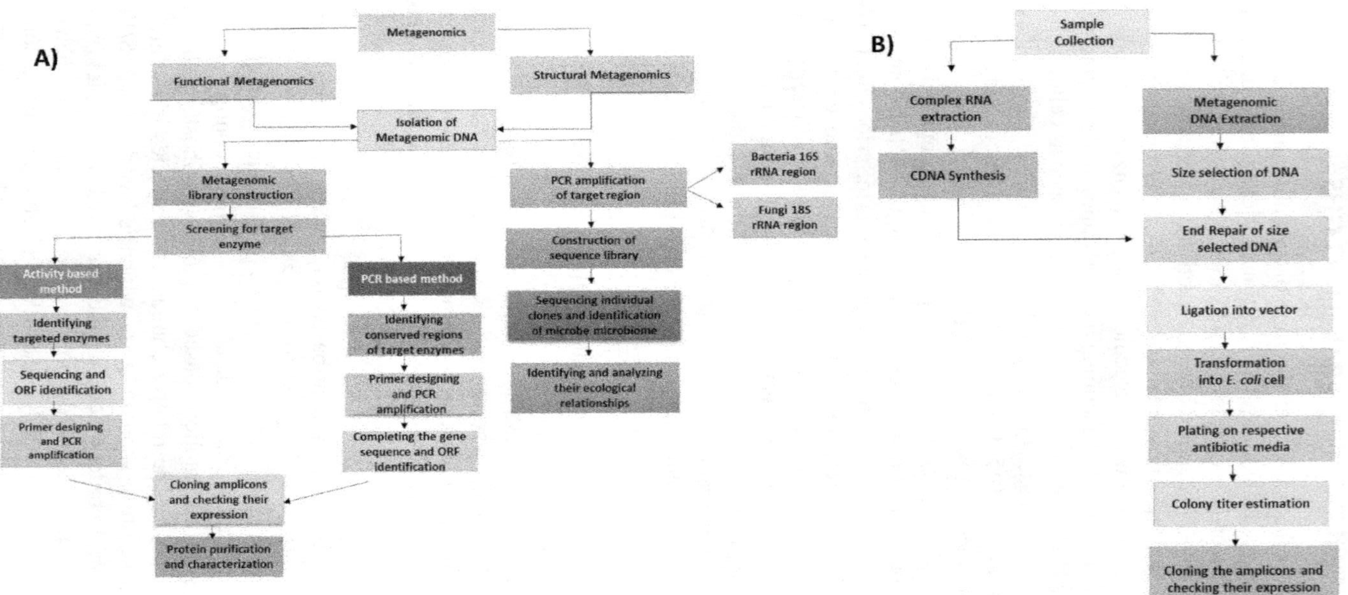

FIGURE 4.2 Different phases in (a) structural and functional metagenomics and (b) metagenomic library preparation (modified from Batista-García et al., 2016; License no. 5164081157475).

revealed the presence of genes encoding for hydrolytic enzymes associated with the degradation of LCB. The *Bacillus thermozeamaize* genome was found to harbor numerous GHs (Lemos et al., 2017). LCB is not easily colonized by microorganisms due to heterogeneity and recalcitrance. Therefore, environments amended with agro-residues are well-suited for unveiling novel biomass-degrading enzymes from unculturable microbes based upon functional metagenomics approaches (Rattanachomsri et al., 2011).

4.8 METAGENOMIC TECHNIQUES FOR BIOPROSPECTING GLYCOSYL HYDROLASES

The metagenome extraction, cloning, and expression of the gene of interest are a challenging task in metagenomics. The recent advent of biotechnological and bioinformatics tools and techniques assists in assessing disparate biocatalysts from extreme environmental niches. The detailed steps involved are detailed in the following subsections.

4.8.1 Enrichment and Nucleic Acid Extraction

Enrichment strategy of a specific gene prior to nucleic acid extraction enhances the probability of sequencing-based screening. Direct extraction technique is more suitable for very negligent genes, as it leads to unbiased genetic representation. Few common methods for selective enrichment includes stable isotope probing (SIP), size-selective filtration, suppression subtractive hybridization (SSH), phage display, differential expression analysis, and affinity capture (Cowan et al., 2005; Crameri & Suter, 1993; Galbraith, Antonopoulos, & White, 2004; Green, Simons, Taillon, & Lewin, 2001; Radajewski, Ineson, Parekh, & Murrell, 2000; Venter et al., 2004). Nevertheless, large inserts with high molecular weight metagenomes are critical for the robust construction of libraries. Gunawardana and his coworkers (Gunawardana et al., 2014) noted that coextractants such as polyphenols and humic acid impede the downstream process, like amplification and restriction digestion of metagenomics DNA.

4.8.2 Construction of Metagenomic Library

The extracted environmental nucleic acid is fragmented and ligated into appropriate vectors, such as plasmids, phages, bacterial artificial chromosomes (BAC), fosmids/cosmids for construction of metagenomics expression libraries. The expression cassette is further transformed into a selective host (*E. coli* or yeast) for assessing the functional expression of the targeted GHs. For library preparation, vector choice depends upon the size of the independent genes or gene clusters. Plasmid or phage vector is preferred for small inserts in the range between 2 and 10 kb, while cosmids (35–45 kb), fosmids (40 kb), and BACs (>200 kb) are suitable for larger gene expression libraries. However, host selection is based on plasmid stability, expression performance, and adaptation according to the screening methods. Other than *E. coli* and yeast, bacteria belonging to *Streptomyces* sp., *Bacillus* sp., *Pseudomonas* sp., etc. are being used for the construction of metagenomic libraries (Gupta & Shukla, 2016).

In general, the genes associated to interrelated metabolic pathways present as gene cassettes like operons/super-operonic clusters in prokaryotes contrast to eukaryotes. Thus, for functional screening, it is important to clone the metagenomic DNA into suitable vectors, like fosmids or cosmids. Further, promoters of a multiple-cloning site (MCS) help bidirectional transcription, which increases the probability of positive clones with the targets (Lämmle et al., 2007). Further, broad host range systems are required to enhance expression successfully and in the identification of targeted genes, as the gene expression significantly depends on the host (Yoon et al., 2013). Although *E. coli* is highly economical, and effective, for production of a wide range of heterologous proteins (Hannig & Makrides, 1998), it poses certain limitations in the number of GHs obtained from metagenomic libraries from an enriched environment (L.-L. Li, McCorkle, Monchy, Taghavi, & van der Lelie, 2009). Furthermore, the application of bacterial host systems reduces the chances of identifying GHs of fungal origin, due to specific codon usage, regulation, and activation of promoter, followed by posttranslational process. These factors impact the expression of eukaryotic genes in prokaryotic systems more specifically at their functional level. Likewise, prokaryotic host systems lack some posttranslational events, thereby affecting their production extracellularly (Flipphi et al., 2013; Juturu & Wu, 2014; L.-L. Li et al., 2009). Since majority of the GHs identified so far belong to prokaryotic systems, alternate host expression systems need to be regarded in other genera. *T. thermophilus* is a successful metagenomic library host in terms of more active clones for screening esterases and expression of xylanases (Angelov, Mientus, Liebl, & Liebl, 2009; Leis et al., 2015). The ongoing projects on the development of eukaryotic host systems like yeast cells (*Pichia pastoris*) will address the limitations of functional expression of eukaryotic LCB-deconstructing enzymes.

4.8.3 Limitations of Metagenomic Library Construction

Besides the techniques available, there exist a few challenges in metagenomic library construction and subsequent functional library screening from LCB-rich niches. The major limitation is attaining high-quality metagenomic DNA (mgDNA), which is critical for metagenomics library construction. While extracting from the lignocellulosic environment, the nucleic acid is often contaminated by humic substances, furans, and phenolic derivatives (Chandel et al., 2014). These co-contaminants denature nucleic acids, act as enzyme inhibitors, and interfere with DNA transformation (Lim et al., 2005). Apart from this, the quality of mgDNA extracted from LCBs may also contain traces of fertilizer and pesticide residues and other chemicals from industrial processes. The mgDNA yield from LCB is also low due to limited microbial colonization of LCB (Chandel et al., 2014; Kanokratana et al., 2013). Consequently, for the aforementioned reasons, optimized mgDNA extraction protocols from LCBs are lacking. However, LCBs supplemented with cellulose for enriching cellulolytic microbial communities would increase the mgDNA yield (Vester, Glaring, & Stougaard, 2015). To separate prokaryotic DNA from the metagenome, centrifugation techniques and size-selective filters are used to ward off the contamination with other genomes of eukaryotes (Mori, Kamei, Hirai, & Kondo, 2014; Venter et al., 2004).

The major limitation in the enrichment studies is the loss of relevance as the microbial populations subjected to enrichment undergo external modifications in the natural biomass ecosystem. The use of silica gel in metagenomic DNA extraction after cell lysis will reduce the shearing to some extent and enhance the mgDNA recovery as well as the quality. However, in the alternate strategy, the large particulate matter is separated by a gentle centrifugation technique (645 times gravity/3,000 rpm) before the mgDNA extraction. The separation step is followed by the addition of a lysis buffer containing CTAB, ethylene diamine, EDTA, NaCl, Tris-HCl, and proteinase K for microbial cell lysis. These strategies are generally employed for unveiling novel genes encoding hydrolytic enzymes, like cellulases, esterases, proteases, lipases, xylanases, etc.

4.8.4 Metagenomic Library Screening

The library constructed from environmental DNA displays huge gene-encoded information, and hence, it is necessary to scout all the probable positive clones expressing the specific targets, say, GHs in our case. Screening metagenomic clones positive for GHs is performed using chromogenic or dye-linked substrates, and the positive clones are spotted out based on staining. Further, enzyme activities are validated by specific quantitative enzymatic assays. For instance, positive clones from metagenomic library for endoglucanase and β-glucosidase are commonly screened on substrates such as CMC, 4-methyl-umbelliferyl-β-D-glucoside (MUG), azurine-linked substrate (AZCL-HE-cellulose and AZCL-β-glucan), and p-nitrophenyl-β-D-glucopyranoside, p-nitrophenyl-β-D-cellobioside, 4-methyl-umbelliferyl β-D-glucoside (MUG), respectively.

Another strategy for screening is using sequence-specific probes or primers. A high-throughput screening technique, called substrate-induced gene expression (SIGEX), based on the induction of catabolic genes is performed for screening positive clones using diverse substrates and analyzed by FACS (T. Uchiyama, Abe, Ikemura, & Watanabe, 2005). The limitation of SIGEX is that it is only used for those substrates which could be relocated into the cell cytoplasm. Since the substrates for screening GHs lack mobility towards cytoplasm, SIGEX cannot be adopted for LCB-deconstructing biocatalysts (Yun & Ryu, 2005).

4.8.5 Massive Sequencing of Glycosyl Hydrolases Using NGS

Metagenomic expression library construction generates only limited genetic information, lacking comprehensive insights on the genetic diversity of GHs. Recent next-generation sequencing technologies allow massive parallel sequencing, which provides novel data on neighbor-joining tree and genetic and metabolism diversity of targeted GHs in the microbiome. The metagenomic 16S rDNA unleashed the potential for unraveling the entire passport details of bacterial species present in the particular niche phylogeny and their putative genes. Novel GHs from divergent niche, like animal sources, are analyzed using next-generation sequencing techniques, like 454 genome sequencer, Ion Torrent, PacBio, Illumina, SOLiD technology, and the recent nanopore technology. The metagenomic surveys encoding various GHs from different environmental samples are represented in Table 4.1. There exists a bias between the gene characterization through culturable or unculturable methods as well as the *de facto* in nature. Also, the diversity of GHs in nature necessitates the understanding of synergism between GHs for efficient valorization of biopolymers and the saccharification process. However, more research is essential to decode the relationship between various GH family proteins and LCB conversion. Certain nonhydrolytic auxiliary enzymes like lytic polysaccharide monooxygenases (LPMOs) are involved in the crystalline cellulose solubilization via oxidative cleavage of glycosidic linkages. The knowledge on new GHs families has the potential for an efficient biomass valorization and saccharification process in biofuel production.

4.8.6 Metagenomic-Derived Functional Glycosyl Hydrolases (GHs)

Several GHs have been isolated from culturable microorganisms; however, common isolation practices and characterization lead to a biased selection of efficient strains. Hence, commercial GHs, such as cellulases produced by *Trichoderma* sp. or *Aspergillus* sp., are unable to perform efficient saccharification as they lack certain accessory enzyme activities (Martinez et al., 2004). Therefore, unraveling metagenomes for more efficient cellulases could open up new vistas in developing robust enzyme cocktails for efficient biomass valorization. For industrial bioconversion of lignocellulosics to value-added, high-value chemicals, robust GHs with more stability and specificity are required. The recent quest on thermostable cellulases from thermophiles and hot springs increases the possibility of increasing the saccharification efficiencies of lignocellulosics (Ganesan et al., 2020; Thankappan et al., 2018; Thankappan, Kandasamy, & Uthandi, 2017). Few cellulases from metagenomic libraries of different environmental samples that are recently identified are presented in Table 4.1.

4.8.6.1 Endoglucanases

Endoglucanase initiates the cellulose degradation process. The complete cellulase genes obtained from environmental samples, including soil, spill water, termite mounds, insect gut, and animal feces, are entitled to the families GH5, GH9, GH12, GH44, and GH45, originated from divergent microbial species. For example: *Clostridium*, *Cellulomonas*, *Cellovibrio*, *Bacillus*, *Vibrio*, and other unidentified microbes. Based on the structure-function relationship and kinetic properties, endoglucanases exhibit variations in the amino acids at 244 to 1005 positions. The functional properties of endoglucanases are largely dependent on the source of genes per se to the environmental niche. Cellulases from biogas digester, sugarcane bagasse composts, rice straw (Geng et al., 2012; Kanokratana, Eurwilaichitr, Pootanakit, & Champreda, 2015; Y. F. Yeh et al., 2013), and hot springs (Thankappan et al., 2018) are thermophilic. However, few enzymes showed discrepancies with the environmental nature that they were isolated from. For instance, endoglucanase from compost

TABLE 4.1
Glycosyl Hydrolases Retrieved on the Basis of Functional Metagenomics

S. N.	Cellulases	Environmental Sample	Type of Library	Average Library Size	No. of Clones Screened	Substrate Used for Screening	Amino Acid Sequence Size	GH Family	Nearest Neighbor (% Identity)	Host/ Expression Vector	Reference
Endoglucanase											
1	Endo-1,4-β-D-glucanase (JX048675, JX048676)	Bovine Rumen	Fosmid	30 kb	10,000	CMC	559 and 554	GH5	Unidentified microorganism (75%)	*E. coli*/pET20b	(KJ Rashamuse et al., 2013)
2	Endoglucanase (DQ139834)	Soil microbial consortia	Cosmid	22	1700	CMC	363	GH5	*Cellvibrio mixtus* (86%)	*E. coli*/pET19b	(Voget, Steele, & Streit, 2006)
3	Endo-1,4-β-D-glucanase (GU585368)	Soil	BAC	50–100	3,024	CMC	443	GH5	*Plesiocystis pacifica* (39%)	*E. coli*/pET29a	(J. Liu et al., 2011)
4	Cellulase (JN98181)	Dairy cow rumen	BAC		6000	CMC	559 and 892	GH5	*Prevotella ruminicola*	*E. coli*/ pET-30a	(Gong et al., 2012)
5	Cellulase (HQ706358)	Biogas digester fed	Fosmid	23–50	100,000	4-MUC	440	GH5	*Clostridium phytofermentans* (48.3%)	*E. coli*/ pET 22r and *T. reesei*/ pDHt/sk-tc2bh	(Geng et al., 2012)
6	Endo-β-1,4-glucanase (JN012243)	Vermicompost	Fosmid	36	89,000	CMC	363		*Cellvibrio mixtus* (88%)	*E. coli*/pET28a	(Yasir et al., 2013)
7	Cellulase	Swamp buffalo rumen	Fosmid		10,000	AZCL-HE-cellulose	525	GH5	Uncultured (79%)	*E. coli*/pET28a	(Nguyen et al., 2012)

(Continued)

TABLE 4.1 *(Continued)*
Glycosyl Hydrolases Retrieved on the Basis of Functional Metagenomics

S. N.	Cellulases	Environmental Sample	Type of Library	Average Library Size	No. of Clones Screened	Substrate Used for Screening	Amino Acid Sequence Size	GH Family	Nearest Neighbor (% Identity)	Host/ Expression Vector	Reference
8	Cellulase	German grassland soil	Fosmid	26.2	364, 568	Dye-labeled hydroxyethyl cellulose (HECred	831	GH9	*Sorangium cellulosum* (50%)	*E. coli*/pET101	(Nacke et al., 2012)
9	Endo-1,4-β-D-glucanase (DQ235086)	Compost soil	Cosmid	33	100,000	CMC	579	GH9	Uncultured bacterium (52%)	*E. coli*/pQE31	(Pang et al., 2009)
10	Endoglucanase (KC618310)	Sugarcane bagasse compost	Fosmid	31.3	120,000	AZCL-β-glucan	807	GH9	*Micromonospora aurantiaca* (66%)	*E. coli*/ pET-28a	(Kanokratana, Eurwilaichitr, Pootanakit, & Champreda, 2015)
11	Endoglucanase (KF509855)	*Bursaph-elenchus xylophilus*	Plasmid	3	5,000	CMC	367	GH8	*Klebsiella oxytoca* (90%)	*E. coli*/ pET-30a	(L. Zhang et al., 2013)
12	Cellulase (HM235800)	Abalone gut	Fosmid	40	3840	CMC	340	GH45	*Vibrio alginolyticus* (100%)	*E. coli*/pCold II	(D. Kim et al., 2011)
13	Cellulase (KF498957)	Sugarcane field land soil	Plasmid	-	-	CMC	428	GH5	*Cellvibrio japonicas* (80%)	*E. coli*/ pET-28a(+)	(Thabata M Alvarez et al., 2013)
14	Cellulase (KF626648)	Leaf-branch compost	Fosmid	38	6000	CMC	244	GH12	*Rhodothermus marinus* (76%)	*E. coli*/pET25b	(Okano et al., 2014)
15	Cellulase (KC684420)	Gut microflora of Hermetiaillucens	Fosmid	-	92,000	CMC	331	GH5	*Dysgonomonas mossii* (72%)	*E. coli*/ pET21a(+)	(Lee, Seo, Kim, & Sim, 2014)

16	Endoglucanase (FJ422812/ FJ422814/ FJ422815)	Spill and water sample	Cosmid	-	3,744	CMC	1005/ 862/ 364	GH5/ GH9/ GH2	*Cellvibrio japonicas* (74%)/ *Cellvibrio japonicas* (73%)/ Uncultured bacterium (93%)	*E. coli*/ pET19b	(Pottkämper et al., 2009)
17	Cellulases (JF895785/ JF826524/ JF826525)	Elephant feces and biogas plant	Fosmid	-	29,000	CMC	604/ 634/ 506	GH9/ GH5/ GH5	*Clostridium Cellulovorans* (41%)/ *Clostridium cellulolyticum* (60%)/ *Fibrobacter Succinogenes* (41%)	*E. coli*/ pET28a/ pMAL-c2	(Ilmberger et al., 2012)
18	Endo-β-1, 4-glucanase (DQ182491)	Rabbit cecum	Cosmid	30–50	32,500	CMC	395	GH5	*Bacillus cellulosilyticus* (46%)	*E. coli*/ pET30a(+)	(Feng et al., 2007)
19	Endoglucanase (JQ081063)	Rice straw composts	Lambda phage	-	2,739	CMC	464	GH12	*Micromonospora aurantiaca* (74%)	*E. coli*/ pBluescript	(Y.-F. Yeh et al., 2013)
20	Cellulase (EU449491)	Buffalo rumens	Cosmid	35	15,000	CMC	546	GH5	Unidentified micro-organism (67%)	*E. coli*/ pET-30a(+)	(Duan et al., 2009)
21	Endoglucanase (KF424270)	Mangrove soil	Fosmid	30	100,000	CMC	648	GH44	*Micromonospora lupini* (48%)	*E. coli*/ pET22b(+)	(Mai et al., 2014)
22	Cellulase	Thermophilic methanogenic digester	Fosmid	36	92,000	CMC	455	GH5	Uncultured bacterium (59%)	*E. coli*/ pET-28a(+)	(M. Wang et al., 2015a)

(Continued)

TABLE 4.1 *(Continued)*
Glycosyl Hydrolases Retrieved on the Basis of Functional Metagenomics

S. N.	Cellulases	Environmental Sample	Type of Library	Average Library Size	No. of Clones Screened	Substrate Used for Screening	Amino Acid Sequence Size	GH Family	Nearest Neighbor (% Identity)	Host/ Expression Vector	Reference
23	Endoglucanase (AY859541)	Soil sample from straw stock	Plasmid	7–10	24,000	CMC	356	GH8	*Rhizobium leguminosarum* (47%)	*E. coli*/ pET-28a(+)	(Xiang et al., 2014)
24	Endoglucanase (KC618310)	Bagasse pile	Fosmid	31.3	100,000	AZCL-HE-cellulose	807	GH9	*Micromonospora aurantiaca* (66%)	*E. coli*/ pET28a(+)	(Kanokratana, Eurwilaichitr, Pootanakit, & Champreda, 2015)
25	Cellulase (KM12345)	Cow rumen	Cosmid	-	-	CMC	756	GH9	*Clostridium cellulovorans* (35.7%)	*E. coli*/ pET-21a(+)	(Kang et al., 2015)
26	Endoglucanase (JX434088) and Endoglucanase (JX434086)	Grasshopper and cutworm gut symbionts	Primer based on the assembled sequences (Illumina Metagenomic Sequences)	-	-	-	357/ 394	GH5/GH8	*Enterobacteriaceae bacterium* (99%) and *Enterobacter cancerogenus* (99%)	*E. coli*/pET161	(W. Shi et al., 2013)
27	Cellulase (EF114228)	Wetland soil sample	Fosmid	40	70,000	CMC	662	GH44	*Cellulomonas pachnodae* (59%)	*E. coli*/ pET21a(+)	(S.-J. Kim et al., 2008; Nam et al., 2008)
28	Endoglucanase (HQ634705)	Yak rumen	BAC	55	76,000	CMC	514	GH5	Unidentified microorganism (77%)	*E. coli*/pET-30a(+)	(Dai et al., 2012)

β-glucosidase										
29	β-glucosidase (EU678637)	alkaline polluted soil	Plasmid	3.5	30,000	Esculin	481		*Plasmodium chabaudi* (21%)	*E. coli*/ pETBlue-2	(Jiang et al., 2009)
30	Glucan 1,4-β-glucosidase (JX163904)	Holstein cow rumen	Fosmid	-	16,896	pNPβG and pNPβC	779	GH3	*Prevotellarum inicola* (86%)	*E. coli*/ pET-41	(Mercedes V Del Pozo et al., 2012)
31	β-glucosidase (HG326254)	Hot spring	Plasmid	4–10	-	Esculin hydrate	495	GH1	*Thermoproteus uzoniensis* (53%)	*E. coli*/ pQE-80L	(Schröder, Elleuche, Blank, & Antranikian, 2014)
32	6-Phospho-β-glucosidase (KC441954)	Pulp wastewater	Plasmid	6–8	60,000	Esculin hydrate	475	GH1	*Clostridium difficile* (89%)	*E. coli*/ pETDuet-1	(Yang et al., 2013)
33	β-glucosidase (JQ844187)	Termite gut	Fosmid	-	10,000	esculin hydrate	455	GH1	*Spirochaeta coccoides* (54%)	*E. coli*/ pET28a (+)	(J. Wang et al., 2012)
34	β-glucosidase (GU647096)	Marine water	BAC	-	-	esculin hydrate	422	GH1	*Alteromonadales bacterium* (98%)	*E. coli*/ pET22b	(Fang et al., 2010)
35	β-glucosidase (GQ849222)	Yak rumen	Cosmid	-	4,000	pNPG	755	GH3	*Parabacteroides distasonis* (48%)	*E. coli*/ pET28a	(Bao et al., 2012)
36	β-glucosidase (DQ842022)	Wetland	Fosmid	35	14,000	MUC	485	GH1	*Chloroflexus aurantiacus* (56%)	*E. coli*/ pET32a(+)	(S.-J. Kim et al., 2007)
37	β-glucosidase (JX566949)	Soil from sugar refinery	Fosmid		90,000	Esculin hydrate	469	GH1	*Streptomyces avermitilis* (87%)	*E. coli*/ pSE380	(Jian Lu et al., 2013)
38	β-glucosidase (JQ957567)	Mangrove soil	Plasmid	5	30,000	Esculin hydrate	422	GH1	*Paenibacillus mucilaginosus* (74%)	*E. coli*/ pET-32a (+)	(G. Li, Jiang, Fan, & Liu, 2012)
39	β-glucosidase (DQ182493)	Rabbit cecum	Cosmid	30–50	32,500	Esculin hydrate	854	GH3	*Clostridium beijerincki* (45%)	*E. coli*/ pET-30a(+)	(Feng et al., 2009)

(*Continued*)

TABLE 4.1 *(Continued)*
Glycosyl Hydrolases Retrieved on the Basis of Functional Metagenomics

S. N.	Cellulases	Environmental Sample	Type of Library	Average Library Size	No. of Clones Screened	Substrate Used for Screening	Amino Acid Sequence Size	GH Family	Nearest Neighbor (% Identity)	Host/ Expression Vector	Reference
40	β-glucosidase (JX962691)	Xiangxi yellow cattle rumen	Fosmid	30	20,160	Esculin hydrate	779	GH3	*Prevotella Bergensis* (62%)	*E. coli/* pET28a (+)	(Y. Li et al., 2014a)
41	β-glucosidases (KF433952/ KF433953)	Amazon soil	Fosmid	-	97,500	MUG	750/ 467	GH3/ GH1	*Corallococcus coralloides* (69%)/ *Methanocella paludicola* (72%)	*E. coli/* pET system	(Bergmann et al., 2014)
42	β-glucosidase	Thermophilic methanogenic digester	Fosmid	36	92,000	Esculin hydrate	462	GH1	Uncultured bacterium (60%)	*E. coli/* pET-28a(+)	(M. Wang et al., 2015a)
43	β-glucosidases (HV348683)	Kusaya Gravy (fermentation food product of dried fish)	Plasmid	-	380,000	pNPGlc	452	GH1	*Clostridiales bacterium* (57%)	*E. coli/* pET29b(+)	(Taku Uchiyama, Yaoi, & Miyazaki, 2015)
44	β-glucosidases (KF660587/ KF660588)	Agricultural soil	Plasmid	5.5–6	45,000	Esculin hydrate	762/ 475	GH3/ GH1	*Rhodococcus rhodochrous* (72%)/ *Sorangium cellulosum* (52%)	*E. coli/* pET-30b(+)	(Biver, Stroobants, Portetelle, & Vandenbol, 2014)
45	β-glucosidase (GQ507800)	Biogas reactor	Cosmid	35	30,000	Esculin hydrate	620	GH1	*Enterobacter* sp. (85%)	*E. coli/* pETBlue-2	(C. Jiang et al., 2010)

Exoglucanase											
46	Exo-1,4-β-glucanase (HQ824368)	*Equus burchelli* Fecal samples	Primer based	-	-	-	318	GH6	*Piromyces equi* (58%)	*E. coli*/pET32a	(Chandrasekharaiah et al., 2012)
Multifunctional cellulases											
47	Endocellulase and Endoxylanase (JQ581599)	Compost	Fosmid	40	251	CMC	111	GH43	Uncultured bacterium (52%)	*E. coli*/pZErO™	(Lee, Seo, Kim, & Sim, 2014)
48	Endocellulase/exocellulase/xylanase (KC963960)	Rumen-fistulated Korean cow (Hanwoo)	Robotic high-throughput screening system	-	-	-	380	GH5	*Prevotella ruminicola* (83%)	*E. coli*/pET-22b(+)	(K.-C. Ko et al., 2013)
49	Endoglucanase/Exoglucanase (GQ849224)	Yak rumen	Cosmid	-	4,000	CMC	336	GH2	Uncultured bacterium (83%)	*E. coli*/pET-21a(+)	(Bao et al., 2011)
50	β-glucosidase/β-xylosidase/α-arabinosidase (KF114873)	Dairy cow rumen	BAC	2–8 kb	6,000	CMC	759	GH3	*Thermotoga neapolitana* (50%)	*E. coli*/pET-30a	(Gruninger, Gong, Forster, & McAllister, 2014)
51	β-glucosidase with lipolytic activity (FJ686869)	Alkaline polluted soil	Plasmid	3.5	30,000	Esculin	172	GH4	*Frankia* sp. (45%) *Clostridium botulinum* (29%)	*E. coli*/pETBlue-2	(C.-J. Jiang et al., 2011)
52	β-glucosidase/xylosidase (GQ324952)	Yak rumen	Cosmid	-	5000	MU-Glc	765	GH3	*Parabacteroides distasonis* (73%)	*E. coli*/pET21a	(Zhou et al., 2012)
53	β-1, 4-xylosidase/β-1,4-endoglucanase (GU132859)	Chinese yak rumen	Lambda Phage	30–40	4,000	CMC	551	GH5	*Prevotella ruminicola* (72%)	*E. coli*/pET21a	(Chang et al., 2011)

(Continued)

TABLE 4.1 *(Continued)*
Glycosyl Hydrolases Retrieved on the Basis of Functional Metagenomics

S. N.	Cellulases	Environmental Sample	Type of Library	Average Library Size	No. of Clones Screened	Substrate Used for Screening	Amino Acid Sequence Size	GH Family	Nearest Neighbor (% Identity)	Host/ Expression Vector	Reference
54	β-xylosidase (β-glucosidase-xylosidase) (GQ324952)	Chinese yak rumen	Cosmid	-	5,000	MU-Glc	898	GH3	*Parabacteroides distasonis* (73%)	*E. coli/* pET21a-RuBGX1	(Zhou et al., 2012)
Xylanolytic enzymes											
55	PersiXyn1	Camel rumen	Direct sequencing	-	-	Xylan	43kDa	GH11	-	NA	(Ariaeenejad et al., 2019)
56	XYL21, XYL38	Compost metagenome	Fosmid	-	-	Xylan	41.9 kDa	GH11	-	Fosmid	(Ellilä et al., 2019)
57	AMOR_GH10A	Arctic mid-ocean ridge vent	Metagenomic data set	-	-	Xylan	-	GH1	-	-	(Fredriksen et al., 2019)
58	XylCMS	Camel rumen metagenome	454 Pyrosequencing	-	-	Xylan	46 kDa	GH11	-	-	(Ghadikolaei et al., 2019)
59	Xyl1	Termite gut metagenome	CopyControl fosmid library production kit	-	-	-	55kDa	GH11	-	Fosmid	(Konanani Rashamuse et al., 2017)
60	-	Kinema soybean metagenome	Whole metagenome sequencing	-	-	-	-	GH39 and GH 43	-	-	(J. Kumar et al., 2019)

61	Multiple xylanolytic enzymes	Goats ruminal liquid metagenome	Copy Control fosmid library production kit	-	-	-	-	GH1 GH5 GH8 GH14 and GH43	-	Fosmid	(Duque et al., 2018)
62	MeXyl31	Soil	CopyControl fosmid library production kit	-	-	-	77kDa	GH31	-	Fosmid	(Matsuzawa, Kimura, Suenaga, & Yaoi, 2016)
63	Biof1_09	Compost	Copy Control fosmid library Production-n kit	-	-	-	-	GH43	-	Fosmid	(Sae-Lee & Boonmee, 2014)
64	AR19M-311–2, AR19M-311–11 and, AR19M-311–21	Hot spring	Direct sequencing	-	-	-	-	GH1 GH3 GH31 and GH43	-	-	(Sato et al., 2017)

soil having optimum activity at 25°C retained its absolute activity at low temperature (10–40°C) (Pang et al., 2009). Besides, thermoalkaliphilic hemicellulases and cellulases were revealed in a multisubstrate approach through functional screening from two wheat straw–degrading microbial consortium (Mukil Maruthamuthu, Jiménez, Stevens, & van Elsas, 2016). However, Cheng et al. (Cheng et al., 2015) also identified a nonspecific endoglucanase from the metagenomic library constructed out of goat rumen (Cheng et al., 2015). To date, approximately 171 GHs (GH-1 to GH-171) with few families like GH-5, GH-13, GH-16, GH-30, and GH- 43 entail large subfamilies of 56, 44, 27, 9, and 37, respectively. These 171 families comprise of 1,030,220 modules, and 20,116 modules are yet to be categorized (CAzy database). Majority of GH-5 families were cellulolytic, whereas GH-93 was arabinan degraders (Talamantes et al., 2016). Among the reported metagenome-derived endoglucanases so far, three families possess carbohydrate-binding domains (CBMs). Endoglucanases with CBM at C-terminal are grouped under type-A CBMs and are capable of cellulose degradation. Subsequent reports were made on the characteristics of metagenome-derived GHs where they differ from their corresponding regular enzymes in their temperature, pH, solvent, and metal tolerance (Pabbathi et al., 2021). Such enzyme characters support further bioprocessing of LCBs and fermentation processes at industrial scale to increase biofuel production.

4.8.6.2 Beta Glucosidase

β-glucosidases are a limiting enzyme that completely converts the LCB into glucose units, including synthesis of many synthetic and catalytic platform chemicals and other high-value molecules, flavor precursors, nutrient supplement, feed additives, and alkyl- or oligosaccharides (Coenen, Schoenmakers, & Verhagen, 1995; J. Lu, L. Du, Y. Wei, Y. Hu, & R. Huang, 2013; Swiegers, Bartowsky, Henschke, & Pretorius, 2005; J. Zhang & Lynd, 2010). Many genes encoding β-glucosidase are isolated from diverse niches like alkaline polluted soil, gut microbiome, hot spring, wastewater, marine water, mangrove soil, agricultural soil, anaerobic digester, and termite gut through high-throughput metagenomic approaches (Tiwari, Nain, Labrou, & Shukla, 2018; Kaushal, Rai, & Singh, 2021; Biver, Stroobants, Portetelle, & Vandenbol, 2014; Jiang et al., 2010; Wang et al., 2015; Schröder, Elleuche, Blank, & Antranikian, 2014). B-glucosidases from metagenomes possess novel properties which are unique to culturable microbial origin. Similarly, Del Pozo et al. (2012) functionally expressed the β-glucosidases from metagenome of cow rumen and validated for corn stover hydrolysis in comparison with a commercial cocktail Celluclast (Novozymes A/S), which is devoid of β-glucosidase. Hydrolysis efficiency and sugar recovery were increased when compared to commercial β-glucosidases (Novozymes 188). The limitation of cow rumen–derived β-glucosidases is end product inhibition, where activity has been inhibited even at low levels (10–50 mM) of glucose. On the contrary, several metagenomic libraries have unveiled glucose-insensitive β-glucosidases also. Expectedly, glucose recovery was enhanced by the supplementation of thermotolerant robust β-glucosidases with commercial cellulases during hydrolysis of pretreated rice straw. Metal ions act as cofactors for β-glucosidase activity. β-glucosidase from alkaline polluted soil and rumen both have been found to be positively influenced by metal ions (C. Jiang et al., 2009; Y. Li et al., 2014b). In recent days, metagenome-derived β-glucosidases are engineered for enhanced substrate binding and catalysis by employing site-directed mutagenesis and genome editing (Yang et al., 2013).

4.8.6.3 Multifunctional Cellulases

Multifunctional enzymes represent cellulases specific to more than one substrate and catalyze different ranges of reactions. The multifunctional nature is due to the presence of multicatalytic domains, combined in a single polypeptide. A metagenome from yak rumen revealed unique bifunctional cellulase with both endo- and exocatalytic activities, hydrolyzing both crystalline and amorphous cellulose with high efficiencies (Bao et al., 2011). The presence of β-1,4-xylosidase/β-1,4-endoglucanase activities favored the one-pot bioconversion of insoluble LCB into xylo-/cello-oligosaccharides. Similarly, a trifunctional cellulase exhibiting endocellulase/exocellulase/xylanase activities from cow rumen metagenome was reported (K. C. Ko et al., 2013). A multifunctional cellulase from alkaline polluted soil possesses β-glucosidase and lipolytic activities (C.-J. Jiang et al., 2011). The multifunctional cellulase displayed significant β-glucosidase activity, albeit the sequence homology showed a close proximity with lipases (Table 4.1).

4.8.7 Functional Screening of Unusual Xylanolytic Extremozymes

Thermo- and acid/alkali-tolerant xylanolytic enzymes, or xylanases, are found to have wide use in pulp, food, animal feed, biofuel, and pharmaceutical companies. Thermostable xylanases uncovered from a pooled environmental metagenome recorded optimum activity at 100°C, which is more unique among existing xylanases. However, xylanases of cultivable *Thermotoga* spp. exhibit optimal xylanase activity at > 80°C only (H. Shi et al., 2014). In general, acidophilic xylanases from compost samples and several metagenomics-derived xylanases harmonized the acidic properties of the compost (Ellilä et al., 2019; Sae-Lee & Boonmee, 2014). Nevertheless, thermostability at 80°C and alkali-stability (pH 9.0–10.0) were also exhibited in xylanases derived from compost (Digvijay Verma, Kawarabayasi, Miyazaki, & Satyanarayana, 2013). Based on hydrophobicity, endoxylanases are grouped into the GH-11 family. Few endoxylanases exhibiting such twin properties are *Geobacillus thermoleovorans* (Sharma, Adhikari, & Satyanarayana, 2007; D. Verma & Satyanarayana, 2012), *Bacillus halodurans* S7 (Mamo, Hatti-Kaul, & Mattiasson, 2006), *B. halodurans* TSEV1 (V. Kumar & Satyanarayana, 2014), *Microcella alkaliphila* JAM-AC0309 (Kuramochi et al., 2016), and *Geobacillus thermodenitrificans* TSAA1 (Anand, Kumar, & Satyanarayana, 2013). Despite the multifunctionality of the endoxylanases from pure culture, low pH and thermal stabilities are hampered in long-run processes. Such limitations are addressed by metagenomics-driven approaches. The unusual xylanases (xyn8) from a metagenomic library show an optimum temperature of 20°C (psychrophilic), which is quite rare, and share properties similar to cold-active xylanolytic bacterial/fungal enzymes (G. Wang et al., 2010; Zhao et al., 2010).

Further, unique halophilic/halotolerant xylanolytic enzymes from camel rumen metagenome get activated in the presence of 5M NaCl by 132%, which is not yet reported in culturables (Ghadikolaei et al., 2019). The carbohydrate-binding domain (CBD) of AMOR_GH10A lacks sequence similarity with CaZy database (Fredriksen

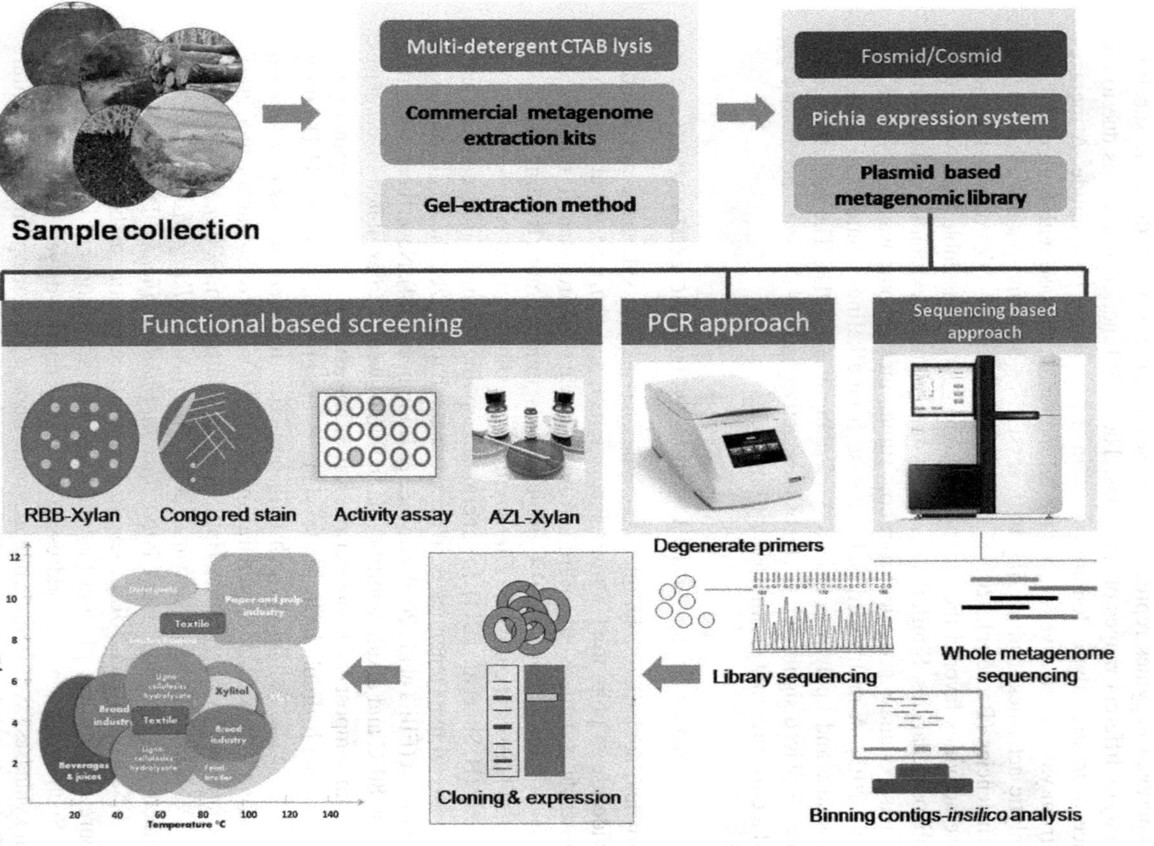

FIGURE 4.3 Schematic representation of the functional screening approaches of xylanolytic enzymes from metagenome.

Source: Adapted from Verma and Satyanarayana (2020).

et al., 2019). Likewise, new candidates of the CBM85 family from metagenomics-derived xylanase (UX66) with promising substrate affinities comprise two catalytic domains and two CBDs (Zhao et al., 2010). Alternatively, GH62 and GH10 families (ferulic acid esterases, arabinofuranosidases, and β-xylanases) obtained from wastewater treatment sludge metagenome exhibited tricatalytic enzyme activities (Holck et al., 2019). The schematic representation of retrieving xylanolytic enzymes using metagenomics is depicted in Figure 4.3.

4.8.7.1 Metagenomic β-Xylosidases

Among xylan-depolymerizing enzymes, Exo-1, 4- β-D xylosidase (E.C. 3.2.1.37) plays a crucial role in the complete degradation of xylans. β-xylosidases are exo-type glycosidases that cleave the glycosidic linkages of short xylo-oligosaccharides, liberating the 5C sugars as an end product (Jain, Kumar, & Satyanarayana, 2015; Shallom & Shoham, 2003). Several limitations associated with existing β-xylosidases include low thermostability, poor efficiency, salt sensitivity, and end product inhibition (Anand, Kumar, & Satyanarayana, 2013; Bao et al., 2012). Therefore, novel candidates of β-xylosidases using metagenomic approaches have been attempted, and of the available β-xylosidases, less than twenty are characterized so far (Rohman, Dijkstra, & Puspaningsih, 2019).

Characterization of metagenomic β-xylosidases exhibited optimum activity at acidic to neutral pH ranges (Douglas B. Jordan et al., 2016; C. Liu, Zou, Yan, & Zhou, 2018; Wagschal, Heng, Lee, & Wong, 2009). Examples: acidic β-xylosidases (AR19M-311–2, AR19M-311–11, and AR19M311–21) and RuBG3A from hot spring and yak rumen metagenomes showed optimum pH of 5.0 and 4.5 on pNPG, respectively (Sato et al., 2017). Metagenomic β-xylosidases (xylM1989) exhibited optimum activity at an alkaline pH of 8.0 (M. Maruthamuthu, Jiménez, & van Elsas, 2017). The temperature tolerance of β-xylosidases (AR19M-311–2, AR19M-311–11, and AR19M311–21) was reported at 90°C, with fair stability at 70°C for an hour (Sato et al., 2017). Thermostable β-xylosidases from a compost soil metagenome showed optimum activity at 60–75°C, retaining 80% of residual activity at 50°C (M. Wang et al., 2015b). A novel GH-31 family metagenomic, α-xylosidase (MeXyl31), from a soil metagenome exhibited stability at an optimum temperature of 45°C (Matsuzawa, Kimura, Suenaga, & Yaoi, 2016). Metagenomic β-xylosidases with bifunctional enzymatic activities which can be used for bioethanol production in combination with endo-xylanases for efficient hydrolysis have been reported (DeCastro, Rodríguez-Belmonte, & González-Siso, 2016; Rohman, Dijkstra, & Puspaningsih, 2019). The hydrolytic efficiency was improved eighty-four-folds by β-xylosidases/arabinofuranosidase (RUM630-BX) due to the stimulation by divalent metal ions like Ca^{2+}, Co^{2+}, Fe^{2+}, Mg^{2+}, Mn^{2+}, and Ni^{2+} by aspartate and histidine residues at the active site (Douglas B. Jordan et al., 2016; D. B. Jordan et al., 2018). A multimeric β-xylosidase from termite gut metagenome possesses the combined catalytic activity of β-glucosidase or β-arabinosidases and is considered as novel due to very low sequence identity/similarity with their homologues (C. Liu, Zou, Yan, & Zhou, 2018).

4.9 CONCLUSION AND FUTURE THRUST

Despite being rich in LCBs, India still lacks industrially viable lignocellulolytic enzymes (GHs) for economic production of 2G biofuels at a commercial scale. In the last decade, next-generation sequencing (NGS) has been recognized due to quick and robust generation of datasets compared with conventional methods. With the advent of metagenomics approaches, multiple applications, such as *de novo* sequencing of metagenomes, quantitation of RNA-seq-based transcript, community structure and/or gene discovery, and rapid markers, have been used to explore the hidden genetic codes in a niche/habitat. The unexplored, uncultivable novel GHs can be unraveled for accelerating 2G biofuel in India. The enormous datasets generated by metagenomic/metatranscriptomics may be further processed in pipelines, such as MEGEN-SEED and KEGG. High-throughput analysis is essential for functional output from the metagenome. Metagenomics has revealed newer GHs with useful characteristics, like pH tolerance, metal tolerance, thermostability, etc., that were inaccessible earlier. Besides metagenomics, metaproteomics is also a useful tool to study functional diversity of noncultured microbes. Certain bottlenecks in metaproteomics-based approaches are bias in protein extraction from the environment, false assignments of MS data to protein, and other downstream bioinformatics analysis to address the structure-function relationship of the environmental microbiome. Hence, bioprospecting novel GHs should be complemented with integrated omics along with actual verification through laboratory-generated data. This will allow novel candidate GH genes of core microbiome, which may provide novel insights into multifaceted applications in lignocellulose-based biorefineries.

REFERENCES

Abbott, D. W., & Boraston, A. B. (2012). Chapter eleven—quantitative approaches to the analysis of carbohydrate-binding module function. In H. J. Gilbert (Ed.), *Methods in enzymology* (Vol. 510, pp. 211–231). Academic Press.

Adrio, J. L., & Demain, A. L. (2014). Microbial enzymes: Tools for biotechnological processes. *Biomolecules*, *4*(1), 117–139. doi: 10.3390/biom4010117.

Agency, I. E. (2017). *A world in transformation. World energy outlook 2017*. https://www.iea.org/news/a-world-in-transformation-world-energy-outlook-2017.

Alain, K., & Querellou, J. (2009). Cultivating the uncultured: Limits, advances and future challenges. *Extremophiles*, *13*(4), 583–594. doi: 10.1007/s00792-009-0261-3.

Alvarez, T. M., Paiva, J. H., Ruiz, D. M., Cairo, J. P. L. F., Pereira, I. O., Paixão, D. A. A., . . . Murakami, M. T. (2013). Structure and function of a novel cellulase 5 from sugarcane soil metagenome. *PLoS One*, *8*(12), e83635. doi: 10.1371/journal.pone.0083635.

Anand, A., Kumar, V., & Satyanarayana, T. (2013). Characteristics of thermostable endoxylanase and β-xylosidase of the extremely thermophilic bacterium Geobacillus thermodenitrificans TSAA1 and its applicability in generating xylooligosaccharides and xylose from agro-residues. *Extremophiles*, *17*(3), 357–366. doi: 10.1007/s00792-013-0524-x.

Angelov, A., Mientus, M., Liebl, S., & Liebl, W. (2009). A two-host fosmid system for functional screening of (meta)genomic libraries from extreme thermophiles. *Systematic and Applied Microbiology*, *32*(3), 177–185. https://doi.org/10.1016/j.syapm.2008.01.003.

Ariaeenejad, S., Hosseini, E., Maleki, M., Kavousi, K., Moosavi-Movahedi, A. A., & Salekdeh, G. H. (2019). Identification and characterization of a novel thermostable xylanase from camel rumen metagenome. *International Journal of Biological Macromolecules*, *126*, 1295–1302. doi: 10.1016/j.ijbiomac.2018.12.041.

Armenta, S., Moreno-Mendieta, S., Sánchez-Cuapio, Z., Sánchez, S., & Rodríguez-Sanoja, R. (2017). Advances in molecular engineering of carbohydrate-binding modules. *Proteins: Structure, Function, and Bioinformatics*, *85*(9), 1602–1617. https://doi.org/10.1002/prot.25327.

Bao, L., Huang, Q., Chang, L., Sun, Q., Zhou, J., & Lu, H. (2012). Cloning and characterization of two β-Glucosidase/Xylosidase enzymes from yak rumen metagenome. *Applied Biochemistry and Biotechnology*, *166*(1), 72–86. doi: 10.1007/s12010-011-9405-x.

Bao, L., Huang, Q., Chang, L., Zhou, J., & Lu, H. (2011). Screening and characterization of a cellulase with endocellulase and exocellulase activity from yak rumen metagenome. *Journal of Molecular Catalysis B: Enzymatic*, *73*(1), 104–110. https://doi.org/10.1016/j.molcatb.2011.08.006.

Batista-García, R. A., del Rayo Sánchez-Carbente, M., Talia, P., Jackson, S. A., O'Leary, N. D., Dobson, A. D. W., & Folch-Mallol, J. L. (2016). From lignocellulosic metagenomes to lignocellulolytic genes: Trends, challenges and future prospects. *Biofuels, Bioproducts and Biorefining*, *10*(6), 864–882. https://doi.org/10.1002/bbb.1709.

Bayer, E. A., Chanzy, H., Lamed, R., & Shoham, Y. (1998). Cellulose, cellulases and cellulosomes. *Current Opinion in Structural Biology*, *8*(5), 548–557. https://doi.org/10.1016/S0959-440X(98)80143-7.

Bergmann, J. C., Costa, O. Y. A., Gladden, J. M., Singer, S., Heins, R., D'haeseleer, P., . . . Quirino, B. F. (2014). Discovery of two novel β-glucosidases from an Amazon soil metagenomic library. *FEMS microbiology letters*, *351*(2), 147–155.

Biver, S., Stroobants, A., Portetelle, D., & Vandenbol, M. (2014). Two promising alkaline β-glucosidases isolated by functional metagenomics from agricultural soil, including one showing high tolerance towards harsh detergents, oxidants and glucose. *Journal of Industrial Microbiology & Biotechnology*, *41*(3), 479–488.

Bromberg, J. S., Fricke, W. F., Brinkman, C. C., Simon, T., & Mongodin, E. F. (2015). Microbiota—implications for immunity and transplantation. *Nature Reviews Nephrology*, *11*(6), 342–353. doi: 10.1038/nrneph.2015.70.

Bušić, A., Marđetko, N., Kundas, S., Morzak, G., Belskaya, H., Ivančić Šantek, M., . . . Šantek, B. (2018). Bioethanol production from renewable raw materials and its separation and purification: A review. *Food Technology and Biotechnology*, *56*(3), 289–311. doi: 10.17113/ftb.56.03.18.5546.

Chandel, A. K., Antunes, F. A. F., Anjos, V., Bell, M. J. V., Rodrigues, L. N., Polikarpov, I., . . . da Silva, S. S. (2014). Multi-scale structural and chemical analysis of sugarcane bagasse in the process of sequential acid—base pretreatment and ethanol production by scheffersomyces shehatae and saccharomyces cerevisiae. *Biotechnology for Biofuels*, *7*(1), 63. doi: 10.1186/1754-6834-7-63.

Chandrasekharaiah, M., Thulasi, A., Bagath, M., Kumar, D. P., Santosh, S. S., Palanivel, C., . . . Sampath, K. T. (2012). Identification of cellulase gene from the metagenome of equus burchelli fecal samples and functional characterization of a novel bifunctional cellulolytic enzyme. *Applied Biochemistry and Biotechnology*, *167*(1), 132–141. doi: 10.1007/s12010-012-9660-5.

Chang, L., Ding, M., Bao, L., Chen, Y., Zhou, J., & Lu, H. (2011). Characterization of a bifunctional xylanase/endoglucanase from yak rumen microorganisms. *Applied Microbiology and Biotechnology*, *90*(6), 1933–1942. doi: 10.1007/s00253-011-3182-x.

Cheng, J., Huang, S., Jiang, H., Zhang, Y., Li, L., Wang, J., & Fan, C. (2015). Isolation and characterization of a non-specific endoglucanase from a metagenomic library of goat rumen. *World Journal of Microbiology and Biotechnology*, *32*(1), 12. doi: 10.1007/s11274-015-1957-4.

Coenen, T. M., Schoenmakers, A. C., & Verhagen, H. (1995). Safety evaluation of beta-glucanase derived from Trichoderma reesei: Summary of toxicological data. *Food and Chemical Toxicology*, *33*(10), 859–866. doi: 10.1016/0278-6915(95)00052-4.

Coughlan, L., Cotter, P., Hill, C., & Alvarez-Ordóñez, A. (2015). Biotechnological applications of functional metagenomics in the food and pharmaceutical industries. *Frontiers in Microbiology*, *6*(672). doi: 10.3389/fmicb.2015.00672.

Couto, G. H., Glogauer, A., Faoro, H., Chubatsu, L. S., Souza, E. M., & Pedrosa, F. O. (2010). Isolation of a novel lipase from a metagenomic library derived from mangrove sediment from the south Brazilian coast. *Genetics and Molecular Research*, *9*(1), 514–523. doi: 10.4238/vol9-1gmr738.

Cowan, D., Meyer, Q., Stafford, W., Muyanga, S., Cameron, R., & Wittwer, P. (2005). Metagenomic gene discovery: Past, present and future. *Trends in Biotechnology*, *23*(6), 321–329. doi: 10.1016/j.tibtech.2005.04.001.

Crameri, R., & Suter, M. (1993). Display of biologically active proteins on the surface of filamentous phages: A cDNA cloning system for selection of functional gene products linked to the genetic information responsible for their production. *Gene*, *137*(1), 69–75. doi: 10.1016/0378-1119(93)90253-y.

Dadheech, T., Shah, R., Pandit, R., Hinsu, A., Chauhan, P. S., Jakhesara, S., . . . Joshi, C. (2018). Cloning, molecular modeling and characterization of acidic cellulase from buffalo rumen and its applicability in saccharification of lignocellulosic biomass. *International Journal of Biological Macromolecules*, *113*, 73–81. https://doi.org/10.1016/j.ijbiomac.2018.02.100.

Dai, X., Zhu, Y., Luo, Y., Song, L., Liu, D., Liu, L., . . . Dong, X. (2012). Metagenomic insights into the fibrolytic microbiome in yak rumen. *PLoS One*, *7*(7), e40430. doi: 10.1371/journal.pone.0040430.

Dashtban, M., Maki, M., Leung, K. T., Mao, C., & Qin, W. (2010). Cellulase activities in biomass conversion: Measurement methods and comparison. *Critical Reviews in Biotechnology*, *30*(4), 302–309. doi: 10.3109/07388551.2010.490938.

DeCastro, M.-E., Rodríguez-Belmonte, E., & González-Siso, M.-I. (2016). Metagenomics of thermophiles with a focus on discovery of novel thermozymes. *Frontiers in Microbiology*, *7*(1521). doi: 10.3389/fmicb.2016.01521.

Del Pozo, M. V., Fernández-Arrojo, L., Gil-Martínez, J., Montesinos, A., Chernikova, T. N., Nechitaylo, T. Y., . . . Golyshin, P. N. (2012). Microbial β-glucosidases from cow rumen metagenome enhance the saccharification of lignocellulose in combination with commercial cellulase cocktail. *Biotechnology for Biofuels*, *5*(1), 73. doi: 10.1186/1754-6834-5-73.

Dennis, R. J., Taylor, E. J., Macauley, M. S., Stubbs, K. A., Turkenburg, J. P., Hart, S. J., . . . Davies, G. J. (2006). Structure and mechanism of a bacterial β-glucosaminidase having O-GlcNAcase activity. *Nature Structural & Molecular Biology*, *13*(4), 365–371. doi: 10.1038/nsmb1079.

Duan, C. J., Xian, L., Zhao, G. C., Feng, Y., Pang, H., Bai, X. L., . . . Feng, J. X. (2009). Isolation and partial characterization of novel genes encoding acidic cellulases from metagenomes of buffalo rumens. *Journal of Applied Microbiology*, *107*(1), 245–256.

Duque, E., Daddaoua, A., Cordero, B. F., Udaondo, Z., Molina-Santiago, C., Roca, A., . . . Ramos, J.-L. (2018). Ruminal metagenomic libraries as a source of relevant hemicellulolytic enzymes for biofuel production. *Microbial Biotechnology*, *11*(4), 781–787. https://doi.org/10.1111/1751-7915.13269.

Ellilä, S., Bromann, P., Nyyssönen, M., Itävaara, M., Koivula, A., Paulin, L., & Kruus, K. (2019). Cloning of novel bacterial xylanases from lignocellulose-enriched compost metagenomic libraries. *AMB Express*, *9*(1), 124. doi: 10.1186/s13568-019-0847-9.

Energy, B. S. R. O. W. (2018). *BP statistical review of world energy* (67th ed., pp. 1–53). BP p.l.c, London.

Energy, B. S. R. O. W. (2019). *BP statistical review of world energy* (68th ed., pp. 1–61). BP p.l.c, London.

Fang, Z., Fang, W., Liu, J., Hong, Y., Peng, H., Zhang, X., . . . Xiao, Y. (2010). Cloning and characterization of a beta-glucosidase from marine microbial metagenome with excellent glucose tolerance. *Journal of Microbiology and Biotechnology*, *20*(9), 1351–1358.

FAOSTAT, F. A. A. O. C. S. D. (2021). Crop. www.fao.org/faostat/en/#data/QC.

Feng, Y., Duan, C.-J., Liu, L., Tang, J.-L., & Feng, J.-X. (2009). Properties of a metagenome-derived β-Glucosidase from the contents of rabbit cecum. *Bioscience, Biotechnology, and Biochemistry*, *73*(7), 1470–1473. doi: 10.1271/bbb.80664.

Feng, Y., Duan, C.-J., Pang, H., Mo, X.-C., Wu, C.-F., Yu, Y., . . . Feng, J.-X. (2007). Cloning and identification of novel cellulase genes from uncultured microorganisms in rabbit cecum and characterization of the expressed cellulases. *Applied Microbiology and Biotechnology*, *75*(2), 319–328.

Flipphi, M., Fekete, E., Ág, N., Scazzocchio, C., & Karaffa, L. (2013). Spliceosome twin introns in fungal nuclear transcripts. *Fungal Genetics and Biology*, *57*, 48–57. https://doi.org/10.1016/j.fgb.2013.06.003.

Fredriksen, L., Stokke, R., Jensen, M. S., Westereng, B., Jameson, J. K., Steen, I. H., & Eijsink, V. G. H. (2019). Discovery of a thermostable GH10 xylanase with broad substrate specificity from the arctic mid-ocean ridge vent system. *Applied and Environmental Microbiology*, *85*(6). doi: 10.1128/aem.02970-18.

Galbraith, E. A., Antonopoulos, D. A., & White, B. A. (2004). Suppressive subtractive hybridization as a tool for identifying genetic diversity in an environmental metagenome: The rumen as a model. *Environmental Microbiology*, *6*(9), 928–937. doi: 10.1111/j.1462-2920.2004.00575.x.

Ganesan, M., Mathivani Vinayakamoorthy, R., Thankappan, S., Muniraj, I., & Uthandi, S. (2020). Thermotolerant glycosyl hydrolases-producing Bacillus aerius CMCPS1 and its saccharification efficiency on HCR-laccase (LccH)-pretreated corncob biomass. *Biotechnology for Biofuels*, *13*(1), 124. doi: 10.1186/s13068-020-01764-2.

Geng, A., Zou, G., Yan, X., Wang, Q., Zhang, J., Liu, F., . . . Zhou, Z. (2012). Expression and characterization of a novel metagenome-derived cellulase Exo2b and its application to improve cellulase activity in Trichoderma reesei. *Applied Microbiology and Biotechnology*, *96*(4), 951–962. doi: 10.1007/s00253-012-3873-y.

Ghadikolaei, K. K., Sangachini, E. D., Vahdatirad, V., Noghabi, K. A., & Zahiri, H. S. (2019). An extreme halophilic xylanase from camel rumen metagenome with elevated catalytic activity in high salt concentrations. *AMB Express*, *9*(1), 86. doi: 10.1186/s13568-019-0809-2.

Ghosh, A., Mehta, A., & Khan, A. M. (2019). Metagenomic analysis and its applications. In S. Ranganathan, M. Gribskov, K. Nakai, & C. Schönbach (Eds.), *Encyclopedia of bioinformatics and computational biology* (pp. 184–193). Academic Press.

Gong, X., Gruninger, R. J., Qi, M., Paterson, L., Forster, R. J., Teather, R. M., & McAllister, T. A. (2012). Cloning and identification of novel hydrolase genes from a dairy cow rumen metagenomic library and characterization of a cellulase gene. *BMC Research Notes*, *5*(1), 566.

Goshadrou, A. (2013). Enhanced NSSF for ethanol production by phosphoric acid pretreatment. The Candaian Society of Bioengineering. Paper No. CSBE13-069.

Gowen, C. M., & Fong, S. S. (2010). Exploring biodiversity for cellulosic biofuel production. *Chemistry & Biodiversity*, *7*(5), 1086–1097. https://doi.org/10.1002/cbdv.200900314.

Gray, K. A., Zhao, L., & Emptage, M. (2006). Bioethanol. *Current Opinion in Chemical Biology*, *10*(2), 141–146. https://doi.org/10.1016/j.cbpa.2006.02.035.

Green, C. D., Simons, J. F., Taillon, B. E., & Lewin, D. A. (2001). Open systems: Panoramic views of gene expression. *Journal of Immunological Methods*, *250*(1–2), 67–79. doi: 10.1016/s0022-1759(01)00306-4.

Gruninger, R., Gong, X., Forster, R., & McAllister, T. (2014). Biochemical and kinetic characterization of the multifunctional β-glucosidase/β-xylosidase/α-arabinosidase, Bgxa1. *Applied Microbiology and Biotechnology*, *98*(7), 3003–3012.

Gunawardana, M., Chang, S., Jimenez, A., Holland-Moritz, D., Holland-Moritz, H., La Val, T. P., . . . Baum, M. M. (2014). Isolation of PCR quality microbial community DNA from heavily contaminated environments. *Journal of Microbiological Methods*, *102*, 1–7. doi: 10.1016/j.mimet.2014.04.005.

Gupta, S. K., & Shukla, P. (2016). Advanced technologies for improved expression of recombinant proteins in bacteria: Perspectives and applications. *Critical Reviews in Biotechnology*, *36*(6), 1089–1098. doi: 10.3109/07388551.2015.1084264.

Handelsman, J., Rondon, M. R., Brady, S. F., Clardy, J., & Goodman, R. M. (1998). Molecular biological access to the chemistry of unknown soil microbes: A new frontier for natural products. *Chemical Biology*, *5*(10), R245–249. doi: 10.1016/s1074-5521(98)90108-9.

Hannig, G., & Makrides, S. C. (1998). Strategies for optimizing heterologous protein expression in Escherichia coli. *Trends in Biotechnology*, *16*(2), 54–60. https://doi.org/10.1016/S0167-7799(97)01155-4.

Henrissat, B. (1991). A classification of glycosyl hydrolases based on amino acid sequence similarities. *Biochemical Journal, 280*(Pt 2), 309–316. doi: 10.1042/bj2800309.

Henrissat, B., & Davies, G. (1997). Structural and sequence-based classification of glycoside hydrolases. *Current Opinion in Structural Biology*, *7*(5), 637–644. https://doi.org/10.1016/S0959-440X(97)80072-3.

Hidaka, M., Kitaoka, M., Hayashi, K., Wakagi, T., Shoun, H., & Fushinobu, S. (2006). Structural dissection of the reaction mechanism of cellobiose phosphorylase. *Biochemical Journal*, *398*(1), 37–43. doi: 10.1042/bj20060274.

Hilton, S. K., Castro-Nallar, E., Pérez-Losada, M., Toma, I., McCaffrey, T. A., Hoffman, E. P., . . . Crandall, K. A. (2016). Metataxonomic and metagenomic approaches vs. Culture-based techniques for clinical pathology. *Frontiers in Microbiology*, *7*(484). doi: 10.3389/fmicb.2016.00484.

Holck, J., Djajadi, D. T., Brask, J., Pilgaard, B., Krogh, K. B. R. M., Meyer, A. S., . . . Wilkens, C. (2019). Novel xylanolytic triple domain enzyme targeted at feruloylated arabinoxylan degradation. *Enzyme and Microbial Technology*, *129*, 109353. https://doi.org/10.1016/j.enzmictec.2019.05.010.

Ilmberger, N., Meske, D., Juergensen, J., Schulte, M., Barthen, P., Rabausch, U., . . . Schmitz, R. A. (2012). Metagenomic cellulases highly tolerant towards the presence of ionic liquids—linking thermostability and halotolerance. *Applied Microbiology and Biotechnology*, *95*(1), 135–146.

Jain, I., Kumar, V., & Satyanarayana, T. (2015). Xylooligosaccharides: An economical prebiotic from agroresidues and their health benefits. *Indian Journal of Experimental Biology*, *53*(3), 131–142.

Jiang, C.-J., Chen, G., Huang, J., Huang, Q., Jin, K., Shen, P.-H., . . . Wu, B. (2011). A novel β-glucosidase with lipolytic activity from a soil metagenome. *Folia Microbiologica*, *56*(6), 563–570. doi: 10.1007/s12223-011-0083-4.

Jiang, C.-J., Hao, Z.-Y., Jin, K., Li, S.-X., Che, Z.-Q., Ma, G.-F., & Wu, B. (2010). Identification of a metagenome-derived β-glucosidase from bioreactor contents. *Journal of Molecular Catalysis B: Enzymatic*, *63*(1), 11–16. https://doi.org/10.1016/j.molcatb.2009.11.009.

Jiang, C., Ma, G., Li, S., Hu, T., Che, Z., Shen, P., Wu, B. (2009). Characterization of a novel β-glucosidase-like activity from a soil metagenome. *The Journal of Microbiology*, *47*(5), 542. doi: 10.1007/s12275-009-0024-y.

Jordan, D. B., Braker, J. D., Wagschal, K., Stoller, J. R., & Lee, C. C. (2016). Isolation and divalent-metal activation of a β-xylosidase, RUM630-BX. *Enzyme and Microbial Technology*, *82*, 158–163. https://doi.org/10.1016/j.enzmictec.2015.10.001.

Jordan, D. B., Stoller, J. R., Kibblewhite, R. E., Chan, V. J., Lee, C. C., & Wagschal, K. (2018). Absence or presence of metal ion activation in two structurally similar GH43 β-xylosidases. *Enzyme and Microbial Technology*, *114*, 29–32. doi: 10.1016/j.enzmictec.2018.03.007.

Juturu, V., & Wu, J. C. (2014). Microbial cellulases: Engineering, production and applications. *Renewable and Sustainable Energy Reviews*, *33*, 188–203. https://doi.org/10.1016/j.rser.2014.01.077.

Kang, Y. M., Kim, M. K., An, J. M., Haque, M. A., & Cho, K. M. (2015). Metagenomics of un-culturable bacteria in cow rumen: Construction of cel9E—xyn10A fusion gene by site-directed mutagenesis. *Journal of Molecular Catalysis B: Enzymatic*, *113*, 29–38. https://doi.org/10.1016/j.molcatb.2014.11.010.

Kanokratana, P., Eurwilaichitr, L., Pootanakit, K., & Champreda, V. (2015). Identification of glycosyl hydrolases from a metagenomic library of microflora in sugarcane bagasse collection site and their cooperative action on cellulose degradation. *Journal of Bioscience and Bioengineering*, *119*(4), 384–391. https://doi.org/10.1016/j.jbiosc.2014.09.010.

Kanokratana, P., Mhuantong, W., Laothanachareon, T., Tangphatsornruang, S., Eurwilaichitr, L., Pootanakit, K., & Champreda, V. (2013). Phylogenetic analysis and metabolic potential of microbial communities in an industrial bagasse collection site. *Microbial Ecology*, *66*(2), 322–334. doi: 10.1007/s00248-013-0209-0.

Kaushal, G., Rai, A. K., & Singh, S. P. (2021). A novel β-glucosidase from a hot-spring metagenome shows elevated thermal stability and tolerance to glucose and ethanol. *Enzyme Microbiology and Technology*, *145*,109764. doi: 10.1016/j.enzmictec.2021.109764.

Ketudat Cairns, J. R., & Esen, A. (2010). β-Glucosidases. *Cellular and Molecular Life Sciences*, *67*(20), 3389–3405. doi: 10.1007/s00018-010-0399-2.

Kim, D., Kim, S.-N., Baik, K. S., Park, S. C., Lim, C. H., Kim, J.-O., . . . Seong, C. N. (2011). Screening and characterization of a cellulase gene from the gut microflora of abalone using metagenomic library. *The Journal of Microbiology*, *49*(1), 141–145.

Kim, S.-J., Lee, C.-M., Han, B.-R., Kim, M.-Y., Yeo, Y.-S., Yoon, S.-H., . . . Jun, H.-K. (2008). Characterization of a gene encoding cellulase from uncultured soil bacteria. *FEMS Microbiology Letters*, *282*(1), 44–51.

Kim, S.-J., Lee, C.-M., Kim, M.-Y., Yeo, Y.-S., Yoon, S.-H., Kang, H.-C., & Koo, B.-S. (2007). Screening and characterization of an enzyme with beta-glucosidase activity from environmental DNA. *Journal of Microbiology and Biotechnology*, *17*(6), 905–912.

Ko, K. C., Lee, J. H., Han, Y., Choi, J. H., & Song, J. J. (2013). A novel multifunctional cellulolytic enzyme screened from metagenomic resources representing ruminal bacteria. *Biochemical and Biophysical Research Communications*, *441*(3), 567–572. doi: 10.1016/j.bbrc.2013.10.120.

Kumar, J., Sharma, N., Kaushal, G., Samurailatpam, S., Sahoo, D., Rai, A. K., & Singh, S. P. (2019). Metagenomic insights into the taxonomic and functional features of kinema, a traditional fermented soybean product of Sikkim Himalaya. *Frontiers in Microbiology*, *10*, 1744–1744. doi: 10.3389/fmicb.2019.01744.

Kumar, V., & Satyanarayana, T. (2014). Production of endoxylanase with enhanced thermostability by a novel polyextremophilic Bacillus halodurans TSEV1 and its applicability in waste paper deinking. *Process Biochemistry*, *49*(3), 386–394. https://doi.org/10.1016/j.procbio.2013.12.005.

Kuramochi, K., Uchimura, K., Kurata, A., Kobayashi, T., Hirose, Y., Miura, T., . . . Horikoshi, K. (2016). A high-molecular-weight, alkaline, and thermostable β-1,4-xylanase of a subseafloor Microcella alkaliphila. *Extremophiles*, *20*(4), 471–478. doi: 10.1007/s00792-016-0837-7.

Lämmle, K., Zipper, H., Breuer, M., Hauer, B., Buta, C., Brunner, H., & Rupp, S. (2007). Identification of novel enzymes with different hydrolytic activities by metagenome expression cloning. *Journal of Biotechnology*, *127*(4), 575–592. https://doi.org/10.1016/j.jbiotec.2006.07.036.

Lee, C. M., Seo, S. H., Kim, S. J., & Sim, J. S. (2014). Screening and characterization of a novel cellulase gene from the gut microflora of Hermetia illucens using metagenomic library. *Journal of Microbiology and Biotechnology*, *24*(9), 1196–1206.

Leis, B., Angelov, A., Mientus, M., Li, H., Pham, V. T. T., Lauinger, B., . . . Liebl, W. (2015). Identification of novel esterase-active enzymes from hot environments by use of the host bacterium Thermus thermophilus. *Frontiers in Microbiology*, *6*(275). doi: 10.3389/fmicb.2015.00275.

Lemos, L. N., Pereira, R. V., Quaggio, R. B., Martins, L. F., Moura, L. M. S., da Silva, A. R., . . . Setubal, J. C. (2017). Genome-centric analysis of a thermophilic and cellulolytic bacterial consortium derived from composting. *Frontiers in Microbiology*, *8*(644). doi: 10.3389/fmicb.2017.00644.

Lepage, P., Leclerc, M. C., Joossens, M., Mondot, S., Blottière, H. M., Raes, J., . . . Doré, J. (2013). A metagenomic insight into our gut's microbiome. *Gut*, *62*(1), 146–158. doi: 10.1136/gutjnl-2011-301805.

Li, G., Jiang, Y., Fan, X.-j., & Liu, Y.-h. (2012). Molecular cloning and characterization of a novel β-glucosidase with high hydrolyzing ability for soybean isoflavone glycosides and glucose-tolerance from soil metagenomic library. *Bioresource Technology*, *123*, 15–22. https://doi.org/10.1016/j.biortech.2012.07.083.

Li, L.-L., McCorkle, S. R., Monchy, S., Taghavi, S., & van der Lelie, D. (2009). Bioprospecting metagenomes: Glycosyl hydrolases for converting biomass. *Biotechnology for Biofuels*, *2*(1), 10. doi: 10.1186/1754-6834-2-10.

Li, Y., Liu, N., Yang, H., Zhao, F., Yu, Y., Tian, Y., & Lu, X. (2014a). Cloning and characterization of a new beta-Glucosidase from a metagenomic library of Rumen of cattle feeding with Miscanthus sinensis. *BMC biotechnology*, *14*(1), 85.

Li, Y., Liu, N., Yang, H., Zhao, F., Yu, Y., Tian, Y., & Lu, X. (2014b). Cloning and characterization of a new β-Glucosidase from a metagenomic library of Rumen of cattle feeding with Miscanthus sinensis. *BMC Biotechnology*, *14*(1), 85. doi: 10.1186/1472-6750-14-85.

Lim, H. K., Chung, E. J., Kim, J.-C., Choi, G. J., Jang, K. S., Chung, Y. R., . . . Lee, S.-W. (2005). Characterization of a forest soil metagenome clone that confers indirubin and indigo production on *Escherichia coli*. *Applied and Environmental Microbiology*, *71*(12), 7768–7777. doi:10.1128/AEM.71.12.7768-7777.2005.

Linton, S. M. (2020). Review: The structure and function of cellulase (endo-β-1,4-glucanase) and hemicellulase (β-1,3-glucanase and endo-β-1,4-mannase) enzymes in invertebrates that consume materials ranging from microbes, algae to leaf litter. *Comparative Biochemistry and Physiology Part B: Biochemistry and Molecular Biology*, *240*, 110354. https://doi.org/10.1016/j.cbpb.2019.110354.

Liu, C., Zou, G., Yan, X., & Zhou, X. (2018). Screening of multimeric β-xylosidases from the gut microbiome of a higher termite, *Globitermes brachycerastes*. *International Journal of Biological Sciences*, *14*(6), 608–615. doi: 10.7150/ijbs.22763.

Liu, J., Liu, W.-D., Zhao, X.-L., Shen, W.-J., Cao, H., & Cui, Z.-L. (2011). Cloning and functional characterization of a novel endo-β-1, 4-glucanase gene from a soil-derived metagenomic library. *Applied Microbiology and Biotechnology*, *89*(4), 1083–1092.

Ljungdahl, L. G. (2008). The cellulase/hemicellulase system of the anaerobic fungusOrpinomycesPC-2 and aspects of its applied use. *Annals of the New York Academy of Sciences*, *1125*(1), 308–321. https://doi.org/10.1196/annals.1419.030.

Lombard, V., Golaconda Ramulu, H., Drula, E., Coutinho, P. M., & Henrissat, B. (2013). The carbohydrate-active enzymes database (CAZy) in 2013. *Nucleic Acids Research*, *42*(D1), D490–D495. doi: 10.1093/nar/gkt1178.

Lorenz, P., Liebeton, K., Niehaus, F., & Eck, J. (2002). Screening for novel enzymes for biocatalytic processes: Accessing the metagenome as a resource of novel functional sequence space. *Current Opinion in Biotechnology*, *13*(6), 572–577. doi: 10.1016/s0958-1669(02)00345-2.

Lu, J., Du, L., Wei, Y., Hu, Y., & Huang, R. (2013). Expression and characterization of a novel highly glucose-tolerant β-glucosidase from a soil metagenome. *Acta Biochim Biophys Sin (Shanghai)*, *45*(8), 664–673. doi: 10.1093/abbs/gmt061.

Lubieniechi, S., Peranantham, T., & Levin, D. B. (2013). Recent patents on genetic modification of plants and microbes for biomass conversion to biofuels. *Recent Pat DNA Gene Seq*, *7*(1), 25–35. doi: 10.2174/1872215611307010005.

Lynd, L. R., Weimer, P. J., Zyl, W. H. V., & Pretorius, I. S. (2002). Microbial cellulose utilization: Fundamentals and biotechnology. *Microbiology and Molecular Biology Reviews*, *66*(3), 506–577. doi:10.1128/MMBR.66.3.506-577.2002.

Mai, Z., Su, H., Yang, J., Huang, S., & Zhang, S. (2014). Cloning and characterization of a novel GH44 family endoglucanase from mangrove soil metagenomic library. *Biotechnology Letters*, *36*(8), 1701–1709. doi: 10.1007/s10529-014-1531-4.

Mamo, G., Hatti-Kaul, R., & Mattiasson, B. (2006). A thermostable alkaline active endo-β-1–4-xylanase from Bacillus halodurans S7: Purification and characterization. *Enzyme and Microbial Technology*, *39*(7), 1492–1498. https://doi.org/10.1016/j.enzmictec.2006.03.040.

Martinez, A., Kolvek, S. J., Yip, C. L. T., Hopke, J., Brown, K. A., MacNeil, I. A., & Osburne, M. S. (2004). Genetically modified bacterial strains and novel bacterial artificial chromosome shuttle vectors for constructing environmental libraries and detecting heterologous natural products in multiple expression hosts. *Applied and Environmental Microbiology*, *70*(4), 2452–2463. doi:10.1128/AEM.70.4.2452-2463.2004.

Maruthamuthu, M., Jiménez, D. J., Stevens, P., & van Elsas, J. D. (2016). A multi-substrate approach for functional metagenomics-based screening for (hemi)cellulases in two wheat straw-degrading microbial consortia unveils novel thermoalkaliphilic enzymes. *BMC Genomics*, *17*(1), 86. doi: 10.1186/s12864-016-2404-0.

Maruthamuthu, M., Jiménez, D. J., & van Elsas, J. D. (2017). Characterization of a furan aldehyde-tolerant β-xylosidase/α-arabinosidase obtained through a synthetic metagenomics approach. *Journal of Applied Microbiology*, *123*(1), 145–158. https://doi.org/10.1111/jam.13484.

Matsuzawa, T., Kimura, N., Suenaga, H., & Yaoi, K. (2016). Screening, identification, and characterization of α-xylosidase from a soil metagenome. *Journal of Bioscience and Bioengineering*, *122*(4), 393–399. https://doi.org/10.1016/j.jbiosc.2016.03.012.

McCarter, J. D., & Stephen Withers, G. (1994). Mechanisms of enzymatic glycoside hydrolysis. *Current Opinion in Structural Biology*, *4*(6), 885–892. https://doi.org/10.1016/0959-440X(94)90271-2.

Milshteyn, A., Schneider, J. S., & Brady, S. F. (2014). Mining the metabiome: Identifying novel natural products from microbial communities. *Chemical Biology*, *21*(9), 1211–1223. doi: 10.1016/j.chembiol.2014.08.006.

Morgan, X. C., & Huttenhower, C. (2012). Chapter 12: Human microbiome analysis. *PLOS Computational Biology*, *8*(12), e1002808. doi: 10.1371/journal.pcbi.1002808.

Mori, T., Kamei, I., Hirai, H., & Kondo, R. (2014). Identification of novel glycosyl hydrolases with cellulolytic activity against crystalline cellulose from metagenomic libraries constructed from bacterial enrichment cultures. *SpringerPlus*, *3*(1), 365. doi: 10.1186/2193-1801-3-365.

Moshi, A. P., Crespo, C. F., Badshah, M., Hosea, K. M. M., Mshandete, A. M., & Mattiasson, B. (2014). High bioethanol titre from Manihot glaziovii through fed-batch simultaneous saccharification and fermentation in automatic gas potential test system. *Bioresource Technology*, *156*, 348–356. https://doi.org/10.1016/j.biortech.2013.12.082.

Nacke, H., Engelhaupt, M., Brady, S., Fischer, C., Tautzt, J., & Daniel, R. (2012). Identification and characterization of novel cellulolytic and hemicellulolytic genes and enzymes derived from German grassland soil metagenomes. *Biotechnology Letters*, *34*(4), 663–675. doi: 10.1007/s10529-011-0830-2.

Nam, K. H., Kim, S. J., Kim, M. Y., Kim, J. H., Yeo, Y. S., Lee, C. M., . . . Hwang, K. Y. (2008). Crystal structure of engineered β-glucosidase from a soil metagenome. *Proteins: Structure, Function, and Bioinformatics*, *73*(3), 788–793.

Nguyen, N. H., Maruset, L., Uengwetwanit, T., Mhuantong, W., Harnpicharnchai, P.,

Champreda, V., . . . Pongpattanakitshote, S. (2012). Identification and characterization of a cellulase-encoding gene from the buffalo rumen metagenomic library. *Bioscience, Biotechnology, and Biochemistry*, *76*(6), 1075–1084. doi: 10.1271/bbb.110786.

Okano, H., Ozaki, M., Kanaya, E., Kim, J.-J., Angkawidjaja, C., Koga, Y., & Kanaya, S. (2014). Structure and stability of metagenome-derived glycoside hydrolase family 12 cellulase (LC-CelA) a homolog of Cel12A from Rhodothermus marinus. *FEBS Open Bio*, *4*, 936–946. doi: 10.1016/j.fob.2014.10.013.

Pabbathi, N. P. P., Velidandi, A., Tavarna, T., Gupta, S., Raj, R. S., Gandam, P. K., & Baadhe, R. R. (2021). Role of metagenomics in prospecting novel endoglucanases, accentuating functional metagenomics approach in second-generation biofuel production: A review. *Biomass Conversion and Biorefinery*. doi: 10.1007/s13399-020-01186-y.

Pang, H., Zhang, P., Duan, C.-J., Mo, X.-C., Tang, J.-L., & Feng, J.-X. (2009). Identification of cellulase genes from the metagenomes of compost soils and functional characterization of one novel endoglucanase. *Current Microbiology*, *58*(4), 404. doi: 10.1007/s00284-008-9346-y.

Pottkämper, J., Barthen, P., Ilmberger, N., Schwaneberg, U., Schenk, A., Schulte, M., . . . Streit, W. R. (2009). Applying metagenomics for the identification of bacterial cellulases that are stable in ionic liquids. *Green Chemistry*, *11*(7), 957–965. doi: 10.1039/b820157a.

Quince, C., Walker, A. W., Simpson, J. T., Loman, N. J., & Segata, N. (2017). Shotgun metagenomics, from sampling to analysis. *Nature Biotechnology*, *35*(9), 833–844. doi: 10.1038/nbt.3935.

Radajewski, S., Ineson, P., Parekh, N. R., & Murrell, J. C. (2000). Stable-isotope probing as a tool in microbial ecology. *Nature*, *403*(6770), 646–649. doi: 10.1038/35001054.

Rashamuse, K., Sanyika Tendai, W., Mathiba, K., Ngcobo, T., Mtimka, S., & Brady, D. (2017). Metagenomic mining of glycoside hydrolases from the hindgut bacterial symbionts of a termite (Trinervitermes trinervoides) and the characterization of a multimodular β-1,4-xylanase (GH11). *Biotechnology and Applied Biochemistry*, *64*(2), 174–186. https://doi.org/10.1002/bab.1480.

Rashamuse, K., Visser, D., Hennessy, F., Kemp, J., Roux-van der Merwe, M., Badenhorst, J., . . . Brady, D. (2013). Characterisation of two bifunctional cellulase—xylanase enzymes isolated from a bovine rumen metagenome library. *Current Microbiology*, *66*(2), 145–151.

Rattanachomsri, U., Kanokratana, P., Eurwilaichitr, L., Igarashi, Y., & Champreda, V. (2011). Culture-independent phylogenetic analysis of the microbial community in industrial sugarcane bagasse feedstock piles. *Bioscience, Biotechnology, and Biochemistry*, *75*(2), 232–239. doi: 10.1271/bbb.100429.

Report, B. P. (2020). *Statistical review of world energy 2020*. BP p.l.c. London, UK.

Rohman, A., Dijkstra, B. W., & Puspaningsih, N. N. T. (2019). β-Xylosidases: Structural diversity, catalytic mechanism, and inhibition by monosaccharides. *International Journal of Molecular Sciences*, *20*(22), 5524.

Rosselló-Mora, R., & Amann, R. (2001). The species concept for prokaryotes. *FEMS Microbiology Reviews*, *25*(1), 39–67. doi: 10.1111/j.1574-6976.2001.tb00571.x.

Sae-Lee, R., & Boonmee, A. (2014). Newly derived GH43 gene from compost metagenome showing dual xylanase and cellulase activities. *Folia Microbiologica*, *59*(5), 409–417. doi: 10.1007/s12223-014-0313-7.

Salehi Jouzani, G., & Taherzadeh, M. J. (2015). Advances in consolidated bioprocessing systems for bioethanol and butanol production from biomass: A comprehensive review. *Biofuel Research Journal*, *2*(1), 152–195. doi: 10.18331/brj2015.2.1.4.

Sato, M., Suda, M., Okuma, J., Kato, T., Hirose, Y., Nishimura, A., . . . Shibata, D. (2017). Isolation of highly thermostable β-xylosidases from a hot spring soil microbial community using a metagenomic approach. *DNA Research*, *24*(6), 649–656. doi: 10.1093/dnares/dsx032.

Schmidt, T. M., DeLong, E. F., & Pace, N. R. (1991). Analysis of a marine picoplankton community by 16S rRNA gene cloning and sequencing. *Journal of Bacteriology, 173*(14), 4371–4378. doi: 10.1128/jb.173.14.4371-4378.1991.

Schröder, C., Elleuche, S., Blank, S., & Antranikian, G. (2014). Characterization of a heat-active archaeal β-glucosidase from a hydrothermal spring metagenome. *Enzyme and Microbial Technology, 57*, 48–54.

Shallom, D., & Shoham, Y. (2003). Microbial hemicellulases. *Current Opinion in Microbiology, 6*(3), 219–228. https://doi.org/10.1016/S1369-5274(03)00056-0.

Sharma, A., Adhikari, S., & Satyanarayana, T. (2007). Alkali-thermostable and cellulase-free xylanase production by an extreme thermophile Geobacillus thermoleovorans. *World Journal of Microbiology and Biotechnology, 23*(4), 483–490. doi: 10.1007/s11274-006-9250-1.

Shi, H., Zhang, Y., Zhong, H., Huang, Y., Li, X., & Wang, F. (2014). Cloning, over-expression and characterization of a thermo-tolerant xylanase from Thermotoga thermarum. *Biotechnology Letters, 36*(3), 587–593. doi: 10.1007/s10529-013-1392-2.

Shi, W., Xie, S., Chen, X., Sun, S., Zhou, X., Liu, L., . . . Yuan, J. S. (2013). Comparative genomic analysis of the endosymbionts of herbivorous insects reveals eco-environmental adaptations: Biotechnology applications. *PLoS Genet, 9*(1), e1003131.

Simon, C., & Daniel, R. (2011). Metagenomic analyses: Past and future trends. *Applied and Environmental Microbiology, 77*(4), 1153–1161. doi: 10.1128/aem.02345-10.

Sobala, L. F., Speciale, G., Zhu, S., Raich, L. s., Sannikova, N., Thompson, A. J., . . . Williams, S. J. (2020). An epoxide intermediate in glycosidase catalysis. *ACS Central Science, 6*(5), 760–770. doi: 10.1021/acscentsci.0c00111.

Sorensen, A., Lübeck, M., Lübeck, P. S., & Ahring, B. K. (2013). Fungal beta-glucosidases: A bottleneck in industrial use of lignocellulosic materials. *Biomolecules, 3*(3), 612–631.

Staley, J. T., & Konopka, A. (1985). Measurement of in situ activities of nonphotosynthetic microorganisms in aquatic and terrestrial habitats. *Annual Review of Microbiology, 39*(1), 321–346. doi: 10.1146/annurev.mi.39.100185.001541.

Swiegers, J. H., Bartowsky, E. J., Henschke, P. A., & Pretorius, I. S. (2005). Yeast and bacterial modulation of wine aroma and flavour. *Australian Journal of Grape and Wine Research, 11*(2), 139–173. https://doi.org/10.1111/j.1755-0238.2005.tb00285.x.

Talamantes, D., Biabini, N., Dang, H., Abdoun, K., & Berlemont, R. (2016). Natural diversity of cellulases, xylanases, and chitinases in bacteria. *Biotechnology for Biofuels, 9*(1), 133. doi: 10.1186/s13068-016-0538-6.

Thankappan, S., Kandasamy, S., Joshi, B., Sorokina, K. N., Taran, O. P., & Uthandi, S. (2018). Bioprospecting thermophilic glycosyl hydrolases, from hot springs of Himachal Pradesh, for biomass valorization. *AMB Express, 8*(1), 168. doi: 10.1186/s13568-018-0690-4.

Thankappan, S., Kandasamy, S., & Uthandi, S. (2017). Deciphering thermostable xylanases from hotsprings: The heritage of Himachal Pradesh for efficient biomass deconstruction. *Madras Agricultural Journal, 104*, 282. doi: 10.29321/maj.2017.000061.

Thuan, N. H., & Sohng, J. K. (2013). Recent biotechnological progress in enzymatic synthesis of glycosides. *Journal of Industrial Microbiology and Biotechnology, 40*(12), 1329–1356. doi: 10.1007/s10295-013-1332-0.

Tiwari, R., Nain, L., Labrou, N. E., & Shukla, P. (2018). Bioprospecting of functional cellulases from metagenome for second generation biofuel production: A review. *Critical Reviews in Microbiology, 44*(2), 244–257. doi: 10.1080/1040841x.2017.1337713.

Uchiyama, T., Abe, T., Ikemura, T., & Watanabe, K. (2005). Substrate-induced gene-expression screening of environmental metagenome libraries for isolation of catabolic genes. *Nature Biotechnology, 23*(1), 88–93. doi: 10.1038/nbt1048.

Uchiyama, T., Yaoi, K., & Miyazaki, K. (2015). Glucose-tolerant β-glucosidase retrieved from the metagenome. *Name: Frontiers in Microbiology, 6*, 548.

Ufarté, L., Potocki-Veronese, G., & Laville, É. (2015). Discovery of new protein families and functions: New challenges in functional metagenomics for biotechnologies and microbial ecology. *Frontiers in Microbiology*, *6*, 563–563. doi: 10.3389/fmicb.2015.00563.

Venter, J. C., Remington, K., Heidelberg, J. F., Halpern, A. L., Rusch, D., Eisen, J. A., . . . Smith, H. O. (2004). Environmental genome shotgun sequencing of the sargasso sea. *Science*, *304*(5667), 66–74. doi: 10.1126/science.1093857.

Verma, D., Kawarabayasi, Y., Miyazaki, K., & Satyanarayana, T. (2013). Cloning, expression and characteristics of a novel alkalistable and thermostable xylanase encoding gene (Mxyl) retrieved from compost-soil metagenome. *PLoS One*, *8*(1), e52459. doi: 10.1371/journal.pone.0052459.

Verma, D., & Satyanarayana, T. (2012). Cloning, expression and applicability of thermo-alkali-stable xylanase of Geobacillus thermoleovorans in generating xylooligosaccharides from agro-residues. *Bioresource Technology*, *107*, 333–338. doi: 10.1016/j.biortech.2011.12.055.

Verma, D., & Satyanarayana, T. (2020). Xylanolytic extremozymes retrieved from environmental metagenomes: Characteristics, genetic engineering, and applications. *Frontiers in Microbiology*, *11*(2232). doi: 10.3389/fmicb.2020.551109.

Vester, J. K., Glaring, M. A., & Stougaard, P. (2015). Improved cultivation and metagenomics as new tools for bioprospecting in cold environments. *Extremophiles*, *19*(1), 17–29. doi: 10.1007/s00792-014-0704-3.

Voget, S., Steele, H., & Streit, W. (2006). Characterization of a metagenome-derived halotolerant cellulase. *Journal of biotechnology*, *126*(1), 26–36.

Vuong, T. V., & Wilson, D. B. (2010). Glycoside hydrolases: Catalytic base/nucleophile diversity. *Biotechnology and Bioengineering, 107*, 195–205.

Wagschal, K., Heng, C., Lee, C. C., & Wong, D. W. S. (2009). Biochemical characterization of a novel dual-function arabinofuranosidase/xylosidase isolated from a compost starter mixture. *Applied Microbiology and Biotechnology*, *81*(5), 855–863. doi: 10.1007/s00253-008-1662-4.

Wang, G., Wang, Y., Yang, P., Luo, H., Huang, H., Shi, P., . . . Yao, B. (2010). Molecular detection and diversity of xylanase genes in alpine tundra soil. *Applied Microbiology and Biotechnology*, *87*(4), 1383–1393. doi: 10.1007/s00253-010-2564-9.

Wang, J., Sun, Z., Zhou, Y., Wang, Q., Ye, J. a., Chen, Z., & Liu, J. (2012). Screening of a xylanase clone from a fosmid library of rumen microbiota in hu sheep. *Animal Biotechnology*, *23*(3), 156–173. doi: 10.1080/10495398.2012.662925.

Wang, M., Lai, G.-L., Nie, Y., Geng, S., Liu, L., Zhu, B., . . . Wu, X.-L. (2015a). Synergistic function of four novel thermostable glycoside hydrolases from a long-term enriched thermophilic methanogenic digester. *Frontiers in Microbiology*, *6*, 509.

Wang, M., Lai, G.-L., Nie, Y., Geng, S., Liu, L., Zhu, B., . . . Wu, X.-L. (2015b). Synergistic function of four novel thermostable glycoside hydrolases from a long-term enriched thermophilic methanogenic digester. *Frontiers in Microbiology*, *6*(509). doi: 10.3389/fmicb.2015.00509.

Whitman, W. B., Coleman, D. C., & Wiebe, W. J. (1998). Prokaryotes: The unseen majority. *Proceedings of the National Academy of Sciences*, *95*(12), 6578. doi: 10.1073/pnas.95.12.6578.

Wu, Z., & Lee, Y. Y. (1998). Nonisothermal simultaneous saccharification and fermentation for direct conversion of lignocellulosic biomass to ethanol. *Applied Biochemistry and Biotechnology*, *70*(1), 479. doi: 10.1007/bf02920161.

Xiang, L., Li, A., Tian, C., Zhou, Y., Zhang, G., & Ma, Y. (2014). Identification and characterization of a new acid-stable endoglucanase from a metagenomic library. *Protein Expression and Purification*, *102*, 20–26.

Xing, M.-N., Zhang, X.-Z., & Huang, H. (2012). Application of metagenomic techniques in mining enzymes from microbial communities for biofuel synthesis. *Biotechnology Advances*, *30*(4), 920–929. https://doi.org/10.1016/j.biotechadv.2012.01.021.

Yang, C., Niu, Y., Li, C., Zhu, D., Wang, W., Liu, X., . . . Xu, P. (2013). Characterization of a novel metagenome-derived 6-Phospho-β-Glucosidase from black liquor sediment. *Applied and Environmental Microbiology*, *79*(7), 2121–2127. doi: 10.1128/AEM.03528-12.

Yasir, M., Khan, H., Azam, S. S., Telke, A., Kim, S. W., & Chung, Y. R. (2013). Cloning and functional characterization of endo-β-1, 4-glucanase gene from metagenomic library of vermicompost. *Journal of Microbiology*, *51*(3), 329–335.

Yeh, Y.-F., Chang, S. C.-y., Kuo, H.-W., Tong, C.-G., Yu, S.-M., & Ho, T.-H. D. (2013). A metagenomic approach for the identification and cloning of an endoglucanase from rice straw compost. *Gene*, *519*(2), 360–366. https://doi.org/10.1016/j.gene.2012.07.076.

Yoon, M. Y., Lee, K.-M., Yoon, Y., Go, J., Park, Y., Cho, Y.-J., . . . Yoon, S. S. (2013). Functional screening of a metagenomic library reveals operons responsible for enhanced intestinal colonization by gut commensal microbes. *Applied and Environmental Microbiology*, *79*(12), 3829–3838. doi: 10.1128/AEM.00581-13.

Yun, J., & Ryu, S. (2005). Screening for novel enzymes from metagenome and SIGEX, as a way to improve it. *Microbial Cell Factories*, *4*(1), 8. doi: 10.1186/1475-2859-4-8.

Zhang, J., & Lynd, L. R. (2010). Ethanol production from paper sludge by simultaneous saccharification and co-fermentation using recombinant xylose-fermenting microorganisms. *Biotechnology and Bioengineering*, *107*(2), 235–244. https://doi.org/10.1002/bit.22811.

Zhang, L., Fan, Y., Zheng, H., Du, F., Zhang, K.-q., Huang, X., . . . Niu, Q. (2013). Isolation and characterization of a novel endoglucanase from a Bursaphelenchus xylophilus metagenomic library. *PLoS One*, *8*(12), e82437. doi: 10.1371/journal.pone.0082437.

Zhao, S., Wang, J., Bu, D., Liu, K., Zhu, Y., Dong, Z., & Yu, Z. (2010). Novel glycoside hydrolases identified by screening a chinese holstein dairy cow rumen-derived metagenome library. *Applied and Environmental Microbiology*, *76*(19), 6701–6705. doi: 10.1128/AEM.00361-10.

Zhou, J., Bao, L., Chang, L., Liu, Z., You, C., & Lu, H. (2012). Beta-xylosidase activity of a GH3 glucosidase/xylosidase from yak rumen metagenome promotes the enzymatic degradation of hemicellulosic xylans. *Letters in Applied Microbiology*, *54*(2), 79–87.

5 Bioprospecting of Lignin Valorization by Microbes and Lignolytic Enzymes for the Production of Value-Added Chemicals

Anamika Sharma[1], *Rameshwar Tiwari*[2], *and Lata Nain*[3]

[1]Department of Postharvest and Food Sciences, Agricultural Research Organization, Volcani Center, Rishon LeZion, Israel

[2]School of Energy and Chemical Engineering, Ulsan National Institute of Science and Technology (UNIST), 50 UNIST-gil, Ulsan 44919, Republic of Korea

[3]ICAR- Indian Agricultural Research Institute, New Delhi, India

CONTENTS

DOI: 10.1201/9781003191247-5

5.1 INTRODUCTION

In response to limited nonrenewable energy sources and continuously increasing industrial demands, focus has been positioned to identify renewable and sustainable feedstocks to replace current petrochemicals (Carlsson et al., 2011). The US Department of Agriculture and the US Department of Energy are aiming to increase bio-based value-added chemicals and materials from 5% in 2005 to 25% in 2030 (Perlack et al., 2011). In this point of view, lignocellulosic biomass is known as the most abundant renewable carbon source (compared with CO_2) and highly functionalized composite feedstock to produce various renewable materials, biofuel molecules, and other value-added chemicals (Jing et al., 2019). The high abundance of biomass on Earth, with an estimated 181.5 billion tons produced annually, makes this feedstock suitable to reduce dependence on petrochemicals (Paul & Dutta, 2018).

Lignocellulosic biomass consists of polysaccharide fraction including cellulose and hemicelluloses and heteropolymer lignin, protein, and ash content. The degree and type of cross-linkage between these cell wall components can vary from the range of diverse biomass (Hatfield, 1993). The polysaccharide portion of the biomass, including cellulose and hemicellulose, is composed of the highest abundant biopolymer on Earth. In lignocellulosic biomass, mostly the highest content is cellulose (30–50%), followed by hemicellulose content (20–35%). Lignin comprises the remaining 15–30% of the biomass and is a complex phenolic polymer structure that is covalently linked with polysaccharides, such as hemicellulose in lignocellulosic biomass (Questell-Santiago et al., 2020). The aromatic heteropolymeric structure of lignin provides rigidity and strength to the plant wall and enables nutrient transport within the plant tissue (Mnich et al., 2020). The polysaccharide consists of a fixed form of monomeric sugars that can be utilized or converted into value-added products. The complete conversion of biomass to "turn waste into a resource" has constraints, like lower productivity and higher cost compared to the petrochemical-based production level (Kumar & Verma, 2020). Hence, solutions for the depolymerization and conversion of each biomass component are essential to achieve sustainable productivity of desired product(s). The conversion of polysaccharides is potentially restricted by the complex biomass structure as well as the recalcitrance of lignin, which acts as a "cement" that packs up the gaps between the polysaccharide fibers in the plant cell wall (Ponnusamy et al., 2019; Yoo et al., 2020). This heterogeneity and complexity of lignin causes natural resistance of the plant cell wall to biological conversion through the microbial attack.

In nature, microbes can degrade a biomass by a set of oxidative and nonoxidative hydrolytic enzymes responsible for complete biomass turnover (Tiwari et al., 2018). The lignin-degradation potential of microbes is limited to converting into its monomeric form (Davis & Moon, 2020). Few microbes can utilize lignin via the "upper pathway" into central intermediates, which can be further converted via the "lower metabolic pathway" into central carbon metabolism as a process called "biological funneling" (Beckham et al., 2016). The complete microbial funneling of lignin to the central metabolic intermediate is limited in most lignin-degrading microbes as well. However, potential microbial sources can be explored to develop lignin valorization bioprocesses with or without coupled chemical or physical methods (Ragauskas et al., 2014). Lignin is composed of various aromatic compounds, and each microbe can utilize those compounds under their metabolic limitations. Mostly, lignin and its monomers, like p-coumaric, coniferyl, and sinapyl alcohols, show a toxic effect on microbial cell growth caused by cell membrane disruption, nucleic acid binding, and other cellular dysfunctions (Lou et al., 2012; Shin et al., 2019). Hence, the presence of lignin and lignin-derived aromatic compounds harms the functionality and efficiency of microbial cell factories in biorefineries. On the other hand, a large amount of lignin has been formed and estimated to be 5–36 × 10^8 tons annually, and 30% of organic carbon in the biosphere, still considered as low value or waste substrate (Abdelaziz et al., 2016). Industrial thermochemical processes have been developed to convert lignin into chemicals, fuel, and electricity, but mostly these processes are energy-intensive and generate toxic and non-eco-friendly by-products (Paone et al., 2020). Therefore, the microbial degradation of lignin and its related aromatic compounds describes research opportunities and challenges to fulfil the aim of lignin removal as well as the complete conversion of lignocellulosic biomass for the development of value-added products (Weng et al., 2021). The scope of this chapter will focus on microbial degradation of lignin or lignin-derived aromatic compounds and their impact on biomass-based biorefineries.

5.2 LIGNIN STRUCTURE AND POLYMERIZATION

The heteropolymer lignin is composed of three main phenylpropanoid monolignols: *p*-coumaryl alcohol, coniferyl alcohol, and sinapyl alcohol. These monolignols are characterized by a phenyl ring decorated with a prop-2-en-1-ol side chain on the C1 atom and a hydroxyl group on the C4 atom. Hence, with the aromatic ring having unmethoxylated, or having one or two methoxyl groups, the three alcohols are called *p*-hydroxyphenyl (H), guaiacyl (G), and syringyl (S), respectively. However, lignin also acquires other monomers, like hydroxycinnamic acids and aldehydes, coniferyl and sinapyl acetates, and coumarates (Grabber et al., 2010; Ralph, 2010; Ralph et al., 2004). The various combinations in both monomer content and chemical bonds are responsible for the heterogeneity in the chemical structure of each isolated lignin from different sources (Watkins et al., 2015). The ratios of each monomer vary based on plant species. For example, grasses mostly have all three monomers, while softwood lignin contains mostly G units and minor amounts of H units, and hardwood lignin contains both G and S units (Dutta et al., 2018).

Lignin polymerization starts with the enzyme-mediated dehydrogenation of lignin monomers and generation of resonance-stabilized free radicals (Tobimatsu & Schuetz, 2019). Afterwards, the reaction is no longer enzymatic, but an arbitrary radical polymerization course begins at the reactive sites. Mostly, lignin linkages are generated with oligomer-oligomer or oligomer-monomer couplings, while monomer-monomer coupling reactions are quite rare (Heitner et al., 2019). Phenoxy oxygen is in highly reactive positions, and the C-β that readily couples into aryl-ether linkages form the most common β-O-4 ether linkage in lignin (van Parijs et al., 2010).

This linkage accounts for 50% and 60% of polymerization reactions in softwood and hardwood, respectively (Adler, 1977). The predominance of β-O-4 ether linkage in lignin makes this a potential target for lignin degradation (Paone et al., 2020). Any enzymatic reaction or chemical process that can break this linkage can be useful for effective lignin degradation in a broad spectrum of biomass. The lignin monomeric coupling can be designated based on ether bonds or C-C bonds. However, β-5, β1, β-β, 5–5, and 4-O-5 linkages are substantially difficult to break and make the complete degradation of lignin difficult in conventional ways (Iram et al., 2021). Various chemical and spectrometric techniques have been explored to analyze the structure of lignin, but interpretation of complete lignin structure is still difficult because of the unrestrained pattern of lignin polymerization.

5.3 CHARACTERIZATION OF LIGNIN CONTENT IN LIGNOCELLULOSIC BIOMASS

The compositional profiling of lignocellulosic biomass is essential to deal with its complex and recalcitrance property, which further leads to the development of biomass processing technologies. For more than 150 years, various protocols have been developed for these estimations from time-consuming conventional digestion-based techniques to evolved, statistical, spectrometric, and chemometric model-based early-composition detection (Krasznai et al., 2018). To study the component analysis of lignocellulosic biomass, and particularly for lignin content, the ideal way is to extract or isolate pure original lignin without any structural or chemical modifications and recover after removal of extractives, cellulose, hemicellulose, and nonlignin constituents (Watkins et al., 2015). Despite various efforts that have been done for lignin isolation, none of the methods is universally accepted as fulfilling all these requirements.

The structure and composition of lignin from biomass mostly have been interpreted based on different chemical digestive methods, including pyrolysis, thioacidolysis, nitrobenzene oxidation, or chemical derivatization (Lourenço & Pereira, 2018). Initially, Henneberg and Stohmann proposed a crude fiber-based method, also called Weende analysis (Henneberg & Stohmann, 1859). However, the complexity of fractionation is a major limitation of this analysis, especially to attain the variability of lignin, and the Weende system has not been used widely in a biomass component analysis. In 1893, a method was published by Klason to estimate black liquor from the kraft pulping process. After forty years, the method was further developed and used for the quantification of lignin from biomass (Marton & Adler, 1996). The mass of the biomass residue leftover after the digestion of the biomass with 64–72% of H_2SO_4 to remove the polysaccharide portion is defined as Klason lignin (Browning, 1967). Lignin estimation through this method was adopted by TAPPI standard and denoted as TAPPI Test Method T222 om-88 (TAPPI, 1999). However, it was also noted that some portion of lignin is acid-soluble and can't be omitted to calculate total lignin content. This limitation was further addressed, and quantification of insoluble lignin was mentioned in TAPPI Test Method T222 om-06 (Tappi, 2002), and recovered acid-soluble lignin was calculated by TAPPI Test Method T250 (UM250, 1985). Generally, Klason lignin and acid-soluble lignin are collectively set up as total lignin from biomass. Further, this method was used to estimate lignin for agriculture, food, and biochemical research (Bunzel et al., 2011; Robertson et al., 2010). However, still this method has limitations, like the overestimation of lignin in samples with high ash and protein content or the presence of furfural and hydroxymethylfurfural formed from reducing sugars (Sjöström & Alén, 1998). Despite these constraints, to date, most of the published research refers to TAPPI methodology and Klason lignin for biomass compositional analysis, and hence an established reference for comparison (Krasznai et al., 2018). With time and advancement of techniques, this method was adopted and modified to achieve early and better detection with some modifications in the Klason method. Lignin was estimated gravimetrically as Klason lignin and ash-free, acid-insoluble residue by the Uppsala method (Theander & Westerlund, 1986). Further, the permanganate method was developed by Tasman and Berzins for lignin based on measuring the Kappa number of woods in pulp and paper biorefinery (Tasman & Berzins, 1957). However, limitations like generic oxidation and the presence of phenolic and other pigments or proteins are responsible for overvaluing lignin percentage by this method. Another method, named as detergent fiber method, was developed for composition analysis, including lignin from lignocellulosic biomass in feed industry (Soest, 1963a; Soest, 1963b). It includes the determination of neutral and acid detergent fiber analyses (NDF and ADF, respectively) (Goering & Van Soest, 1970). The limitation of this method is the underestimation of total lignin by acid detergent lignin (ADL) compared with the Klason lignin method (Dien et al., 2006). Despite these wet chemical methods, acetyl bromide, or derivatization followed by reductive cleavage (DFRC) method, was developed to estimate and analyze lignin and its structure (Fukushima & Hatfield, 2004). In this method, derivatization, or cell wall solubilization, is achieved with acetyl bromide in acetic acid. The reductive cleavage of resulting β-bromo ethers utilizes zinc in an acidic medium following acetylation using acetic anhydride and pyridine. The generated acetylated lignin monomers are further quantified and analyzed through spectroscopic and chromatographic techniques. However, again, overestimation of lignin is a major bottleneck of this method due to minimal xylan digestibility (Hatfield et al., 1999). Few modifications have been done in the hydrolysis step of the DFRC method to remove the xylan interference (Del Río et al., 2007). In recent years, laboratory analytical procedures (LAPs) have been established for lignocellulosic biomass compositional analysis by the National Renewable Energy Laboratory (NREL) (Sluiter et al., 2008). These LAPs were designed based on similar previous techniques with some alteration, mainly in Uppsala method and ASTM E1758–01 (ASTM, 2003). The complete compositional analysis of the lignocellulosic biomass mentioned as NREL LAP TP-510–42618 has already been referred by various research groups (Chinga-Carrasco et al., 2019; Chinga-Carrasco et al., 2018; Johnston et al., 2020; Kamireddy et al., 2013). Here, acid-soluble and insoluble lignin were analyzed by spectroscopy and gravimetric techniques, respectively. The advantage of this method is the use of advanced methods to detect extractives and protein content, which can the lignin estimation indirectly. There is a large database available in the public domain for the composition of various plant residues using the NREL LAPs, which helps compare the results and give confidence to building globally standardized procedures used for various biorefineries. Other than these digestion-based methods, nondestructive methods, such as Fourier transform infrared (FTIR) (Gelbrich et al., 2012), near-infrared spectroscopy (NIRS) (Ramadevi et al., 2010), and nuclear magnetic resonance (NMR) (Fu et al., 2015), have also been explored to characterize the lignin content in biomass.

5.4 LIGNIN VALORIZATION BY MICROBES

5.4.1 Fungal Valorization of Lignin

Fungi are the most efficient and evolved for lignin depolymerization in nature by secreting various lignin-degrading enzymes (Li et al., 2019a). Fungi can degrade and metabolize lignin mainly through extracellular oxidases, like lignin, manganese, versatile, and dye-decolorizing peroxidases (LiP, MnP, VP, and DyP) and laccase (Mäkelä et al., 2016). These oxidases create free radicals which cleave chemical linkages between the heteropolymer structure of lignin and generate various aromatic compounds which are further assimilated by microbes through metabolic pathways (Abdel-Hamid et al., 2013). Other accessory oxidases secreted from fungal sources also lead to H_2O_2 generation and are further used by peroxidases and involved in lignin depolymerization (Pollegioni et al., 2015). In fungal sources, the mechanism of ligninolytic enzyme action can be explained by laccase and peroxidases, which oxidize small intermediates and catalyze the generation of electrons from the lignin aromatic structure. Fungal mycelia also helps in lignin depolymerization through a nonenzymatic Fenton reaction (Kerem et al., 1999; Sista Kameshwar & Qin, 2020). As it is difficult for lignolytic enzymes to penetrate bulky cell walls, the generation of free radicals and mycelial structure of fungus favours to pervading through the cell wall which sequentially helps in effective action of the lignolytic enzyme system (Iram et al., 2021; Tanaka et al., 1999). After the action of the fungal ligninolytic enzyme, some lignin monomers enter fungal hyphae and are further consumed through metabolic pathways (Martínez et al., 2005). Moreover, the fungal mycelial system helps create pores and increase surface area to penetrate as well as increase enzymatic attacks (Blanchette, 1984; Gregorio et al., 2006). It was also observed that different inhibitors were also found to be less effective against fungal metabolism.

Lignolytic fungi mostly include three types, namely, white-rot, brown-rot, and soft-rot fungi. White-rot fungi have a complete set of enzymes to degrade lignin into CO_2 and H_2O (Del Cerro et al., 2021). White-rot fungi mostly include basidiomycetes and a few species of ascomycetes. Few white-rot fungal strains, like *Phanerochaete chrysosporium*, *Pleurotus ostreatus*, *Sporotrichum* sp., *Trametes versicolor*, and *T. hirsuta*, have the capacity to degrade lignin, cellulose, and hemicellulose from biomass (Kumar & Chandra, 2020). *P. chrysosporium* is a well-studied and efficient white-rot fungus for delignification (Gold & Alic, 1993). Therefore, this capacity of *P. chrysosporium* was utilized by various biotransformation processes, like biological pretreatment of biomass, composting, and lignolytic enzyme production, as well as bioremediation of environmental pollutants (Chen et al., 2018; Seo et al., 2018; Sharma et al., 2016). Mostly, *P. chrysosporium* degrades lignin through a non-specific oxidative approach by secreting highly effective MnP and LiP enzymes. Moreover, the involvement of a peroxidase-independent oxidoreductase, the cytochrome P450 monooxygenase system, has also been identified and displays a unique catalytically diverse oxidoreductase enzyme mechanism which supports lignin degradation (Sakai et al., 2018). Pregenomic and postgenomic identification of *P. chrysosporium*, combined with transcriptional and proteomics surveys–based functional characterization, also elaborated its delignification and biodegradable potential (Arntzen et al., 2020; Bak, 2015; Gold & Alic, 1993).

Brown-rot fungi represent 7% of wood-rotting basidiomycetes and are mostly grown on softwoods (Mäkelä et al., 2020). Brown-rot fungi–governed decay can be identified through wood shrinks and shows a brown discoloration due to oxidized lignin and cracks into roughly cubical pieces (Chandra & Madakka, 2019). The fungal group can hydrolyze cellulose and hemicellulose efficiently but displays partial lignin degradation capacity. Lignin degradation by brown-rot fungi is initiated by the generation of hydroxyl radicals via Fenton oxidation chemistry. The main characteristic of this group is to reduce Fe^{3+} of Fe-oxalate complex to Fe^{2+} by extracellular hydroquinones, which further react with H_2O_2 and produce hydroxyl radical ($Fe^{2+} + H_2O_2 \rightarrow Fe^{3+} + -OH +. OH$). The redox cycling further converts oxidized quinone into hydroquinone (Mäkelä et al., 2020). Few brown-rot fungi, like *Gloeophyllum trabeum*, *Laetiporus portentosus*, and *Fomitopsis lilacinogilva*, have already been explored for lignin depolymerization (Abdel-Hamid et al., 2013; Arimoto et al., 2015).

The other group of lignolytic fungi is named soft-rot fungi, which are mostly known to act on the syringyl units of lignin (Zabel & Morrell, 2012). This fungal group is mostly found in a dry environment, where the growth of other ligninolytic fungal groups, including white- and brown-rot fungi, is inhibited due to high moisture and temperature, low aeration, and the presence of preservatives (Brischke & Alfredsen, 2020). Soft-rot fungi are known for modifying lignin through demethoxylation rather than mineralizing the lignin, as reports on the presence of lignolytic enzymes are very limited (Filley et al., 2002). The most studied soft-rot fungi belong to the *Aspergillus* and *Trichoderma* genera, which have strong hydrolytic enzyme systems (Ray & Behera, 2017).

5.4.2 Bacterial Valorization of Lignin

Instead of fungi, bacteria also possess the ability to degrade lignin with high efficiency. However, fungi have higher lignin-degradation capacity due to the lignolytic enzyme system compared to bacteria, but the major problem with fungal ligninolytic enzymes is the expression in active form to attain large-scale delignification (de Gonzalo et al., 2016). Few bacterial strains are known to produce efficient lignolytic enzymes which not only depolymerize lignin but also have metabolic pathways to utilize aromatic compounds generated from lignin. Lignolytic bacterial strains mostly belong to *Actinobacteria*, *Proteobacteria*, and *Firmicutes* (Chauhan, 2020). These bacteria can utilize lignin in aerobic conditions by secreting diverse oxidative enzymes. *Actinobacteria*, including strains of *Streptomyces* and *Rhodococcus* genera, are known for having lignin-degradation capacity. *Streptomyces viridosporus* T7A secretes lignolytic oxidases for lignin degradation and causes reduced guaiacyl units of lignin. *S. viridosporus* T7A and *S. setonii* 75Vi2 were found to remove 30–45% of lignin from different biomasses within twelve weeks (Antai & Crawford, 1981). These Streptomyces strains generate lignin-derived, acid-precipitable polymeric lignin (APPL) when grown on lignocellulosic biomass (Zeng et al., 2013). *Streptomyces griseorubens* ssr38 was also shown in the depolymerization of lignin up to a high extent, and around 25% of depolymerized lignin was recovered as value-added APPL (Saritha et al., 2013). *Rhodococcus* is another robust bacterial genus that shows tolerance as well as hydrolytic capacity for toxic aromatic metabolites. *R. jostii* RHA1 is a most promising candidate for lignin degradation in this genus, which was initially known for polychlorinated biphenyl (PCB) degradation (Seto et al., 1995). However, *R. jostii* RHA1 was further explored for not only utilizing kraft lignin and wheat straw but also producing value-added chemicals, like aromatic dicarboxylic acids and vanillin (Mycroft et al., 2015; Sainsbury et al., 2013). The DyP-type peroxidase DypB from *R. jostii* RHA1 can oxidize both polymeric lignin and a lignin model compound and appears to have both MnII and lignin oxidation sites (Ahmad et al., 2011). PCB-degrading *R. erythropolis* strain also exhibited lignin metabolizing activity on nitrated lignin of wheat straw (Chung et al., 1994).

In *Proteobacteria*, members of *Pseudomonas*, *Pandoraea*, and *Comamonas* genus are also known to degrade lignin. The most promising member of this group is *Pseudomonas putida* KT2440, derived from *P. putida* mt-2, which can degrade 30% of lignin derived from alkali-pretreated liquor (Salvachúa et al., 2015). Recently, the genome annotation of *P. putida* KT2440 has displayed highly efficient and diverse degradative pathways, including lignin-derived aromatic compounds, and therefore can be a good candidate for the development of metabolically engineered strain for efficient lignin degradation (Belda et al., 2016). *P. putida* NX-1, a novel isolate, also utilized kraft lignin by secreting lignolytic enzymes (Xu et al., 2018). *Pandoraea* sp. B-6 was another strain that can degrade kraft lignin and further needs to be explored for improved lignin-degradation efficiency (Liu et al., 2018).

Many strains from the Bacillus genus of *Firmicutes* can decompose lignin polymer (Tian et al., 2014). Genome and proteomic analysis revealed three lignin degradation pathways in *Bacillus*, including a gentisate, benzoic acid, and the β-ketoadipate pathway (Weng et al., 2021). *Bacillus ligniniphilus* L1 was found to utilize alkaline lignin as a single carbon or energy source, and fifteen single-phenol ring aromatic compounds were identified after its degradation (Zhu et al., 2017). *Bacillus amyloliquefaciens* SL-7 was isolated based on its property to utilize tobacco straw lignin as the sole carbon source (Mei et al., 2020). This strain can secrete lignolytic enzymes and depolymerize the lignin by changing its morphology. Bacterial strain belonging to the *Bacillus*-derived genera *Paenibacillus glucanolyticus* was isolated from pulping waste black liquor enriched with lignin content. The *P. glucanolyticus* SLM1 strain can hydrolyze cellulose and hemicellulose for growth, as well as degrade lignin under aerobic and anaerobic conditions (Mathews et al., 2016).

Lignin degradation was also proposed by *Citrobacter* and *Klebsiella* genus. *Citrobacter freundii* and another *Citrobacter* sp. were cocultured and displayed around 62% decolorization of kraft lignin (Chandra & Bharagava, 2013). Absorbance at 280 nm (A280) was reduced by 23.8, 28.5, and 19.4% during the seven-day incubation with *K. pneumoniae* NX-1, *P. putida* NX-1, and *O. tritici* NX-1, respectively (Xu et al., 2018). Despite these strains, *Novosphingobium* (Chen et al., 2012), *Cupriavidus basilensis* (Shi et al., 2017), *Nocardia* (Malarczyk et al., 1994), and *Pandoraea* (Liu et al., 2018) species have also shown lignin degradation and are used for various applications.

5.5 LIGNOLYTIC ENZYMES

The complete degradation of lignin is mainly involved in two steps, namely, depolymerization and mineralization (Li et al., 2019a). During lignin depolymerization, a collection of ligninolytic enzymes oxidatively degrades lignin polymer to generate smaller aromatic compounds. Lignin mineralization is a consequent step where these aromatic compounds are metabolized through the advanced catabolic potential of the microbial cell (Bugg et al., 2011). The knowledge of enzymology and metabolic pathways responsible for complete lignin degradation has evolved over time. The enzyme functions, properties, catalytic behavior, and structural features are essential to explore the biological mechanism of lignin degradation.

5.5.1 Enzymes for Lignin Depolymerization

5.5.1.1 Lignolytic Peroxidases

Heme-containing peroxidases, including Lip, MnP, and VP, mostly known as fungal class II peroxidases, are the major enzymes responsible for lignin depolymerization using H_2O_2 as a cosubstrate (Kumar & Chandra, 2020). This group of enzymes initiates the attack on lignin content of plant biomass, and further hydrolysis of polysaccharide is facilitated. Therefore, this group was classified into auxiliary activities family 2 (AA2) in the Carbohydrate-Active Enzyme (CAZy) database (Levasseur et al., 2013).

Lip has a helicoidal structure with calcium ions for structural stability and a heme group that is connected to protein by hydrogen bridges. The heme iron is associated with His amino acid and the high redox potential of the enzyme. This enzyme redox potential increases with the distance between each heme group and His amino acid and establishes an electronic deficiency in the porphyrin ring of the iron (Li et al., 2019a). MnP enzyme–active site contains peroxisomal His, H-bonded to an Asp, and peroxidase-binding pocket containing catalytic His and Arg (Martin, 2002).

These peroxidases degrade not only lignin-containing compounds with β-O-4-linked but also several nonphenolic lignin derivatives to homologous ketones or aldehydes and hydroxylation of benzylic methylene groups of aromatic ring cleavages (Shin et al., 2005). The peroxidase enzyme action involves several steps and is initiated by forming two highly oxidized intermediates, compound I and compound II. The reaction is initiated with the interaction between H_2O_2 and the Fe^{3+} of the heme group, producing the short-lived intermediate transient ferric hydroperoxide (compound 0; Fe^{4+}-O-O-H). Further, compound 0 is converted into the compound I ($[Fe^{4+}O]^{+}$) by the cleavage of the O-O bond facilitated by protonation of the distal oxygen. Further, compound I is reduced into compound II, where iron is still in the Fe^{4+} state, but porphyrin radical has been extinguished. Finally, compound III consists of a complex in which an oxygen molecule is bound to the ferrous iron in peroxidase, and this compound is formed when there is an excess of hydrogen peroxide (Kumar & Chandra, 2020). All the peroxidases have a mechanism where H_2O_2 and small reducing substrates activate heme and turn out into a substrate for further action (Figure 5.1).

In the same way, MnP and VP can oxidize Mn^{2+}, which is ample in woody biomass, into Mn^{3+} state. The complete mechanism of peroxidases has a long route of electron transfer from lignin or nonlignin molecule to the heme group. Hence, this

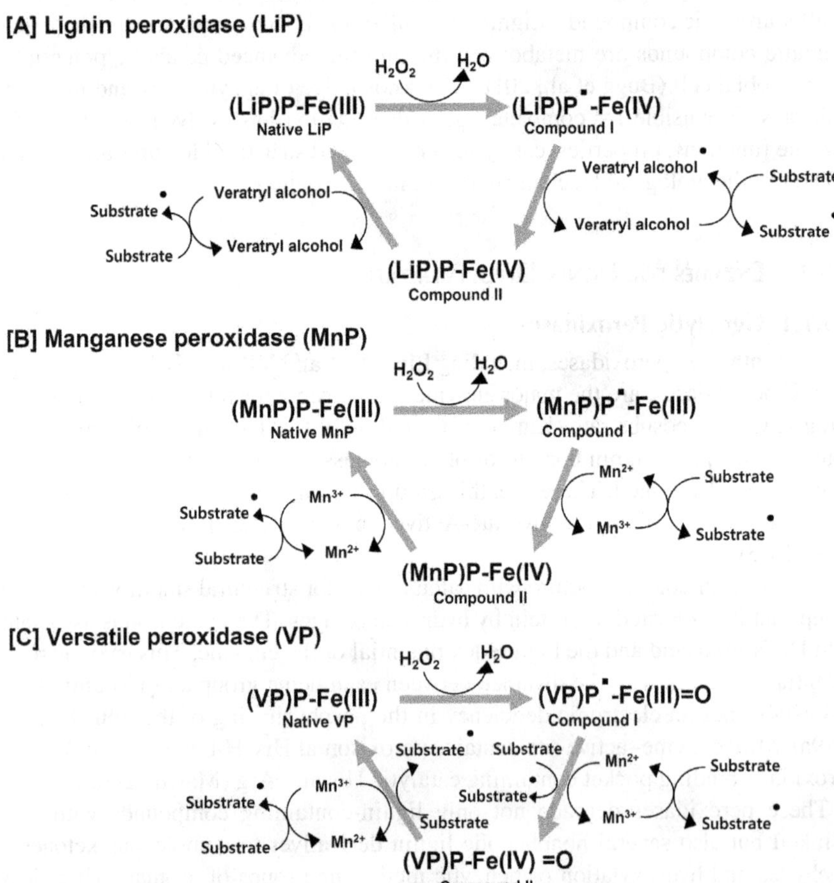

FIGURE 5.1 Catalytic mechanism of peroxidases-mediated lignin degradation. (A) Lignin Peroxidase (LiP); (B) manganese peroxidase (MnP); (C) versatile peroxidase (VP).

mechanism requires enzymes with biphasic mechanisms and high binding efficiency for the substrate, which are the main characteristics of ligninolytic peroxidases.

5.5.1.2 Laccases

Laccases belong to the largest group of polyphenol oxidases that appear in the rhizosphere and can oxidize polyphenols, methoxy-substituted phenols, diamines, and other compounds (Janusz et al., 2020). Laccase belongs to the family of multicopper oxidase that uses O_2 as the terminal electron acceptor for the oxidation of aromatic and nonaromatic compounds of lignin polymer. Generally, laccases are multicopper oxidases, distributed mainly in fungus but also in both bacteria and archaea as well (Patel et al., 2019). Laccases share mechanical similarities with other enzymes containing different atoms, like Cu, Mn, Zn, or Fe atoms, in their combinations. Laccases are grouped in the CAZy family AA1 with their involvement in the delignification of plant biomass (Levasseur et al., 2013). The structure of laccases mostly contains three domains with a β-barrel topology, which is stabilized by two disulfide bridges bond, but in some cases with three disulfide bridges bond (Weng et al., 2021). Laccases retain the active site through three types of copper atoms based on their coordination with each other, known as a mononuclear (T1Cu) and a trinuclear cluster (T2Cu and T3Cu). The mononuclear (T1Cu) featured blue color at the reduced resting state, also called the substrate-reducing site, shows a diverse triangular planar coordination, while T2Cu displays coordination with two His residues and a water molecule. T3Cu, the two Cu molecules with six His, is present in two groups of three active sites to share oxygen molecules that are reduced into a water molecule. This oxidizing characteristic of laccase over a broad range of substrates using oxygen and split-out water molecules as the only by-product makes it a 'green catalyst' (Su et al., 2018). However, low redox potential is the major limitation for laccase to act effectively on a substrate like lignocellulosic biomasses due to their complex size and low accessibility. This problem can be eliminated by the involvement of a redox mediator which can continuously control the cyclic redox process. For example, hydroxycinnamic acids, which were released from lignin by the enzymatic action of feruloyl esterase, were used as a natural mediator of laccase (Nghi et al., 2012). Laccase makes the mediator oxidized, and its small size facilitates penetration into the pores of the lignocellulosic biomass (Paz et al., 2019). Due to the limitation of this low redox potential and the use of mediators, the oxidation mechanism of laccases is classified into two types: direct and indirect substrate oxidation (Ozer et al., 2020). In the first type of direct oxidation, the substrate gets oxidized with direct contact of laccase's copper cluster (Matera et al., 2008). In another case, the low redox potential of laccase cannot activate substrate oxidation and requires a mediator to oxidize in the first step. This is called an indirect oxidation mechanism, where a further mediator oxidizes the substrate (Agrawal et al., 2018). In an indirect mechanism, the mediator must possess some qualities, like high specificity towards laccase in both oxidizing and reducing forms without inhibiting the enzyme kinetics, and its conversion must be cyclic (Figure 5.2).

Laccases perform an important role in delignification by oxidizing β-O-4 lignin dimers in the presence of a mediator. This oxidation involves Cα-oxidation, Cα-Cβ cleavage, β-ether cleavage, and aromatic ring cleavage of lignin or lignin-derived

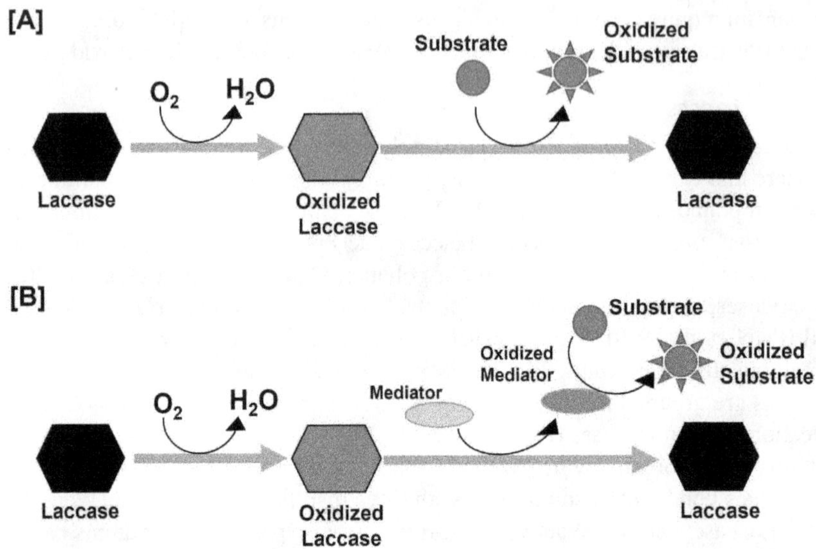

FIGURE 5.2 Catalytic mechanism of laccase-mediated lignin degradation. The substrate is oxidized by direct (A) and indirect (B) oxidation in the presence of a mediator.

chemicals like vanillyl glycol, 4,6-di (t-butyl) guaiacol, and syringaldehyde to generate phenoxy radicals (Wong, 2009). Therefore, laccases have been extensively used not only for the depolymerization of lignin from plant biomass but also for the detoxification of lignin residues and achieve better efficiency of following biorefinery steps, like enzymatic saccharification and the fermentation process (Heap et al., 2014; Moreno et al., 2016).

5.5.2 Enzymes for Lignin Mineralization

Nonradical ligninolytic enzymes, like Cα-dehydrogenase and β-etherases, are the good alternatives of lignin cleavage and valorization (Zakzeski et al., 2010). These enzymes also have the capacity to cleave the most abundant β-O-4 aryl ether linkages between phenylpropane units. Cα-dehydrogenase (LigD) and β-etherases (LigF) from *Pseudomonas paucimobilis* SYK-6 are two model enzymes which were illustrated for their catalytic activity for lignin mineralization (Picart et al., 2015). In mechanical reactions, Cα-dehydrogenase initially oxidizes the Cα of guaiacylglycerol-β-guaiacyl ether by consuming NAD^+, and subsequently, β-etherases mediated the nucleophilic attack of glutathione to the common β-O-4-aryl ether motif in lignin (Husarcíková et al., 2018). Although the enzyme activities are quite low towards lignin mineralization in high amounts.

Dioxygenases, nonheme, mononuclear, iron-containing enzyme families, are another class of intracellular enzymes that can cleave aromatic ring structure and facilitate its mineralization through central metabolic pathways (Eltis & Bolin, 1996). Protecatechuate and catechol dioxygenases are highly widespread natural enzymes that catalyze intradiol cleavage of the aromatic ring (Fritsche & Hofrichter, 2005). Dioxygenases are further classified as extradiol and intradiol based on mechanism and iron oxidation state (Vaillancourt et al., 2006). In aerobic organisms, aromatic molecules, such as benzoate and *p*-coumarate, are degraded aerobically through intermediates catechol and protocatechuate. Further, O_2-dependent dioxygenase enzymes cleave the catechol or protocatechuate that exhibit either ortho (intradiol) or meta (extradiol) cleavage between or adjacent to the hydroxyl groups, respectively (Knoot et al., 2015). The cleavage products are degraded through distinct pathways that yield different molecules for entry into central metabolism as different combinations of acetyl-CoA, succinate, and pyruvate. This concept of biological funneling-based lignin valorization has a high impact on biomass refineries (Figure 5.3). The screening of potential candidates of dioxygenases combined with advanced metabolic engineering techniques can promote efficient lignin mineralization.

5.6 LIGNIN VALORIZATION TO PRODUCE VALUE-ADDED CHEMICALS

For many years, lignocellulosic biomass-based biorefineries have been focusing on the conversion of polysaccharide components, including cellulose and hemicellulose, into value-added products while downgrading the lignin as a waste product for biorefineries (Kumar & Verma, 2020). Nevertheless, the native phenolic components derived from lignin are now considered as a substrate for sustainable production of green aromatic value-added products (Cao et al., 2019). With technical advancement, biorefineries are changing their objective with "lignin-first" or lignin-based biorefinery by using a reductive catalytic fractionation approach to produce value-added products from depolymerized lignin (Paone et al., 2020). In recent years, chemical-based lignin biorefinery has been developed, where chemical catalytic reactions depolymerize the extracted lignin from biomass and stabilize the obtained phenolic units, which are further transformed into value-added chemicals, fuels, polymers, and pharmaceutical ingredients (Figure 5.3). However, biochemical processes involved with microbes or lignolytic enzymes have their own benefits, like sustainability, mild biological process, eco-friendliness, and selectivity of products (Shin et al., 2019).

The Laccase-mediator system has been widely used in the pulp and paper industry (Virk et al., 2012). Delignification of wood pulps and bleaching of kraft pulps are essential and incorporate value and quality in the pulp and paper industry. Moreover, laccase is also involved in food industries with decolorizing food and beverages by eliminating phenolic compounds (Singh & Gupta, 2020). LiP and MnP enzymes are also being used in pulp and paper industries for delignification (Rajwar et al., 2016). These enzymes are also involved in bioremediation by degrading xenobiotic compounds and polyaromatic hydrocarbons (PAH) (Sharma et al., 2018). Laccases are also used for detoxification during saccharification and fermentation of pretreated biomass, as cellulases and hemicellulases have high affinity towards lignin molecules (Heap et al., 2014). Moreover, the growth of fermentation microorganisms like yeast or bacterial strains is mostly inhibited due to lignin-derived aromatic compounds (Klinke et al., 2004). The leftover lignin or lignin-derived content after pretreatment of biomass can reduce the efficiency of hydrolytic enzyme action and metabolic capacity of fermentation microbes (Chandel et al., 2013). Therefore, laccase can remove these lignin monomers and its negative impact on hydrolytic enzymes or inhibition effect on fermentation microorganisms.

Despite lignolytic enzymes, whole microbial cell factories were also developed to produce value-added products from lignin (Becker & Wittmann, 2019). *Rhodococcus*, having an aromatic ring opening β-ketoadipate pathway, can convert lignin-derived aromatic compounds into fatty acids (Li et al., 2019b). The β-ketoadipate pathway transforms lignin-derived aromatic compounds into a precursor for fatty acid synthesis, i.e., acetyl-CoA. Oleaginous fungi *Trichosporon cutaneum* was also explored to produce lipid from lignin model compounds like 4-Hydroxybenzaldehyde. Polyhydroxyalkanoates (PHAs), often referred to as "carbonosomes," are the bio-based polymers of hydroxyalkanoic acid and can replace conventional plastic due to its biodegradable nature (Hu et al., 2018). Microbes produce PHAs as energy and carbon reserves under harsh environmental conditions. Few aerobic microbes can produce PHAs from lignin-derived aromatic compounds in five different ways (Li et al., 2019a). *p*-coumaric acid and ferulic acid derived from lignin have been shown to be converted into protocatechuic acid, followed by acetyl-CoA, and then converted into PHAs. *C. basilensis* B-8 (Si et al., 2018), *Burkholderia* strain ISTR5 (Morya et al., 2021), *Pseudomonas* sp. (Wang et al., 2018), and *Pandoraea* sp. ISTKB (Kumar et al., 2017) have already been explored to produce PHAs from lignin or lignin-derived chemicals.

Vanillin is a high-value market compound that is used predominantly in food, perfumery, and pharmaceutical industries (Sainsbury et al., 2013). Other than a conventional substrate for vanillin production like guaiacol, it can also be produced from lignin as well. Mostly in industries, kraft lignin waste is valorized into vanillin through the oxidation process (Banerjee & Chattopadhyay, 2019). However, various lignin mineralizing microbes, like *Bacillus subtilis* and *Pediococcus acidilactici*, and microbial consortium have been explored to produce lignin from lignocellulosic biomass, particularly from lignin. The low production yield is the major problem that was further taken care of by genetic manipulations. Ferulic acid is metabolized by engineered *Escherichia coli* for vanillin production with a high yield of 94.3% (Yang et al., 2013).

Cis, cis-Muconic acid (MA) is another valuable platform chemical used as a precursor molecule to produce various polymers and drugs, including adipic acid and terephthalic acid (Choi et al., 2020). Bio-based MA production, especially from biomass and derived lignin, has also drawn interest by using microbes that can tolerate and metabolize aromatic compounds. The MA branch in the β-ketoadipate pathway inside *Corynebacterium glutamicum* can convert aromatic compounds to catechol and further enters the TCA cycle through acetyl-CoA and succinyl-CoA (Becker et al., 2018). Metabolically engineered *C. glutamicum* strain, which lacks catB and enhanced expression of catA, can produce 85 g/L of MA from catechol and 1.8 g/L of MA from hydrothermally depolymerized softwood lignin (Becker et al., 2018). Similar to *C. glutamicum*, the β-ketoadipate pathway was engineered in *P. putida* to produce MA from aromatic compound 4-hydroxybenzoic acid (4-HBA) and vanillin

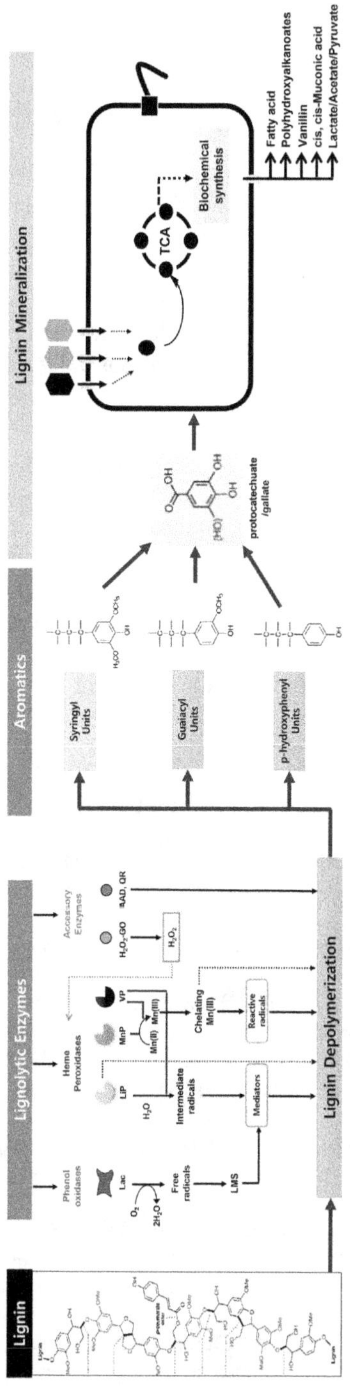

FIGURE 5.3 Schematic representation of lignin valorization, including depolymerization and mineralization, to produce value-added compounds.

(Sonoki et al., 2014; Sonoki et al., 2018). Other microbes like *Amycolatopsis* spp. (Barton et al., 2018) and *Enterobacter cloacae* (Johnson et al., 2016) are also used to produce MA from lignin or lignin-derived compounds.

5.7 CONCLUSION AND FUTURE PROSPECTS

Lignocellulosic biomass–based biorefineries have a broad perspective to produce high value-added products as well as replacement of petrochemicals. In a few years, lignin valorization has been integrated into this biorefinery model with improved productivity as well as economic viability. Though effective lignin depolymerization/mineralization through microbial systems is still a challenge that complemented high process costs, low yields, and unsustainable performance. The bioconversion of lignin from lignolytic microbes may produce various value-added products relevant to biofuel, biomedical, bioplastic, textile, and food industries. These products are mostly produced from simple aromatic compounds like phenol, cresol, guaiacol, and anisole or extracted lignin polymer. However, the efficiency and catalytic upgradation is now essential to utilize all major components, including cellulose, hemicellulose, as well as lignin in a “single-pot” manner instead of in separate or extracted components. Therefore, searching for new or metabolic upgradation of existing microbial cell factories is required to tolerate as well as depolymerize and metabolize a broad range of aromatic compounds derived from lignin. Overall, the prosperous lignin-based economy is achievable only with the cooperative attempts of fractionation, depolymerization, and upgrading steps of lignin.

REFERENCES

Abdel-Hamid, A.M., Solbiati, J.O., Cann, I.K. 2013. Insights into lignin degradation and its potential industrial applications. *Advances in Applied Microbiology*, **82**, 1–28.

Abdelaziz, O.Y., Brink, D.P., Prothmann, J., Ravi, K., Sun, M., Garcia-Hidalgo, J., Sandahl, M., Hulteberg, C.P., Turner, C., Lidén, G. 2016. Biological valorization of low molecular weight lignin. *Biotechnology Advances*, **34**(8), 1318–1346.

Adler, E. 1977. Lignin chemistry—past, present and future. *Wood Science and Technology*, **11**(3), 169–218.

Agrawal, K., Chaturvedi, V., Verma, P. 2018. Fungal laccase discovered but yet undiscovered. *Bioresources and Bioprocessing*, **5**(1), 1–12.

Ahmad, M., Roberts, J.N., Hardiman, E.M., Singh, R., Eltis, L.D., Bugg, T.D. 2011. Identification of DypB from *Rhodococcus jostii* RHA1 as a lignin peroxidase. *Biochemistry*, **50**(23), 5096–5107.

Antai, S.P., Crawford, D.L. 1981. Degradation of softwood, hardwood, and grass lignocelluloses by two *Streptomyces* strains. *Applied and Environmental Microbiology*, **42**(2), 378–380.

Arimoto, M., Yamagishi, K., Wang, J., Tanaka, K., Miyoshi, T., Kamei, I., Kondo, R., Mori, T., Kawagishi, H., Hirai, H. 2015. Molecular breeding of lignin-degrading brown-rot fungus *Gloeophyllum trabeum* by homologous expression of laccase gene. *AMB Express*, **5**(1), 1–7.

Arntzen, M.Ø., Bengtsson, O., Várnai, A., Delogu, F., Mathiesen, G., Eijsink, V.G. 2020. Quantitative comparison of the biomass-degrading enzyme repertoires of five filamentous fungi. *Scientific Reports*, **10**(1), 1–17.

ASTM, E1758-01. 2003. Standard test method for determination of carbohydrates in biomass by high performance liquid chromatography. In: *American Society for Testing and Materials,* Annual Book of ASTM Standards, Vol. 11.06, West Conshohocken, PA.

Bak, J.S. 2015. Lignocellulose depolymerization occurs via environmentally adapted metabolic cascades in the wood-rotting basidiomycete Phanerochaete chrysosporium. *Microbiologyopen*, **4**(1), 151–166.

Banerjee, G., Chattopadhyay, P. 2019. Vanillin biotechnology: The perspectives and future. *Journal of the Science of Food and Agriculture*, **99**(2), 499–506.

Barton, N., Horbal, L., Starck, S., Kohlstedt, M., Luzhetskyy, A., Wittmann, C. 2018. Enabling the valorization of guaiacol-based lignin: Integrated chemical and biochemical production of cis, cis-muconic acid using metabolically engineered *Amycolatopsis* sp ATCC 39116. *Metabolic Engineering*, **45**, 200–210.

Becker, J., Kuhl, M., Kohlstedt, M., Starck, S., Wittmann, C. 2018. Metabolic engineering of *Corynebacterium glutamicum* for the production of cis, cis-muconic acid from lignin. *Microbial Cell Factories*, **17**(1), 1–14.

Becker, J., Wittmann, C. 2019. A field of dreams: Lignin valorization into chemicals, materials, fuels, and health-care products. *Biotechnology Advances*, **37**(6), 107360.

Beckham, G.T., Johnson, C.W., Karp, E.M., Salvachúa, D., Vardon, D.R. 2016. Opportunities and challenges in biological lignin valorization. *Current Opinion in Biotechnology*, **42**, 40–53.

Belda, E., Van Heck, R.G., José Lopez-Sanchez, M., Cruveiller, S., Barbe, V., Fraser, C., Klenk, H.P., Petersen, J., Morgat, A., Nikel, P.I. 2016. The revisited genome of *Pseudomonas putida* KT2440 enlightens its value as a robust metabolic chassis. *Environmental Microbiology*, **18**(10), 3403–3424.

Blanchette, R.A. 1984. Screening wood decayed by white rot fungi for preferential lignin degradation. *Applied and Environmental Microbiology*, **48**(3), 647–653.

Brischke, C., Alfredsen, G. 2020. Wood-water relationships and their role for wood susceptibility to fungal decay. *Applied Microbiology and Biotechnology*, **104**(9), 3781–3795.

Browning, B.L. 1967. *Methods of Wood Chemistry*. Volumes I & II. Wiley.

Bugg, T.D., Ahmad, M., Hardiman, E.M., Singh, R. 2011. The emerging role for bacteria in lignin degradation and bio-product formation. *Current Opinion in Biotechnology*, **22**(3), 394–400.

Bunzel, M., Schüßler, A., Tchetseubu Saha, G.R. 2011. Chemical characterization of Klason lignin preparations from plant-based foods. *Journal of Agricultural and Food Chemistry*, **59**(23), 12506–12513.

Cao, Y., Chen, S.S., Zhang, S., Ok, Y.S., Matsagar, B.M., Wu, K.C.-W., Tsang, D.C. 2019. Advances in lignin valorization towards bio-based chemicals and fuels: Lignin biorefinery. *Bioresource Technology*, **291**, 121878.

Carlsson, A.S., Yilmaz, J.L., Green, A.G., Stymne, S., Hofvander, P. 2011. Replacing fossil oil with fresh oil—with what and for what? *European Journal of Lipid Science and Technology*, **113**(7), 812–831.

Chandel, A.K., Da Silva, S.S., Singh, O.V. 2013. Detoxification of lignocellulose hydrolysates: Biochemical and metabolic engineering toward white biotechnology. *BioEnergy Research*, **6**(1), 388–401.

Chandra, M.R.G.S., Madakka, M. 2019. Comparative biochemistry and kinetics of microbial lignocellulolytic enzymes. In: *Recent Developments in Applied Microbiology and Biochemistry*, Elsevier, pp. 147–159.

Chandra, R., Bharagava, R.N. 2013. Bacterial degradation of synthetic and kraft lignin by axenic and mixed culture and their metabolic products. *Journal of Environmental Biology*, **34**(6), 991.

Chauhan, P.S. 2020. Role of various bacterial enzymes in complete depolymerization of lignin: A review. *Biocatalysis and Agricultural Biotechnology*, **23**, 101498.

Chen, Y., Chai, L., Tang, C., Yang, Z., Zheng, Y., Shi, Y., Zhang, H. 2012. Kraft lignin biodegradation by *Novosphingobium* sp. B-7 and analysis of the degradation process. *Bioresource Technology*, **123**, 682–685.

Chen, Y., Chen, Y., Li, Y., Wu, Y., Zhu, F., Zeng, G., Zhang, J., Li, H. 2018. Application of Fenton pretreatment on the degradation of rice straw by mixed culture of *Phanerochaete chrysosporium* and *Aspergillus niger*. *Industrial Crops and Products*, **112**, 290–295.

Chinga-Carrasco, G., Ehman, N.V., Filgueira, D., Johansson, J., Vallejos, M.E., Felissia, F.E., Håkansson, J., Area, M.C. 2019. Bagasse—A major agro-industrial residue as potential resource for nanocellulose inks for 3D printing of wound dressing devices. *Additive Manufacturing*, **28**, 267–274.

Chinga-Carrasco, G., Ehman, N.V., Pettersson, J., Vallejos, M.E., Brodin, M.W., Felissia, F.E., Hakansson, J., Area, M.A.C. 2018. Pulping and pretreatment affect the characteristics of bagasse inks for three-dimensional printing. *ACS Sustainable Chemistry & Engineering*, **6**(3), 4068–4075.

Choi, S., Lee, H.-N., Park, E., Lee, S.-J., Kim, E.-S. 2020. Recent advances in microbial production of cis, cis-muconic acid. *Biomolecules*, **10**(9), 1238.

Chung, S.-Y., Maeda, M., Song, E., Horikoshij, K., Kudo, T. 1994. A Gram-positive polychlorinated biphenyl-degrading bacterium, *Rhodococcus erythropolis* strain TA421, isolated from a termite ecosystem. *Bioscience, Biotechnology, and Biochemistry*, **58**(11), 2111–2113.

Davis, K., Moon, T.S. 2020. Tailoring microbes to upgrade lignin. *Current Opinion in Chemical Biology*, **59**, 23–29.

de Gonzalo, G., Colpa, D.I., Habib, M.H., Fraaije, M.W. 2016. Bacterial enzymes involved in lignin degradation. *Journal of Biotechnology*, **236**, 110–119.

Del Cerro, C., Erickson, E., Dong, T., Wong, A.R., Eder, E.K., Purvine, S.O., Mitchell, H.D., Weitz, K.K., Markillie, L.M., Burnet, M.C. 2021. Intracellular pathways for lignin catabolism in white-rot fungi. *Proceedings of the National Academy of Sciences*, **118**(9).

Del Río, J.C., Marques, G., Rencoret, J., Martínez, Á.T., Gutiérrez, A. 2007. Occurrence of naturally acetylated lignin units. *Journal of Agricultural and Food Chemistry*, **55**(14), 5461–5468.

Dien, B.S., Jung, H.-J.G., Vogel, K.P., Casler, M.D., Lamb, J.F., Iten, L., Mitchell, R.B., Sarath, G. 2006. Chemical composition and response to dilute-acid pretreatment and enzymatic saccharification of alfalfa, reed canarygrass, and switchgrass. *Biomass and Bioenergy*, **30**(10), 880–891.

Dutta, T., Papa, G., Wang, E., Sun, J., Isern, N.G., Cort, J.R., Simmons, B.A., Singh, S. 2018. Characterization of lignin streams during bionic liquid-based pretreatment from grass, hardwood, and softwood. *ACS Sustainable Chemistry & Engineering*, **6**(3), 3079–3090.

Eltis, L.D., Bolin, J.T. 1996. Evolutionary relationships among extradiol dioxygenases. *Journal of Bacteriology*, **178**(20), 5930–5937.

Filley, T., Cody, G., Goodell, B., Jellison, J., Noser, C., Ostrofsky, A. 2002. Lignin demethylation and polysaccharide decomposition in spruce sapwood degraded by brown rot fungi. *Organic Geochemistry*, **33**(2), 111–124.

Fritsche, W., Hofrichter, M. 2005. Aerobic degradation of recalcitrant organic compounds by microorganisms. In: *Environmental Biotechnology Concepts and Applications*, Wiley, pp. 203–227.

Fu, L., McCallum, S.A., Miao, J., Hart, C., Tudryn, G.J., Zhang, F., Linhardt, R.J. 2015. Rapid and accurate determination of the lignin content of lignocellulosic biomass by solid-state NMR. *Fuel*, **141**, 39–45.

Fukushima, R.S., Hatfield, R.D. 2004. Comparison of the acetyl bromide spectrophotometric method with other analytical lignin methods for determining lignin concentration in forage samples. *Journal of Agricultural and Food Chemistry*, **52**(12), 3713–3720.

Gelbrich, J., Mai, C., Militz, H. 2012. Evaluation of bacterial wood degradation by Fourier Transform Infrared (FTIR) measurements. *Journal of Cultural Heritage*, **13**(3), S135–S138.

Goering, H.K., Van Soest, P.J. 1970. *Forage Fiber Analyses (Apparatus, Reagents, Procedures, and Some Applications)*, US Agricultural Research Service.

Gold, M.H., Alic, M. 1993. Molecular biology of the lignin-degrading basidiomycete Phanerochaete chrysosporium. *Microbiological Reviews*, **57**(3), 605–622.

Grabber, J.H., Schatz, P.F., Kim, H., Lu, F., Ralph, J. 2010. Identifying new lignin bioengineering targets: 1. Monolignol-substitute impacts on lignin formation and cell wall fermentability. *BMC Plant Biology*, **10**(1), 1–13.

Gregorio, A.P.F., Da Silva, I.R., Sedarati, M.R., Hedger, J.N. 2006. Changes in production of lignin degrading enzymes during interactions between mycelia of the tropical decomposer basidiomycetes Marasmiellus troyanus and Marasmius pallescens. *Mycological Research*, **110**(2), 161–168.

Hatfield, R.D. 1993. Cell wall polysaccharide interactions and degradability. *Forage Cell Wall Structure and Digestibility*, 285–313.

Hatfield, R.D., Grabber, J., Ralph, J., Brei, K. 1999. Using the acetyl bromide assay to determine lignin concentrations in herbaceous plants: Some cautionary notes. *Journal of Agricultural and Food Chemistry*, **47**(2), 628–632.

Heap, L., Green, A., Brown, D., van Dongen, B., Turner, N. 2014. Role of laccase as an enzymatic pretreatment method to improve lignocellulosic saccharification. *Catalysis Science & Technology*, **4**(8), 2251–2259.

Heitner, C., Dimmel, D., Schmidt, J. 2019. *Lignin and Lignans: Advances in Chemistry*, CRC Press.

Henneberg, W., Stohmann, F. 1859. Uber das Erhaltungsfutter volljahrigen Rindviehs. *J. Landwirtsch*, **3**, 485–551.

Hu, M., Wang, J., Gao, Q., Bao, J. 2018. Converting lignin derived phenolic aldehydes into microbial lipid by Trichosporon cutaneum. *Journal of Biotechnology*, **281**, 81–86.

Husarcíková, J., Voß, H., de María, P.D., Schallmey, A. 2018. Microbial β-etherases and glutathione lyases for lignin valorisation in biorefineries: Current state and future perspectives. *Applied Microbiology and Biotechnology*, **102**(13), 5391–5401.

Iram, A., Berenjian, A., Demirci, A. 2021. A review on the utilization of lignin as a fermentation substrate to produce lignin-modifying enzymes and other value-added products. *Molecules*, **26**(10), 2960.

Janusz, G., Pawlik, A., Świderska-Burek, U., Polak, J., Sulej, J., Jarosz-Wilkołazka, A., Paszczyński, A. 2020. Laccase properties, physiological functions, and evolution. *International Journal of Molecular Sciences*, **21**(3), 966.

Jing, Y., Guo, Y., Xia, Q., Liu, X., Wang, Y. 2019. Catalytic production of value-added chemicals and liquid fuels from lignocellulosic biomass. *Chem*, **5**(10), 2520–2546.

Johnson, C.W., Salvachúa, D., Khanna, P., Smith, H., Peterson, D.J., Beckham, G.T. 2016. Enhancing muconic acid production from glucose and lignin-derived aromatic compounds via increased protocatechuate decarboxylase activity. *Metabolic Engineering Communications*, **3**, 111–119.

Johnston, P.A., Zhou, H., Aui, A., Wright, M.M., Wen, Z., Brown, R.C. 2020. A lignin-first strategy to recover hydroxycinnamic acids and improve cellulosic ethanol production from corn stover. *Biomass and Bioenergy*, **138**, 105579.

Kamireddy, S.R., Li, J., Abbina, S., Berti, M., Tucker, M., Ji, Y. 2013. Converting forage sorghum and sunn hemp into biofuels through dilute acid pretreatment. *Industrial Crops and Products*, **49**, 598–609.

Kerem, Z., Jensen, K.A., Hammel, K.E. 1999. Biodegradative mechanism of the brown rot basidiomycete *Gloeophyllum trabeum*: Evidence for an extracellular hydroquinone-driven fenton reaction. *FEBS Letters*, **446**(1), 49–54.

Klinke, H.B., Thomsen, A., Ahring, B.K. 2004. Inhibition of ethanol-producing yeast and bacteria by degradation products produced during pre-treatment of biomass. *Applied Microbiology and Biotechnology*, **66**(1), 10–26.

Knoot, C.J., Purpero, V.M., Lipscomb, J.D. 2015. Crystal structures of alkylperoxo and anhydride intermediates in an intradiol ring-cleaving dioxygenase. *Proceedings of the National Academy of Sciences*, **112**(2), 388–393.

Krasznai, D.J., Champagne Hartley, R., Roy, H.M., Champagne, P., Cunningham, M.F. 2018. Compositional analysis of lignocellulosic biomass: Conventional methodologies and future outlook. *Critical Reviews in Biotechnology*, **38**(2), 199–217.

Kumar, A., Chandra, R. 2020. Ligninolytic enzymes and its mechanisms for degradation of lignocellulosic waste in environment. *Heliyon*, **6**(2), e03170.

Kumar, B., Verma, P. 2020. Biomass-based biorefineries: An important architype towards a circular economy. *Fuel*, 119622.

Kumar, M., Singhal, A., Verma, P.K., Thakur, I.S. 2017. Production and characterization of polyhydroxyalkanoate from lignin derivatives by *Pandoraea* sp. ISTKB. *ACS Omega*, **2**(12), 9156–9163.

Levasseur, A., Drula, E., Lombard, V., Coutinho, P.M., Henrissat, B. 2013. Expansion of the enzymatic repertoire of the CAZy database to integrate auxiliary redox enzymes. *Biotechnology for Biofuels*, **6**(1), 1–14.

Li, C., Chen, C., Wu, X., Tsang, C.-W., Mou, J., Yan, J., Liu, Y., Lin, C.S.K. 2019a. Recent advancement in lignin biorefinery: With special focus on enzymatic degradation and valorization. *Bioresource Technology*, **291**, 121898.

Li, X., He, Y., Zhang, L., Xu, Z., Ben, H., Gaffrey, M.J., Yang, Y., Yang, S., Yuan, J.S., Qian, W.-J. 2019b. Discovery of potential pathways for biological conversion of poplar wood into lipids by co-fermentation of Rhodococci strains. *Biotechnology for Biofuels*, **12**(1), 1–16.

Liu, D., Yan, X., Zhuo, S., Si, M., Liu, M., Wang, S., Ren, L., Chai, L., Shi, Y. 2018. Pandoraea sp. B-6 assists the deep eutectic solvent pretreatment of rice straw via promoting lignin depolymerization. *Bioresource Technology*, **257**, 62–68.

Lou, Z., Wang, H., Rao, S., Sun, J., Ma, C., Li, J. 2012. p-Coumaric acid kills bacteria through dual damage mechanisms. *Food Control*, **25**(2), 550–554.

Lourenço, A., Pereira, H. 2018. Compositional variability of lignin in biomass. *Lignin—Trends and Applications*, **10**.

Mäkelä, M.R., Bredeweg, E.L., Magnuson, J.K., Baker, S.E., De Vries, R.P., Hildén, K. 2016. Fungal ligninolytic enzymes and their applications. *Microbiology Spectrum*, **4**(6), 4.6.16.

Mäkelä, M.R., Hilden, K., Kuuskeri, J. 2020. Fungal lignin-modifying peroxidases and H_2O_2-producing enzymes. In: *Encyclopedia of Mycology*, Elsevier.

Malarczyk, E., Rogalski, J., Leonowicz, A. 1994. Transformation of ferulic acid by soil bacteria *Nocardia* provides various valuable phenolic compounds. *Acta Biotechnologica*, **14**(3), 235–241.

Martin, H. 2002. Review: Lignin conversion by manganese peroxidase (MnP). *Enzyme and Microbial Technology*, **30**, 454–466.

Martínez, Á.T., Speranza, M., Ruiz-Dueñas, F.J., Ferreira, P., Camarero, S., Guillén, F., Martínez, M.J., Gutiérrez Suárez, A., Río Andrade, J.C.D. 2005. Biodegradation of lignocellulosics: Microbial, chemical, and enzymatic aspects of the fungal attack of lignin. *International Microbiology*, **8**(3).

Marton, J., Adler, E. 1996. *Preface, Dedication*, ACS Publications.

Matera, I., Gullotto, A., Tilli, S., Ferraroni, M., Scozzafava, A., Briganti, F. 2008. Crystal structure of the blue multicopper oxidase from the white-rot fungus *Trametes trogii* complexed with p-toluate. *Inorganica Chimica Acta*, **361**(14–15), 4129–4137.

Mathews, S.L., Grunden, A.M., Pawlak, J. 2016. Degradation of lignocellulose and lignin by *Paenibacillus glucanolyticus*. *International Biodeterioration & Biodegradation*, **110**, 79–86.

Mei, J., Shen, X., Gang, L., Xu, H., Wu, F., Sheng, L. 2020. A novel lignin degradation bacteria-*Bacillus amyloliquefaciens* SL-7 used to degrade straw lignin efficiently. *Bioresource Technology*, **310**, 123445.

Mnich, E., Bjarnholt, N., Eudes, A., Harholt, J., Holland, C., Jørgensen, B., Larsen, F.H., Liu, M., Manat, R., Meyer, A.S. 2020. Phenolic cross-links: Building and de-constructing the plant cell wall. *Natural Product Reports*, **37**(7), 919–961.

Moreno, A.D., Ibarra, D., Alvira, P., Tomás-Pejó, E., Ballesteros, M. 2016. Exploring laccase and mediators behavior during saccharification and fermentation of steam-exploded wheat straw for bioethanol production. *Journal of Chemical Technology & Biotechnology*, **91**(6), 1816–1825.

Morya, R., Sharma, A., Kumar, M., Tyagi, B., Singh, S.S., Thakur, I.S. 2021. Polyhydroxyalkanoate synthesis and characterization: A proteogenomic and process optimization study for biovalorization of industrial lignin. *Bioresource Technology*, **320**, 124439.

Mycroft, Z., Gomis, M., Mines, P., Law, P., Bugg, T.D. 2015. Biocatalytic conversion of lignin to aromatic dicarboxylic acids in *Rhodococcus jostii* RHA1 by re-routing aromatic degradation pathways. *Green Chemistry*, **17**(11), 4974–4979.

Nghi, D.H., Bittner, B., Kellner, H., Jehmlich, N., Ullrich, R., Pecyna, M.J., Nousiainen, P., Sipilä, J., Huong, L.M., Hofrichter, M. 2012. The wood rot ascomycete Xylaria polymorpha produces a novel GH78 glycoside hydrolase that exhibits α-L-rhamnosidase and feruloyl esterase activities and releases hydroxycinnamic acids from lignocelluloses. *Applied and Environmental Microbiology*, **78**(14), 4893–4901.

Ozer, A., Sal, F.A., Belduz, A.O., Kirci, H., Canakci, S. 2020. Use of feruloyl esterase as laccase-mediator system in paper bleaching. *Applied Biochemistry and Biotechnology*, **190**(2), 721–731.

Paone, E., Tabanelli, T., Mauriello, F. 2020. The rise of lignin biorefinery. *Current Opinion in Green and Sustainable Chemistry*, **24**, 1–6.

Patel, N., Shahane, S., Majumdar, R., Mishra, U. 2019. Mode of action, properties, production, and application of laccase: A review. *Recent Patents on Biotechnology*, **13**(1), 19–32.

Paul, S., Dutta, A. 2018. Challenges and opportunities of lignocellulosic biomass for anaerobic digestion. *Resources, Conservation and Recycling*, **130**, 164–174.

Paz, A., Outeiriño, D., Guerra, N.P., Domínguez, J.M. 2019. Enzymatic hydrolysis of brewer's spent grain to obtain fermentable sugars. *Bioresource Technology*, **275**, 402–409.

Perlack, R.D., Eaton, L.M., Turhollow Jr, A.F., Langholtz, M.H., Brandt, C.C., Downing, M.E., Graham, R.L., Wright, L.L., Kavkewitz, J.M., Shamey, A.M. 2011. *US Billion-ton Update: Biomass Supply for a Bioenergy and Bioproducts Industry*, Bioenergy Technologies Office.

Picart, P., Dominguez de Maria, P., Schallmey, A. 2015. From gene to biorefinery: Microbial β-etherases as promising biocatalysts for lignin valorization. *Frontiers in Microbiology*, **6**, 916.

Pollegioni, L., Tonin, F., Rosini, E. 2015. Lignin-degrading enzymes. *The FEBS Journal*, **282**(7), 1190–1213.

Ponnusamy, V.K., Nguyen, D.D., Dharmaraja, J., Shobana, S., Banu, J.R., Saratale, R.G., Chang, S.W., Kumar, G. 2019. A review on lignin structure, pretreatments, fermentation reactions and biorefinery potential. *Bioresource Technology*, **271**, 462–472.

Questell-Santiago, Y.M., Galkin, M.V., Barta, K., Luterbacher, J.S. 2020. Stabilization strategies in biomass depolymerization using chemical functionalization. *Nature Reviews Chemistry*, **4**(6), 311–330.

Ragauskas, A.J., Beckham, G.T., Biddy, M.J., Chandra, R., Chen, F., Davis, M.F., Davison, B.H., Dixon, R.A., Gilna, P., Keller, M. 2014. Lignin valorization: Improving lignin processing in the biorefinery. *Science*, **344**(6185).

Rajwar, D., Joshi, S., Rai, J. 2016. Ligninolytic enzymes production and decolorization potential of native fungi isolated from pulp and paper mill sludge. *Nature Environment & Pollution Technology*, **15**(4).

Ralph, J. 2010. Hydroxycinnamates in lignification. *Phytochemistry Reviews*, **9**(1), 65–83.

Ralph, J., Lundquist, K., Brunow, G., Lu, F., Kim, H., Schatz, P.F., Marita, J.M., Hatfield, R.D., Ralph, S.A., Christensen, J.H. 2004. Lignins: Natural polymers from oxidative coupling of 4-hydroxyphenyl-propanoids. *Phytochemistry Reviews*, **3**(1), 29–60.

Ramadevi, P., Meder, R., Varghese, M. 2010. Rapid estimation of kraft pulp yield and lignin in Eucalyptus camaldulensis and *Leucaena leucocephala* by diffuse reflectance near-infrared spectroscopy (NIRS). *Southern Forests*, **72**(2), 107–111.

Ray, R.C., Behera, S.S. 2017. Solid state fermentation for production of microbial cellulases. In: *Biotechnology of Microbial Enzymes*, Elsevier, pp. 43–79.

Robertson, J.A., I'Anson, K.J., Treimo, J., Faulds, C.B., Brocklehurst, T.F., Eijsink, V.G., Waldron, K.W. 2010. Profiling brewers' spent grain for composition and microbial ecology at the site of production. *LWT-Food Science and Technology*, **43**(6), 890–896.

Sainsbury, P.D., Hardiman, E.M., Ahmad, M., Otani, H., Seghezzi, N., Eltis, L.D., Bugg, T.D. 2013. Breaking down lignin to high-value chemicals: The conversion of lignocellulose to vanillin in a gene deletion mutant of *Rhodococcus jostii* RHA1. *ACS Chemical Biology*, **8**(10), 2151–2156.

Sakai, K., Matsuzaki, F., Wise, L., Sakai, Y., Jindou, S., Ichinose, H., Takaya, N., Kato, M., Wariishi, H., Shimizu, M. 2018. Biochemical characterization of CYP505D6, a self-sufficient cytochrome P450 from the white-rot fungus *Phanerochaete chrysosporium. Applied and Environmental Microbiology*, **84**(22), e01091–18.

Salvachúa, D., Karp, E.M., Nimlos, C.T., Vardon, D.R., Beckham, G.T. 2015. Towards lignin consolidated bioprocessing: Simultaneous lignin depolymerization and product generation by bacteria. *Green Chemistry*, **17**(11), 4951–4967.

Saritha, M., Arora, A., Singh, S., Nain, L. 2013. Streptomyces griseorubens mediated delignification of paddy straw for improved enzymatic saccharification yields. *Bioresource Technology*, **135**, 12–17.

Seo, H., Kim, K.-J., Kim, Y.H. 2018. In silico-designed lignin peroxidase from Phanerochaete chrysosporium shows enhanced acid stability for depolymerization of lignin. *Biotechnology for Biofuels*, **11**(1), 1–13.

Seto, M., Kimbara, K., Shimura, M., Hatta, T., Fukuda, M., Yano, K. 1995. A novel transformation of polychlorinated biphenyls by *Rhodococcus* sp. strain RHA1. *Applied and Environmental Microbiology*, **61**(9), 3353–3358.

Sharma, A., Singh, S.B., Sharma, R., Chaudhary, P., Pandey, A.K., Ansari, R., Vasudevan, V., Arora, A., Singh, S., Saha, S. 2016. Enhanced biodegradation of PAHs by microbial consortium with different amendment and their fate in in-situ condition. *Journal of Environmental Management*, **181**, 728–736.

Sharma, B., Dangi, A.K., Shukla, P. 2018. Contemporary enzyme based technologies for bioremediation: A review. *Journal of Environmental Management*, **210**, 10–22.

Shi, Y., Yan, X., Li, Q., Wang, X., Xie, S., Chai, L., Yuan, J. 2017. Directed bioconversion of Kraft lignin to polyhydroxyalkanoate by *Cupriavidus basilensis* B-8 without any pretreatment. *Process Biochemistry*, **52**, 238–242.

Shin, K.-S., Kim, Y.H., Lim, J.-S. 2005. Purification and characterization of manganese peroxidase of the white-rot fungus *Irpex lacteus. Journal of Microbiology*, **43**(6), 503–509.

Shin, S.K., Ko, Y.J., Hyeon, J.E., Han, S.O. 2019. Studies of advanced lignin valorization based on various types of lignolytic enzymes and microbes. *Bioresource Technology*, **289**, 121728.

Si, M., Yan, X., Liu, M., Shi, M., Wang, Z., Wang, S., Zhang, J., Gao, C., Chai, L., Shi, Y. 2018. In situ lignin bioconversion promotes complete carbohydrate conversion of rice straw by *Cupriavidus basilensis* B-8. *ACS Sustainable Chemistry & Engineering*, **6**(6), 7969–7978.

Singh, D., Gupta, N. 2020. Microbial Laccase: A robust enzyme and its industrial applications. *Biologia*, **75**(8), 1183–1193.

Sista Kameshwar, A.K., Qin, W. 2020. Systematic metadata analysis of brown rot fungi gene expression data reveals the genes involved in Fenton's reaction and wood decay process. *Mycology*, **11**(1), 22–37.

Sjöström, E., Alén, R. 1998. *Analytical Methods in Wood Chemistry, Pulping, and Papermaking*, Springer Science & Business Media.

Sluiter, A., Hames, B., Ruiz, R., Scarlata, C., Sluiter, J., Templeton, D., Crocker, D. 2008. Determination of structural carbohydrates and lignin in biomass. *Laboratory Analytical Procedure*, **1617**(1), 1–16.

Soest, P.V. 1963a. Use of detergents in the analysis of fibrous feeds. I. Preparation of fiber residues of low nitrogen content. *Journal of the Association of Official Agricultural Chemists*, **46**(5), 825–829.

Soest, P.V. 1963b. Use of detergents in the analysis of fibrous feeds. II. A rapid method for the determination of fiber and lignin. *Journal of the Association of Official Agricultural Chemists*, **46**(5), 829–835.

Sonoki, T., Morooka, M., Sakamoto, K., Otsuka, Y., Nakamura, M., Jellison, J., Goodell, B. 2014. Enhancement of protocatechuate decarboxylase activity for the effective production of muconate from lignin-related aromatic compounds. *Journal of Biotechnology*, **192**, 71–77.

Sonoki, T., Takahashi, K., Sugita, H., Hatamura, M., Azuma, Y., Sato, T., Suzuki, S., Kamimura, N., Masai, E. 2018. Glucose-free cis, cis-muconic acid production via new metabolic designs corresponding to the heterogeneity of lignin. *ACS Sustainable Chemistry & Engineering*, **6**(1), 1256–1264.

Su, J., Fu, J., Wang, Q., Silva, C., Cavaco-Paulo, A. 2018. Laccase: A green catalyst for the biosynthesis of poly-phenols. *Critical Reviews in Biotechnology*, **38**(2), 294–307.

Tanaka, H., Itakura, S., Enoki, A. 1999. Hydroxyl radical generation by an extracellular low-molecular-weight substance and phenol oxidase activity during wood degradation by the white-rot basidiomycete *Trametes Versicolor*. *Journal of Biotechnology*, **75**(1), 57–70.

Tappi, T. 1999. *TEST METHOD T 222 om-88, Acid-Insoluble Lignin in Wood and Pulp*, Tappi Press Atlanta.

Tappi, T. 2002. 222 om-02: Acid-insoluble lignin in wood and pulp. *2002–2003 TAPPI Test Methods*.

Tasman, J., Berzins, V. 1957. The permanganate consumption of pulp materials. *Tappi*, **40**(9), 691.

Theander, O., Westerlund, E.A. 1986. Studies on dietary fiber. 3. Improved procedures for analysis of dietary fiber. *Journal of Agricultural and Food Chemistry*, **34**(2), 330–336.

Tian, J.-H., Pourcher, A.-M., Bouchez, T., Gelhaye, E., Peu, P. 2014. Occurrence of lignin degradation genotypes and phenotypes among prokaryotes. *Applied Microbiology and Biotechnology*, **98**(23), 9527–9544.

Tiwari, R., Nain, L., Labrou, N.E., Shukla, P. 2018. Bioprospecting of functional cellulases from metagenome for second generation biofuel production: A review. *Critical Reviews in Microbiology*, **44**(2), 244–257.

Tobimatsu, Y., Schuetz, M. 2019. Lignin polymerization: How do plants manage the chemistry so well? *Current Opinion in Biotechnology*, **56**, 75–81.

UM250, T. 1985. *Acid-Soluble Lignin in Wood and Pulp*, TAPPI Press.

Vaillancourt, F.H., Bolin, J.T., Eltis, L.D. 2006. The ins and outs of ring-cleaving dioxygenases. *Critical Reviews in Biochemistry and Molecular Biology*, **41**(4), 241–267.

van Parijs, F.R., Morreel, K., Ralph, J., Boerjan, W., Merks, R.M. 2010. Modeling lignin polymerization. I. Simulation model of dehydrogenation polymers. *Plant Physiology*, **153**(3), 1332–1344.

Virk, A.P., Sharma, P., Capalash, N. 2012. Use of laccase in pulp and paper industry. *Biotechnology Progress*, **28**(1), 21–32.

Wang, X., Lin, L., Dong, J., Ling, J., Wang, W., Wang, H., Zhang, Z., Yu, X. 2018. Simultaneous improvements of *Pseudomonas* cell growth and polyhydroxyalkanoate production from a lignin derivative for lignin-consolidated bioprocessing. *Applied and Environmental Microbiology*, **84**(18), e01469–18.

Watkins, D., Nuruddin, M., Hosur, M., Tcherbi-Narteh, A., Jeelani, S. 2015. Extraction and characterization of lignin from different biomass resources. *Journal of Materials Research and Technology*, **4**(1), 26–32.

Weng, C., Peng, X., Han, Y. 2021. Depolymerization and conversion of lignin to value-added bioproducts by microbial and enzymatic catalysis. *Biotechnology for Biofuels*, **14**(1), 1–22.

Wong, D.W. 2009. Structure and action mechanism of ligninolytic enzymes. *Applied Biochemistry and Biotechnology*, **157**(2), 174–209.

Xu, Z., Qin, L., Cai, M., Hua, W., Jin, M. 2018. Biodegradation of kraft lignin by newly isolated Klebsiella pneumoniae, *Pseudomonas putida*, and Ochrobactrum tritici strains. *Environmental Science and Pollution Research*, **25**(14), 14171–14181.

Yang, W., Tang, H., Ni, J., Wu, Q., Hua, D., Tao, F., Xu, P. 2013. Characterization of two *Streptomyces* enzymes that convert ferulic acid to vanillin. *PloS One*, **8**(6), e67339.

Yoo, C.G., Meng, X., Pu, Y., Ragauskas, A.J. 2020. The critical role of lignin in lignocellulosic biomass conversion and recent pretreatment strategies: A comprehensive review. *Bioresource Technology*, **301**, 122784.

Zabel, R.A., Morrell, J.J. 2012. *Wood Microbiology: Decay and Its Prevention*, Academic press.

Zakzeski, J., Bruijnincx, P.C., Jongerius, A.L., Weckhuysen, B.M. 2010. The catalytic valorization of lignin for the production of renewable chemicals. *Chemical Reviews*, **110**(6), 3552–3599.

Zeng, J., Singh, D., Laskar, D., Chen, S. 2013. Degradation of native wheat straw lignin by *Streptomyces viridosporus* T7A. *International Journal of Environmental Science and Technology*, **10**(1), 165–174.

Zhu, D., Zhang, P., Xie, C., Zhang, W., Sun, J., Qian, W.-J., Yang, B. 2017. Biodegradation of alkaline lignin by *Bacillus ligniniphilus* L1. *Biotechnology for Biofuels*, **10**(1), 1–14.

6 Microbial Valorization of Food Industry Wastes for Production of Nutraceutical Molecules

Ranjitha K., Vijay Rakesh Reddy, and Harinder Singh Oberoi
Division of Post Harvest Technology and Agricultural Engineering, ICAR-Indian Institute of Horticultural Research, Hessaraghatta Lake Post, Bangalore, India-560089

CONTENTS

6.1 INTRODUCTION

The global food industry is a rapidly growing complex network of businesses that supply raw and processed foods to the global population. Large volumes of organic wastes are generated during the functioning of these industries. Disposal of these wastes has become an increasingly problematic issue due to the potential pollution; at the same time, many of these wastes act as a source of valuable biomass, high-value biochemicals, and nutrients. Industrial food loss represents 5% of the total food loss (Redcorn et al., 2018). Being generated in high volumes at specific points (industry

DOI: 10.1201/9781003191247-6

sites) and homogeneous, these wastes can be targeted as economically feasible substrates for further value addition.

Food industry wastes often represent a source of various nutrients and biologically active compounds called nutraceuticals. Thus, these wastes represent a gold mine for biochemical compounds obtained either as direct extraction or through microbe-mediated (fermentation) methods. These microbes help in either better extraction of the existing bioactive metabolites, *de novo* synthesis during the growth, or biotransformation of certain molecules to more biologically active forms. In the present scenario, most food industry waste valorization strategies mediated through microbes are available only in scientific literature, and their industrial application is very limited due to the associated legal issues and economic feasibility problems. The present chapter is a comprehensive outlook on the possible biologically active molecule production from food industry wastes through fermentation techniques.

6.2 COMPOSITION OF FOOD INDUSTRY WASTES

The biochemical nature of the waste generated depends on the source material and the end product intended in that particular food industry unit. For example, fruit juice production units produce solid wastes like pomace, seeds, and peels as waste, while pickling units and dairy industries generate liquid wastes, such as brine and whey, respectively. The chemical composition of common food industry wastes is presented in Table 6.1. It is important to mention here that with the advancement in fermentation and separation technologies, some by-products, such as whey, have attained the status of a main raw material for commercially feasible products preparation.

6.3 CONCEPT OF BIOACTIVES AND NUTRACEUTICALS

The term *bioactive* is commonly used in nutrition science and pharmacology. In a broader perspective, the term represents any molecule that affects a living organism, tissue, or cell. In the field of nutrition, these are extra nutritional components present in foods, providing health benefits beyond the nutritional value of the food. Nutraceutical bioactives, in broad, are specific, nonnutritional biochemical components of food with a significant role in modifying and maintaining normal physiological function to maintain good health. Nutraceuticals have therapeutic potential due to their influence on energy intake, anti-inflammatory properties, ability to prevent oxidative stress and metabolic disorders. Due to these effects, nutraceuticals slow down the risk of physiological disorders, such as cardiovascular diseases, cancer, diabetes, cataract, and neurodegenerative diseases. This concept has been proposed as a modern approach to food science, for possibly using food constituents as "*beyond the diet, but before the drugs*" molecules. Nutraceuticals act through modulating metabolic processes, acting as antioxidants, receptor activity inhibitors, enzyme activity modulators, and regulators of gene expression. The major categories of nutraceutical molecules are dietary fiber and prebiotics, polyunsaturated fatty acids, antioxidant vitamins, carotenoids, polyphenols, and bioactive peptides (Wildman and Wallace, 2016).

TABLE 6.1
General Characteristics of Industrial Food Wastes (Redcorn et al., 2018)

Industry	Resource	Total Solids	Digestible Carbohydrates	Soluble and Insoluble Fibers			Proteins	Lipids	Lignin
				Cellulose	Hemi- cellulose	Pectin			
Potato processing	Potato peels	9%	63%	—	—	—	17%	1%	10%
Coffee roasting	Coffee grounds	20%	—	9%	37%	—	14%	17%	—
Pomegranate juice	Pomegranate peels	92%	—	5%—20%	—	5%—11%	—	—	9%
	Pomegranate seeds	13%—19%	—	19%	—	—	—	12%—20%	21%
Citrus processing	Citrus peels	19%	1%	22%—37%	5%—17%	23%	16%—23%	1%—4%	7%—9%
Pineapple processing	Peel	15%	-	18%		1.3%			
Rice processing	Husk	>90%	-	28.%—43.3%	22.0%—29.%	-	-	-	19.%—24.%
Soybean processing	Okara	20%-24%	25%		15%		25%	10%	-
Fish processing	Fish tails, skin, heads, and bones	20%—50%	—	—	—	—	15%—30%	0–25%	—
Cheese processing	Whey	6%	78%	—	—	—	11%	7%	—
Sugar beet Industry	Sugar beet pulp	87%		26.3	18.5				2.5
Cane sugar industry	Bagasse	91%		30.2%	56%	13%			

6.4 MICROBIAL VALORIZATION OF FOOD INDUSTRY WASTES FOR NUTRACEUTICALS

Production and extraction of nutraceutical molecules by fermentation method is a fascinating alternative to the presently followed solvent extraction methods. Biological methods of bioactive extraction provide superior-quality extracts as the process precludes the residual toxicity of organic solvent extraction. In certain fermentation processes, most often bioactive compounds are obtained as secondary metabolites produced by microorganisms in the stationary phase of growth, typically during growth-limitation situations, such as exhaustion of key nutrients such as carbon, nitrogen, and phosphate (Barrios-Gonzalez et al., 2005). The microbial processes help in better extraction due to the liberation of the bioactive compounds, *de novo* synthesis of the bioactives, or biotransformation of the precursors to molecules with higher biological activity level.

6.4.1 Polyphenolic Antioxidants

Phenolic compounds are widely present in the plant kingdom. In plants, they act as pigments, play important roles in attracting pollinators, form the part of structural polymers called lignin, enhance plant defense against different stresses, and act as plant signal molecular chemical defense mechanism against UV radiation and insect attack.

The major nutraceutical value of plant phenolics is due to their antioxidant value, thus potentially reducing the risks of coronary heart diseases and other degenerative disorders (Brewer, 2011). Phenolic nutraceuticals include simple phenolics (e.g., coumarins and phenolic acids) and polyphenolics. The major polyphenolic compounds in the plant system are flavonoids (e.g., flavonols, flavones, flavan-3-ols, anthocyanidins, flavanones, and isoflavones) and nonflavonoids (tannins, lignans, and stilbenes).

Lignocellulosic wastes from foods of plant origin are potential sources of phenolic acids. Bagasse, stover, cobs, and husks represent the most popular of such sources. Fungi are by far most useful for this purpose due to the secretion of polysaccharide hydrolases and oxidative ligninolytic enzymes, which degrade lignin and open phenyl rings, increasing the free phenolic content (Sánchez, 2009). The key extracellular enzymes catalyzing lignin degradation are lignin peroxidase, manganese peroxidase, and laccase. Besides the above enzymes, other hydrolytic enzymes, such as β-glucosidase, tannin acyl hydrolase, ellagitannin acyl hydrolase, α-amylase, laccase, etc., have an important role in releasing nutraceutical phenolics during fermentation from lignocellulosic-rich agro-wastes (Robledo et al., 2008; Zheng and Shetty, 2000).

The lignin degradation releases mainly phenolic acids, such as ferulic, *p*-coumaric, syringic, vanillic, and *p*-hydroxybenzoic acids (Mussatto et al., 2007). The lignin conversion to phenolic acids is most often achieved through solid-state fermentation (SSF) of the raw material. Pretreatments are given to substrates for improving their fermentability. Chemical or physical pretreatments and sterilization of the substrate, addition of nutrients for the acceleration of microbial growth, maintenance of optimum water activity, inoculum levels, temperature, pH, agitation, aeration, *etc.* have a significant effect on the productivity of solid-state fermentation processes (nee'Nigam et al., 2009). Among the different process parameters, moisture and available water content (a_w) have an important role in SSF. Available water (a_w) indicates available or accessible water for the growth of the microorganism and affects their growth, metabolism, and mass transfer processes during SSF. Generally, substrates with 30–85% moisture levels are used for SSF. Desiccation and lower available water levels induce sporulation of the microorganism, but higher moisture levels lead to undesirable bacterial growth and contamination (Pérez-Guerra et al., 2003).

Among the different fungal species, *Phanerochaete chrysosporium*, *Trametes versicolor*, *T. hirsuta*, and *Bjerkandera adusta* are notable lignin degraders. During their growth, fungi utilize the polysaccharides released from the lignin degradation process (Dey et al., 2016). For example, Barbosa et al. (2008) reported that green coconut husk is a good source of ferulic acid, which can be converted to vanillin through SSF using *Phanerochaete chrysosporium*.

Fruits and vegetable wastes possess only negligible levels of lignin but are rich in simple forms of polyphenolic compounds, such as tannins, flavonoids, anthocyanins, *etc.* with high antioxidant potential (Babbar et al., 2011,2015). Pomegranate wastes, for example, contain significant levels of phenolic compounds, such as anthocyanins, hydrolyzable tannins such as catechin, epicatechin, punicalin, pedunculagin, punicalagin, gallic, and ellagic acid, etc., and several lignans such as isolariciresinol, medioresinol, matairesinol, pinoresinol, syringaresinol, and secoisolariciresinol. Pomegranate peel can be effectively converted to ellagic acid through the activity of the strain *Aspergillus niger* GH1. The process yielded as high as 8 kg ellagic acid per ton waste, with high economic potential, considering the commercial value of this product and availability of raw material at low costs (Gonzalez-Aguilar et al., 2008; Robledo et al., 2008).

Similar to pomegranate waste, cranberry pomace, the major by-product of the cranberry juice industry, is also an extremely good source of the antioxidant phenolic ellagic acid. Bioprocessing of cranberry pomace with *Lentinus edodes* increases the ellagic acid content (Vattem and Shetty, 2003). Teri pod (*Caesalpinia digyna*) residue, the solid waste obtained after oil recovery, is a widely available food industry waste in South Asian countries. Teri pod residue is abundant in tannins, from which gallic acid may be industrially produced through bioconversion using the fungus *Rhizopus oryzae* (Kar et al., 1999).

Pineapple canning is a very popular fruit industry in the tropics. Wastes were tested for their potential to yield vanillic acid and vanillin by a fermentation process using *Aspergillus niger* I-1472 and *Pycnoporus cinnabarinus* MUCL 39533. Under optimal conditions, these wastes produced 2,800 mgL^{-1} of vanillin per gram of pineapple cannery waste (Lun et al., 2014). Olive mill waste and winery wastes were found to be potentially useful for increased phenolic extraction by SSF using *Rhizopus oryzae* (Leite et al., 2019).

Soy whey is a liquid nutritional by-product of soybean manufacture. This is rich in nutraceuticals like peptides, oligosaccharides, and isoflavones. *Cordyceps militaris* SN-18 fermentation could further increase the contents of all the aforementioned compounds, thus indicating its functional properties (Dai et al., 2021).

Similarly, citrus wastes represent a source for an array of polyphenolics. *Paecilomyces variotii* is a good candidate organism for improving the functional property of citrus waste polyphenolics (Madeira Jr et al., 2014). A yield increase of 900%, 1,400%, and 1,330% hesperidin, naringenin, and ellagic acid was reported with a concomitant increase of 73% in the antioxidant capacity during the process. This is strong proof of the potential of improving the biological activity of polyphenol mixtures through microbial biotransformation.

Lactobacillus rhamnosus fermentation increased the amounts of free caffeic acid and 4-hydroxybenzoic acid through SSF of wheat bran–whey permeate. Bioaccessibility of free total phenolic acids was also improved by more than 40% due to fermentation. Further, the inclusion of *Saccharomyces cerevisiae* as coculture with *L. rhamnosus* increased recovery of phenolics and antioxidant activity compared to monoculture. These findings provide new insights for commercial exploration to improve the bioactive properties of wheat bran using mixed-culture fermentation (Bertsch et al., 2020).

Research interest to microbially valorize fruit industry wastes is on the rise now, consequent to which varying substrates are tested for their usefulness to improve their antioxidant yield by solid-state fermentation (Table 6.2).

TABLE 6.2
Phenolics Production through Solid-State Fermentation

Substrate	Culture	Inference	Reference
Soybean hull	*Aspergillus oryzae*	Increase in phenolic acids, flavonoids, and isoflavone	Cabezudo et al., 2021
Mango seed	*Aspergillus niger*	Increase in antioxidant polyphenolics	Torres-León et al., 2019
Chestnut (*Castanea sativa mill.*) Waste	*A. japonicus*	Sixfold increase in ellagic acid	Gulsunoglu-Konuskan, 2021
Tamarind seed flour (SF) and mixed peel and seed flour (PSF)	*Rhizopus* spp	524% increase in total phenols	Santos et al., 2020
Apple waste	*Aspergillus* spp	Taxifolin production by *A. aculeatus* and *A. japonicus* Eriodictyol and catechin by *A. niger* and *A. tubingensis*.	Gulsunoglu et al., 2020
Black plum seed powder	*Aspergillus niger*	Gave highest yield of gallic acid (13.31 mg/g of substrate)	Saeed et al., 2021
Chokeberry (*Aronia melanocarpa*) pomace	*Aspergillus niger* and *Rhizopus oligosporus* strains	1.5-fold increase in total flavonoids	Dulf et al., 2018
Grape pomace and wheat bran	*Aspergillus* niger 3T5B8	Higher proanthocyanidins extraction	Teles et al., 2019

6.4.2 Carotenoid Antioxidants

Carotenoids are fat-soluble natural isoprenoid pigments imparting yellow, orange, or red color to fruits and vegetables. Their structural diversity comes from variations in the number of units and chemical modifications. Many carotenoids can act as antioxidants through their free radical–scavenging activity. Beta-carotene is the most vital carotenoid due to its provitamin A activity and antioxidant activity. Nonprovitamin carotenoids, like α-carotene, lutein, lycopene, astaxanthin, cryptoxanthin, *etc*., also contribute to a good level of antioxidant activities, thus protecting against degenerative diseases.

A wide range of microbes also possesses the ability to act as commercial sources of carotenoids (Table 6.3). Similarly, a wide variety of food industry wastes, such as winery waste, glucose syrup, molasses from beet and sugar, soybean flour extract, whey, technical glycerol, *etc*., are low-cost substrates for carotenoid production by microbes (De Carvalho et al., 2013, 2014). The microbial valorization of food wastes for carotenoids is basically about fulfilling the nutritional requirement for *de novo* synthesis of these nutraceuticals rather than providing the precursor molecule/release of existing bound forms.

The Mucorales fungus *Blakeslea trispora*, is one of the best-known sources of microbial carotenoids. Carotenoid pigments are produced by this fungus during sexual reproduction. The physiological function of the carotenoids in *B. trispora* is to act as precursors for the synthesis of a sex-specific hormone called trisporic acid (TSA) to enable the fungus to participate in sexual reproduction. The enzyme carotene oxygenase mediates the synthesis of trisporic acid (TSA). Trisporic acid carotene acts as a signaling molecule to stimulate sexually complementary cells to make contact with each other, thus initiating and controlling sexual reproduction. Therefore, the use of two mating types in the same culture medium is the key to the production of large amounts of carotenoids. Another specialty is the positive feedback control of β-carotene in the carotenogenesis process, which in turn further stimulates carotenogenesis and the production of trisporoid. Furthermore, it also acts as a hormone stimulator of its biosynthesis. Thus, the organism requires a threshold concentration of TSA sufficient to activate carotenogenesis (production

TABLE 6.3
Antioxidant Carotenoids Produced by Microbes

Name of the Carotenoid	Source Organism
Lutein	*Spongiococcum excentricum, Muriellopsis* sp
Astaxanthin	*Haematococcus pluvialis, Xanthophyllomyces dendrorhous Paracoccus carotinifaciens*
β-carotene (orange)	*Blakeslea trispora, Rhodotorula glutinis, Dunaliella salina*
Lycopene (red)	*Blakeslea trispora*
Toruline	*Rhodotorula glutinis*
Torularhardine	*Sporobolomyces ruberrimus*
Canthaxanthin	*Rhodotorula glutinis, Dietzia natronolimnaea*

of carotenoids). This level is approximately equivalent to 0.5% of the spore dry weight, which is usually stored in zygospores of *B. trispora* (Papaioannou et al., 2012; Avalos et al., 2017).

Fruit and vegetable waste (orange, carrot, and papaya peels) could produce as high as 0.127 mg/mL carotenoids using microbial strain *Blakeslea trispora*(+) MTCC 884 in solid-state fermentation, out of which 76% was β-carotene. Horticultural wastes such as cabbage, watermelon rind, and peach peels can be added as the main carbon source into submerged *B. trispora* cultures for improved biomass of this fungus. In an experimental trial, a series of agro-food wastes was used with corn steep liquor and thiamine, which yielded a biomass of 10.2 g/L and carotenoid levels 230 mg/L, respectively, with approximately 76% beta-carotene fraction, while the corresponding in a synthetic medium was 9 mg/L and 46 mg/L, respectively. This study proved the efficiency of *B. trispora* to use diverse agro-food wastes for carotenoid production (Papaioannou et al., 2012).

Rhodotorula glutinis and *Rhodotorula mucilaginosa* are two yeast species with high carotenoid production potential. Besides carotene, *Rhodotorula* produces two other carotenoids, *viz*., toruline and torulohodin, both of which have antioxidant properties (Kot et al., 2018). The proportions of these three carotenoids vary depending on the growth medium constitution as well as environmental factors. For example, under high-intensity light, torulohodin production is more than β-carotene. Carotenogenesis in *Rhodotorula* is highly regulated by environmental factors, *viz*., light, aeration, and temperature. The presence of organic solvents such as ethanol, methanol, glycol-reactive oxygen species, osmolytes, *etc*. imparts mild stress to induce carotenogenesis. Various metal ions like Ba, Fe, Mg, Ca, Zn, and Co were also demonstrated to stimulate carotenoid production by *R. glutinis*. The stimulation effect is differential; for instance, Al^{3+} and Zn^{2+} stimulate β-carotene and γ-carotene production, while Zn^{2+} and Mn^{2+} inhibit torulene and torularhodin production. These effects of trace elements are due to their role as cofactors to activate specific carotenogenic enzymes (Avalos et al., 2017). Growth and carotenoid production by *R. glutinis* is already proven with different waste- and residue-based fermentation substrates such as flour extracts, grape must, molasses, radish brine, and whey (Frengova et al., 2009). *Rhodotorula* yeasts initiate carotenoid synthesis during the late logarithmic phase and maximize it in the stationary phase. Substrates such as hydrolyzed mustard waste isolates, hemicellulosic hydrolyzates of eucalyptus globules wood, hydrolyzed mung bean waste flour, sugarbeet molasses, corn hydrolyzate, milk whey, *etc*. have already been explored as carbon substrates for carotenoid production. Rhodotorula yeasts grown in medium with potato wastewater supplemented with 3–5% or 5% glycerol also synthesize 230 μg/g carotenoids (Kot et al., 2019).

6.4.3 Prebiotics

Prebiotics are food components that enhance the growth and activity of beneficial gut microbiota. Dietary prebiotics is typically nondigestible fiber compounds which reach the intestine to stimulate the gut bacteria, largely with beneficial properties. They were first identified and named by Marcel Roberfroid in 1995. Fructans, galactans, resistant starch, pectin, beta-glucans, xylooligosaccharides, etc. fit well in the definition of prebiotics.

6.4.3.1 Fructooligosaccharides (FOS)

Among the different oligosaccharides, FOS has additional benefits of being low-calorie sweeteners and is expected to be a popular ingredient in food and beverages aimed at the functional food sector. FOS also has other nutraceutical properties, like reducing cholesterol levels and helping in mineral absorption. These are conventionally produced from sucrose using β-fructofuranosidase (FFase; 3.2.1.26) and fructosyltransferase (FTase; 2.4.1.9) enzymes. Fungal strains of *Aspergillus japonicus*, *A. ibericus*, *A. oryzae*, *A. niger*, *Aureobasidium pullulans*, *Rhizopus stolonifer*, and *Penicillium citreonigrum* are the most popular microbial sources of FOS-producing enzymes.

Sucrose-rich food industry wastes, such as molasses, fruit peels, bagasse, pomaces, and coffee processing by-products (coffee pulp, coffee husk, and coffee spent grain), are successfully used for FOS production through fermentation. Agro-food waste and microorganisms used in the production of fructooligosaccharides are listed in Table 6.4.

Screening of suitable strains for improved FOS production has been a backbone of research. Kestose, Nystose, neokestose, etc. are commonly produced by different

TABLE 6.4
Agro-Food Waste and Microorganisms Used in Production of Fructooligosaccharides

Agro-Food Waste	Microorganisms	Enzymes Produced	Reference
Molasses	*Aspergillus*	β-fructofuranosidase	Reddy et al., 2010
	Saccharomyces cerevisiae	Fructofuranosidase	Bali et al., 2015
	Aspergillus japonicus-FCL 119T	β-Fructosyltransferase	Bali et al., 2015
Sugarcane bagasse	*Aspergillus flavus* NFCCI 2364	Fructosyltransferase	Ganaie et al., 2017
Banana peel/leaf	A. flavus NFCCI 2364	Fructosyltransferase	Ganaie et al., 2017
	Chrysonilia sitophila	β-fructofuranosidase	Patil et al., 2011
	Saccharomyces cerevisiae GVT263	β-D-fructofuranosidase	Goud et al., 2013
Agave mead	*Aspergillus oryzae DIA-MF*	Fructosyltransferase	Muñiz-Márquez et al., 2016
Coffee by-products	*Aspergillus japonicus*	β-fructofuranosidase	Mussatto andTeixeira, 2010
	Aspergillus japonicus	β-fructofuranosidase	Mussatto et al., 2013
Cassava wastes	*Rhizopus stolonifer LAU 07*	Fructosyltransferase	Lateef and Gueguim-Kana, 2012
Apple pomace	*Aspergillus versicolor*	β-fructofuranosidase	Arfelli et al., 2016
Wheat bran fructosyltransferase	*Fusarium graminearum* *Aspergillus awamori* GHRTS	β-D-fructofuranosidase	Sathish and Prakasham, 2013
Spent osmotic solutions	*Aspergillus oryzae* N74	Fructosyltransferase	Ruiz et al., 2014
Date by-products	Aspergillus awamori NBRC 4033	β-D-fructofuranosidase	Smaali et al., 2012

fungal strains during their growth in a sucrose-rich medium. Zambelli et al. (2014) isolated the filamentous fungi *Penicillium sizovae* (CK1) and *Cladosporium cladosporioides* (CF215) from molasses and jams and tested their FOS yield. *C. cladosporioides* synthesized mainly 1-kestose, nystose, blastoise, fructosyl nystose, 6-kestose, and neokestose, while *P. sizovae* produced mainly 1- kestose, with traces of neoFOS and levan-type FOS.

Lactic acid bacteria species such as *L. mesenteroides* are reported to produce as high as 124 g/kg fructooligosaccharides from cane molasses (Kaprasob et al., 2018). Also, a functional monosaccharide, D-psicose, was produced in the fermentation media by epimerization of D-fructose (Sharma et al., 2016). In a process optimization study, Dorta et al. (2006) found that additives like peptone to the molasses-based media (2–5%) improve cellular growth, β-fructosyltransferase, and FOS production by around 60% by *A. japonicus*-FCL 119T and *A. niger* ATCC 20611. The strain *A. pullulans* KCCM 12017 yielded 0.5 g/FOS per gram molasses with twenty-four-hour incubation at 55°C in a molasses-based medium.

Aguamiel is a potential fermentable by-product. It is the sugar-rich sap that oozes out after cutting off the inflorescence stalk of agave (Santos-Zea et al., 2012). Augmiel has 11.5% dry matter, out of which 75% are sugars (represented by fructose, glucose, and sucrose) and of FOS (10%), protein (3% wt), and amino acids (0.3% wt) and minerals (Ortiz-Basurto et al., 2008; Santos-Zea et al., 2012; Silos-Espino et al., 2007; Tovar et al., 2008). Aguamiel is therefore a very rich source of essential nutrients for FOS production through microorganisms. FOS production by *A. oryzae* DIA-MF using aguamiel is twofold higher than that of the synthetic medium, with a yield level of 0.84 g FOS L^{-1} h^{-1} (Muñiz-Márquez et al., 2019).

6.4.3.2 Galactooligosaccharides (GOS)

Galactooligosaccharides (GOS) are important prebiotics synthesized conventionally from lactose by a special enzyme called *β*-galactosidase (EC 3.2.1.23). *β*-galactosidases are very commonly produced enzymes across microbial groups. This enzyme hydrolytically splits the lactose as well as catalyzes transgalactosylation to produce galactooligosaccharides (Prenosil et al., 1987). Instead of using pure enzymes, GOS production through microbial cell factories is explored. Recently, GOS production from dairy effluents using mixed cultures of *Bacillus singularis* and *Saccharomyces* sp. was patented in the USA (US9139856B2—process for the production of galactooligosaccharides [GOS], Google Patents). The system is claimed to have more economic advantage by making possible the repeated use of cell biomass and by obtaining pure GOS without interference from galactose.

6.4.3.3 Xylooligosaccharides

Xylooligosaccharides (XOSs) are yet another group of oligosaccharides with prebiotic properties. These oligomers are typified by two to seven xylose units linked through β-(1,4)-xylosidic bonds and with acetyl groups, uronic acids, and arabinose units in the side chains, giving out a ramified structure (Kumar and Satyanarayana, 2011). Xylan is the main constituent of hemicelluloses and the second most abundant carbohydrate in the lignocellulosic biomass (Nieto-Domínguez et al., 2017). Xylan is the precursor of XOS production through fermentation. The origin of the biomass and the production process decide the structure of XOS (Samanta et al., 2016). Several agro-food residues, such as sugarcane bagasse, corncobs, rice husks, olive pits, barley straw, tobacco stalk, cotton stalk, sunflower stalk, wheat straw, etc., are already found in their use as potential sources of XOS. The XOS-yielding hydrolytic treatments conventionally are acid hydrolysis, alkaline hydrolysis, auto-hydrolysis, and enzymatic hydrolysis. Like in FOS production, the mediators for XOS are also fungal strains, such as *Trichoderma harzianum*, *T. reesei Cellulosimicrobium cellulans*, *Penicillium janczewskii*, *P. echinulatu*, and *A. awamori*. These microbes predominantly harbor the enzyme complex comprising endoxylanase and side groups–splitting enzymes. The bacterium *Bacillus subtilis KCX006* is also remarkably useful due to its ability to synthesize endoxylanases and xylan-debranching enzymes without β-xylosidase activity with a constitutive expression of the genes. The process produced xylobiose to xylotetraose as the major products in wheat bran and groundnut oilcake substrate to give a final yield of 2158 IU and 24.92 per gram dry weight respectively (Reddy and Krishnan, 2016). Brewers' spent grain (BSG) as substrate was found to be highly promising for the single-step production of XOS by *Trichoderma reesei*. Within three days' growth in BSG (20 g/L) under optimal conditions, 38 mg/g xylose equivalents/g of BSG was achieved. The profiling studies showed these as arabino-xylooligosaccharides with two to five monomeric units. This reported productivity is comparable with that of commercial enzyme-mediated processes (Amorim et al., 2019a).

Similar to all microbe-mediated biotechnological processes, recombinant DNA technology is a powerful strategy to improve XOS production too (Milessi et al., 2016). Cloning of *xyn2* gene from *Trichoderma reesei* with appropriate secretory tag in *B. subtilis* 3610 secreted xylanase enzyme to obtain as high306 mg XOS/g^{-1} beechwood xylan. The fermentation medium was optimized to contain 2.5 g L^{-1} of xylan and pH 6.0. Fermentation was carried out at 42.5°C for eight hours (Amorim et al., 2019b; Rashid and Sohail, 2021).

6.4.4 Polyunsaturated Fatty Acids

Polyunsaturated fatty acids (PUFAs) are key biological molecules due to their structural and functional properties with wide use in biomedical and nutraceutical fields. They primarily control the structural design, dynamics, fluidity, and permeability of biological membranes and influence the activity of membrane-bound proteins. Besides this, PUFAs act as essential precursors of hormones and metabolites, such as prostaglandins, leukotrienes, and hydroxy-fatty acids.

γ-linolenic acid (C18:3 n-6; GLA) is a PUFA synthesized from the linoleic acid by the activity of the key enzyme Δ6-desaturase. Diseases such as atopic dermatitis, diabetes, arthritis, *etc.* diminish this conversion process, and dietary supplementation is essential for these patients. The main commercial source of polyunsaturated fatty acids (mostly ω-3) is from various types of fish. Oleaginous microbes are also reliable sources of PUFAs. The production of γ-linolenic acid from food industry wastes with fermentation processes is given in Table 6.5. Among the oleaginous fungi, the Zygomycetes fungi have a remarkable ability to produce GLA from agrifood wastes. Genera such as *Cunninghamella*, *Mortierella*, *Mucor*, *Rhizopus*, and *Thamnidium*

TABLE 6.5
Microbial Production of Ɣ- Linolenic Acid (GLA) from Food Industry Wastes

Microorganism	Substrate	GLA yield g/kg BP
Thamnidium elegans	Spent flakes/SMG	07.2
Thamnidium elegans	Crushed corn	10.0
Thamnidium elegans	Wheat bran/SMG/SO	10.0
Thamnidium elegans	Wheat bran/SMG/SO/plant extracts	20.0
Mortierella isabellina	Pear pomace	2.9
Cunninghamella elegans	Barley/SMG/peanut oil	14.2
Cunninghamella echinulata	Orange peel	1.4
Mucor rouxii	Rice bran	6.0

SMG: spent malt grains; SO: sunflower oil BP: bioproduct/fermented mass.
Source: Mechmeche et al. (2017)

can form GLA during their growth on diverse food industry waste substrates such as cereal bran, soybean meal, spent malt grain, fruit peels, and pomaces. The GLA yield depends greatly on substrate-strain interaction in the fermentation process. In a study, *Mortierella isabellina* grown on various wastes produced 18 to 2.9 g of GLA/kg dry fermented mass. Rice bran is a suitable substrate for GLA production using *Mucor rouxii*, with a yield level of 6 g/kg of fermented mass. *Thamnidium elegans* is also capable of utilizing a variety of cereals to yield up to 5 g GLA/kg of fermented mass (Čertík et al., 2012).

Acid-hydrolyzed molasses serve as an excellent carbon source for PUFA production by fungi such as *Mucor recurves*, *Cunninghamella echinulata*, and *Mortierella isabellina* (Chatzifragkou et al., 2011). *Schizochytrium* sp. *and Aurantiochytrium* sp. were able to produce docosahexaenoic acid (DHA) during growth on sugarcane bagasse hydrolysate and similar substrates (Iwasaka et al., 2013; Yin et al., 2019). Other waste substrates like tomato hydrolysate enriched with glucose as well as starch waste from the potato industry can also be used for the growth of *C. echinulate* (Fakas et al., 2008, Kothri et al., 2020). Distillery wastes from shochu distillery and brewery industry have been used as substrates for DHA production using *Schizochytrium* sp. and *Aurantiochytrium* sp. (Yamasaki et al., 2006; Ryu et al., 2013). Coconut water is a good growth medium for the growth of the marine protist *Schizochytrium mangrovei*, a microbial cell factory for DHA production (Unagul et al., 2007). Orange peel and the peel extract are ideal substrates for DHA production through *Aurantiochytrium*. Empty fruit bunches after palm oil extraction are also suitable for *Aurantiochytrium* sp. cell growth and lipid production rich in DHA (Park et al., 2018).

6.4.5 Bioactive Peptides

Bioactive peptides are generally protein hydrolysates possessing approximately two to twenty amino acids, and their physiological effects vary with amino acid sequences (Capriotti et al., 2015; Xue et al., 2015). Antihypertensive, antioxidative, antimicrobial, antidiabetic, and immune-modulatory properties are some of the

TABLE 6.6
Microbial Production of Bioactive Peptides through Fermentation of Food Industry Wastes

Fermentation method	Fermentative organism	Substrate	Bioactivity of the peptides	References
Solid state fermentation	*Bacillus subtilis*	Soybean meal	ACE inhibitory	Wang et al., 2013
Submerged	*Bacillus subtilis*	Defatted wheat germ	Antioxidant	Niu et al, 2013
Submerged	*Aspergillus oryzae*	Turbot skin	Antioxidant	Fang et al., 2017
Submerged	*Bacillus subtilis*	Tomato seeds	ACE inhibitory, antioxidant	Moayedi et al., 2017
Submerged	*Lactobacillus plantarum*	Soy whey	Antioxidant	Xiao et al., 2015

nutraceutical properties conferred by these short peptides (Liu et al., 2010). Protein foods such as milk, meat, *etc.* inherently possess certain amounts of bioactive peptides (Salampessy et al., 2010). Microbial metabolism converts even other substrates to bioactive peptides. For example, *Lactobacillus plantarum* can be used to obtain high-value peptides from tomato seed meal extract substrate. This organism could metabolize tomato seed proteins into bioactive peptides with antioxidant activity (Mechmeche et al., 2017). Corn gluten meal with 46% moisture and fortified with 5% peptone could be used for solid-state fermentation with *Bacillus subtilis* MTCC5480 to obtain bioactive peptides in five days (Jiang et al., 2020). Commendable work done on bioactive peptide preparation through fermentation of food industry wastes is mentioned in Table 6.6.

6.5 CONCLUSION AND FUTURE PERSPECTIVES

Food industry wastes provide an excellent substrate for bioconversion to extract, *de novo* synthesis, and biotransformation of bioactive molecules. Production of nutraceuticals like polyphenolics, prebiotics, carotenoid antioxidants, polyunsaturated fatty acids (PUFA), and bioactive peptides has already been studied. Solid-state fermentation is the most common fermentation system used for polyphenolic production from agro-food industry waste, while submerged fermentation is used for bioactive peptide production from liquid by-products of the dairy industry. *Aspergillus* spp. is the most commonly explored microbe for polyphenols extraction. *Blakeslea trispora* and *Rhodotorula spp.* are well-known to grow on food wastes and produce beta-carotene. Prebiotic oligosaccharides are formed by an array of fungi and bacteria. Oleaginous yeasts and zygomycetes fungi have shown potential for PUFA production from several agri-food wastes. The present limitation is the nonavailability of scale-up studies, with most of the scientific information restricted to academic interest. The downstream processing aspects remain as the major bottleneck for exploring commercial production. Moreover, strain improvement, specifically for their ability to grow and yield the desirable nutraceuticals with a negligible quantity of interfering and undesirable metabolites, needs to be addressed for commercialization of the wide array of technical knowledge generated so far.

REFERENCES

Amorim, Cláudia, Sara C. Silvério, and Lígia R. Rodrigues. "One-step process for producing prebiotic arabino-xylooligosaccharides from brewer's spent grain employing *Trichoderma* species." *Food Chemistry* 270 (2019a): 86–94. doi:10.1016/j.foodchem.2018.07.080.

Amorim, Cláudia, Sara C. Silvério, Raquel F. S. Gonçalves, Ana C. Pinheiro, Soraia Silva, Elisabete Coelho, Manuel A. Coimbra, Kristala L. J. Prather, and Lígia R. Rodrigues. "Downscale fermentation for xylooligosaccharides production by recombinant *Bacillus subtilis* 3610." *Carbohydrate Polymers* 205 (2019b): 176–183.

Arfelli, V., C. Henn, T. B. Dapper, V. C. Arfelli, C. Henn, and M. R. Simões. "Fructofuranosidase production by *Aspergillus versicolor* isolated from Atlantic forest and grown on apple pomace β-Fructofuranosidase production by *Aspergillus versicolor* isolated from Atlantic forest and grown on apple pomace." *African Journal of Microbiology Research* 10, no. 25 (2016): 938–948. https://doi.org/10.5897/AJMR2016.8038.

Avalos, Javier, Steffen Nordzieke, Obdulia Parra, Javier Pardo-Medina, and M. Carmen Limon. "Carotenoid production by filamentous fungi and yeasts." In *Biotechnology of yeasts and filamentous fungi*, pp. 225–279. Springer, Cham, 2017. doi:10.1007/978-3-319-58829-2-8.

Babbar, N., H. S. Oberoi, and S. K. Sandhu. "Sandhu therapeutic and nutraceutical potential of bioactive compounds extracted from fruit residues." *Critical Reviews in Food Science and Nutrition* 55, no. 3 (2015): 319–337. doi:10.1080/10408398.2011.653734.

Babbar, N., H. S. Oberoi, D. S. Uppal, and R. T. Patil. "Total phenolic content and antioxidant capacity of extracts obtained from six important fruit residues." *Food Research International* 44 (2011): 391–396.

Bali, Vandana, Parmjit S. Panesar, Manab B. Bera, and Reeba Panesar. "Fructo-oligosaccharides: Production, purification and potential applications." *Critical Reviews in Food Science and Nutrition* 55, no. 11 (2015): 1475–1490.

Barbosa, dos Santos, Elisabete, Daniel Perrone, Ana Lúcia do Amaral Vendramini, and Selma Gomes Ferreira Leite. "Vanillin production by Phanerochaete chrysosporium grown on green coconut agro-industrial husk in solid state fermentation." *BioResources* 3, no. 4 (2008): 1042–1050.

Barrios-Gonzalez, J., F. J. Fernandez, A. Tomasini, and A. Mejia. "Secondary metabolites production by solid-state fermentation." *Malaysian Journal of Microbiology* 1, no. 1 (2005): 1–6.

Bertsch, Annalisse, Denis Roy, and Gisèle LaPointe. "Fermentation of wheat bran and whey permeate by mono-cultures of Lacticaseibacillus rhamnosus strains and co-culture with yeast enhances bioactive properties." *Frontiers in Bioengineering and Biotechnology* 8 (2020): 956. doi:10.3389/fbioe.2020.00956.

Brewer, M. S. "Natural antioxidants: Sources, compounds, mechanisms of action, and potential applications." *Comprehensive Reviews in Food Science and Food Safety* 10, no. 4 (2011): 221–247. doi:10.1111/j.1541-4337.2011.00156.x.

Cabezudo, Ignacio, María-Rocío Meini, Carla C. Di Ponte, Natasha Melnichuk, Carlos E. Boschetti, and Diana Romanini. "Soybean (*Glycine max*) hull valorization through the extraction of polyphenols by green alternative methods." *Food Chemistry* 338 (2021): 128131. doi:10.1016/j.foodchem.2020.128131.

Capriotti, Anna Laura, Giuseppe Caruso, Chiara Cavaliere, Roberto Samperi, Salvatore Ventura, Riccardo Zenezini Chiozzi, and Aldo Laganà. "Identification of potential bioactive peptides generated by simulated gastrointestinal digestion of soybean seeds and soy milk proteins." *Journal of Food Compositionand Analysis* 44 (2015): 205–213.

Čertík, Milan, Zuzana Adamechová, and Kobkul Laoteng. "Microbial production of γ-linolenic acid: Submerged versus solid-state fermentations." *Food Science and Biotechnology* 21, no. 4 (2012): 921–926. doi:10.1007/s10068-012-0121-2.

Chatzifragkou, Afroditi, Anna Makri, Aikaterini Belka, Stamatina Bellou, Marilena Mavrou, Maria Mastoridou, Paraskevi Mystrioti, Grace Onjaro, George Aggelis, and Seraphim Papanikolaou. "Biotechnological conversions of biodiesel derived waste glycerol by yeast and fungal species." *Energy* 36, no. 2 (2011): 1097–1108.

Dai, Yiqiang, Jianzhong Zhou, Lixia Wang, Mingsheng Dong, and Xiudong Xia. "Biotransformation of soy whey into a novel functional beverage by Cordyceps militaris SN-18." *Food Production, Processing and Nutrition* 3, no. 1 (2021): 1–11. doi:10.1186/s43014-021-00054-0.

De Carvalho, Ana Flávia Azevedo, Pedro de Oliva Neto, Douglas Fernandes Da Silva, and Gláucia Maria Pastore. "Xylo-oligosaccharides from lignocellulosic materials: Chemical structure, health benefits and production by chemical and enzymatic hydrolysis." *Food Research International* 51, no. 1 (2013): 75–85. doi:10.1016/j.foodres.2012.11.021.

De Carvalho, Júlio C., Lígia C. Cardoso, Vanessa Ghiggi, Adenise Lorenci Woiciechowski, Luciana Porto de Souza Vandenberghe, and Carlos Ricardo Soccol. "Microbial pigments." In *Biotransformation of waste biomass into high value biochemicals*, pp. 73–97. Springer, New York, 2014.

Dey, Tapati Bhanja, Subhojit Chakraborty, Kavish Kr Jain, Abha Sharma, and Ramesh Chander Kuhad. "Antioxidant phenolics and their microbial production by submerged and solid state fermentation process: A review." *Trends in Food Science & Technology* 53 (2016): 60–74. doi:10.1016/j.tifs.2016.04.007.

Dorta, Claudia, Rubens Cruz, Pedro de Oliva-Neto, and Danilo José Camargo Moura. "Sugarcane molasses and yeast powder used in the fructooligosaccharides production by *Aspergillus japonicus*-FCL 119T and *Aspergillus niger* ATCC 20611." *Journal of Industrial Microbiology and Biotechnology* 33, no. 12 (2006): 1003.

Dulf, Francisc Vasile, Dan Cristian Vodnar, Eva-Henrietta Dulf, Zoriţa Diaconeasa, and Carmen Socaciu. "Liberation and recovery of phenolic antioxidants and lipids in chokeberry (*Aronia melanocarpa*) pomace by solid-state bioprocessing using *Aspergillus niger* and *Rhizopus oligosporus* strains." *LWT* 87 (2018): 241–249. doi:10.1016/j.lwt.2017.08.084.

Fakas, Stylianos, Milan Čertik, Seraphim Papanikolaou, George Aggelis, Michael Komaitis, and Maria Galiotou-Panayotou. "γ-Linolenic acid production by *Cunninghamella echinulata* growing on complex organic nitrogen sources." *Bioresource Technology* 99, no. 13 (2008): 5986–5990.

Fang, B., J. Sun, P. Dong, C. Xue, and X. Mao. "Conversion of turbot skin wastes into valuable functional substances with an eco-friendly fermentation technology." *Journal of Cleaner Production* 156 (2017): 367–377.

Frengova, Ginka I., and Dora M. Beshkova. "Carotenoids from *Rhodotorula* and Phaffia: Yeasts of biotechnological importance." *Journal of Industrial Microbiology and Biotechnology* 36, no. 2 (2009): 163. doi:10.1007/s10295-008-0492-9.

Ganaie, Mohd Anis, Hemant Soni, Gowhar Ahmad Naikoo, Layana Taynara Santos Oliveira, Hemant Kumar Rawat, Praveen Kumar Mehta, and Narendra Narain. "Screening of low cost agricultural wastes to maximize the fructosyltransferase production and its applicability in generation of fructooligosaccharides by solid state fermentation." *International Biodeterioration & Biodegradation* 118 (2017): 19–26.

Gonzalez-Aguilar, G., R. M. Robles-Sánchez, M. A. Martínez-Téllez, G. I. Olivas, E. Alvarez-Parrilla, and L. A. De La Rosa. "Bioactive compounds in fruits: Health benefits and effect of storage conditions." *Stewart Postharvest Review* 4, no. 3 (2008): 1–10.

Goud, K. Gnaneshwar, K. Chaitanya, and Gopal Reddy. "Enhanced production of β-D-fructofuranosidase by Saccharomyces cerevisiae using agro-industrial wastes as substrates." *Biocatalysis and Agricultural Biotechnology* 2, no. 4 (2013): 385–392.

Gulsunoglu-Konuskan, Zehra, Funda Karbancioglu-Guler, and Meral Kilic-Akyilmaz. "Development of a bioprocess for production of ellagic acid from chestnut (*Castanea sativa* mill.) waste by fermentation with *Aspergillus* spp." *Food Bioscience* (2021): 101058. doi:10.1016/j.fbio.2021.101058.

Gulsunoglu-Konuskan, Zehra, Randy Purves, Funda Karbancioglu-Guler, and Meral Kilic-Akyilmaz. "Enhancement of phenolic antioxidants in industrial apple waste by fermentation with *Aspergillus* spp." *Biocatalysis and Agricultural Biotechnology* 25 (2020): 101562. doi:10.1016/j.bcab.2020.101562.

Iwasaka, Hiroaki, Tsunehiro Aki, Hirofumi Adachi, Kenshi Watanabe, Seiji Kawamoto, and Kazuhisa Ono. "Utilization of waste syrup for production of polyunsaturated fatty acids and xanthophylls by Aurantiochytrium." *Journal of Oleo Science* 62, no. 9 (2013): 729–736. doi:10.5650/jos.62.729. PMID: 24005017.

Jiang, Xin, Ziqi Cui, Lihua Wang, Hongjian Xu, and Yonggen Zhang. "Production of bioactive peptides from corn gluten meal by solid-state fermentation with *Bacillus subtilis* MTCC5480 and evaluation of its antioxidant capacity *in vivo*." *LWT* 131 (2020): 109767. doi:10.1016/j.lwt.2020.109767.

Kaprasob, Ratchadaporn, Orapin Kerdchoechuen, Natta Laohakunjit, and Promluck Somboonpanyakul. "B vitamins and prebiotic fructooligosaccharides of cashew apple fermented with probiotic strains *Lactobacillus* spp., *Leuconostoc mesenteroides* and *Bifidobacterium longum*." *Process Biochemistry* 70 (2018): 9–19.

Kar, B., R. Banerjee, and B. C. Bhattacharyya. "Microbial production of gallic acid by modified solid state fermentation." *Journal of Industrial Microbiology and Biotechnology* 23, no. 3 (1999): 173–177. doi:10.1038/sj.jim.2900713.

Kot, Anna M., Stanisław Błażejak, Iwona Gientka, Marek Kieliszek, and Joanna Bryś. "Torulene and torularhodin: 'New' fungal carotenoids for industry?" *Microbial Cell Factories* 17, no. 1 (2018): 1–14. doi:10.1186/s12934-018-0893-z.

Kot, Anna M., Stanisław Błażejak, Marek Kieliszek, Iwona Gientka, and Joanna Bryś. "Simultaneous production of lipids and carotenoids by the red yeast *Rhodotorula* from waste glycerol fraction and potato wastewater." *Applied Biochemistry and Biotechnology* 189, no. 2 (2019): 589–607. doi:10.1007/s12010-019-03023-z.

Kothri, Maria, Maria Mavrommati, Ahmed M. Elazzazy, Mohamed N. Baeshen, Tarek A. A. Moussa, and George Aggelis. "Microbial sources of polyunsaturated fatty acids (PUFAs) and the prospect of organic residues and wastes as growth media for PUFA-producing microorganisms." *FEMS Microbiology Letters* 367, no. 5 (2020): fnaa028. https://doi.org/10.1093/femsle/fnaa028.

Kumar, Vikash, and T. Satyanarayana. "Applicability of thermo-alkali-stable and cellulase-free xylanase from a novel thermo-halo-alkaliphilic *Bacillus halodurans* in producing xylooligosaccharides." *Biotechnology Letters* 33, no. 11 (2011): 2279–2285.

Lateef, A., and E. B. Gueguim-Kana. "Utilization of cassava wastes in the production of fructosyltransferase by *Rhizopus stolonifer* LAU 07." *Romanian Biotechnological Letters* 17, no. 3 (2012): 7309–7316.

Leite, Paulina, Cátia Silva, José Manuel Salgado, and Isabel Belo. "Simultaneous production of lignocellulolytic enzymes and extraction of antioxidant compounds by solid-state fermentation of agro-industrial wastes." *Industrial Crops and Products* 137 (2019): 315–322. doi:10.1016/j.indcrop.2019.04.044.

Liu, J. B., Y. Wang, Y. Guo, H. L. Xu, S. Y. Lin, and Y. G. Yin. "Optimization of bioactive peptides derived from egg white protein." *Jilin Daxue Xuebao (Gongxueban)/Journal of Jilin University (Engineering and Technology Edition)* 40 (2010): 389–394.

Lun, Ong Khai, Tan Bee Wai, and Liew Siew Ling. "Pineapple cannery waste as a potential substrate for microbial biotranformation to produce vanillic acid and vanillin." *International Food Research Journal* 21, no. 3 (2014): 953.

Madeira Jr, Jose Valdo, Vania Mayumi Nakajima, Juliana Alves Macedo, and Gabriela Alves Macedo. "Rich bioactive phenolic extract production by microbial biotransformation of Brazilian Citrus residues." *Chemical Engineering Research and Design* 92, no. 10 (2014): 1802–1810. doi:10.1016/j.cherd.2014.07.014.

Mechmeche, Manel, Faten Kachouri, Hamida Ksontini, and Moktar Hamdi. "Production of bioactive peptides from tomato seed isolate by *Lactobacillus plantarum* fermentation and enhancement of antioxidant activity." *Food Biotechnology* 31, no. 2 (2017): 94–113. doi:10.1080/08905436.2017.1302888.

Milessi, Thais S. S., Willian Kopp, Mayerlenis J. Rojas, Anny Manrich, Alvaro Baptista-Neto, Paulo W. Tardioli, Roberto C. Giordano, Roberto Fernandez-Lafuente, Jose M. Guisan, and Raquel LC Giordano. "Immobilization and stabilization of an endoxylanase from *Bacillus subtilis* (XynA) for xylooligosaccharides (XOs) production." *Catalysis Today* 259 (2016): 130–139.

Moayedi, A., L. Mora, M. C. Aristoy, et al. "ACE-inhibitory and antioxidant activities of peptide fragments obtained from tomato processing by-products fermented using *Bacillus subtilis*: Effect of amino acid composition and peptides molecular mass distribution." *Applied Biochemistry and Biotechnology* 181 (2017): 48–64. https://doi.org/10.1007/s12010-016-2198-1.

Muñiz-Márquez, Diana B., Juan C. Contreras, Raúl Rodríguez, Solange I. Mussatto, José A. Teixeira, and Cristóbal N. Aguilar. "Enhancement of fructosyltransferase and fructooligosaccharides production by *A. oryzae* DIA-MF in Solid-State Fermentation using aguamiel as culture medium." *Bioresource Technology* 213 (2016): 276–282.

Muñiz-Márquez, Diana B., José A. Teixeira, Solange I. Mussatto, Juan C. Contreras-Esquivel, Raúl Rodríguez-Herrera, and Cristóbal N. Aguilar. "Fructo-oligosaccharides (FOS) production by fungal submerged culture using aguamiel as a low-cost by-product." *LWT* 102 (2019): 75–79.

Mussatto, Solange I., Lina F. Ballesteros, Silvia Martins, Dulce A. F. Maltos, Cristóbal N. Aguilar, and José A. Teixeira. "Maximization of fructooligosaccharides and β-fructofuranosidase production by *Aspergillus japonicus* under solid-state fermentation conditions." *Food and Bioprocess Technology* 6, no. 8 (2013): 2128–2134.

Mussatto, Solange I., Giuliano Dragone, and Inês Conceicao Roberto. "Ferulic and p-coumaric acids extraction by alkaline hydrolysis of brewer's spent grain." *Industrial Crops and Products* 25, no. 2 (2007): 231–237. doi:10.1016/j.indcrop.2006.11.001.

Mussatto, Solange I., and José A. Teixeira. "Increase in the fructooligosaccharides yield and productivity by solid-state fermentation with *Aspergillus japonicus* using agro-industrial residues as support and nutrient source." *Biochemical Engineering Journal* 53, no. 1 (2010): 154–157.

nee'Nigam, Poonam Singh, and Ashok Pandey. "Solid-state fermentation technology for bioconversion of biomass and agricultural residues." In *Biotechnology for agro-industrial residues utilisation*, pp. 197–221. Springer, Dordrecht, 2009.

Nieto-Domínguez, Manuel, Laura I. de Eugenio, María J. York-Durán, Barbara Rodríguez-Colinas, Francisco J. Plou, Empar Chenoll, Ester Pardo, Francisco Codoñer, and María Jesús Martínez. "Prebiotic effect of xylooligosaccharides produced from birchwood xylan by a novel fungal GH11 xylanase." *Food Chemistry* 232 (2017): 105–113.

Niu, L. Y., S. T. Jiang, and L. J. Pan. "Preparation and evaluation of antioxidant activities of peptides obtained from defatted wheat germ by fermentation." *The Journal of Food Science and Technology* 50, no. 1 (2013 February): 53–61. doi:10.1007/s13197-011-0318-z.

Ortiz-Basurto, Rosa Isela, Gérald Pourcelly, Thierry Doco, Pascale Williams, Manuel Dornier, and Marie-Pierre Belleville. "Analysis of the main components of the aguamiel produced by the maguey-pulquero (*Agave mapisaga*) throughout the harvest period." *Journal of Agricultural and Food Chemistry* 56, no. 10 (2008): 3682–3687.

Papaioannou, Emmanouil H., and Maria Liakopoulou-Kyriakides. "Agro-food wastes utilization by *Blakeslea trispora* for carotenoids production." *Acta Biochimica Polonica* 59, no. 1 (2012).

Park, Won-Kun, Myounghoon Moon, Sung-Eun Shin, Jun Muk Cho, William I. Suh, Yong Keun Chang, and Bongsoo Lee. "Economical DHA (Docosahexaenoic acid) production from *Aurantiochytrium* sp. KRS101 using orange peel extract and low cost nitrogen sources." *Algal Research* 29 (2018): 71–79. https://doi.org/10.1016/j.algal.2017.11.017.

Patil, P. R., G. S. N. Reddy, and M. B. Sulochana. "Production, optimization and characterization of-fructofuranosidase by *Chrysonilia sitophila* PSSF84—A novel source." *Indian Journal of Biotechnology* 10 (2011): 56–64.

Pérez-Guerra, N., A. Torrado-Agrasar, C. López-Macias, and L. Pastrana. "Main characteristics and applications of solid substrate fermentation." *Electronic Journal of Environmental, Agricultural and Food Chemistry* 2, no. 3 (2003).

Prenosil, J. E., E. Stuker, and J. R. Bourne. "Formation of oligosaccharides during enzymatic lactose: Part I: State of art." *Biotechnology and Bioengineering* 30, no. 9 (1987): 1019–1025.

Rashid, Rozina, and Muhammad Sohail. "Xylanolytic Bacillus species for xylooligosaccharides production: A critical review." *Bioresources and Bioprocessing* 8, no. 1 (2021): 1–14.

RedCorn, Raymond, Samira Fatemi, and Abigail S. Engelberth. "Comparing end-use potential for industrial food-waste sources." *Engineering* 4, no. 3 (2018): 371–380. doi:10.1016/j.eng.2018.05.010.

Reddy, P. P., G. S. N. Reddy, and M. B. Sulochana. "Screening of β-fructofuranosidase producers with high transfructosylation activity and its 32 experimental run studies on reaction rate of enzyme." *Journal of Biological Sciences* 10, no. 3 (2010): 237–241.

Reddy, Shyam Sunder, and Chandraraj Krishnan. "Production of xylooligosaccharides in SSF by *Bacillus subtilis* KCX006 producing β-xylosidase-free endo-xylanase and multiple xylan debranching enzymes." *Preparative Biochemistry and Biotechnology* 46, no. 1 (2016): 49–55. doi:10.1080/10826068.2014.970694.

Robledo, Armando, Antonio Aguilera-Carbó, Raúl Rodriguez, José Luis Martinez, Yolanda Garza, and Cristobal N. Aguilar. "Ellagic acid production by *Aspergillus niger* in solid state fermentation of pomegranate residues." *Journal of Industrial Microbiology and Biotechnology* 35, no. 6 (2008): 507–513. doi:10.1007/s10295-008-0309-x.

Ruiz, Yolanda, Bernadette Klotz, Juan Serrato, Felipe Guio, Jorge Bohórquez, and Oscar F. Sánchez. "Use of spent osmotic solutions for the production of fructooligosaccharides by *Aspergillus oryzae* N74." *Food Science and Technology International* 20, no. 5 (2014): 365–372.

Ryu, Byung-Gon, Kyochan Kim, Jungmin Kim, Jong-In Han, and Ji-Won Yang. "Use of organic waste from the brewery industry for high-density cultivation of the docosahexaenoic acid-rich microalga, *Aurantiochytrium* sp. KRS101." *Bioresource Technology* 129 (2013): 351–359. https://doi.org/10.1016/j.biortech.2012.11.049.

Saeed, Shagufta, Sofia Aslam, Tahir Mehmood, Rahat Naseer, Sadia Nawaz, Huma Mujahid, Sehrish Firyal, Aftab Ahmed Anjum, and Aeysha Sultan. "Production of gallic acid under solid-state fermentation by utilizing waste from food processing industries (2021)." *Waste and Biomass Valorization* 12, no. 1 (2021): 155–163. doi:10.1007/s12649-020-00980-z.

Salampessy, Junus, Michael Phillips, Saman Seneweera, and Kasipathy Kailasapathy. "Release of antimicrobial peptides through bromelain hydrolysis of leatherjacket (*Meuchenia* sp.) insoluble proteins." *Food Chemistry* 120, no. 2 (2010): 556–560.

Samanta, Ashis Kumar, A. P. Kolte, A. V. Elangovan, A. Dhali, S. Senani, M. Sridhar, K. P. Suresh, N. Jayapal, C. Jayaram, and Sohini Roy. "Value addition of corn husks through enzymatic production of xylooligosaccharides." *Brazilian Archives of Biology and Technology* 59 (2016).

Sánchez, Carmen. "Lignocellulosic residues: Biodegradation and bioconversion by fungi." *Biotechnology Advances* 27, no. 2 (2009): 185–194.

Santos, Tacila Rayane Jericó, Alessandra Gabrielly Santos Vasconcelos, Luciana Cristina Lins de Aquino Santana, Nayjara Carvalho Gualberto, Paula Ribeiro Buarque Feitosa, and Airla Carla Pires de Siqueira. "Solid-state fermentation as a tool to enhance the polyphenolic compound contents of acidic Tamarindus indica by-products." *Biocatalysis and Agricultural Biotechnology* 30 (2020): 101851.

Santos-Zea, Liliana, Ana Maria Leal-Diaz, Enrique Cortes-Ceballos, and Janet Alejandra Gutierrez-Uribe. "Agave (*Agave* spp.) and its traditional products as a source of bioactive compounds." *Current Bioactive Compounds* 8, no. 3 (2012): 218–231.

Sathish, Thadikamala, and Reddy Shetty Prakasham. "Intensification of fructosyltransferases and fructo-oligosaccharides production in solid state fermentation by *Aspergillus awamori* GHRTS." *Indian Journal of Microbiology* 53, no. 3 (2013): 337–342.

Sharma, Manisha, Satya Narayan Patel, Kusum Lata, Umesh Singh, Meena Krishania, Rajender S. Sangwan, and Sudhir P. Singh. "A novel approach of integrated bioprocessing of cane molasses for production of prebiotic and functional bioproducts." *Bioresource Technology* 219 (2016): 311–318. doi:10.1016/j.biortech.2016.07.131.

Silos-Espino, G., N. González-Cortés, A. Carrillo-López, F. Guevaralara, M. E. Valverde-González, and O. Paredes-López. "Chemical composition and *in vitro* propagation of *Agave salmiana* 'gentry'." *The Journal of Horticultural Science and Biotechnology* 82, no. 3 (2007): 355–359.

Smaali, Issam, Souhir Jazzar, Asma Soussi, Murielle Muzard, Nathalie Aubry, and M. Nejib Marzouki. "Enzymatic synthesis of fructooligosaccharides from date by-products using an immobilized crude enzyme preparation of β-D-fructofuranosidase from *Aspergillus awamori* NBRC 4033." *Biotechnology and Bioprocess Engineering* 17, no. 2 (2012): 385–392.

Teles, Aline S. C., Davy W. H. Chávez, Raul A. Oliveira, Elba P. S. Bon, Selma C. Terzi, Erika F. Souza, Leda M. F. Gottschalk, and Renata V. Tonon. "Use of grape pomace for the production of hydrolytic enzymes by solid-state fermentation and recovery of its bioactive compounds." *Food Research International* 120 (2019): 441–448. doi:10.1016/j.foodres.2018.10.083.

Torres-León, Cristian, Nathiely Ramírez-Guzmán, Juan Ascacio-Valdes, Liliana Serna-Cock, Maria T. dos Santos Correia, Juan C. Contreras-Esquivel, and Cristóbal N. Aguilar. "Solid-state fermentation with *Aspergillus niger* to enhance the phenolic contents and antioxidative activity of Mexican mango seed: A promising source of natural antioxidants." *LWT* 112 (2019): 108236. doi:10.1016/j.lwt.2019.06.003.

Tovar, Luis Raul, Manuel Olivos, and Ma Eugenia Gutierrez. "Pulque, an alcoholic drink from rural Mexico, contains phytase. Its *in vitro* effects on corn tortilla." *Plant Foods for Human Nutrition* 63, no. 4 (2008): 189–194.

Unagul, Panida, Caetharin Assantachai, Saranya Phadungruengluij, Manop Suphantharika, Morakot Tanticharoen, and Cornelis Verduyn. "Coconut water as a medium additive for the production of docosahexaenoic acid (C22: 6 n3) by *Schizochytrium mangrovei* Sk-02." *Bioresource Technology* 98, no. 2 (2007): 281–287. https://doi.org/10.1016/j.biortech.2006.01.013.

Vattem, Dhiraj A., and Kalidas Shetty. "Ellagic acid production and phenolic antioxidant activity in cranberry pomace (Vaccinium macrocarpon) mediated by Lentinus edodes using a solid-state system." *Process Biochemistry* 39, no. 3 (2003): 367–379. doi:10.1016/s0032-9592(03)00089-x.

Wang, Haikuan, Shanting Zhang, Yan Sun, and Yujie Dai. "ACE-inhibitory peptide isolated from fermented soybean meal as functional food." *International Journal of Food Engineering* 9, no. 1 (2013): 1–8. https://doi.org/10.1515/ijfe-2012-0207.

Wildman, Robert E.C., and Taylor C. Wallace. *Handbook of nutraceuticals and functional foods*. Boca Raton, Florida: CRC Press, 2016.

Xue, Zhaohui, Haichao Wen, Lijuan Zhai, Yanqing Yu, Yanni Li, Wancong Yu, Aiqing Cheng, Cen Wang, and Xiaohong Kou. "Antioxidant activity and anti-proliferative effect of a bioactive peptide from chickpea (*Cicer arietinum* L.)." *Food Research International* 77 (2015): 75–81.

Yamasaki, Takashi, Tsunehiro Aki, Masami Shinozaki, Masahiro Taguchi, Seiji Kawamoto, and Kazuhisa Ono. "Utilization of Shochu distillery wastewater for production of polyunsaturated fatty acids and xanthophylls using thraustochytrid." *Journal of Bioscience and Bioengineering* 102, no. 4 (2006): 323–327. https://doi.org/10.1263/jbb.102.323.

Yin, Feng-Wei, Si-Yu Zhu, Dong-Sheng Guo, Lu-Jing Ren, Xiao-Jun Ji, He Huang, and Zhen Gao. "Development of a strategy for the production of docosahexaenoic acid by *Schizochytrium* sp. from cane molasses and algae-residue." *Bioresource Technology* 271 (2019): 118–124. https://doi.org/10.1016/j.biortech.2018.09.114.

Zambelli, Paolo, Lucía Fernandez-Arrojo, Diego Romano, Paloma Santos-Moriano, María Gimeno-Perez, Ana Poveda, Raffaella Gandolfi, María Fernández-Lobato, Francesco Molinari, and F. J. Plou. "Production of fructooligosaccharides by mycelium-bound transfructosylation activity present in *Cladosporium cladosporioides* and *Penicilium sizovae*." *Process Biochemistry* 49, no. 12 (2014): 2174–2180.

Zheng, Zuoxing, and Kalidas Shetty. "Solid-state bioconversion of phenolics from cranberry pomace and role of *Lentinus edodes* β-glucosidase." *Journal of Agricultural and Food Chemistry* 48, no. 3 (2000): 895–900. doi:10.1021/jf990972u.

7 Sustainable Production of Polyhydroxyalkanoate (PHA) from Food Wastes

Sunanda Joshi, Monika Chaudhary, Varsha Upadhayay, and Arindam Kuila
Department of Bioscience and Biotechnology, Banasthali Vidyapith, Rajasthan-304022

CONTENTS

DOI: 10.1201/9781003191247-7

7.1 INTRODUCTION

Based on the current United Nations Food and Agriculture Organization (FAO), over a third of all food produced is wasted each year around the world. Food waste can originate from a range of sources, including home, municipal, industrial, and agricultural garbage, and its content varies greatly depending on the source and kind of trash (1).

PHAs are one-of-a-kind polyesters formed spontaneously inside the cellular structure of many bacteria when they are faced with growth-limiting conditions, such as a shortage of food, electron donor, or acceptor (2). Food waste is a worldwide issue that impacts the full cycle of food production, from cultivation to storage, packing, and disposal (3). Another factor to consider when it comes to food waste is the amount of energy expended in the waste's manufacturing, processing, and transportation.

If an alternative method of converting food waste into value-added products can be identified, energy can be effectively turned into usable things (Figure 7.1). The conversion of food waste into tangible by-products has sparked a lot of interest, with systems being developed to produce biofuels, polymers, and a number of additional chemical feedstocks (4).

Food waste is a strong option for an economical carbon source because of its extensive availability and potential to solve major waste concerns when used to make PHAs (Figure 7.2). Food waste is a global concern that impacts all aspects

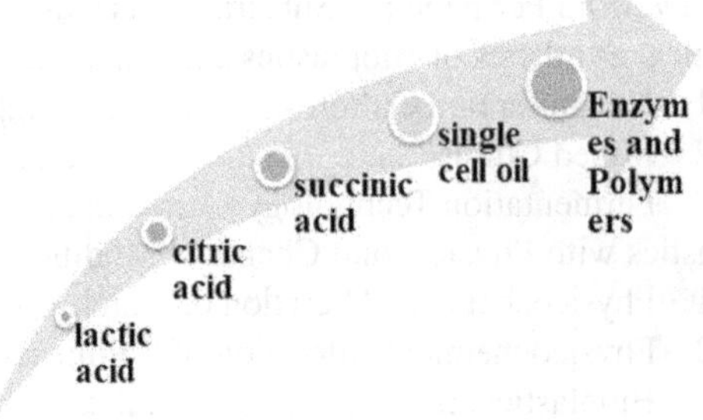

FIGURE 7.1 Food waste used in the production of high-value compounds.

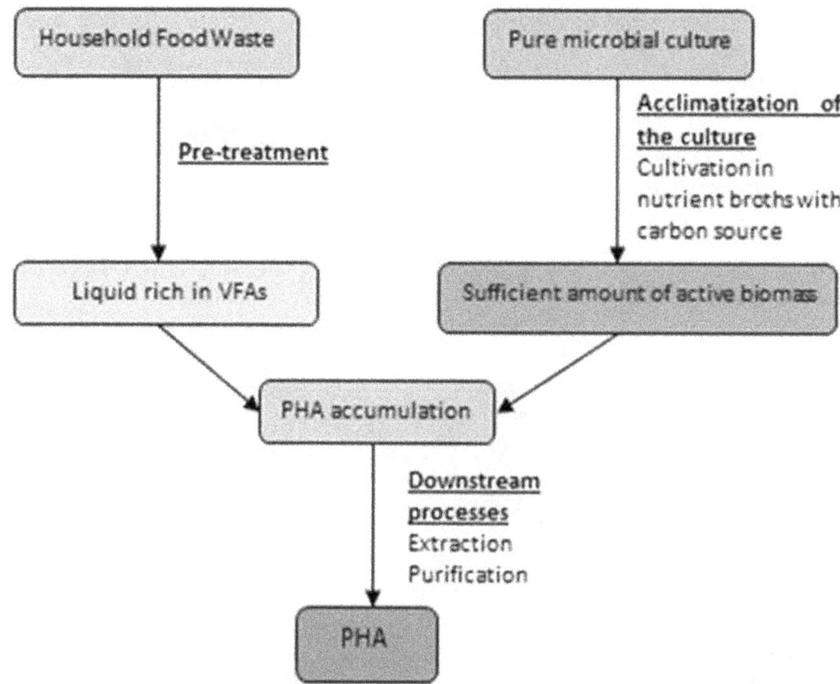

FIGURE 7.2 Food-waste-to-PHA production process.

of the food supply chain, from cultivation through warehousing, processing, and landfill (3). PHA and its copolymers are now being studied for use in a variety of industries and fields.

For example, "poly-(3-hydroxybutyrate) (P3HB) is a toxicologically safe homopolymer of PHAs with excellent biocompatibility in mammalian cells, making it ideal for pharmaceutical and food applications" (5–7).

Food waste is estimated to equal 88 million tons in Europe, with 57 million tons originating from families and restaurants (8). Almost 37 million tons of food were disposed of in US municipal solid waste systems in 2013, accounting for nearly 14% of total waste created in the country. PHA is gaining appeal as a "green plastic" that is bio-based, biocompatible, and biodegradable. Food waste (FW) is characterized as wasted food at the end of the supply chain, which is influenced by consumer preferences, perceived value, and retailer business models. Landfilling, composting, and fermentation are used to dispose of the majority of FW. "Food waste is an organic matrix rich in key components such as starch, cellulose, hemicelluloses, lignin, proteins, lipids, and organic acids that can be managed in a more sustainable and cost-effective manner by using it as a raw resource in bulk chemical manufacture."

These polymers are composed of monomers with wide variations in chain length (9–11). The PHB (Figure 7.3) is well-known since it is produced by various wild-type and environmental isolates when cultured on a variety of carbon sources (11–14). Bacteria that make biodegradable plastic are a major social and environmental issue.

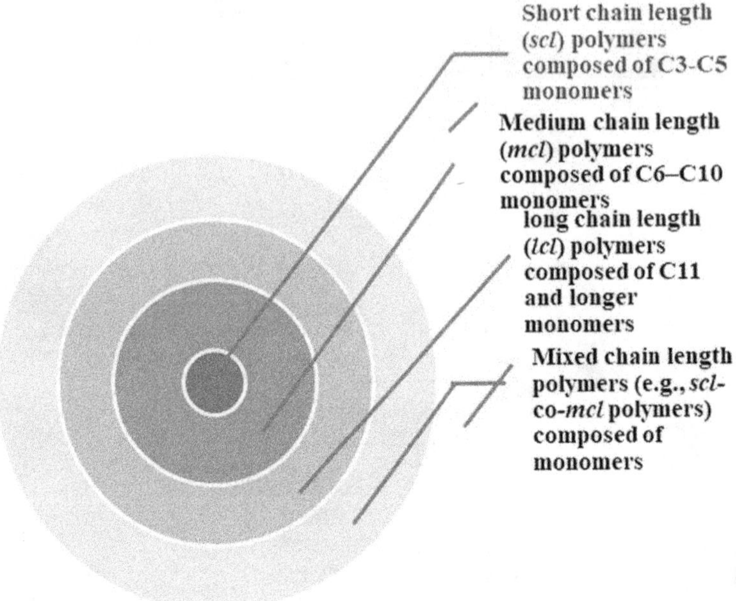

FIGURE 7.3 Polymers are divided into four categories.

The cost of making this bioplastic via conventional methods, on the other hand, is significantly greater than petrochemical-based polymers.

Sugar-rich feedstocks such as whey, molasses, starch, and lignocellulosic biomass are available as by-products of the food processing industry and are suitable for scl-PHA synthesis.

Because of their vulnerability to genetic modification, the bacteria are model microorganisms for studying the PHA synthesis-degradation cycle (gene knockouts and other DNA alterations). *C. necator* can also manufacture PHAs at a rate of up to 90% (w w-1) of their cell dry weight (CDW) (15–17).

PHAs are bio-based polyesters that have long been touted as excellent alternatives to petroleum-based polymers. These molecules are found in granules inside microbial species and have a variety of activities that have just recently been discovered (18), providing carbon storage and equivalent reductions (19). PHAs are biocompatible and biodegradable, having properties similar to those of traditional polymers. As a result, PHAs are now mass-produced in factories and employed in a wide range of items, such as packaging and medical devices.

This is a significant impediment to the commercial manufacture of PHA bioplastic for nonmedical applications.

In the recent literature, agricultural wastes, agro-industrial wastes, food wastes, and other wastes (Figure 7.4) (6) have been widely used for PHA production, especially when combined with pure microbial cultures. Food waste is a high-quality organic material that can give bacteria the nutrients they need to survive, thrive, and accumulate PHAs. It is a significant waste item that generates aromas, pollutes water,

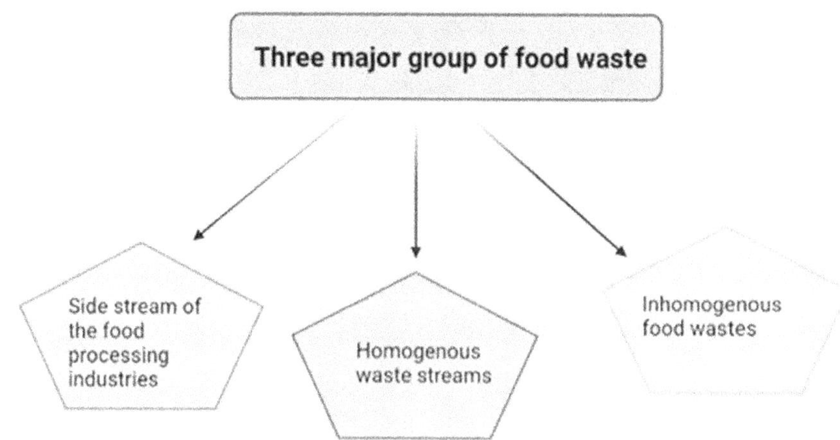

FIGURE 7.4 Major groups of food waste.

and emanates greenhouse emissions due to its high relative humidity and environmental friendliness (7).

7.2 FOOD WASTE AS RESOURCE OF PHA

Any food processing end product that is not consumed, recycled, or used for other purposes is considered as waste. Every year, almost one billion tons of food waste is produced worldwide (20, 21). It is worthy to point out the differences between industrial and household food wastes, especially if the purpose is to use these wastes as feedstocks in bioprocesses. Production of bioplastics like PHA is a great way to get rid of FW. Bioplastics made from FW are created using a long-term, renewable process that uses carbon-neutral resources (Figure 7.5).

Such bioplastics should comply with the standards and performance objectives of international norms for recyclable plastics/products. PHAs are nontoxic to humans and the environment. PHA decomposition is favored in settings where microbial growth is high, such as soil and sewage sludge.

For a successful biorefinery, the process must be adapted to the market needs and bulk biochemical productions driven by supply/demand issues as well as scales of the economy. PHAs are competing with relatively inexpensive, traditional petroleum-based polymers as bio-based, biodegradable thermoplastics. Therefore, the development of scalable, inexpensive PHA production processes is necessary to compete with these traditional thermoplastics.

Huschner et al. developed a high cell density cultivation method to produce P(HB-*co*-HV) from mixed organic acids using a dual acid-acid salt-feeding method (22). Fermentative lactic acid (LA) and polylactate (PLA) production strategies have been described using municipal food wastes as carbon sources.

Venus et al. investigated LA synthesis using two different production approaches of a one-step approach using simultaneous hydrolysis of macromolecules found in food wastes and fermentation of hydrolysates, and another two-step approach that

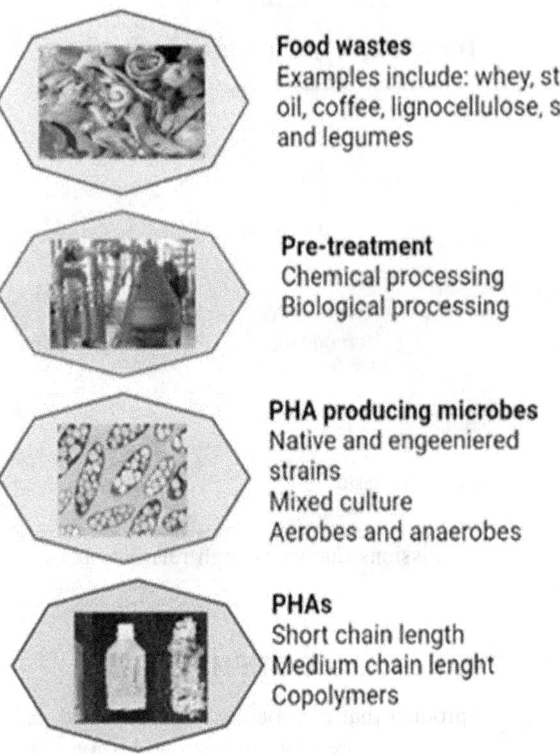

FIGURE 7.5 Food waste as resource of PHA.

separated the hydrolysis of waste carbons and subsequent fermentation of bioavailable carbons into separate steps.

7.2.1 PHAs from Whey

Whey is a by-product of the cheese-making methodology that carries a variety of nutrients, such as "lactose, proteins, lipids, water-soluble vitamins, mineral salts, and other minerals," all of which are necessary for microbial development. Regardless of the fact that they may be utilized to make lactose, casein, and protein powder, half of them are likely to be discarded (23).

Whey is a nutrient-dense media that promotes microbial growth. They're a form of energy-rich storage fluid that granulates in cells in difficult situations, like inadequate intake and so on. PHAs are seen as a particularly appealing and viable substitute for traditional plastics due to mechanical properties similar to fossil-derived polymers (such as polypropylene), biocompatibility, and complete low toxicity (24, 25).

Whey is a dairy industry by-product produced by precipitating and extracting milk casein. Whey is generated in vast quantities every year all around the world, but only about half of it is used to make human and animal nutrition products.

While genetic modification is a versatile and potential way of increasing PHA production, it does need the use of more closely monitored production lines. Indeed, widespread concern with genetically engineered microorganism regarding GMMs being mistakenly discharged into the environment and subsequently transmitting genes to microbiological and nonmicroorganism habitats has been raised in scientific research facilities and the biotechnology sector (26).

For whey-based PHA production, mixed microbial cultures (MMCs) have recently become a popular alternative to single pathogenic microorganisms (Figure 7.6). MMCs are frequently collected from domestic waste and modified for PHA synthesis via expanding the number of strains that can produce PHA-based carbon storage.

Because whey permeates or supernatants are sterile, uniform, and transparent, they are easier to deal with; nonetheless, their use increases manufacturing costs. Additionally, when entire whey and whey supernatants were used as carbon sources in a complex mineral environment, biomass development and PHA accumulation in *Methylobacterium* sp. ZP24 were revealed to be divergent (27). As a conclusion, interpreting results from whey penetrates or supernatants to total whey should be interpreted cautiously.

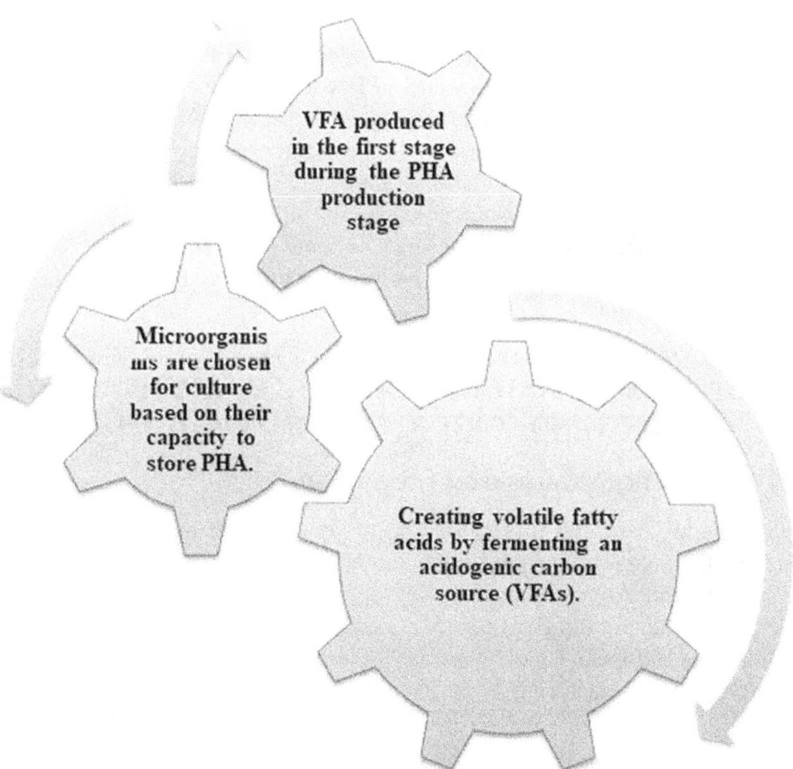

FIGURE 7.6 PHA production through three-step process using MMC.

Source: Duque et al. (2014); Colombo et al. (2016).

In PHA study, whole whey should be used more regularly. Enzymatic hydrolysis raises the cost of a prospective PHA generation technique dramatically. Whey lactose hydrolysates were also produced by chemical lactose hydrolysis using HCl, which might be used for microbiological PHA synthesis. Whey for PHA production could make it more cost-effective and environmentally friendly.

7.2.2 Waste Cooking Oil

Cooking oils, i.e., waste cooking oils, have been considered as a carbon source for the generation of PHA.

> Triacylglycerides (TAGS) are a promising alternative feedstock for the synthesis of mcl-PHAs and lcl-PHAs. They are generated from waste oil fatty acids and are a promising alternative feedstock for the synthesis of mcl-PHAs and lcl-PHAs. An investigation using crude glycerol as a nitrogen source in combination with meat and bone meals produced 5.9 gl-1 PHAs and a co-polymer PHBV without the use of a media precursor, saving even more expenditure. (28)

It has been proven that adding additives like salts or nitrogen sources to whey is necessary for obtaining optimal PHA production. As a result and as previously said, if these additions could be made from other food wastes, it would be tremendously advantageous. Whey has been utilized as a PHA supplement manufacture, while the use of whey as an addition for PHA production has yet to be fully researched. When *C. necator* was employed as a complex nitrogen source, adding protease-hydrolyzed whey to the mix increased PHA synthesis (29).

As a result, mixing waste sources opens up virtually endless possibilities for increasing PHA manufacturing efficiency without the use of costly chemicals. Low-cost additions, as well as alternative waste sources, could be investigated for increased PHA production.

7.3 THE CURRENT STATE OF BIOPLASTIC SYNTHESIS AND THE GATHERING AND SORTING OF FOOD WASTE FEEDSTOCK

7.3.1 Bioplastic Synthesis from Food Waste

Ravindran and Jaiswal (2016) reported postproduction, handling/storage, manufacturing, wholesale/retail, and consumption are all waste-producing processes in the food supply chain. Approximately 30% of food is classified as FW (30). The US generated 39.6 million tons of FW in 2015, but only 5.3% of it was utilized for anaerobic digestion. The European Union generates 90 million tons of FW per year. The food processing and manufacturing industries account for 38% of the total (31).

7.3.2 Current Status of Bioplastic Production

In the biodegradable polymer economy, PHA is one of the most important drivers. Chen et al. (2016) and Briassoulis and Giannoulis (2018) reported that PHA is a large polymer family that has been in research for a long time but is ready to go commercial, with manufacturing capacity projected to treble in the coming five years. Because of its biodegradability and rubbery properties, it has shown great promise as a replacement for traditional plastics (32, 33).

7.3.3 Technologies for Collecting and Classifying Food Waste as a Feedstock

Most components of FW can be recycled (if separated), reducing trash disposal costs. Food waste recycling rates, in particular, must be enhanced. Collecting and sorting garbage at the source can lower the rate of later processes, allowing for a well-thought-out approach to increase output and profit while lowering environmental impact and improving material recycling productivity.

Considering the enormous amounts of FW generated by industry and agriculture is extremely basic. The makeup of domestic FW, on the other hand, is considerably different.

7.4 FOOD-WASTE-TO-FERMENTABLE-SUBSTRATE CONVERSION TECHNOLOGIES

FW is a wonderful bioplastics starting material, but it must be altered to improve and change its physicochemical and biological properties. The most prevalent pretreatment, treatment, and even some cellulase-mediated saccharification are described in this section, and also their influence on bioplastic manufacturing output.

Barisik et al. (2016) and Kim (2018) reported that the voluntary or involuntary emancipation of monomers from FW (e.g., lignocellulosic components), combined with increased accessibility of proteins, lipids, and polysaccharides (e.g., starch and cellulose) for subsequent enzymatic hydrolysis and fermentation, results in efficient conversion methods (34, 35).

7.4.1 Fermented Products Obtained from Food Wastes through Biological Route

The fungus that causes white rot aids delignification by increasing the rate and productivity of enzymatic saccharification (36). Isroi et al. (2011) reported that fungi have been used as a pretreatment method for FW in several studies (37).

Cianchetta et al. (2014) examined the consequences of five distinct fungi on wheat straw enzymatic hydrolysis and identified white-rot fungi as potential wheat straw carbohydrate producers (38). *Ceriporiopsis subvermispora* produced the maximum net carbohydrate output while reducing weight and cellulose losses, based on the results of a biomass combination of the best fungal strains (39).

7.4.2 Fermentable Sugars Obtained from Food Waste through Chemical Route

Mussoline et al. (2013) stated that it is common to pretreat FW, particularly lignocellulosic materials (40).

> Alkaline treatment makes low-cost and abundant feedstocks more accessible to hydrolytic enzymes while reducing the amount of cell proliferation inhibitors (41). Acid-treated lignocellulose can be transformed into lignocellulose-derived byproducts, which can block or deactivate enzymes and affect fermentation bacteria activity.

Monavari et al. (2011), Barisik et al. (2016), and Kim (2018) reported that among the inhibitors are furan derivatives (such as furfural and hydroxymethylfurfural [HMF]), lignin-derived phenolics (such as phenols, which block cellulolytic enzymes), and several weak acids (41, 42, 43).

7.4.3 Food Waste Is Converted to Fermentable Organic Molecules via Mechanical and Thermal Processes

Physical preparation is accomplished by mechanical and thermal conversion techniques. Sasmal et al. (2012) and Pagliaccia et al. (2016) reported that outer layer, segregation efficiency, cellulase-mediated hydrolysis, or soluble substrates should all be increased (including sugar, amino, fats, and carbohydrates) (44, 45). PHB is used in a laboratory-scale PHA manufacturing technique, made from jambool seeds that were desiccated in a 60°C oven to reduce humidity content before being crushed into minute particles. Agbor et al. (2011) and Pielhop et al. (2016) reported that removing lignin and hemicellulose from cellulose via steam-explosion processing of lignocellulosic waste (160 to 260°C and 0.7 to 4.8 MPa) is a typical approach (46–49).

7.4.4 Cellulase-Mediated Saccharification of Food Waste

The most common method for turning polymers into monomers and/or intermediates is hydrolysis. Enzymatic hydrolysis improves FW's hydrolytic ability while lowering volatile suspended solids levels.

Although enzyme hydrolysis can convert cellulose and hemicellulose into fermentable sugars, lignin, due to its phenylpropanoid units, is one of the most resistant substances.

The authors discovered that among enzymes, carbohydrates, proteases, and lipases, protease had the highest rate of reduction of volatile suspended particles and that the combination-enzyme treatment was more effective than the single-enzyme treatment. Heng et al. (2017) reported utilization of *Burkholderia cepacia* to optimize PHA manufacturing method from rice husks through alkali treatment, enzymatic hydrolysis, and biosynthetic synthesis (50).

7.5 Bioplastics Made from Fermentable Substrates Such as Food Waste

7.5.1 Biological Synthesis of Bioplastics

For commercial PHA production, only a few microorganisms have been used, despite the identification of 250 different species of natural PHA producers. Among the bacteria discovered that convert diverse carbon sources into PHA are "*Alcaligenes latus*, *Bacillus megaterium*, *Cupriavidus necator*, and *Pseudomonas oleovorans*" (51). Although marine microorganisms have shown significant promise in the production of bioplastics, little attention has been paid to their usage in PHA production (52).

The bacterium *Halomonas hydrothermalis*, *H. campaniensis LS21* can grow in both artificial seawater and FW-like mixed substrates comprising carbohydrates, proteins, triglycerides, and essential fats, according to this study (53).

Pandian et al. (2010) investigated B. megaterium SRKP-3 for production of PHB from dairy waste and seawater (54).

7.5.1.1 Unadulterated Culture

PHA is made in two stages, starting with pure culture (55). The first is about cell growth in relation to nutrition availability. Furthermore, PHA biosynthesis is widespread in the second phase.

> Bacteria that produce PHA in response to growth, such as *A. latus* and recombinant *E.coli*, on the other hand, do not necessitate a significant amount of resources. However, in a fed-batch mode, the nutrient-feeding technique can be used to generate PHA. A nutrient-feeding method is needed for high-yield PHA synthesis in fed-batch cultures of growth-associated PHA-producing bacteria.

Because cell growth and PHA synthesis happen at the same time, this shows that both can be enhanced.

7.5.1.2 Mixed Culture

Microorganisms that may live in the very same broth culture are known as diversified microbial cultures (MMC). Samorì et al. (2015) and Hilliou et al. (2016a, 2016b) reported that when uptake of nutrients is limited, three methods for generating PHA from mixed cultures are used (56–58).

Ben et al. (2016) evaluated a number of mixed-culture modes, including a brewery effluent–fed aerobic sequencing batch reactor (SBR) (59).

It was previously found that the consortia formed when PHA was produced from FW fermentation utilizing a system of aeration tanks.

Proteobacteria (39%) and uncultured bacteria (16%) were the most frequent bacterial groupings according to the results of the gene expression studies. Pure culture reduces the requirement for aseptic conditions and lowers operational expenses.

7.5.1.3 Fermentation Technology

It has been established that the type of fermentation used by PHA-producing bacteria plays a critical role in the synthesis of bioplastics, thanks to developments in culture techniques used in significant PHA manufacture (Figure 7.7).

Fed-batch technology produces more PHA than batch culture methods. Because the percentage of N/P in this process is limited, cell concentration can be easily controlled by changing the carbon source's feed intake velocity. As a result, a sufficiently high fraction of carbon sources can be avoided in order to provide PHA producers with substantial osmotic pressure.

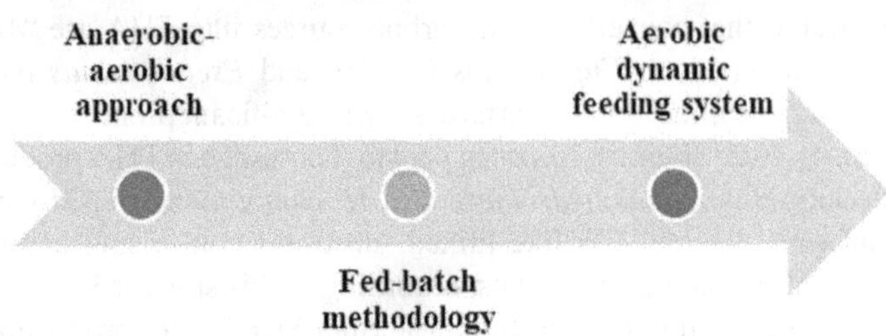

FIGURE 7.7 Methodology to generate PHA from mixed cultures.

7.5.2 Bioplastics with Physical and Chemical Modifications

The classic biodegradable plastic synthesis technique comprises physical melting and/or chemical cross-linking to entirely employ feedstock or a single entity (e.g., fiber, fat). Recent research has focused on employing industrial chemical biotechnology to convert biomass into novel polymers by separating monomers or oligomers.

Hemiacetal or hemiketal connections can be found in polysaccharides. Polymeric repeating units or short oligosaccharide sequences are produced by these compounds and are coupled to other bioplastics. Polysaccharides make up between 22 and 37% of all resources domestic waste (60).

7.5.2.1 Physicochemical Alteration of Starch-Based Bioplastics

There is a lot of information out there about chemically altering starch. Substantial alkaline treatment of crystalline parts of starch granules can create starch nanocrystals. Around the world, bread waste is produced at a pace of up to 27 million kilograms each year. Because starch is the main constituent in bread, bioplastics made from a succinic acid monomer are a viable choice for making bioplastics from it.

Hydrothermal treatment can change the configuration of starch without causing thermal loss of the original shape. Dextrose from maize or other carbohydrate sources can be chemically converted to PLA. Lactic acid is produced by fermenting dextrose, which is subsequently polycondensed with lactic acid monomers or lactide.

7.5.2.2 Physicochemical Alteration of Cellulose-Based Bioplastics

Often, chemically altered cellulose is utilized. Cellulose is the main polymer made composed of three hydroxyl groups in the monomer and anhydro D-glucopyranose repeating units. High-molecular-weight cellulose has a very high crystalline structure. Due to the strong intermolecular and intramolecular hydrogen interactions that occur between and within the individual chains, cellulose is poorly soluble in aqueous conditions (61).

Some researchers have concentrated on generating covalent-active patches on the cellulose surface to improve adhesion between "polar OH cellulose fibre groups and non-polar polymer backbones."

7.5.2.3 Physicochemical Alteration of Chitin-Based Biopolymers

Chitin is a biopolymer that can be found in the exoskeletons of crustaceans, insects, mushrooms, and yeasts. Shell waste, which is a prominent chitin source, is produced annually in the amount of 18 Tg. Because of its limited solubility, chitin cannot be employed as an initial feedstock. Chitin, on the other hand, can be transformed to chitosan through a chemical process (62, 63). The equivalent main amino functional group is produced via deacetylation. It is measured as a percentage of acetyl glucosamine to glucosamine conversion. Chitin's physical, chemical, and biological properties are all affected by deacetylation

7.5.2.4 Physicochemical Alteration of Caprolactone-Based Bioplastics

An essential step in the manufacture of PCL is the ring-opening polymerization of caprolactone. The polymer chain has been found to have segmental mobility as a result of the low percolation threshold, which should aid in the movement of ions (64).

The chemical treatment of saccharides, which consists of a two-step conversion, is the most significant implementation (Figure 7.8).

These products are usually combined with other organic substances to produce high-quality biodegradable polymers at an affordable price.

7.5.3 Food Waste Is Used as a Substrate for the Manufacture of PHA

The possibility of generating PHA utilizing food waste as a substrate has been comprehensively explored (Figure 7.9), including data on maximal PHA production rate, storage yield, and accumulation capacity (65).

7.5.3.1 Food Waste as a Source of Carbon

Carbon substrates for the PHA fermentation process usually include fruits and vegetables. Half of the fresh fruit is extracted, and the remainder is discarded in the production of some fruit products.

The sugar concentration of these wastes is high, but the protein level is minimal. Citrus wastes are used to make "citric acid, succinic acid, dietary fibre, prebiotic oligosaccharides, and natural antioxidant enzymes (especially pectinase)," all of which

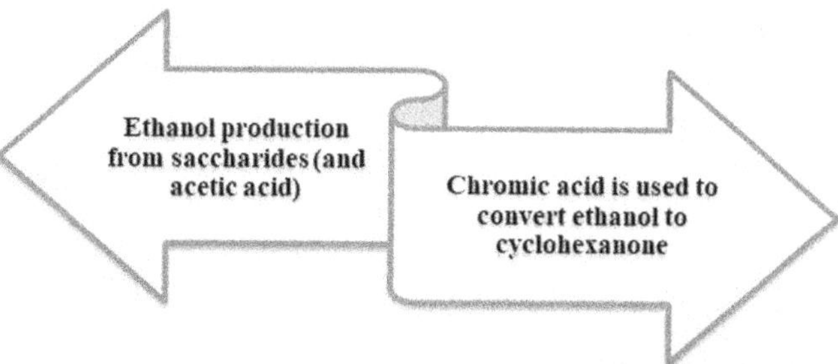

FIGURE 7.8 A Baeyer-Villiger reaction converts caprolactone to cyclohexanone.

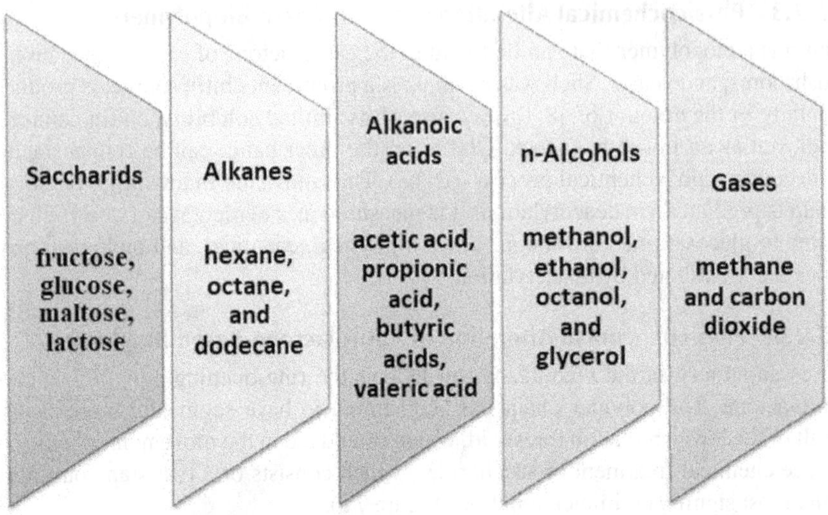

FIGURE 7.9 Key carbon sources for biosynthesis of PHA.

can help increase profit (66) in septic systems. In addition, simultaneous fermentation of potato wastes was carried out (67).

7.6 CASE STUDIES OF PHAs MADE FROM FOOD AND AGRICULTURAL WASTE

A number of factors influence the efficiency of bioplastic synthesis (e.g., the kind of waste produced in the target location, operational cost, and feedstock accessibility, transportation costs, raw material costs, and other typical industrial costs). In addition to great productivity, industrial bioplastic production must adhere to government pollution guidelines.

Seasonal oscillations may have an impact on bioplastics manufacturing, limiting the delivery of particular raw resources for a phase of time, due to the different types of waste produced in each season.

However, in industrial-scale manufacturing, technoeconomic profitability, access to raw materials, and industrial products are all important factors. Markets must all be assessed or analyzed.

7.6.1 Palm Tree Biomass–Based Refineries: Case Studies

A range of influences impact the efficient, premium, and economically safe production of bioplastics. The most essential aspect in the start-up of bioplastic manufacturing is the availability of substitutes.

> The site of a biorefinery should provide for a consistent supply of raw materials from adjacent places and in 2006, Malaysia was the world's leading exporter and producer

of palm oil (producing 15.88 million tonnes, accounting for 43 percent of total global supply).

Milling for palm oil production and crops resulted in a large aquifer. Palms cover millions of hectares in Malaysia, for example.

For instance, consider the amount of waste biomass, such as EFB. One hectare of palm oil plantation can yield 50–70 tons of biomass waste, according to estimates. As a result, biofuels can be made from waste biomass

7.6.2 Biomass-Based Refineries That Use Bananas as a Feedstock: Case Studies

In 2012, 6.67% of total banana harvests were squandered and 26.46% were rejected. (68). In equal amounts, these waste agricultural and food products can be used to generate biofuels, sugars, and PHB (68). The starch content of banana peels is usually about 40%; after ripening, it may be turned to sugars and used to make a range of products (68).

PHB was a one-of-a-kind product; it was made from hydrolyzed starch derived from banana pulp. A biorefinery generated glucose, ethanol, and PHB.

The final scenario proposed that the processes be integrated in terms of mass and energy. In scenarios 1, 2, and 3, the costs for PHB, glucose, and ethanol were estimated to be 2.7/2.3/1.6, -/0.9/0.7, and -/1.3/0.6 USD kg1, respectively. Colombian labor costs of "USD 2.14 h1 and USD 4.29 h1" were used in the cost analysis for operators and supervisors, respectively. Furthermore, comparing revealed corresponding economic margins of 22/43/106%, -/0/22.2%, and -/18.2/45.5%. The manufacturing of PHB from leftover banana can cut energy and water consumption by 30.6% and 35%, respectively.

7.6.3 Case Studies of Biomass-Based Sugarcane Refineries

The cost of sugarcane used to create PHA is ten times lower than the cost of pure sugar (i.e., sugarcane 0.04 USD kg1 vs. glucose 0.44 USD kg1) (69) and the cost of creating P (3HB) was calculated to be USD 3.44 kg1 P (3HB) in another case study, which is 41% less than the cost of generating commercial glucose. (70)

Three scenarios based on Cambodian conditions were used to make the technoeconomic assessment (Figure 7.10).

7.7 CONCLUSIONS

Plastic manufacture and accumulation have disastrous environmental consequences, necessitating the development of environmentally acceptable plastic alternatives. PHAs seemed to be viable alternatives to distinctive plastics in this context since they are biodegradable, are biocompatible, and can be manufactured biologically. Bioplastic is a natural polymeric material with outstanding biocompatibility, biodegradability, and material qualities that has witnessed substantial development in the previous two decades.

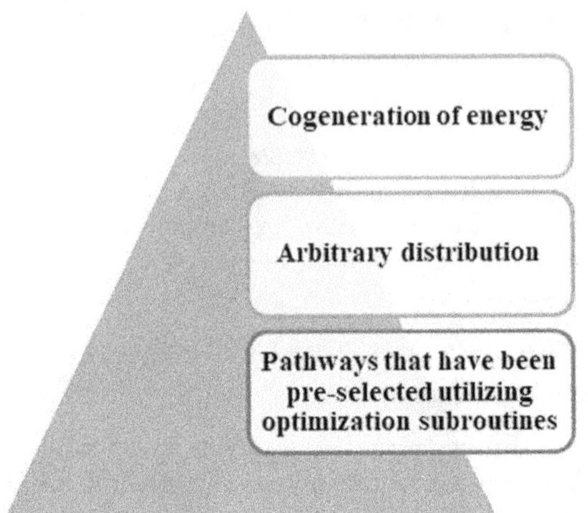

FIGURE 7.10 Scenarios based on Cambodian conditions to make the technoeconomic assessment.

The interconnection of biotechnology processes in the manufacturing of bioplastics is a major technique for maximizing the usage of food waste while also boosting the overall bioprocessing chain's potential profitability. Given the inevitability of FW generation, the environmental costs of waste dumping should be kept to a minimum. As an outcome, the perspective of FW for bioplastic manufacturing to address critical environmental challenges was established. As a result, the physical, thermochemical, and biological procedures for preparing bioplastic raw materials from FW have been presented. It was also emphasized that PHA should be produced using unadulterated/mixed culture and fermentation processes. Most significantly, governments, regulators, corporations, the public, and customers should work together to eliminate the present environmental implications of FW disposal.

REFERENCES

1. Xue, L., Liu, G., Parfitt, J., Liu, X., Van Herpen, E., Stenmarck, Å., O'Connor, C., Östergren, K., Cheng, S. Missing food, missing data? A critical review of global food losses and food waste data. *Environ. Sci. Technol.*, 51 (2017), pp. 6618–6633.
2. Serafim, L.S., Lemos, P.C., Albuquerque, M.G.E., Reis, M.A.M. Strategies for PHA production by mixed cultures and renewable waste materials. *Appl. Microbiol. Biotechnol.*, 81 (2008), pp. 615–628. Doi:10.1007/s00253-008-1757-y.
3. Parfitt, J., Barthel, M., Macnaughton, S. Food waste within food supply chains: Quantification and poten-tial for change to 2050. *Philos. Trans. R Soc. Lond. B Biol. Sci.*, 365 (2010), pp. 3065–3081.
4. Matharu, A.S., de Melo, E.M., Houghton, J.A. Opportunity for high value-added chemicals from food supply chain wastes. *Bioresour Technol.*, 215 (2016), pp. 123–130.
5. Tan, G.Y.A., Chen, C.L., Li, L., Ge, L., Wang, L., Razaad, I.M.N., Li, Y., Zhao, L., Mo, Y., Wang, J.Y. Start a research on biopolymer polyhydroxyalkanoate (PHA): A review. *Polymers*, 6(3) (2014), pp. 706–754.

6. Shen, L., Haufe, J., Patel, M.K. *PRO-BIP* (2009), pp. 111–112. www.chem.uu.nl/nws/www/research/e&e/PROBIP2009%20Final%20June%202009.pdf.
7. Bugnicourt, E., Cinelli, P., Lazzeri, A., Alvarez, V. Polyhydroxyalkanoate (PHA): Review of synthesis, characteristics, processing and potential applications in packaging. *Express. Polym. Lett.*, 8(11) (2014), pp. 791–808.
8. Stenmarck, A., Jensen, C., Quested, T., Moates, G. Estimates of European food waste levels. *IVL Swedish Environmental Research Institute* 79 (2016).
9. Riedel, S.L., Lu, J., Stahl, U., Brigham, C.J. Lipid and fatty acid metabolism in Ralstonia eutropha: Relevance for the biotechnological production of value-added products. *Appl. Microbiol. Biotechnol.*, 98 (2014), pp. 1469–1483. Doi:10.1007/s00253-013-5430-8.
10. Sudesh, K., Abe, H., Doi, Y. Synthesis, structure and properties of polyhydroxyalkanoates: Biological polyesters. *Prog Polym Sci.* 25(10) (2000), pp. 1503–1555. Doi:10.1016/S0079-6700(00)00035-6.
11. Reinecke, F., Steinbuchel, A. Ralstonia eutropha strain H16 as model organism for PHA metabolism and for biotechnological production of technically interesting biopolymers. *J Mol Microbiol Biotechnol.*, 16 (2008), pp. 91–108. Doi:10.1159/000142897.
12. Wilde, E. Untersuchungen uber Wachstum und Speich- erstoffsynthese von Hydrogenomonas. *Arch fur Mikrobiol.*, 43 (1962), pp. 109–137. Doi:10.1007/BF00406429.
13. James, B.W., Mauchline, W.S., Dennis, P.J., Keevil, W., Wait, R., Keevil, C.W. Poly-3-Hydroxybutyrate in Legionella pne-umophila, an energy source for survival in low nutrient environments poly-3-hydroxybutyrate in Legionella pne-umophila, an energy source for survival in low-nutrient environments. *Appl. Environ. Microbiol.*, 65 (1999), pp. 822–827.
14. Qi, Q., Rehm, B.H.A. Polyhydroxybutyrate biosynthesis in Caulobacter crescentus: Molecular characterization of the polyhydroxybutyrate synthase. *Microbiol.*, 47 (2001), pp. 3353–3358. Doi:10.1099/00221287-147-12-335.
15. Steinbuchel, A. Polyhydroxyalkanoic acids. *Biomater. Nov. Mater.* Springer; Germany, 1991, pp. 123–213. Doi:10.1007/978-1-349-11167-1_3 3. Brigham, C.J., Budde, C.F., Holder, J.W., Zeng, Q., Mahan, A.E., Rha, C.K., Sinskey, A.J. Elucidation of β-oxidation pathways in Ralstonia eutropha H16 by examination of global gene expression. *J. Bacteriol.*, 192 (2010), pp. 5454–5464. Doi:10.1128/JB.00493-10.
16. Brigham, C.J., Budde, C.F., Holder, J.W., Zeng, Q., Mahan, A.E., Rha, C.K., Sinskey, A.J. Elucidation of β-oxidation pathways in Ralstonia eutropha H16 by examination of global gene expression. *J. Bacteriol.*, 192 (2010), pp. 5454–5464. Doi:10.1128/JB.00493-10.
17. Riedel, S.L., Bader, J., Brigham, C.J., Budde, C.F., Yusof, Z.A.M., Rha, C., Sinskey, A.J. Production of poly(3-hydroxybutyrate-co-3-hydroxyhexanoate) by Ralstonia eutropha in high cell density palm oil fermentations. *Biotechnol Bioeng.*, 109 (2012), pp. 74–83. Doi:10.1002/bit.23283.
18. Koller, M., Maršálek, L., de Sousa Dias, M.M., Braunegg, G. Producing microbial polyhydroxyalkanoate (PHA) biopolyesters in a sustainable manner. *New Biotechnol.*, 37 (2017), pp. 24–38. Doi:10.1016/j.nbt.2016.05.001.
19. Madison, L.L., Huisman, G.W. Metabolic engineering of poly(3-hydroxyalkanoates): From DNA to plastic. *Microbiol. Mol. Biol. Rev.*, 63 (1999), pp. 21–53.
20. Venus, J., Fiore, S., Demichelis, F., Pleissner, D. Centralized and decentralized utilization of organic residues for lactic acid production. *J. Clean Prod.* 172 (2018), pp. 778–785. Doi:10.1016/j.jclepro.2017.10.259.
21. Kwan, T.H., Hu, Y., Lin, C.S.K. Techno-economic analysis of a food waste valorization process for lactic acid, lactide and poly(lactic acid) production. *J. Clean. Prod.*, 181 (2018), pp. 72–87. Doi:10.1016/j.jclepro.2018.01.179.
22. Huschner, F., Grousseau, E., Brigham, C.J., Plassmeier, J., Popovic, M., Rha, C., Sinskey, A.J. Development of a feeding strategy for high cell and PHA density fed-batch fermentation of Ralstonia eutropha H16 from organic acids and their salts. *Process Biochem.*, 50 (2015), pp. 165–172. Doi:10.1016/j.procbio.2014.12.004.

23. Pescuma, M., Valdez, G.F. De., Mozzi, F. Whey-derived valuable products obtained by microbial fermentation. *Appl. Microbiol. Biotechnol.*, 99 (2015), pp. 6183–6196.
24. Nath, A., Dixit, M., Bandiya, A., Chavda, S., Desai, A.J. Enhanced PHB production and scale up studies using cheese whey in fed batch cultures of *Methylobacterium* sp ZP24. *Biores Techn.*, 99(13) (2008), pp. 5749–5755.
25. Gahlawat, G., Srivastava, A.K. Model-based nutrient feeding strategies for the increased production of polyhydroxybutyrate (PHB) by *Alcaligenes latus. Appl. Biochem. Biotechnol.*, 183(2) (2017), pp. 530–542.
26. Colombo, B., Favini, F., Scaglia, B., Sciarria, T.P., D'Imporzano, G., Pognani, M., et al. Enhanced polyhydroxyalkanoate (PHA) production from the organic fraction of municipal solid waste by using mixed microbial culture. *Biotechnol. Biofuels.*, 10 (2017), pp. 1–15. Doi:10.1186/s13068-017-0888-8.
27. Yellore, V., Desai, A. Production of poly-3-hydroxybutyrate from lactose and whey by *Methylobacterium* sp. ZP24. *Lett. Appl. Microbiol.*, 26 (1998), pp. 391–394. Doi:10.1046/j.1472–765x.1998.00362.x.
28. Koller, M., Bona, R., Braunegg, G., Hermann, C., Horvat, P., Kroutil, M., Martinz, J., Neto, J., Pereira, L., Varila, P. Production of polyhydroxyalkanoates from agricultural waste and surplus materials. *Biomacromolecules.*, 6 (2005), pp. 561–565.
29. Obruca, S., Benesova, P., Petrik, S., Oborna, J., Prikryl, R., Marova, I. Production of polyhydroxyalkanoates using hydrolysate of spent coffee grounds. *Process Biochem.*, 49 (2014), pp. 1409–1414. Doi:10.1016/j.procbio.2014.05.013.
30. Ravindran, R., Jaiswal, A. Exploitation of food industry waste for high-value products. *Trends Biotechnol.*, 34 (2016), pp. 58–69.
31. Pfaltzgraff, L.A., De bruyn, M., Cooper, E.C., Budarin, V., Clark, J.H. Food waste biomass: A resourcefor high-value chemicals. *Green Chem.*, 15 (2013), pp. 307–314.
32. Chen, G.Q., Jiang, X.R., Guo, Y. Synthetic biology of microbes synthesizing polyhydroxyalkanoates (PHA). *Synth. Syst. Biotechnol.*, 1(4) (2016), pp. 236–242.
33. Briassoulis, D., Giannoulis, A. Evaluation of the functionality of bio-based food packaging films. *Polym. Test.*, 69 (2018), pp. 39–51.
34. Barisik, G., Isci, A., Kutlu, N., Bagder Elmaci, S., Akay, B. Optimization of organic acid pretreatment of wheat straw. *Biotechnol. Prog.*, 32 (2016), pp. 1487–1493.
35. Kim, D. Physico-chemical conversion of lignocellulose: Inhibitor effects and detoxification strategies: A mini review. *Molecules.*, 23 (2018), p. 309.
36. Kalyani, D., Lee, K.M., Kim, T.S., Li, J., Dhiman, S.S., Kang, Y.C., Lee, J.K. Microbial consortia for saccharification of woody biomass and ethanol fermentation. *Fuel.*, 107 (2013), pp. 815–822.
37. Isroi, I., Millati, R., Niklasson, C., Cayanto, C., Taherzadeh, M.J., Lundquist, K. Biological pretreatment of lignocelluloses with white-rot fungi and its applications: Review. *Bioresour.*, 6 (2011), pp. 5224–5259.
38. Cianchetta, S., Di Maggio, B., Burzi, P.L., Galletti, S. Evaluation of selected whiterot fungal isolates for improving the sugar yield from wheat straw. *Appl. Biochem. Biotechnol.*, 173 (2014), pp. 609–623.
39. Saha, B.C., Qureshi, N., Kennedy, G.J., Cotta, M.A. Biological pretreatment of corn stover with white-rot fungus for improved enzymatic hydrolysis. *Int. Biodeterior. Biodegrad.*, 109 (2016), pp. 29–35.
40. Mussoline, W., Esposito, G., Giordano, A., Lens, P. The anaerobic digestion of rice straw: A review. *Crit. Rev. Environ. Sci. Technol.*, 43 (2013), pp. 895–915.
41. Monavari, S., Galbe, M., Zacchi, G. The influence of ferrous sulfate utilization on the sugar yields from dilute-acid pretreatment of softwood for bioethanol production. *Bioresour. Technol.*, 102 (2011), pp. 1103–1108.
42. Barisik, G., Isci, A., Kutlu, N., Bagder Elmaci, S., Akay, B. Optimization of organic acid pretreatment of wheat straw. *Biotechnol. Prog.*, 32 (2016), pp. 1487–1493.

43. Kim, D. Physico-chemical conversion of lignocellulose: Inhibitor effects and detoxification strategies: A mini review. *Molecules.*, 23 (2018), p. 309.
44. Sasmal, S., Goud, V.V., Mohanty, K. Ultrasound assisted lime pretreatment of lignocellulosic biomass toward bioethanol production. *Energy and Fuels.*, 26 (2012), pp. 3777–3784.
45. Pagliaccia, P., Gallipoli, A., Gianico, A., Montecchio, D., Braguglia, C.M. Single stage anaerobic bioconversion of food waste in mono and co-digestion with olive husks: Impact of thermal pretreatment on hydrogen and methane production. *Int. J. Hydrogen Energy.*, 41 (2016), pp. 905–915.
46. Agbor, V.B., Cicek, N., Sparling, R., Berlin, A., Levin, D.B. Biomass pretreatment: Fundam entals toward application. *Biotechnol. Adv.*, 26 (2011), pp. 675–685.
47. Pielhop, T., Amgarten, J., von Rohr, P.R., Studer, M.H. Steam explosion pretreatment of softwood: The effect of the explosive decompression on enzymatic digestibility. *Biotechnol. Biofuels.*, 9(1) (2016), p. 152.
48. Colombo, B., Sciarria, T.P., Reis, M., Scaglia, B., Adani, F. Polyhydroxyalkanoates (PHAs) production from fermented cheese whey by using a mixed microbial culture. *Bioresour. Technol.*, 218 (2016), pp. 692–699. Doi:10.1016/j.biortech.2016.07.024.
49. Duque, A.F., Oliveira, C.S.S., Carmo, I.T.D., Gouveia, A.R., Pardelha, F., Ramos, A.M., et al. Response of a three-stage process for PHA production by mixed microbial cultures to feedstock shift: Impact on polymer composition. *New Biotechnol.*, 31 (2014), pp. 276–288. Doi:10.1016/j.nbt.2013.10.010.
50. Heng, K.S., Hatti-Kaul, R., Adam, F., Fukui, T., Sudesh, K. Conversion of rice husks to polyhydroxyalkanoates (PHA) via a three-step process: Optimized alkaline pretreatment, enzymatic hydrolysis, and biosynthesis by *Burkholderia emicell* USM (JCM 15050).
51. Reddy, C.S.K., Ghai, R., Kalia, V.C.C. Polyhydroxyalkanoates: An overview. *Bioresour. Technol.*, 87 (2003), pp. 137–146.
52. Takahashi, R., Castilho, N., Silva, M., Miotto, M., Lima, A. Prospecting for marine bacteria for polyhydroxyalkanoate production on low-cost substrates. *Bioeng.*, 4 (2017), p. 60.
53. Yue, H., Ling, C., Yang, T., Chen, X., Chen, Y., Deng, H., Wu, Q., Chen, J., Chen, G.Q. A seawater-based open and continuous process for polyhydroxyalkanoates production by recombinant Halomonas campaniensis LS21 grown in mixed substrates. *Biotechnol. Biofuels.*, 7 (2014), p. 108.
54. Pandian, R.K., et al. Optimization and fed-batch production of PHB utilizing dairy waste and sea water as nutrient sources by *Bacillus megaterium* SRKP-3. *Bioresour. Technol.*, 101 (2010), pp. 705–711. Doi: 10.1016/j.biortech.2009.08.040.
55. Chen, G.Q., Jiang, X.R., Guo, Y. Synthetic biology of microbes synthesizing polyhydroxyalkanoates (PHA). *Synth. Syst. Biotechnol.*, 1(4) (2016), pp. 236–242.
56. Hilliou, L., Machado, D., Oliveira, C.S.S., Gouveia, A.R., Reis, M.A.M., Campanari, S., Villano, M., Majone, M. Impact of fermentation residues on the thermal, structural, and rheological properties of polyhydroxy(butyrate-co-valerate) produced from cheese whey and olive oil mill wastewater. *J. Appl. Polym. Sci.*, 133(2) (2016), p. 42818.
57. Hilliou, L., Teixeira, P.F., Machado, D., Covas, J.A., Oliveira, C.S.S., Duque, A.F., Reis, M.A.M. Effects of fermentation residues on the melt processability and thermomechanical degradation of PHBV produced from cheese whey using mixed microbial cultures. *Polym. Degrad. Stab.*, 128 (2016), pp. 269–277.
58. Samorì, C., Abbondanzi, F., Galletti, P., Giorgini, L., Mazzocchetti, L., Torri, C., Tagliavini, E. Extraction of polyhydroxyalkanoates from mixed microbial cultures: Impact on polymer quality and recovery. *Bioresour. Technol.*, 189 (2015), pp. 195–202.
59. Ben, M., Kennes, C., Veiga, M.C. Optimization of polyhydroxyalkanoate storage using mixed cultures and brewery wastewater. *J. Chem. Technol. Biotechol.*, 91 (2016), pp. 2817–2826.
60. Tommonaro, G., Pejin, B., Iodice, C., Tafuto, A., De Rosa, S. Further in vitro biological activity evaluation of amino-, thio- and ester-derivatives of avarol. *J. Enzyme Inhib. Med. Chem.*, 30 (2016), pp. 333–335.

61. Sandhya, M., Aravind, J., Kanmani, P. Production of polyhydroxyalkanoates from *Ralstonia eutropha* using paddy straw as cheap substrate. *Int. J. Environ. Sci. Technol.*, 10 (2013), pp. 47–54.
62. Kumar, M.N.R. A review of chitin and chitosan applications. *React. Funct. Polym.*, 46 (2000), pp. 1–27.
63. Rinaudo, M. Chitin and chitosan: Properties and applications. *Prog. Polym. Sci.*, 31 (2006), pp. 603–632.
64. Ravi, M., Song, S., Wang, J., Nadimicherla, R., Zhang, Z. Preparation and characterization of biodegradable poly (ε-caprolactone)-based gel polymer electrolyte films. *Ionics*, 22(5) (2016), pp. 661–670.
65. Santhanam, S., Sasidharan, R. Microbial production of polyhydroxy alkanotes (PHA) from *Alcaligens* spp. And *Pseudomonas oleovorans* using different carbon sources. *African J Biotechnol.*, 9 (2010), pp. 3144–3150.
66. Matsumoto, K., Taguchi, S. Enzyme and metabolic engineering for the production of novel biopolymers: Crossover of biological and chemical processes. *Curr. Opin. Biotechnol.*, 24 (2013), pp. 1054–1060.
67. Smerilli, M., Neureiter, M., Wurz, S., Haas, C., Frühauf, S., Fuchs, W. Direct fermentation of potato starch and potato residues to lactic acid by *Geobacillus stearothermophilus* under non-sterile conditions. *J. Chem. Technol. Biotechnol.*, 90 (2015), pp. 648–657.
68. Quinaya, S.H.D., Alzate, C.A.C. Plantain and banana fruit as raw material for glucose production. *J. Biotechnol.*, 185 (2014), p. S34.
69. Suwannasing, W., Imai, T., Kaewkannetra, P. Cost-effective defined medium for the production of polyhydroxyalkanoates using agricultural raw materials. *Bioresour. Technol.*, 194 (2015), pp. 67–74.
70. Zahari, M.A.K.M., Ariffin, H., Mokhtar, M.N., Salihon, J., Shirai, Y., Hassan, M.A. Case study for a palm biomass biorefinery utilizing renewable non-food sugars from oil palm frond for the production of poly(3-hydroxybutyrate) bioplastic. *J. Clean. Prod.*, 87 (2015), pp. 284–290.

8 New Insights into Feruloyl Esterase

The Enzyme with Potential Biotechnological Applications

Meghna Diarsa and Akshaya Gupte
Natubhai V. Patel College of Pure and Applied Sciences, Vallabh Vidyanagar-388120

CONTENTS

8.1 INTRODUCTION

The plant cell wall contains lignocelluloses such as cellulose, hemicellulose, lignin, and pectin. Arabinoxylan is a most abundant hemicellulose. The structure of

DOI: 10.1201/9781003191247-8

arabinoxylan consists of a linear chain of β-1-4-linked D-xylopyranose units attached at O2&/or O3 with L-arabinofuranose, acetate, or 4-O-methyl glucuronic acid. These linkages are a main component of the plant cell wall. It is established by cinnamic acid, which joins to the arabinofuranosyl side chains (Prates et al., 2001). Ferulic acid links hemicellulose and lignin through ester bonds, since feruloyl esterase is a key enzyme for the hydrolysis of the ester bond present between the arabinofuranoside and ferulic acid (Wang et al., 2020). During the hydrolysis process, feruloyl esterase liberates the hemicellulose and lignin and also provides a free polysaccharide product. These polysaccharide products are more susceptible to degradation by other biocatalysts, such as endo-β-1–4 xylanases and arabinofuranosidases. Figure 8.1 shows the ester bond of ferulic acid linkages with arabinoxylan-lignin.

The hydrolysis of ester bonds in the plant cell wall produces many commercially interesting products. Products such as ferulic acid and ρ-coumaric acid are produced, which are mainly used in pharma industries and food industries (vanillin as flavoring agent). Feruloyl esterase enzyme is classified on the basis of amino acid sequences and substrate specificity for aromatic moieties (Koseki et al., 2009a). The sequences of organisms facilitate the powerful tools in enzyme discovery. Aspergillus species have been widely studied for the feruloyl esterase enzyme. Food industries widely use *Aspergillus niger*, *Aspergillus awamori*, and *Aspergillus kawachii* (Koji mold) for the production of feruloyl esterase enzyme (Koseki et al., 2009).

8.2 SOURCES OF FERULIC ACID ESTERASE

Ferulic acid esterase, releasing the hydroxycinnamic acid from the wheat bran, was first detected in the culture of *Streptomyces olivochromogens* (Wang et al., 2014). Feruloyl esterase is an extracellular enzyme (E.C.3.1.1.73), and it is a subclass of carboxylic acid esterase (E.C.3.1.1.1). Feruloyl esterase is also known as FA esterase, cinnamic acid esterase, or cinnamoyl esterase and ferulic acid esterase (Koseki et al., 2009). The molecular weight of the feruloyl esterase enzyme is found in the range of 11–211 Kda (Rumbold et al., 2003). Feruloyl esterase has been isolated from a number of fungi, bacteria, and yeast (Faulds et al., 1997). *Aspergillus niger* has been reported for the two feruloyl esterases FaeA and FaeB (Gopalan, et al., 2016; Faulds et al., 1997). Anaerobic fungi *Neocallimastix* strain MC-2 produce high level of FAE and ρ-coumaroyl esterase (Borneman et al., 1992; Faulds and Williamson, 1994). *Penicillium funiculosum* (Kroon et al., 2000), *Penicillium expansum* (Donaghy and Mckay, 1997), and black yeast *Aureobasidium pullulans* (Rumbold et al., 2003) are also reported for the production of feruloyl esterase. Several mushrooms have been reported for feruloyl esterase activity, such as *Russula virescens* (Wang et al., 2014), *Schizophyllum commune* (Mackenzie and Bilous, 1988), and *Pleurotus sapidus* (Linke et al., 2013).

8.3 PRODUCTION OF FERULOYL ESTERASE ENZYME

Many bacterial and fungal species are capable of producing feruloyl esterase enzymes in submerged as well as solid-state fermentation (McAuley et al., 2004). The production media contains agro-industrial waste (wheat bran, sugar beet pulp, and oat spelt) as a carbon source. The media also contains soluble ferulic acid esters.

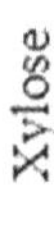

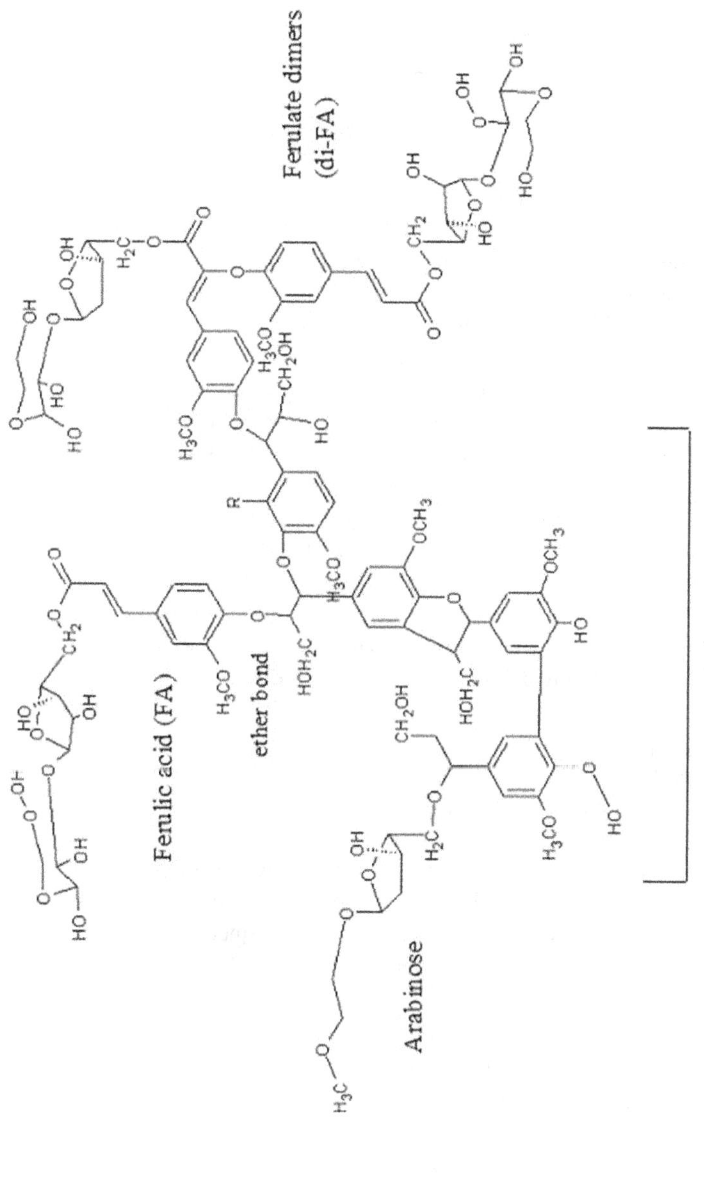

FIGURE 8.1 The ester linkages of ferulic acid-arabinoxylan-lignin in monocots.

Source: Adapted from Antonopoulou (2017).

The successful production of feruloyl esterase depends upon the use of the appropriate substrates, because it also works as an inducer for microorganisms. The production of feruloyl esterase is not supported by mono- and disaccharides like glucose, lactose, maltose, and xylose. Complex carbon sources like destarched wheat bran, sugar beet pulp, and maize bran have been employed effectively for the production of feruloyl esterase.

8.3.1 Solid-State and Submerged Fermentation

Many researchers study the use of solid-state and submerged fermentation for the production of feruloyl esterase enzymes. Solid-state fermentation is a fermentation process in which microorganisms grow on the solid substrate in the absence of free liquid. Solid-state fermentation is more preferable compared to submerged fermentation. Solid-state fermentation has many advantages, like higher productivity, lower risk of contamination, less catabolic repression, and it also increases enzyme stability (Gopalan et al., 2015). In solid-state fermentation, solid substrates are used as a carbon source and energy source. Agricultural, forestry, and food processing of lignocellulosic waste is mostly used in solid-state fermentation.

On the other hand, submerged fermentation is done in liquid production media. In submerged fermentation, the risk of contamination is high, and the stability of the enzyme is lower as compared to solid-state fermentation. Kumar et al. (2011) examined the production of feruloyl esterase by *Aspergillus terreus* strain GA2 that grows in solid-state fermentation. They obtained maximum activity 1,162 U/gram of dry substrate on the seventh day. Sachan et al. (2014) studied the production of feruloyl esterase by *Trichophytonajelloi* MTCC 4878 in submerged fermentation. They found the highest feruloyl esterase activity of 397 U/mg on the sixth day. Shin and Chen (2006) also obtained 334.6 U/l feruloyl esterase activity on the fifth day in submerged fermentation by *Fusarium proliferatum* NRRL 26517. *Fusarium proliferatum* produces Type B feruloyl esterase enzymes. These comparative studies clearly indicate that solid-state fermentation has more advantages compared to submerged fermentation. The production of feruloyl esterase depends on the choice of fermentation technique, substrate, and microorganisms used.

Table 8.1 shows some properties of feruloyl esterase enzymes of different microorganisms. The first purified feruloyl esterase was reported in *Streptomyces olivochromogenes* by Faulds and Williamson in 1991. Since then, feruloyl esterase

TABLE 8.1
Properties of Feruloyl Esterase Enzyme

Species	Mol. Mass (Kda)	$pH_{opt.}$	T_{opt} (°C)	$pH_{stability}$	$T_{stability}$ (°C)	References
Aspergillus awamori	35	5.0	45	4.0–11.0	50	Koseki et al., 2005
Aspergillus nidulans	130	7.0	45	4.0–9.5	45	Shin and Chen, 2007
Aspergillus niger	36	5.0	60	-	-	Faulds and Williamson, 1994

TABLE 8.1 *(Continued)*
Properties of Feruloyl Esterase Enzyme

Species	Mol. Mass (Kda)	$pH_{opt.}$	T_{opt} (°C)	$pH_{stability}$	$T_{stability}$ (°C)	References
Fusarium proliferatum	31	6.5–7.5	50	6–8	50	Shin and Chen, 2006
Fusarium oxysporum	27	7.0	45	7.0–9.0	45	Topakas et al., 2003b
Neocallimastix	11	7.2	40	-	-	Borneman et al., 1991
Penicillium chrysogenum	62	6+7	50	-	-	Sakamoto et al., 2005
Sporotricum thermophile	27	8	60	6.0–8.0	4	Topakas et al., 2003
Talaromyces stipitatus	66	6–7	60	-	-	Crepin et al., 2003b

have been purified and characterized from many organisms, such as *Aspergillus* spp., *Butyrivibrio fibrisolvens*, *Cellvibrio japonicus*, *Clostridium thermocellum*, *Lactobacillus acidophilus*, *Penicillium funiculosum*, *Piromyces equi*, *Neocallimastix* MC-2, and *Talaromyces stipitatus*. The physicochemical characteristics (molecular weight, isoelectric point, optimum pH, temperature, and stability) of purified feruloyl esterase enzyme vary from species to species.

8.4 TYPES OF FERULOYL ESTERASE ENZYME

Feruloyl esterase enzymes are classified into four different classes (Types A, B, C, and D), depending on their sequence homologies for substrate specificity (Wu et al., 2017; Gao et al., 2016). Table 8.2 depicts the classification of feruloyl esterase on the basis of substrate specificity. Feruloyl esterase enzymes can be characterized by both natural and synthetic substrates, such as methyl ferulate (MFA), methyl sinapate (MSA), methyl caffeate (MCA), and methyl ρ-coumarate (MpCA) (Qi et al., 2011).

Type A feruloyl esterase prefers the substrate that contains the methoxy group substitutions at C-3/ or C-5 of the phenolic ring, as occurs in ferulic acid and sinapic acid (Dilokpimol et al., 2016; Topakas et al., 2007). It shows specificity against ρ-coumaric acid, ferulic and sinapic acid, but not against caffeic acid. It releases the di form of hydroxycinnamic acid from 1-s ester-linked feruloylated arabinose. The feruloyl esterase enzyme of *Aspergillus niger*, *Aspergillus tubingensis*, and *Aspergillus awamori* belongs to the type A enzyme (Koseki et al., 2005, 2009). The sequence shows similarity with fungal lipase. Type A FAE shows activity against methyl ferulate, methyl sinapate, and methyl ρ-coumarate (Wong, 2006).

Type B feruloyl esterase enzymes are able to release the ferulic acid linked to C-2 of feruloylated arabinose or C-6 of galactose (Crepin et al., 2004). It is not able to release the di form of ferulic acid. Type B feruloyl esterase shows preference with the substrates which contain one or two hydroxyl substrates, found in caffeic acid and

TABLE 8.2
Classification of Feruloyl Esterase (Crepin et al., 2004)

Types of FAE	ρ- Coumaric acid	Caffeic acid	Ferulic acid	Sinapic acid	Di form	Sequence similarity
A	+	No	++/+++	+++	Yes	Lipase
B	++/+++	++/+++	+	No	No	Acetyl xylan esterase
C	Yes	Yes	Yes	Yes	No	Tannase
D	Yes	Yes	Yes	Yes	Yes	Xylanase

ρ-coumaric acid. Type B FAE also acts on 1–5 ester-linked hydroxycinnamic acid, but it is comparatively lower than the Type A ferulic acid esterase. Type B shows activity against methyl caffeate, methyl ferulate, and methyl ρ-coumaric acid (Crepin et al., 2004). It includes *Penicillium funiculosum* and *Neurospora crassa* (Wong, 2006). Type B FAE enzymes contain less hydrolytic activity for authentic and natural acetylated substrates such as α-naphthyl acetate and acetylated xylan.

Type C and Type D show specificity against all four synthetic hydroxycinnamic acids, i.e., ρ-coumaric acid, caffeic acid, ferulic acid, and sinapic acid (Topakas et al., 2007; Romero-Borbón et al., 2018). There are only two differences between Type C and Type D feruloyl esterase. Type D is able to hydrolyze the di form of ferulic acid. Type C is not able to hydrolyze (Dilokpimol et al., 2016; Kühnel et al., 2012). *Aspergillus niger* FaeB, *Talaromyces stipitatus* FaeC (Garcia-Conesa et al., 2004), *Fusarium oxysporum* FaeC (Moukouli et al., 2008), and *Aspergillus oryzae* FaeC (Koseki et al., 2009) are classified as Type C, and their sequences are similar to fungal tannases. The study reported that *Aspergillus niger* FaeB and *Aspergillus oryzae* FaeB and FaeC have hydrolytic activity for chlorogenic acid.

On the basis of functionality, feruloyl esterase is further classified into twelve families of different subgroups. Feruloyl esterase of Type A is classified in subfamily FAE12A. Feruloyl esterase of *Aspergillus nidulans*, *P. chrysogenum*, and *Aspergillus niger* is classified into Type B, which is classified into the subfamily FAE-4A. However, other sequences of Type B feruloyl esterase from *P. funiculosum, N. crassa, and Aspergillus oryzae* were included in the subfamily FEF5B, FEF6A, and FEF12B, respectively. Feruloyl esterase of Type C were classified into subfamily FEF4B. The other families of feruloyl esterase contain the mixture sequence of microorganisms and plantae. It indicates that during evolution, the related sequences of feruloyl esterase might have coevolved together from a common ancestor into different families.

8.5 PRIMARY STRUCTURE OF FERULOYL ESTERASE

The amino acid sequence of feruloyl esterase reveals that the sequence of feruloyl esterase is similar with the sequence of lipase, acetyl xylan esterase, tannase, and xylanase. On the basis of the sequence of amino acids, feruloyl esterase is classified into four subclasses, i.e., Types A, B, C, and D. The sequences of the *Aspergillus niger* and *Aspergillus tubingensis* enzymes align, and the protein structure shows 281 and 280 amino acids are present, respectively. The sequences of *Aspergillus awamori* FaeA contain 281 amino acids, and it shows 91 and 96% identity and similarity to the sequences of *Aspergillus niger.* The sequences of *Aspergillus niger, Aspergillus tubingensis,* and *Aspergillus awamori* show 36% and 50% sequence identity and sequence similarity with the sequence of lipase. Feruloyl esterase of *N. crassa* and *Penicillium funiculosum* is a Type B feruloyl esterase. The amino acid sequence of the Type B Fae of *N. crassa* and *P. funiculosum* is aligned, and the protein structure shows that it contains 292 and 359 amino acids, respectively, with 18 single-peptide amino acids (Crepin et al., 2003). The protein structure shows that 276 residues of the N-terminal domain and 39 residues of C-terminal carbohydrate-binding domain are responsible for cellulose bonding. The N- and C-terminal domains are linked through the 20 amino acids which are rich in Thr/ser/pro. Feruloyl esterase-B of *Aspergillus ficuum, Aspergillus awamori, Aspergillus oryzae,* and *P. purpurogenum* exhibits 60% sequence similarity with acetyl xylan esterase. The feruloyl esterase of *T. stiptatus* is included in the class of Type C (Crepin et al., 2004). The protein structure shows that 530 amino acids with 25 residues of a single peptide are present. *P. equi* and *C. japonicas* have Type D feruloyl esterase. The sequence of *P. equi* is aligned, and the protein structure is made up of 536 amino acids with 55.5 Kda molecular weight. *C. japonicus* has slightly higher molecular weight (58.5 Kda) than the *P. equi* (Fillingham et al., 1999).

8.6 3D STRUCTURE OF FERULOYL ESTERASE

The crystal structure of the *Aspergillus niger* (AnFaeA) is shown in Figure 8.2. The structure is globular in shape, with 65A° × 57A° × 48A° diameter. The protein structure is based on the α/β fold and contains nine major and two minor strands with a β-sheet surrounded by the seven helices. Amongst the nine major strands, eight strands

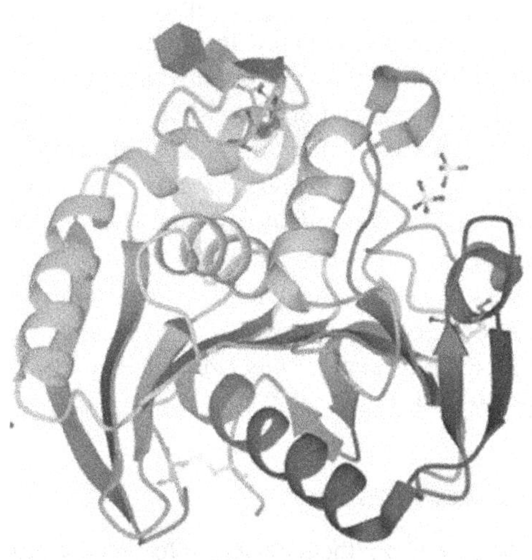

FIGURE 8.2 3D structure of *Aspergillus niger* (AnFaeA).

are arranged in the parallel order. However, the N-terminal strand present next to the C-terminal strand is arranged in the antiparallel manner. The β-sheet is a core of the esterase and exhibits superhelical twists at 90° angle with first β and last β_2 strands of the protein. Three disulfide bridges are present in the structure, which generate a core compact fold. Amongst these three disulfide bridges, Cys91-Cys94 is present near the flap region, which generates a strong tight turn as two Cys residues are only two residues apart. The disulfide bridge (Cys29-Cys258) protects the end of the C-terminal, which is near the core structure of the protein. It creates a hydrophobic pocket which protects the C-terminal Trp. residue. The last disulfide bridge (Cys22-Cys234) is attached to β-9, which extends the C-terminal. The protein structure reveals that Cys235 is not involved in the development of the disulfide bridge (Hermoso et al., 2004).

The protein structure of the feruloyl esterase of *Penicillium piceum* was predicted by Gao et al. in 2017, and it showed 29.79% similarity with the *Aspergillus oryzae* feruloyl esterase. The SWISS-MODEL program was used to create the protein structure of *Penicillium piceum* feruloyl esterase. The protein structure of *Penicillium piceum* FAE contains two domains (catalytic domain and lid domain). The catalytic domain contains Ser191, Asp405, and His441, which are equivalent to Ser203, Asp417, and His457 of *A. oryzae*. The lid domain prepares a subbinding pocket which protects the catalytic domain. The disulfide bond maintains the integrity of the enzyme's active site (Suzuki et al., 2014).

8.7 QUALITATIVE AND QUANTITATIVE METHODS TO DETECT FERULIC ACID ESTERASE

8.7.1 Plate Assay

Plate assay is an easy technique to screen a large number of microorganisms for different primary and secondary metabolites. Lai et al. (2012) incorporated 0.1% ethyl ferulate as a model substrate for feruloyl esterase enzyme into the de Man, Rogosa, and Sharpe (MRS) plate medium for the screening of a *Lactobacillus* strain. The appearance of the halo/clear zone around the inoculated colony is due to the hydrolysis of ethyl ferulate to show the presence of feruloyl esterase activity. Donaghy et al. (1998) also used de Man, Rogosa, and Sharpe (MRS) plate medium and screened 80 *Bacillus* strain and 50 gram bacterial strains such as *Brevibacillus brevis*, *Bacillus cereus*, *Bacillus circulans*, *Bacillus firmus*, *Bacillus licheniformis*, *Bacillus megaterium*, *Bacillus mycoides*, *Bacillus sphaericus*, and *Bacillus subtilis*. Furthermore, gram-positive bacteria, including *Enterococcus faecium*, *E. divas*, *Enterococcus faecalis*, *L. farciminis*, *L. delbrueckii*, *Lactobacillus fermentum*, *Lactobacillus reuteri*, *Lactobacillus sakei*, *Lactococcus spp.*, and *Leuconostoc spp.*, were screened on agar plate incorporated with ethyl ferulate. A clear zone around the colonies suggests the presence of feruloyl esterase activity.

8.7.2 Thin-Layer Chromatography

Thin-layer chromatography is another technique to measure the activity of ferulic acid esterase enzymes. The qualitative analysis of feruloyl esterase is done on a

Merck silica gel 60 F_{254} plates. TLC plate method has several advantages, such as it is a very easy and quick method, has faster development of drops, has auto sampler, and has a constant amount of sample dispenser.

8.7.3 High-Performance Liquid Chromatography (HPLC)

HPLC is the most widely used technique to quantify ferulic acid liberated from the substrates. The feruloyl esterase activity from isolation and characterized microorganisms is measured by a method which is based on the extraction of ferulic acid and coumaric acid. This method is a very sensitive method which detects a very low amount of sample, and it does not require any prior concentration and purification of the sample. Abokitse et al. (2010) extracted ferulic acid from triticale bran. Phenolic acid was quantitatively determined by the NX 3 μm C18 column (150 × 4.6 mm) with diode array detector.

8.7.4 Spectrophotometric Analyses

Spectrophotometric methods are easy to perform and quantify the activity of feruloyl esterase. Destarched wheat bran is used as a substrate to produce feruloyl esterase by *Fusarium oxysporum*. The assay is done by using 100 mM MOPS buffer at pH-6 (Topakas et al., 2003). Esterase activity is detected using various substrates, such as p-NP and methyl ester. DMSO or isopropanol are used to dissolve the substrates. The release of p-nitrophenol is recorded at 420 nm in a UV spectrophotometer (Abokitse et al., 2010). Donnelly and Crawford (1988) used p-nitrophenyl butyrate to quantify the amount of extracellular enzyme by spectrophotometric method. Gopalan et al. (2016) used ethyl ferulate as a substrate for enzyme assay.

8.8 APPLICATIONS OF FERULOYL ESTERASE

Feruloyl esterase enzyme acts as an accessory enzyme which breaks down the cell wall. It is also used in the production of biofuels and in pulp and paper industries (Deng et al., 2019). Industrial researchers mainly focus on the hydrolytic activity of FAE; it releases saccharides as an additional enzyme in biofuel production. Biofuels are alternatives to petroleum. The production of ethanol is done by microorganisms (bacteria and fungi) which are capable of producing enzymes like cellulase and hemicellulase from food sources, for instance, corn (Koseki et al., 2009).

8.8.1 Application in Pulp and Paper Industries

The quality and whiteness of paper is a major concern of paper-making industries. The bleaching step is required to remove lignin from the wood pulp. This process requires extensive toxic and costly chemicals, such as chlorine dioxide and H_2O_2. This problem can be avoided by using an eco-friendly method. Enzymes are alternatives to hazardous and costly chemicals, which reduce water pollution and cleaning cost. Some research studies show the use of feruloyl esterase enzymes in paper and pulp industry for bleaching. Pulp and paper industries use ferulic acid esterase and xylanase as oxidoreductase in the bleaching process. The enzymatic treatment reduces energy consumption and decreases the level of COD (chemical oxygen demand) in wastewater (Dilokpimol et al., 2016). The activities of xylanase and laccase combine with recombinant FAEA produced from *Aspergillus niger* to enhance the delignification of wheat straw and oilseed flax straw pulps. The enzymes increase the chemical and mechanical properties of paper and pulp. It also removes the fine particles from the pulp wastewater (Antonopoulou, 2017).

8.8.2 Application in Animal Feed

Food digestion is an important part of animal feed. Poor digestion affects the growth of the animal and increases immunological stress, which decreases feed ration conversion ratio in livestock; hence, it affects the beneficiality of farmers. Feruloyl esterase improves in situ digestion by degrading hemicellulose in animal feed. Enzymes can also help in gaining body weight, increasing digestibility, reducing immunological stress, providing phytonutrients, and also enhancing the milk yield in cattle and sheep (Howard et al., 2003; Yu et al., 2005).

8.8.3 Application in Food Industries

Food industries use feruloyl esterase enzymes as a flavoring agent and to remove undesirable odor and flavors in alcoholic beverages. Flavors and odors are the most important parameters in fermentation and beverage industries. FAEs are widely used in the baking of bread. It solubilizes the arabinoxylan in dough, which increases the volume and quality of bread (Butt et al., 2008). Enzymes are also involved in the clarification of juice. Other applications include the use of enzyme in the saccharification process (Topakas et al., 2005). The combined action of xylanases and cellulases enhances the extraction of ferulic acid from the cell wall of rice and cereal grains. It can be transformed to aromatic derivatives during the process of aging and fermentation. FAE not only degrades the plant biomass but also incorporates novel bioactive compounds to expand food materials.

8.8.4 Application of FAE in the Production of Ferulic Acid

Ferulic acid is a hydroxycinnamic acid. The first time, ferulic acid was obtained from the plant *Ferula fetida Reg.* Ferulic acid and other phenolic derivatives are present in the plant cell wall, which are mainly used in food, cosmetic, and pharmaceutical industries. Ferulic acid is known to have antioxidant activity (Cheng et al., 2007) (de Oliveira Silva and Batista, 2017). It has the capability to neutralize free radicals. Food industries add ferulic acid into food products to enhance the quality of foods. Ferulic acid is a carrier of vitamins C and E, which protect the skin from UV radiation (Antonopoulou, 2017). Ferulic acid also acts as an antimicrobial, antifungal, anti-inflammatory, antidiabetic, anticancer, and antithrombosis agent (Zduńska et al., 2018; Xu et al., 2019). Commercially, FA produced from rice oil is known as a γ-oryzanol.

8.8.5 Feruloyl Esterase as Biosynthetic Tool

Natural antioxidants and flavoring agents such as ferulic acid and other hydroxycinnamic acids have widespread potential by virtue of their antioxidant property. The solubility of natural antioxidants in aqueous media is very low, which reduces usefulness in food processing (oil based), pharmaceutical, and cosmetic products (Antonopoulou, 2017; Faulds, 2010). Esterification or transesterification is a common alternative to enhance the solubility of natural oxidants. The synthesis of ester-linked hydroxycinnamic acid through transesterification reaction is done by interchanging the organic group of an ester (donor) with the organic group of alcohol (acceptor) (Dilokpimol et al., 2016). The first transesterification activity of feruloyl esterase enzyme was investigated from the *Sporotrichum thermophile* (StFaeC) by using arabinose and arabinobiose as acceptor (Topakas et al., 2005). Transesterification has several advantages over hydrolysis, such as enhancing the availability of acceptor molecules, which can differ from different carbohydrates, alcohols, and glycerol.

8.9 CONCLUSION AND FUTURE PERSPECTIVE

This chapter provides information in the aspect of biotechnological applications of feruloyl esterase and the release of ferulic acid from the plant cell wall. A number of bacteria and fungi have been isolated, purified, and characterized for the production of feruloyl esterase. Over the past year, there has been an increase in the range of various industrial applications, conversion of agro-waste into valuable products, and health and cosmetic applications. Phenolic-acting enzymes include feruloyl esterase, peroxidases, laccases, and polyphenol oxidase. Out of these enzymes, feruloyl esterase is the least-understood member of the family. In recent times, research interest in enhancing health-improving components, such as hydroxycinnamates, in traditional food has grown, and along with it, demands for new techniques to utilize lignocellulosic biomass has also been keenly investigated. Together, these can surely lead to a better understanding of microbial interaction with plant components and how we can exploit it for our health benefits. We need to find a new novel feruloyl esterase enzyme which may be classified in a distinct group by phylogenetic analysis. These new FAEs require further study to know the structure and substrate specificity, which may improve knowledge in biochemical characterization and enhance their applications as biocatalysts.

REFERENCES

Abokitse, K. Wu, M. Bergeron, H. Grosse, S. Lau, P. C. 2010. Thermostable feruloyl esterase for the bioproduction of ferulic acid from triticale bran. *Appl Microbiol Biotechnol.* 87:195–203.

Antonopoulou, I. 2017. Use of feruloyl esterases for chemoenzymatic synthesis of bioactive compounds (Doctoral dissertation, Luleå University of Technology).

Borneman, W. S. Ljungdahl, L. G. Hartley, R. D. Akin, D. E. 1991. Isolation and characterization of p-coumaroyl esterase from the anaerobic fungus *Neocallimastix* strain MC-2. *Appl Environ Microbiol.* 57:2337–2344.

Borneman, W. S. Ljungdahl, L. G. Hartley, R. D. Akin, D. E. 1992. Purification and partial characterization of two feruloyl esterases from the anaerobic fungus *Neocallimastix* strain MC-2. *Appl Environ Microbiol.* 58:3762–3766.

Butt, M. S. Tahir-Nadeem, M. Ahmad, Z. Sultan, M. T. 2008. Xylanases and their applications in baking industry. *Food Technol Biotechnol.* 46:22–31.

Cheng, J. C. Dai, F. Zhou, B. Yang, L. Liu, Z. L. 2007. Antioxidant activity of hydroxycinnamic acid derivatives in human low density lipoprotein: Mechanism and structure—activity relationship. *Food Chem.* 104:132–139.

Crepin, V. F. Faulds, C. B. Connerton, I. F. 2003. Production and characterization of the *Talaromyces stipitatus* feruloyl esterase FAEC in Pichia pastoris: Identification of the nucleophilic serine. *Protein Expr Purifi.* 29:176–184.

Crepin, V. F. Faulds, C. B. Connerton, I. F. 2004. Functional recognition of new classes of feruloyl esterase. *Appl Microbiol Biotechnol.* 63:647–652.

De Oliveira Silva, E. Batista, R. 2017. Ferulic acid and naturally occurring compounds bearing a feruloyl moiety: A review on their structures, occurrence, and potential health benefits. *Compr Rev Food Sci Food Saf.* 16:580–616.

Deng, H. Jia, P. Jiang, J. Bai, Y. Fan, T. P. Zheng, X. Cai, Y. 2019. Expression and characterization of feruloyl esterases from *Lactobacillus fermentum* JN248 and release of ferulic acid from wheat bran. *Int J Biol Macromol.* 138:272–277.

Dilokpimol, A. Mäkelä, M. R. Aguilar-Pontes, M. V. Benoit-Gelber, I. Hildén, K. S. de Vries, R. P. 2016. Diversity of fungal feruloyl esterases: Updated phylogenetic classification, properties, and industrial applications. *Biotechnol Biofuels.* 9:1–18.

Donaghy, J. Kelly, P. F. McKay, A. M. 1998. Detection of ferulic acid esterase production by *Bacillus* spp. And *lactobacilli*. *Appl Microbiol Biotechnol.* 50:257–260.

Donaghy, J. McKay, A. M. 1997. Purification and characterization of a feruloyl esterase from the fungus *Penicillium expansum*. *J Appl Microbiol.* 83:718–726.

Donnelly, P. K. Crawford, D. L. 1988. Production by Streptomyces viridosporus T7A of an enzyme which cleaves aromatic acids from lignocellulose. *Appl Environ Microbiol.* 54:2237–2244.

Faulds, C. B. 2010. What can feruloyl esterases do for us? *Phytochem Rev.* 9:121–132.

Faulds, C. B. DeVries, R. P. Kroon, P. A. Visser, J. Williamson, G. 1997. Influence of ferulic acid on the production of feruloyl esterases by *Aspergillus niger*. *FEMS Microbiol Letters.* 157:239–244.

Faulds, C. B. Williamson, G. 1991. The purification and characterization of 4-hydroxy-3-methoxycinnamic (ferulic) acid esterase from *Streptomyces olivochromogenes. Microbiol.* 137:2339–2345.

Faulds, C. B. Williamson, G. 1994. Purification and characterization of a ferulic acid esterase (FAE-III) from *Aspergillus niger*: Specificity for the phenolic moiety and binding to microcrystalline cellulose. *Microbiol.* 140:79–787.

Fillingham, I. J. Kroon, P. A. Williamson, G. Gilbert, H. J. Hazlewood, G. P. 1999. A modular cinnamoyl ester hydrolase from the anaerobic fungus *Piromyces equi* acts synergistically with xylanase and is part of a multiprotein cellulose-binding cellulase—hosphorizat complex. *Biochem J.* 343:215–224.

Gao, L. Wang, M. Chen, S. Zhang, D. 2016. Biochemical characterization of a novel feruloyl esterase from *Penicillium piceum* and its application in biomass bioconversion. *J Mol Catal B Enzym.*133:S388–S394.

Garcia-Conesa, M. T. Crepin, V. F. Goldson, A. J. Williamson, G. Cummings, N. J. Connerton, I. F. Kroon, P. A. 2004. The feruloyl esterase system of *Talaromyces stipitatus*: Production of three discrete feruloyl esterases, including a novel enzyme, TsFaeC, with a broad substrate specificity. *J Biotechnol.* 108:227–241.

Gopalan, N. Nampoothiri, K. M. Szakacs, G. Parameswaran, B. Pandey, A. 2016. Solid-state fermentation for the production of biomass valorizing feruloyl esterase. *Biocataly Agric Biotechnol.* 7:7–13.

Gopalan, N. Rodríguez-Duran, L. V. Saucedo-Castaneda, G. Nampoothiri, K. M. 2015. Review on technological and scientific aspects of feruloyl esterases: A versatile enzyme for biorefining of biomass. *Bioresour Technol.* 193:534–544.

Hermoso, J. A. Sanz-Aparicio, J. Molina, R. Juge, N. González, R. Faulds, C. B. 2004. The crystal structure of feruloyl esterase A from *Aspergillus niger* suggests evolutive functional convergence in feruloyl esterase family. *J Mol Biol.* 338:495–506.

Howard, R. L. Abotsi, E. L. J. R. Van Rensburg, E. J. Howard, S. 2003. Lignocellulose biotechnology: Issues of bioconversion and enzyme production. *Afr J Biotechnol.* 2:602–619.

Koseki, T. Fushinobu, S. Shirakawa, H. Komai, M. 2009. Occurrence, properties, and applications of feruloyl esterases. *Appl Microbiol Biotechnol.* 84:803–810.

Koseki, T. Hori, A. Seki, S. Murayama, T. Shiono, Y. 2009. Characterization of two distinct feruloyl esterases, AoFaeB and AoFaeC, from *Aspergillus oryzae. Appl Microbiol Biotechnol.* 83:689–696.

Koseki, T. Takahashi, K. Fushinobu, S. Iefuji, H. Iwano, K. Hashizume, K. Matsuzawa, H. 2005. Mutational analysis of a feruloyl esterase from *Aspergillus awamori* involved in substrate discrimination and pH dependence. *Biochim Biophy Acta (BBA)-General Sub.* 1722:200–208.

Kroon, P. A. Williamson, G. Fish, N. M. Archer, D. B. Belshaw, N. J. 2000. A modular esterase from *Penicillium funiculosum* which releases ferulic acid from plant cell walls and binds crystalline cellulose contains a carbohydrate binding module. *Europ J Biochem.* 267:6740–6752.

Kühnel, S. Pouvreau, L. Appeldoorn, M. M. Hinz, S. W. A. Schols, H. A. Gruppen, H. 2012. The ferulic acid esterases of *Chrysosporium lucknowense* C1: Purification, characterization and their potential application in biorefinery. *Enzyme Microb Technol.* 50:77–85.

Kumar, C. G. Kamle, A. Mongolla, P. Joseph, J. 2011. Parametric optimization of feruloyl esterase production from *Aspergillus terreus* strain GA2 isolated from tropical agro-ecosystems cultivating sweet sorghum. *J Microbiol Biotechnol.* 21:947–953.

Lai, K. K. Vu, C. Valladares, R. B. Potts, A. H. Gonzalez, C. F. 2012. Identication and characterization of feruloyl esterases produced by probiotic bacteria. In: Ahmad, R. (ed.). *Protein Purification.* IntechOpen, pp. 151–166.

Linke, D. Matthes, R. Nimtz, M. Zorn, H. Bunzel, M. Berger, R. G. 2013. An esterase from the basidiomycete *Pleurotus sapidus* hydrolyzes feruloylated saccharides. *Appl Microbiol Biotechnol.* 97:7241–7251.

MacKenzie, C. R. Bilous, D. 1988. Ferulic acid esterase activity from *Schizophyllum commune. Appl Environ Microbiol.* 54:1170–1173.

McAuley, K. E. Svendsen, A. Patkar, S. A. Wilson, K. S. 2004. Structure of a feruloyl esterase from *Aspergillus niger. Acta Crystallogr D: Biol Crystallogr.* 60:878–887.

Moukouli, M. Topakas, E. Christakopoulos, P. 2008. Cloning, characterization and functional expression of an alkalitolerant type C feruloyl esterase from *Fusarium oxysporum. Appl Microbiol Biotechnol.* 79:245–254.

Prates, J. A. Tarbouriech, N. Charnock, S. J. Fontes, C. M. Ferreira, L. M. Davies, G. J. 2001. The structure of the feruloyl esterase module of xylanase 10B from *Clostridium thermocellum* provides insights into substrate recognition. *Structure.* 9:1183–1190.

Qi, M. Wang, P. Selinger, L. B. Yanke, L. J. Forster, R. J. McAllister, T. A. 2011. Isolation and characterization of a ferulic acid esterase (Fae1A) from the rumen fungus *Anaeromyces mucronatus. J Appl Microbiol.* 110:1341–1350.

Romero-Borbón, E. Grajales-Hernández, D. Armendáriz-Ruiz, M. Ramírez-Velasco, L. Rodríguez-González, J. A. Cira-Chávez, L. A. Estrada-Alvarado M. I., Mateos-Díaz, J. C. 2018. Type C feruloyl esterase from *Aspergillus ochraceus*: A butanol specific biocatalyst for the synthesis of hydroxycinnamates in a ternary solvent system. *Electron. J. Biotechnol.* 35:1–9.

Rumbold, K. Biely, P. Mastihubová, M. Gudelj, M. Gübitz, G. Robra, K. H. Prior, B. A. 2003. Purification and properties of a feruloyl esterase involved in lignocellulose degradation by *Aureobasidium pullulans*. *Appl Environ Microbiol*. 69:5622–5626.

Sachan, A. Mishra, S. Singh, S. Nigam, V. K. 2014. Production of ferulic acid esterase from *Trichophyton ajeolli* MTCC 4878. *Int J Basic Appl Biol*. 2:13–17.

Sakamoto, T. Nishimura, S. Kato, T. Sunagawa, Y. Tsuchiyama, M. Kawasaki, H. 2005. Efficient extraction of ferulic acid from sugar beet pulp using the culture supernatant of *Penicillium chrysogenum*. *J Appl Glycosci*. 52:115–120.

Shin, H. D. Chen, R. R. 2006. Production and characterization of a type B feruloyl esterase from *Fusarium proliferatum* NRRL 26517. *Enzyme Microb Technol*. 38:478–485.

Shin, H. D. Chen, R. R. 2007. A type B feruloyl esterase from *Aspergillus nidulans* with broad pH applicability. *Appl Microbiol Biotechnol*. 73:1323–1330.

Suzuki, K. Hori, A. Kawamoto, K. Thangudu, R. R. Ishida, T. Igarashi, K. Fushinobu, S. 2014. Crystal structure of a feruloyl esterase belonging to the tannase family: A disulfide bond near a catalytic triad. *Proteins: Struct Funct Bioinform*. 82:2857–2867.

Topakas, E. Vafiadi, C. Stamatis, H. Christakopoulos, P. 2005. *Sporotrichum thermophile* type C feruloyl esterase (StFaeC): Purification, characterization, and its use for phenolic acid (sugar) ester synthesis. *Enzyme Microb Technol*. 36:729–736.

Topakas, E. Kalogeris, E. Kekos, D. Macris, B. J. Christakopoulos, P. 2003a. Production and partial hosphorization of feruloyl esterase by *Sporotrichum thermophile* in solid-state fermentation. *Process Biochem*. 38:1539–1543.

Topakas, E. Stamatis, H. Biely, P. Kekos, D. Macris, B. J. Christakopoulos, P. 2003b. Purification and characterization of a feruloyl esterase from *Fusarium oxysporum* catalyzing esterification of phenolic acids in ternary water—organic solvent mixtures. *J Biotechnol*. 102:33–44.

Topakas, E. Stamatis, H. Mastihubova, M. Biely, P. Kekos, D. Macris, B. J. Christakopoulos, P. 2003c. Purification and characterization of a *Fusarium oxysporum* feruloyl esterase (FoFAE-I) catalysing transesterification of phenolic acid esters. *Enzyme Microb Technol*. 33:729–737.

Wang, L. Zhang, R. Ma, Z. Wang, H. Ng, T. 2014. A feruloyl esterase (FAE) characterized by relatively high thermostability from the edible mushroom *Russula virescens*. *Appl Biochem Biotechnol*. 172:993–1003.

Wang, R. Yang, J. Jang, J. M. Liu, J. Zhang, Y. Liu, L. Yuan, H. 2020. Efficient ferulic acid and xylo-oligosaccharides production by a novel multi-modular bifunctional xylanase/feruloyl esterase using agricultural residues as substrates. *Bioresour Technol*. 297:122487.

Wong, D. W. 2006. Feruloyl esterase. *Appl Biochem Biotechnol*. 133:87–112.

Wu, H. Li, H., Xue, Y. Luo, G. Gan, L. Liu, J. Long, M. 2017. High efficiency co-production of ferulic acid and xylooligosaccharides from wheat bran by recombinant xylanase and feruloyl esterase. *Biochem Eng J*. 120:41–48.

Xu, Z. Wang, T. Zhang, S. 2019. Extracellular secretion of feruloyl esterase derived from *Lactobacillus crispatus* in *Escherichia coli* and its application for ferulic acid production. *Bioresour Technol*. 288:121526.

Yu, P. McKinnon, J. J. Christensen, D. A. 2005. Hydroxycinnamic acids and ferulic acid esterase in relation to biodegradation of complex plant cell walls. *Can J Anim Sci*. 85:255–267.

Zduńska, K. Dana, A. Kolodziejczak, A. Rotsztejn, H. 2018. Antioxidant properties of ferulic acid and its possible application. *Skin Pharmacol Physiol*. 31:332–336.

9 Production of Biomass-Based Butanol
Strategies and Challenges

Dhanya M. S.
Department of Environmental Sciences and Technology, Central University of Punjab, Bathinda (India) 151401

CONTENTS

DOI: 10.1201/9781003191247-9

9.1 INTRODUCTION

The more complete oxidation of n-butanol makes it the future biofuel in comparison to ethanol with efficient operation, pollutant mitigation, and CO_2 reduction (Han et al. 2013). The biomass as feedstock for biobutanol balances carbon dioxide emissions from its combustion as they release only carbon taken up during photosynthesis. Biobutanol has the potential to be produced from various domestic feedstocks and create jobs and thereby increase energy security.

9.2 WHY BUTANOL IS CONSIDERED AS A RENEWABLE AND SUSTAINABLE BIOFUEL

Butanol, with a higher number of carbon atoms than ethanol, produces nearly 30% more energy (Dürre 2007). Butanol is less corrosive and distributed through existing pipelines. Butanol has low Reid vapor pressure, which reduces evaporation and explosion. Butanol is a good alternative fuel as a biofuel extender or as a replacement because of its more superior properties than ethanol and is comparable to gasoline (Sakuragi et al. 2015). The presence of high oxygen content in butanol helps in clean combustion (Örs et al. 2020). The properties of n-butanol are given in Table 9.1.

TABLE 9.1
Physicochemical Properties of Butanol

Properties	n-butanol
Formula	C_4H_9OH
Density (g/L)	0.8098
Fusion point (°C)	–89.3
Boiling point (°C)	118
Auto ignition temperature (°C)	385
Energy density (MJ/kg)	33.1
Air-fuel ratio	11.2
Heat of vaporization (MJ/kg)	0.43
Kinematic viscosity at 40 °C (mm^2/s)	2.63
Flash point (°C)	35
Pour point (°C)	11
Research octane number	96
Motor octane number	78

Source: Compiled from Dürre 2007; Gholizadeh 2009; Si et al. 2014

Butanol has higher heating value, high mileage, low hygroscopicity, low volatility, low vapor pressure, and it has the ability to be used without engine modifications (Dürre 2007; Sakuragi et al. 2015). Si et al. (2014) compared the energy density of n-butanol with gasoline and ethanol and found that it is 10–20% lower than gasoline and around 38% higher than ethanol. The lower Reid vapor pressure results in reduced emissions from relatively low volatilization and evaporation. Butanol is able to produce the same power from less fuel, good ignition capability, less heat requirement, and better mixing ability in internal combustion engines in comparison to ethanol and methanol (Atmanli 2016).

9.2.1 Benefits in Blending of Biobutanol in Transportation Fuel

In comparison to the blends used in gasoline, butanol has high calorific value of 29.2–32.5 MJ/kg (Yoruklu et al. 2019). The high miscibility, solubility, low volatility, high heat of combustion, and less ignition problems of butanol help for improvements in fuel properties from blending butanol with petrol or diesel engines as well as in biodiesel blends (Ndaba et al. 2020).

Many researchers have evaluated the performance and emission pattern of butanol addition to different biodiesel-diesel blends. Butanol is considered as a viable additive with biodiesel-diesel blends with its low carbon monoxide, hydrocarbons, and particulate matter emissions in comparison to other alcohols (Atmanli 2016). Rahman et al. (2017) confirmed lower butanol addition (5%) to diesel-macadamia biodiesel, rice bran biodiesel, and waste cooking oil biodiesel blends and decreased carbon monoxide, nitrogen oxide, and particulate matter emissions. This has also been able to lower heat release rate and lower ignition delay more than does conventional diesel. But the increase in butanol blending to 20% has increased CO and HC emissions in contrast to lower CO emissions from 5 and 10% blending.

9.3 FEEDSTOCKS FOR BIOBUTANOL

Butanol is produced from sugars, starch, and cellulose by ABE fermentation process. Based on the sources of feedstock, butanol is classified into first-, second-, and third-generation butanol (Figure 9.1).

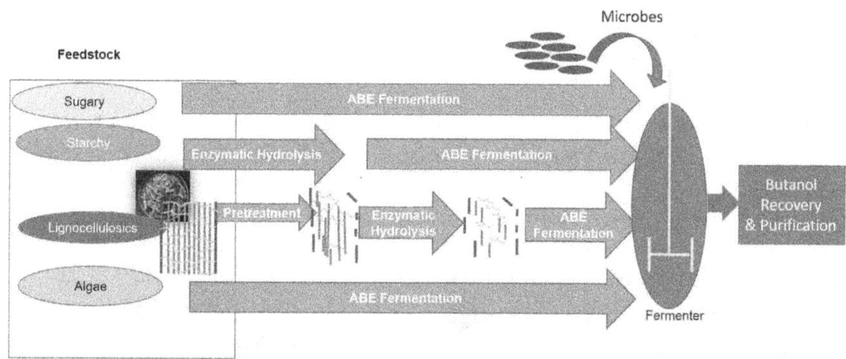

FIGURE 9.1 Processes involved in butanol production from different feedstock.

9.3.1 First-Generation Butanol

First-generation butanol is produced mainly from sugary and starchy crops. Starch grains from maize, wheat, rice, and cassava undergo hydrolysis, followed by saccharification with glucoamylase for glucose conversion. The higher yields of first-generation butanol have been reported from different studies.

The continuous fermentation of glucose by *Clostridium acetobutylicum* DSM 1731 produces butanol productivity of 0.41 g/L/h (Dolejš et al. 2014). The mixture of glucose, mannose, arabinose, and xylose improves ABE fermentation by *C. acetobutylicum* DSM 792 (Raganati et al. 2020). Sugary feedstock such as sugar beets and sugarcane are commonly used in biobutanol production (Qureshi et al. 2010).

C. acetobutylicum is also reported to directly ferment starch to ABE without any hydrolysis. Maize meal is used also by *C. beijerinckii* for butanol (Ezeji et al. (2007). The sugary sap of palm oil produces a butanol yield of 0.35 g/g by *C. acetobutylicum* DSM 1731 (Komonkiat and Cheirsilp 2013). Cassava starch–based butanol production by *C. beijerinckii tyrobutyricum* and *C. acetobutylicum* DP 217, respectively, has been studied by Li et al. (2013) and Li et al. (2014).

Soy molasses is used by *Clostridium beijerinckii* BA101 for butanol production (Qureshi et al. 2001). *Clostridium saccharoperbutylacetonicum* N1-4 produces butanol from cassava (Thang et al. 2010). ABE fermentation from Jerusalem artichokes was commercially optimized by Marchal et al. (1985). Gelatinized sago starch is converted to butanol by *Clostridium acetobutylicum* (Madihah et al. 2001).

9.3.2 Second-Generation Butanol

The need of the hour with respect to food security has forced a shift from edible sources to nonedible sources for butanol production (Birgen et al. 2019). Many researchers have demonstrated butanol production by ABE fermentation from different lignocellulosic materials, like rice straw, wheat straw, cornstalk, corncob, switchgrass, and eucalyptus (Qureshi et al. 2010; Zhang et al. 2013; Liu et al. 2015; Cai et al. 2016). Abd-Alla and El-Enany (2012) reported butanol production of 21.56 g/L from spoilage date palm fruits by *Clostridium acetobutylicum* ATCC 824 and *Bacillus subtilis* DSM 4451. Market-refused vegetables improve butanol production by *C. acetobutylicum* DSM 792 (Survase et al. 2013). Butanol production by *C. sporogenes* BE01 from rice straw was reported by Gottumukkala et al. (2013).

Various researchers have worked on potential lignocellulosic substrates for butanol production by *Clostridium beijerinckii*. Butanol from agricultural residues is also produced by *C. beijerinckii* (Qureshi et al. 2010; Huang et al. 2010). Butanol is produced by *C. beijerinckii* CECT 508 from coffee silverskin (a by-product of coffee roasting industries) with 7.02 g/L butanol with a yield of 0.269 g/g (Hijosa-Valsero et al. 2018). Moon et al. (2015) used *C. beijerinckii optinoii* for improvement of butanol production from sugarcane molasses. Bellido et al. (2015) used sugar beet pulp for butanol yield by *C. beijerinckii* DSM 6422. Wechgama et al. (2016) produced butanol of 12.55 g/L from sugarcane molasses with *C. beijerinckii* TISTR 1461.

9.3.3 Third-Generation Biobutanol

Algae are another good, promising option as butanol feedstock. The requirement of a large area for cultivation, processing, and storage for feedstocks of first- and second-generation biobutanol is overcome by third-generation butanol from algae (Cheng et al. 2015; Ndaba et al. 2020). The fast-growing characteristics of algae also add to the benefits of third-generation biofuel. The carbohydrate content in both micro- and macroalgae (seaweeds) is used for biobutanol production. The less lignin content in algae decreases cost, time, and easy conversion. The brown seaweed *Laminaria digitata* is reported to produce a butanol yield of 0.42 g/g (Hou et al. 2017). Jernigan et al. (2013) studied butanol from steam pretreatment of wastewater algae by *C. saccharobutylacetonicum*. The acid hydrolysis of algae was studied by many researchers for cellulose and hemicellulose (Castro et al. 2015). Wang et al. (2017) produced butanol from acid pretreated biomass of *Neochloris aquatica* CL-M1 grown in swine wastewater with 12 g L^{-1} of butanol yields and high COD removal. The sulfuric acid pretreatment of *Ulva lactuca* by *C. beijerinckii* NCIMB 8052 yields more ABE of 0.35 g/g sugar in comparison to *C. acetobutylicum* ATCC 824 (van der Wal et al. 2013). Wang et al. (2016) enhanced butanol production from *Chlorella vulgaris* JSC-6 by sequential alkali pretreatment and acid hydrolysis. The residue from algae after extracting oil for biodiesel was reported for butanol production by *C. acetobutylicum* with 3.86 g/L (Cheng et al. 2015).

Fu et al. (2021) investigated butanol production from *Saccharina japonica* using engineered *Clostridium tyrobutyricum* and reported ultrasonic-assisted acid hydrolysate had the highest butanol yield of 0.26 g/g. The engineered strain Ct-pMA12G showed improved butanol yield of 0.34 g/g by overexpression of heat-shock protein. Some butanol-producing algae with their potential are given in Table 9.2.

TABLE 9.2
Butanol Production from Algal Biomass by *Clostridia* sp.

Algae	*Clostridia* sp.	Butanol Production (g/L)	References
Nannochloropsis sp.	*C. acetobutylicum* B1787	13.2	Efremenko et al. 2012
Arthrospira sp.		9.1	
Ulva lactuca	*C. beijerinckii* ATCC 55025 & *C. saccharoperbutylacetonicum*	4	Potts et al. 2012
Wastewater microalgae	*C. saccharoperbutylacetonicum* N1–4	7.8	Ellis et al. 2012
Ulva lactuca	*C. beijerinckii* NCIMB 8052	3	van der Wal et al. 2013
	C. acetobutylicum ATCC 824	0.8	
Wastewater microalgae	*C. saccharoperbutylacetonicum* N1–4	3.74	Castro et al. 2015
Chlorella vulgaris JSC-6	*C. acetobutylicum* ATCC 824	13.1	Wang et al. 2016
Neochloris aquatica CL-M1	*C. acetobutylicum* ATCC 824	12.0	Wang et al. 2017
Laminaria digitata	*C. beijerinckii* DSM-6422	7.16	Hou et al. 2017
Lyngbya limnetica and *Oscillatoria obscura*	*C. beijerinckii* ATCC 35702.	8.87	Kushwaha et al. 2020
Saccharina japonica	*C. tyrobutyricum*	12.15	Fu et al. 2021

Some industrial wastes have the potential as inexpensive raw materials for producing butanol. Glycerol, a major by-product of biodiesel production, is used for butanol by *Clostridium pasteurianum* ATCC 6013 (Taconi et al. 2009), *Clostridium acetobutylicum* KF158795 (Yadav et al. 2014), and *Clostridium sp.* Strain CT7 (Chen et al. 2019). Cannilla et al. (2020) evaluated zeolite-assisted etherification of glycerol with butanol for biodiesel-oxygenated additives production. Butanol production of 5 g/L/h by immobilized *C. pasteurianum* is reported from biodiesel-derived crude glycerol (Khanna et al. 2013). Another important feedstock for butanol production is syngas. Syngas is used for butanol production by *C. carboxidivorans* P7T (Bruant et al. 2010).

9.4 PRETREATMENT IN BUTANOL PRODUCTION FROM LIGNOCELLULOSIC BIOMASS

Butanol production from lignocellulosics by ABE fermentation generally has three sequential stages of the process, namely, pretreatment, enzymatic hydrolysis, and fermentation. The different types of pretreatments, like physical, chemical, and biological methods, alter the structure of the biomass. The lignocellulose requires pretreatment to remove lignin, reduce polymerization and cellulose crystallinity, and increase surface area, porosity, and enzymatic accessibility (Wyman et al. 2005).

The acid hydrolysates of willow biomass are able to produce butanol by fermentation (Han et al. 2013). Zhang et al. (2013) developed a wet-disk milling for improved butanol production from corncob. Butanol yield from acid-hydrolyzed Brewer's spent grain at pH 1 produced by *Clostridium beijerinckii* DSM 6422 was 75 g butanol/kg (Plaza et al. 2017).

Cai et al. (2016) reported the dilute alkaline NaOH treatment of cob, leaf, and stem hydrolysate, flower, and husk of corn produced 9.4 g/L, 7.6 g/L, 7.5 g/L, and 7 g/L butanol respectively by *C. acetobutylicum* ABE 1301. Valles et al. (2021) optimized butanol production from rice straw by NaOH pretreatment (0.75% w/v) at 134°C for twenty minutes by *Clostridium beijerinckii*. Butanol production by *C. acetobutylicum* NRRL B-591 was also improved with organosolv pretreatment of rice straw by aqueous ethanol 75% (v/v) added with 1% w/w sulfuric acid at 180°C for thirty minutes (Amiri et al. 2014). The combination of dilute acid and oxidative ammonolysis increased enzymatic hydrolysis in sugarcane bagasse by *C. acetobutylicum* CH02. Butanol production was improved to 7.9 g/L from acid hydrolysates of barley straw by *C. acetobutylicum* DSM 1731 with polyethylene glycol PEG 4000 surfactant-assisted xylanase and cellulase treatment (Yang et al. 2017). Higher butanol was produced by *Clostridium beijerinckii* DSM 6422 from hydrolysates of wheat straw by steam explosion and ozone treatment (Plaza et al. 2017). *Clostridium saccharobutylicum* DSM 13864 produced butanol efficiently from pretreatment of corn stover with ten times recycled ionic liquid [Bmim][Cl] (Ding et al. 2015). Butanol production by microwave-assisted alkali pretreatment and enzymatic hydrolysis was studied by Valles et al. (2020).

9.5 BUTANOL-PRODUCING MICROBES

9.5.1 Clostridium spp.

The most common natural fermentative microorganism in butanol fermentation is *Clostridium* spp. The genus of this bacteria is Firmicutes, and family is Clostridiaceae. *Clostridium* spp. is a rod-shaped, gram-positive, obligate anaerobic bacteria. Louis Pasteur in 1861 reported for the first time about butanol production by *Vidrion butyrique*, which was actually a mixed culture (Dürre 2008). Later, Weizmann isolated *Clostridium acetobutylicum* (Jones and Woods 1986).

The four major groups of clostridia, *C. acetobutylicum*, *C. beijerinckii*, *C. saccharoperbutylacetonicum*, and *C. saccharobutylicum*, are butanol producers (Lee et al. 2016). The most widely studied butanol-producing bacteria is *C. acetobutylicum* (Jones and Woods 1986). *Clostridium acetobutylicum* has utilized different carbon sources, like glucose, xylose, sucrose, lactose, and starch, for ABE fermentation. Acetone, butanol, and ethanol production by *Clostridium acetobutylicum* from ABE fermentation has a ratio of 3:6:1 respectively (Jones and Woods 1986). Table 9.3 summarizes butanol-producing Clostridia sp. with butanol yield from different carbon sources.

Other species of clostridia with butanol production capacity are *Clostridium therrnopapyrolyticum* sp. nov., *Clostridium cellulolyticum*, *Clostridium butylicum*, *C. aurantibutyricum*, and *C. pasteurianum*, *Clostridium thermocellum*, and *Clostridium tyrobutyricum*.

Bruant et al. (2010) reported *C. carboxidivorans* strain P7^T with butanol production from syngas utilizing carbon monoxide following Wood-Ljungdahl pathway for CO and CO_2 fixation and conversion to acetyl-CoA. Virunanon et al. (2008)

TABLE 9.3
Butanol-Producing *Clostridia* sp. with Butanol Yield from Different Carbon Sources

Substrate	*Clostridium* sp.	Butanol Production (g/L)	Butanol Yield (g/g)	Reference
Starch	*C. acetobutylicum* P262	16.0	0.21	Madihah et al. 2001
	C. acetobutylicum GX01	18.8	0.26	Li et al. 2015
	C. acetobutylicum NRRL B-591	15.3	0.21	Kheyrandish et al. 2015
	C. acetobutylicum SE25	15.15	0.36	Li et al. 2016
	C. beijerinckii BA101	13.4	0.11	Ezeji et al. 2007
	C. saccharoperbutylacetonicum N1–4	16.9	0.33	Thang et al. 2010
Corn stover	*C. beijerinckii* P260	14.5	0.44	Qureshi et al. 2010
Corncob	*C. saccharobutylicum* DSM 13864	12.27	0.35	Gao et al. 2014
Wheat straw	*C. beijerinckii* P260	8.09	0.32	Qureshi et al. 2010
Barley straw	*C. beijerinckii* P260	18.01	0.43	Qureshi et al. 2010
Rice bran	*C. saccharoperbutylacetonicum* N1–4	7.72	0.44	Al-Shorgani et al. 2012

isolated fifteen solventogenic-cellulolytic clostridia with 83–100% similarity range from decomposed agricultural wastes, cow feces, and dry grass in Thailand based on screening by three selected criteria, namely, endospore formation, sulfite-reducing ability, and metabolic products, and also 16S rDNA identification. Butanol production by *Clostridium saccharoperbutylacetonicum* was improved with increase in consumption rate of lactic acid and acetic acid (Zhou et al. 2018). A non-acetone-producing *Clostridium sporogenes* BE01 improved biobutanol production to 5.52 g/L from acid hydrolyzed rice straw (Gottumukkala et al. 2013).

9.5.2 Lactobacillus spp.

Russmayer et al. (2019) reported on the naturally occurring, butanol-producing *Lactobacillus diolivorans* from glucose by two-step microbial process with *Serratia marcescens*. Liu et al. (2021) identified dysregulated proteins in butanol-tolerant *Lactobacillus mucosae* BR0713–33 by proteomic analysis. Butanol-tolerant *Lactobacillus* sp. was also reported by Knoshaug and Zhang (2009). Liu et al. (2012) isolated and adapted eight *Lactobacillus* sp., namely, *Lactobacillus mucosae* strains BR0605–3, BR0605-B15, BR0713–18, BR0713–20, BR0713–30, BR0713–33, *Lactobacillus amylovorus* strain NE-L 0206–19, *Lactobacillus crispatus* NE-L 0206–47, with butanol tolerance of 3–4%. A higher butanol tolerance was also exhibited by *L. brevis* and *L. delbrueckii* than the *Clostridium* sp. (Li et al. 2010).

9.5.3 Other Butanol-Producing Microbes

Some research efforts for butanol fermentation were found successful by alternative microbes with genetic modifications. Shen and Liao (2008) reported the potential of *E. coli* in butanol fermentation. Shen et al. (2011) reported enhancement in butanol synthesis in *E. coli* with metabolic pathways designed with the help of enzyme mechanisms. Steen et al. (2008) reported the butanol production capability of *Saccharomyces cerevisiae*. Berezina et al. (2010) engineered *Lactobacillus brevis* with n-butanol metabolic pathway.

9.6 METABOLIC PATHWAY IN BUTANOL SYNTHESIS OF CLOSTRIDIUM ACETOBUTYLICUM

Butanol production in two common butanol-producing microbes, *C. acetobutylicum* and *C. beijerinckii*, consists of acidogenesis and solventogenesis phases. The starchy and lignocellulosic substrates are converted to glucose directly or by pretreatments before entering the acidogenic phase in the metabolic pathway of *C. acetobutylicum*. The first stage is the acidogenesis with organic acid formation and continuous ATP production at low pH. Then, solvents are produced from organic acid with shift to the second phase of solventogenesis with production of acetone, butanol, and ethanol as the main products (Dürre 2007, 2008).

The existing pathways used for 1-butanol production are the ketoacid and fermentative pathway. *Clostridium acetobutylicum* has commonly followed the metabolic pathway (Figure 9.2) that uses glucose and conversion to organic acids and organic

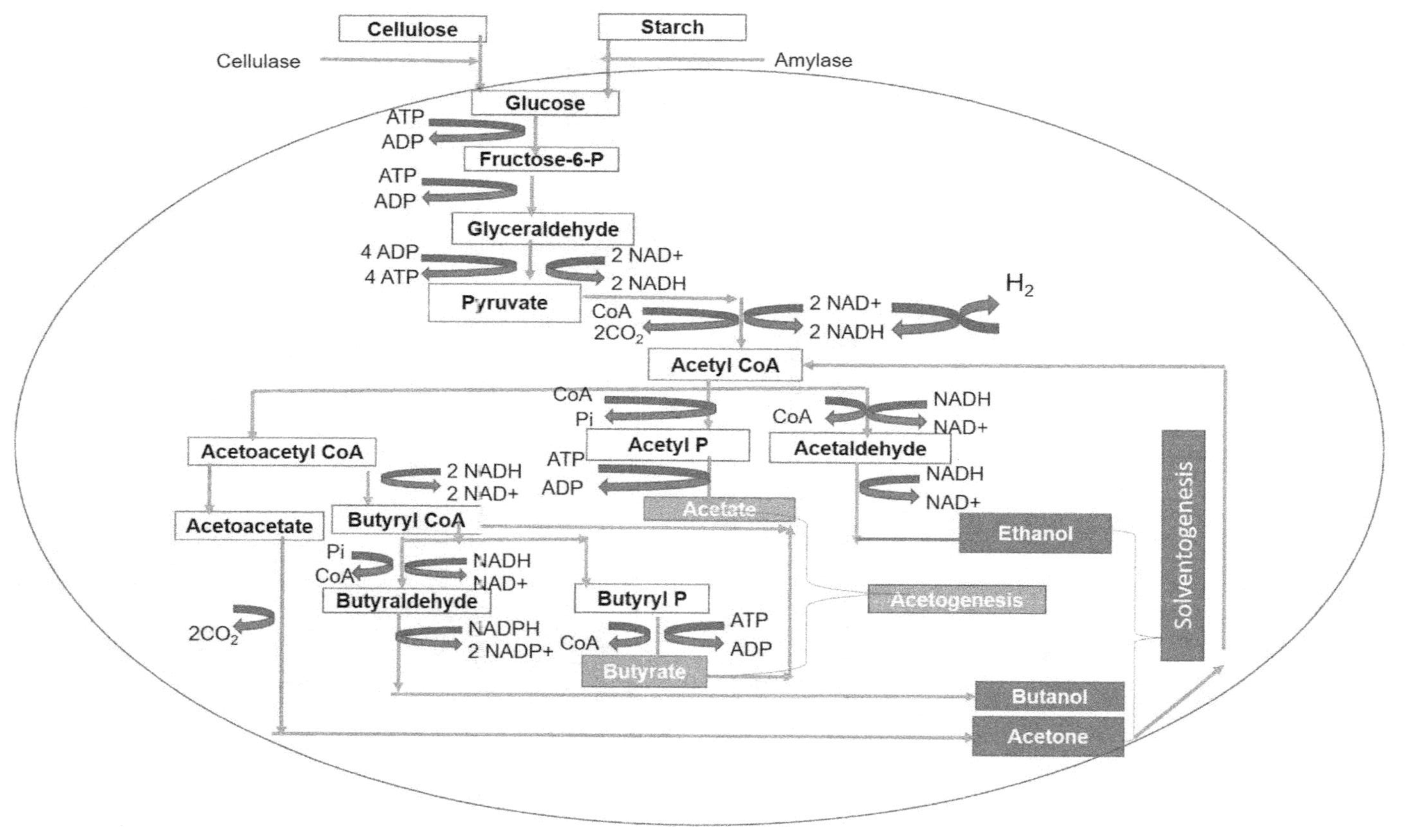

FIGURE 9.2 Metabolic route in ABE fermentation by *Clostridium* spp.

solvents by ABE fermentation (Ezeji et al. 2010; Huang et al. 2010). Pentoses and hexoses are converted first to 2 moles of pyruvate by the Embden-Meyerhof pathway, along with 2 adenosine triphosphate molecules and 2 nicotinamide adenine dinucleotide molecules. Further, each mole of pyruvate is converted to acetyl-CoA and carbon dioxide. The acetyl CoA finally forms acetone, ethanol, and butanol (Stanier et al. 1970; Dürre 2008; Li et al. 2010).

$$1\ C_6H_{12}O_6 \longrightarrow 2\ C_3H_3O_3^- + 2\ ATP + 2\ NADH \ldots\ldots(1)$$

$$1\ C_3H_3O_3^- \longrightarrow Acetyl\text{-}CoA + CO_2 \ldots\ \ldots\ldots(2)$$

Acetyl-CoA ⟶ acetaldehyde/butyraldehyde ⟶ (acetone, acetate) + butanol + ethanol

Girbal and Soucaille (1994) reported the role of NADH/NAD ratio and ATP pool on the regulation of *Clostridium acetobutylicum* metabolism.

9.7 ENZYMES INVOLVED IN BUTANOL-PRODUCING BACTERIA

Hartmanis and Gatenbeck (1984) explained the enzymes associated with production of acetate and butyrate by intermediary metabolism in *Clostridium acetobutylicum*. The enzymes involved in butanol production in the metabolic pathway are acetoacetyl-CoA:acyl-CoA transferase and acetoacetyl-CoA:acetate/butyrate CoA transferase. The CoA derivatives of acetate and butyrate are produced by the acetoacetyl-CoA:acyl-CoA transferase. Ethanol formation from acetone and acetaldehyde is mediated by the acetoacetyl-CoA: acetate/butyrate: CoA transferase (CoAT) pathway. The butaraldehydes formed from butyryl-CoA is further converted to ethanol and butanol (Jones and Woods 1986; Huang et al. 2010; Schadeweg and Boles 2016).

Vasconcelos et al. (1994) demonstrated that metabolism of *Clostridium acetobutylicum* at neutral pH has failed to have butyraldehyde dehydrogenase activity and has only low activities of coenzyme A-transferase, butanol, and ethanol dehydrogenase. Mitchell and Tangney (2005) reported butanol production from carbohydrate uptake is linked to phosphotransferases system proteins with cytoplasmic proteins. The genes ptsI and ptsH regulate enzyme I and HPr, a heat-stable histidine-phosphorylatable protein, respectively, along with sugar-specific enzyme II complexes involved in butanol synthesis.

9.8 BOTTLENECKS IN COMMERCIAL BIOBUTANOL PRODUCTION

Butanol production depends on feedstock cost, performance of microbes, mode of fermentation, and process of product recovery. Microbial inhibitors generated during the pretreatment process, ineffective techniques followed in genetic engineering, lack of proper knowledge on metabolic pathways, and energy-intensive separation process also limit the commercialization of butanol fermentation.

9.8.1 Why Butanol Is Not Economically Viable

Biobutanol production has challenges in commercialization. The lack of efficient microbial strain limits the commercial production of butanol. This results in low productivity, low yield of butanol, and selectivity in ABE fermentation. The lack of cost-effective feedstock also adds to the challenges in butanol commercialization. The cost factor related to pretreatment and enzymatic hydrolysis also limits butanol fermentation from lignocellulosic feedstock. The substrate cost accounts to nearly 60% of the total cost of butanol production (Qureshi and Blaschek 2000). The pretreatment technology covers 18–20% of the total cost in biofuel production (Qureshi et al. 2020). Paniagua-García et al. (2018) also reported high cost of pretreatment and low hydrolysis efficiency as major problems in lignocellulosic butanol production. The limitations in pretreatment process also include butanol inhibition and low butanol yield (Jones and Woods 1986). Economic feasibility can be attained, and high proportion of feedstock cost can be reduced, with the utilization of cheap and sustainable alternate feedstock (Qureshi et al. 2010).

9.8.2 Butanol Feedback Inhibition

The solventogenic bacteria has toxicity by the solvents produced from the fermentation, and it limits the process or alters the cell (Isken and de Bont 1998). Butanol inhibition in ABE fermentation is the major challenge in commercialization of butanol (Bowles and Ellefson 1985; Li et al. 2010). The toxicity of butanol also negatively affects the time of ABE fermentation, yield, and productivity.

Bowles and Ellefson (1985) explained disruption of the function of embedded membrane proteins and causing alterations in membrane fluidity on *Clostridium acetobutylicum* by butanol. Solvent accumulation affects the permeability of cell membrane, thereby resulting in passive flux of ATP, protons, ions, and macromolecules, such as RNA and proteins (Sikkema et al. 1995). The transfer of the phosphoryl group from PEP occurrs in sequence with enzyme I, heat-stable histidine phosphorylatable protein, enzyme II, and translocation of substrate through the cell membrane. Further solutes are transferred into the cytoplasm by a non-PTS system. Glycolysis in butanol-producing microorganisms is affected by butanol accumulation from hexose and pentose catabolism (Ezeji et al. 2010).

9.8.3 Microbial Inhibitors from Pretreatment Process

The inhibitors produced from the pretreatment of lignocellulosics, namely, organic acids, furan derivatives, and phenolic compounds, reduce ABE fermentation and butanol production (Díaz and Tost 2018). The inhibitory products from the fermentation of lignocellulosic hydrolysates are summarized in Figure 9.3.

Díaz and Tost (2018) reported simultaneous in situ recovery of inhibitors from prehydrolysates without any additional total energy requirements and recovery of value-added chemicals like butyric acid, vanillin, and furfuryl alcohol.

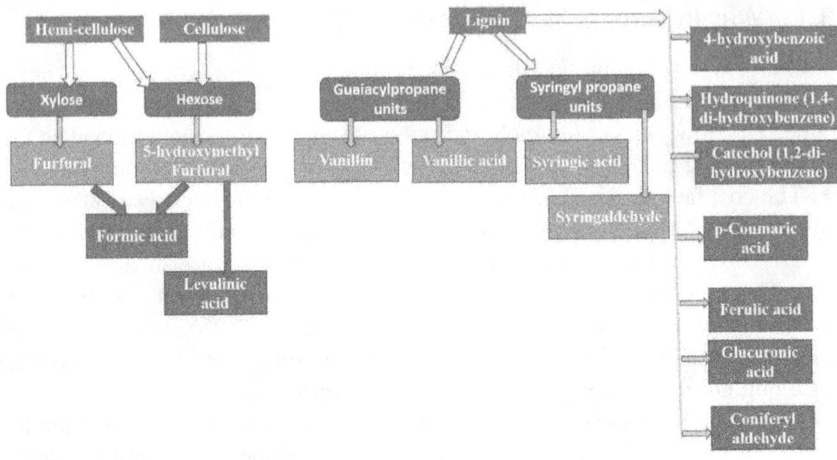

FIGURE 9.3 Microbial inhibitors produced from pretreatment of lignocellulosics that affect butanol fermentation.

Source: Palmqvis and Hahn-Hagerdal (2000); Ezeji et al. (2007); Zautsen et al. (2009)

9.9 STRATEGIES FOR IMPROVING EFFICIENCY OF UPSTREAM PROCESSES

Improvement in upstream processes is expected in the following:

- Design of process technologies
- Integration of saccharification and fermentation processes
- Integration of fermentation and recovery
- Continuous fermentation of immobilized cells
- Advanced fermentation techniques
- Optimization of the fermentation process

9.9.1 Modes of Butanol Fermentation

Batch, fed-batch, and continuous ABE fermentation with pH control for butanol generation was evaluated by Li et al. (2011). Enhanced reactor productivity is observed in fed-batch and continuous fermentation with decreased reactor size, low capital cost, and use of concentrated substrates. Batch fermentation has advantages of simplicity in operation and easiness in maintenance with minimum control requirements (Vees et al. 2020). Enhancement in butanol to 13.82 g/L from optimized batch fermentation by *C. acetobutylicum* YM1 was observed by Al-Shorgani et al. (2018). The product inhibition, dead time periods required for medium preparation, and sterilization limit the batch fermentation, resulting in lower butanol productivity (Vees et al. 2020).

The reduction in hydraulic load and the generation of wastewater result from the utilization of highly concentrated substrate by fed-batch fermentation than batch and continuous fermentations (Qureshi and Blaschek 2000; Ezeji et al. 2007). Fed-batch fermentation reduces substrate inhibition more than batch fermentation (Chen et al. 2017). Niglio et al. (2019) improved butanol production in *C. saccharobutylicum* from corn syrup by fed-batch fermentation. Fed-batch fermentation resulting in high butanol production by immobilized *C. saccharoperbutylacetonicum* N1–4 was also observed by Darmayanti et al. (2018). Darmayanti et al. (2020) also improved butanol yield from high cell density fermentation in a large extractant volume.

Commercial butanol production commonly uses continuous mode of fermentation due to its more productivity. The shortcomings of batch fermentation are overcome by the continuous mode, which avoids butanol accumulation, and hence the toxicity. The yield and productivity of butanol have increased with constant rate due to continuous adding up of fresh medium and removal of the product. Girbal and Soucaille (1994) studied the mixed-substrate steady-state continuous culture of *Clostridium acetobutylicum* in butanol metabolism.

Ezeji et al. (2007) reported that the continuous mode of fermentation led to improvement in butanol from degermed corn by *C. beijerinckii*. Ni et al. (2012) investigated butanol fermentation by *C. saccharobutylicum* DSM 13864 in continuous fermentation of cane molasses and corn stover hydrolysates. Cane molasses produced butanol production of 13.75 g/L, and corn stover hydrolysate with 7.81 g/L, in a four-stage continuous fermentation. The biphasic nature of ABE fermentation, phage contamination, and flocculation of bacterial growth leading to unsteady fermentation are the limitations of the continuous mode of fermentation. Butanol yield by *C. acetobutylicum* CICC 8012 was found to be improved in a continuous and closed-circulating fermentation system with a membrane bioreactor (Chen et al. 2017).

9.9.2 Detoxification and Cell Immobilization

Methods of detoxification, like evaporation, overliming, peroxidase treatment, adsorption, and alkaline peroxide treatment, remove inhibitors (Paniagua-García et al. 2018). Liu et al. (2015) observed that activated carbon detoxification increased butanol production to 11 g/L from switchgrass. Yang et al. (2017) reported enhanced butanol production to 14.7 g/L from lignocellulosic hydrolysates of *Salix schwerinii* by supplementing barley grain starchy slurry with improvement in xylose utilization from hemicelluloses. Ana Maria et al. (2021) reported that the cofermentation of sugarcane bagasse hydrolysate and molasses by *Clostridium saccharoperbutylacetonicum* with furan derivatives lower than 0.1 g/L produced butanol of 7.8 g/L with 0.28 g/g.

The cell immobilization process has improved the productivity and stability in butanol fermentation (Survase et al. 2013). Immobilization helps in cell protection from toxicity and to avoid bleeding during continuous fermentation (Gholizadeh and Baroghi 2009). Butanol yield from waste starch is increased by immobilized *Clostridium acetobutylicum* (Kheyrandish et al. 2015). Qureshi et al. (2000) demonstrated 2.5 times higher butanol yield from continuous fermentation using immobilized *Clostridium beijerinckii* BA101 on clay brick. The coculturing of immobilized *Clostridium beijerinckii* and *Clostridium tyrobutyricum* increased butanol yield from cassava starch and cane molasses (Li et al. 2013). Gallego-Morales et al. (2015) reported butanol yield of 4,500 L/ha/year from king grass (*Pennisetum hybridum*) by *C. acetobutylicum* ATCC 824 in simultaneous and separated hydrolysis and fermentation. Butanol yield was improved from Ca^{2+}- and NAD/NADPH-mediated cassava hydrolysate by *Clostridium* species strain BOH3 (Li et al. 2015).

9.10 ADAPTATION OF BUTANOL-PRODUCING STRAINS AGAINST STRESSES

The different means of adaptation to butanol tolerance in bacteria have been studied by different researchers.

9.10.1 Butanol-Tolerant Microbes

The most commonly used *Clostridium* strains are less tolerant to butanol concentration higher than 2% (Borden and Papoutsakis 2007). Butanol-tolerant microbes have the potential for high butanol productivity. Naturally butanol-tolerant microorganisms can act as a host resource for engineering tolerant strains to increase butanol productivity.

Knoshaug and Zhang (2009) screened twenty-four nonclostridia strains of *Escherichia coli*, *Zymomonas mobilis*, *Saccharomyces cerevisiae*, and non-Saccharomyces yeasts (*Candida and Pichia*) and *Lactobacillus* for butanol tolerance. *Lactobacillus brevis* and *Lactobacillus delbrueckii* have been found to have potential to tolerate 24.3 g/L butanol, while others have growth limited in 1–2%. Kataoka et al. (2011) reported *Bacillus subtilis* GRSW2-B1 with butanol tolerance up to 5% v/v.

Kanno et al. (2013) isolated sixteen bacteria with butanol tolerance higher than 2% (v/v) belonging to nine genera under phyla Firmicutes and Actinobacteria. The genera reported were *Bacillus*, *Lysinibacillus*, *Rummeliibacillus*, *Brevibacillus*, *Coprothermobacter*, *Caloribacterium*, *Enterococcus*, *Hydrogenoanaerobacterium*, and *Cellulosimicrobium*. Three bacterial strains, namely CM4A (from grease-contaminated soil), SK5A from freshwater sediment, both with close similarity to *Enterococcus faecalis*, and GK12 (from thermophilic anaerobic digester), with similarity to *Eubacterium cylindroides*, showed tolerance of 3.5%, 3%, and 3% respectively. Goyal et al. (2019) isolated *Staphylococcus sciuri* KM16 from soil with butanol tolerance up to 2.25% (v/v). Rühl et al. (2009) reported three solvent-tolerant *Pseudomonas putida* strains, namely, *Pseudomonas putida* DOT-T1E, *Pseudomonas putida* S12, and *Pseudomonas* sp. Strain VLB120 can tolerate 6% (vol/vol) butanol. Tashiro et al. (2007) reported improved butanol fermentation utilizing butyrate by *Clostridium saccharoperbutylacetonicum* N1–4 (ATCC 13564) with the help of an electron carrier, methyl viologen.

9.10.2 Alteration in Cellular Membrane

Ezeji et al. (2010) explained the transport of nutrients to the microbial cytoplasm from the culture media with the feedstock as a basic requisite for cell metabolism and, thereby, growth and development of microbial cells. Microorganisms depend on various passive and active transport mechanisms for nutrient uptake through cell membrane and accumulation in cytoplasm. Osmosis, diffusion, and facilitated diffusion are the common passive-transport mechanisms with no energy consumption involved in bacteria for intake and deposition. Sugar transportation by active mechanisms is mainly associated with adenosine triphosphate, phosphoenolpyruvate, or ion gradient system. The solventogenic clostridia mainly has a PEP-dependent phosphotransferase transport system (PTS) for cell metabolism (Mitchell and Tangney 2005).

Genetically engineered microbes with systematic and mutagenic techniques help in improving the productivity of butanol and developing tolerant strains.

9.11 ENGINEERING IDEAL MICROORGANISMS FOR BUTANOL PRODUCTION

The hosts for butanol production in metabolic engineering must be able to tolerate product toxicity and also fermentation inhibitors. The various techniques involved in engineering butanol-producing and tolerant microbes are:

- Recombinant technology
- Engineering genes or enzymes
- Construction of metabolic pathway
- Metabolic engineering
  - Enhancing butanol production
  - Increasing butanol selectivity
  - Improving butanol tolerance

Ranganathan and Maranas (2010) reported low butanol yield from all the seven pathways of pyruvate to butanoyl-CoA metabolism in E. coli. The new butanol synthesis pathway of thiobutanoate pathway with reduction by decarboxylation of methylthiobutanoate to 1 butanol was reported by Atsumi et al. (2008). Boynton et al. (1996) demonstrated cloning, sequencing, and expression of genes, namely pta and ack encoding phosphotransacetylase and acetate kinase, respectively, were involved in the transformation of acetyl coenzyme A to acetate in *Clostridium acetobutylicum* ATCC 824. The route for butanol synthesis in engineered *E. coli* was dependent on end products of glycolysis (Atsumi and Liao 2008). Tomas et al. (2003) reported the overexpression of groESL also resulted in increased butanol yield and tolerance in *Clostridium acetobutylicum* ATCC 824.

Alsaker et al. (2004) used DNA microarray-based gene expression analysis to examine responses to butanol, butyrate, and acetate stresses in *Clostridium acetobutylicum*. The upregulation of stress genes and downregulation of butyryl-CoA, butyrate, and acetate formation genes were observed. The solvent-tolerant genes for *Clostridium acetobutylicum* were identified by Borden and Papoutsakis (2007).

Bruant et al. (2010) reported the sequencing of P7T genome for metabolic engineering strategies to enhance butanol by *C. carboxidivorans* capacity from syngas. Xiao et al. (2012) demonstrated metabolically engineered *Clostridium beijerinckii* strain 8052xylR-xylT(ptb) with D-xylose consumption for butanol from corncob.

Russmayer et al. (2019) reported a high butanol generation utilizing meso-2,3-BTD conversion from glucose by an engineered *L. diolivorans* strain with the help of overexpression of alcohol dehydrogenase pduQ.

9.11.1 Genes Involved in Butanol Tolerance

The genes responsible for butanol production, such as hbd, crt, adhE1, adhE2, and bdhB, were isolated from *Clostridium acetobutylicum* and transformed to other microorganisms (Yoo et al. 2016). Metabolic engineering mainly depends on overexpression, insertion, gene knocking out, and gene knocking down of the enzymes involved in acetone, ethanol, and butanol fermentation (Huang et al. 2010). The genes coded for heat-shock proteins, operon, transcription, etc. were also used for developing butanol-tolerant microbes by recombinant technology (Huang et al. 2010). Butanol producing recombinant *Escherichia coli* was developed by metabolic engineering (Atsumi and Liao 2008).

Nielsen et al. (2009) also engineered *Pseudomonas putida* and *Bacillus subtilis* as microbial hosts for butanol fermentation. Steen et al. (2008) demonstrated and compared metabolic engineering of *Saccharomyces cerevisiae* with a number of isozymes involved in the butanol synthesis pathway of different microorganisms. Many researchers have listed genes used in designing *Saccharomyces cerevisiae* for n-butanol production (Azambuja and Goldbeck 2020). The metabolic pathway for butanol synthesis was also reconstructed in *Lactobacillus brevis* (Berezina et al. 2010, Li et al. 2020)

Harris et al. (2002) reported the role of overexpression of the spo0A gene results towards butanol stress, resulting in high number of cells in either the swollen clostridial form or in endospore formation. Schadeweg and Boles (2016) reported an increase in butanol production by *Saccharomyces cerevisiae* by providing excess acetyl CoA or NADH by limiting ethanol formation. CoA synthesis was achieved by overexpression of the pantothenate kinase coaA gene from *Escherichia coli.*

9.11.2 Metabolic Engineering

Enhancement in butanol production by gene manipulation techniques like recombinant technology and antisense RNA techniques is evident in nonnative microbial hosts (Nawab et al. 2020). The spo0A regulates the stationary phase of ABE fermentation and is essential for the transcription of solvent formation genes in *Clostridium acetobutylicum* and in *C. beijerinckii* (Ravagnani et al. 2000). Harris et al. (2002) reported that the spo0A promoted conversion of acetate and butyrate into acetone, butanol, and ethanol with the help of fast expression of solventogenic transcript *aad-ctfAB* in *Clostridium acetobutylicum* ATCC 824 (pMSPOA). Alsaker et al. (2004) reported that overexpression of *spo0A* helped increase butanol tolerance and continued fermentation in 824 (pMSPOA). The response of butanol stress was also related to different genes involved in the enhancement of butanol tolerance.

Tomas et al. (2003) carried out DNA array–based transcriptional analysis of two non-spore-forming and nonsolventogenic strains of *Clostridium acetobutylicum* SKO1 strain with inactivated spo0A gene (Harris et al. 2002) and M5 strain with megaplasmid pSOL1. Alsaker et al. (2005) developed full genome microarrays for *Clostridium acetobutylicum* and validated with mutant (M5) strain lacking the pSOL1 178-gene megaplasmid compared to wild-type strain. Sillers et al. (2009) reported aldehyde-alcohol dehydrogenase and/or thiolase overexpression coupled with CoA transferase downregulation for high butanol production and selectivity in *Clostridium acetobutylicum.* Guo et al. (2011) developed *Clostridium beijerinckii* mutant by atmospheric pressure glow discharge. Khanna et al. (2013) developed a mutant strain of *C. pasteurianum* with high tolerance to crude glycerol with a butanol productivity of 1.80 g/L/h.

Gao et al. (2012) isolated *C. acetobutylicum* CICC 8012 strain F2- GA after two times of genome shuffling resulted in ABE production of 22.21 g/L with 14.15 g butanol. The genome editing by CRISPR-Cas has benefits over other conventional techniques. Cell engineering with the help of CRISPR-Cas-based biotechnology tool helps in improvement of butanol production (Xue et al. 2016). Goyal et al. (2019) reported that the contribution of four possible genes was responsible for butanol tolerance in *Staphylococcus sciuri* KM16.

9.12 EFFECTIVE RECOVERY OF BUTANOL

ABE production is not economically viable due to severe product inhibition. Many researchers have combined fermentation by clostridia with butanol tolerance with effective butanol recovery as a solution to butanol inhibition (Ezeji et al. 2007). Butanol separation techniques reduce inhibition and enhancement in fermentation efficiency by concentrating solvents and also reducing energy cost of distillation (Zheng et al. 2009). The different technologies associated with recovery of butanol are pervaporation, perstraction, adsorption, gas stripping, liquid-liquid extraction, and reverse osmosis (Wagner et al. 2019; Roussos et al. 2019; Raganati et al. 2020). These techniques reduce butanol toxicity and improve productivity. The most common butanol recovery techniques are given in Figure 9.4.

9.12.1 Adsorption

Butanol recovery by adherence of solvents is possible with adsorbents such as silicalite resin, clay, and activated carbon. Adsorption is easy and requires less energy. Butanol recovery by adsorption is possible using Dowex Optipore SD-2 resin from ABE fermentation of *C. acetobutylicum* ATCC 824 (Nielsen and Prather 2009). Raganati et al. (2020) demonstrated Amberlite XAD-7 as an effective adsorbent for recovery of butanol.

9.12.2 Pervaporation

The pervaporation technique uses a hydrophilic or hydrophobic membrane which selectively separates butanol from ABE fermentation media and thus decreases inhibitory effect on solventogenic clostridia (Li et al. 2010; Zhu et al. 2020). The bioreactor used for pervaporation has a feed pump, membrane module, condenser, and vacuum pump that help in keeping butanol under threshold level of toxicity (Liu et al. 2021).

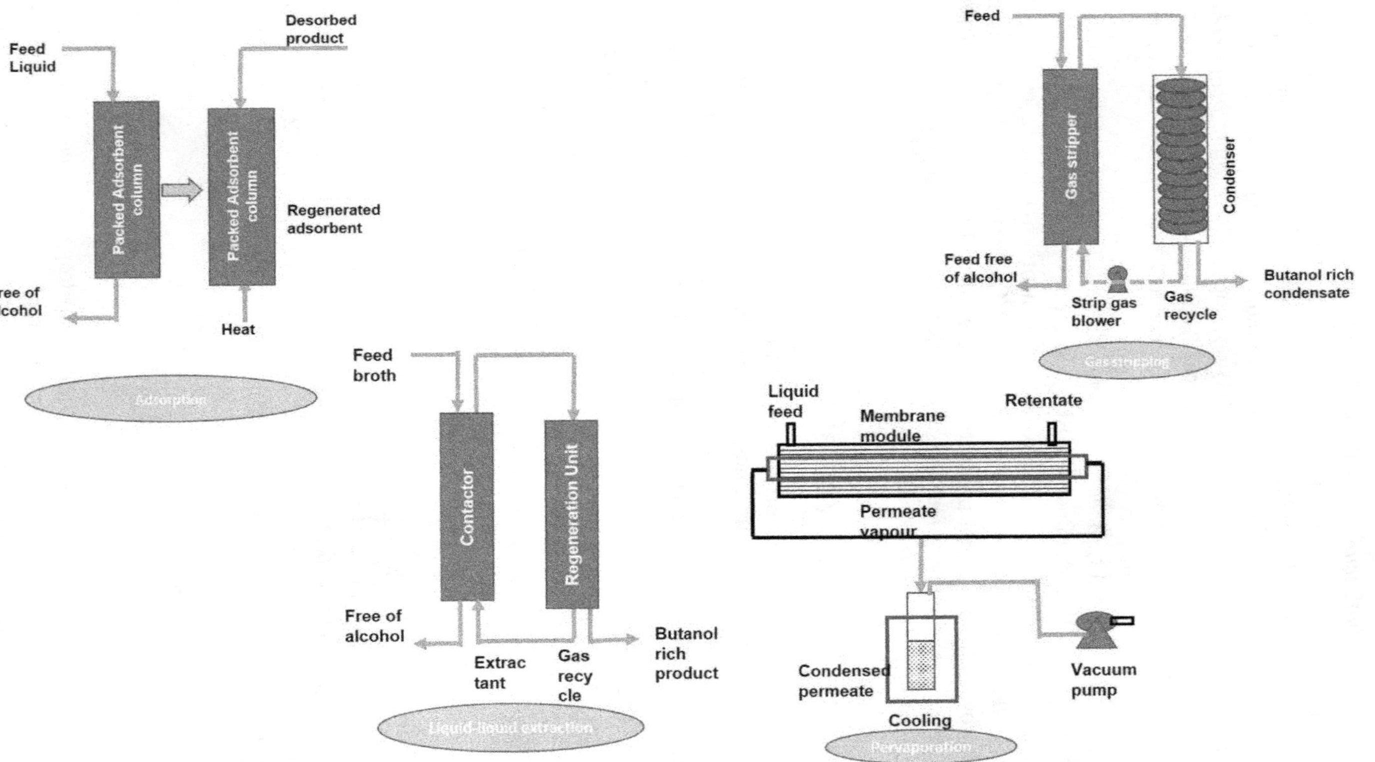

FIGURE 9.4 Diagrammatic representation of common in situ butanol recovery techniques.

Source: Adapted from Lu (2011).

The steps in pervaporation are solvent adsorption on membrane surface, followed by solvent diffusion through membrane, and further, absorption into permeate vapor. Li et al. (2014) demonstrated fermentation-pervaporation (PV) coupled process for butanol generation from cassava and reduction in butanol toxicity to the *Clostridium acetobutylicum* DP217. The most common membrane used in pervaporation is polydimethylsiloxane with butanol-water separation factor of 40 to 60 (Vane 2008). The other membranes used in pervaporation are polyether block amide, hydrophobic zeolite membranes, and silicalite-filled silicone (Li et al. 2004). Cai et al. (2013) reported the pervaporation-detoxified hydrolysate from sweet sorghum bagasse and butanol production of 12.3 g/L. Zhu et al. (2020) demonstrated pervaporation with fluorinated polydimethylsiloxane membrane coupled with fed-batch acetone-butanol-ethanol (ABE) fermentation process improved ABE productivity by 51%.

9.12.3 Gas Stripping

The simplicity in gas stripping process makes it economical and the most preferred option in butanol recovery, reducing inhibition (Qureshi and Blaschek 2000; Zheng et al. 2009; Qureshi et al. 2010). Gas stripping works with the principle of bubbling oxygen-free nitrogen or carbon dioxide or hydrogen into the fermentation media and further cooling and condensation of the gas along with acetone butanol and ethanol. Finally, collecting the condensed, enriched gas with solvents in a collector (Xue et al. 2016; Qureshi et al. 2020; Rochón et al. 2020). The gas is further recycled for repeated ABE fermentation.

This technique selectively recovers volatile substances such as butanol and acetone but is not able to remove nutrients (Dürre 2008). Gas stripping helps in maintaining butanol production below the limits of toxicity level with full sugar utilization (Ezeji et al. 2010). Lu et al. (2012) reported gas stripping recovered ABE solvents from cassava bagasse hydrolysate fermented by *C. acetobutylicum* JB200 with higher amount of butanol and very low concentration of ethanol. The phase separation further improved butanol concentration.

9.12.4 Liquid-Liquid Extraction

This technique uses organic extractant which is immiscible in water by adding to the aqueous fermentation broth, and butanol is extracted and back-recovered by using an extractant or distillation (Teke and Pott 2021). The extractants should be nontoxic in nature and should have high-quality extraction. The commonly used extractant is oleyl alcohol (Motghare et al. 2019; Verma et al. 2018). Roffler et al. (1987) demonstrated high butanol extraction and improved volumetric butanol productivity from extraction by oleyl alcohol and benzyl benzoate. Ezeji et al. (2007) listed that the limitations of liquid-liquid extraction system are cell inhibition caused by extractant, formation of emulsion, volatility of extractant, and rag layer formation. Ha et al. (2010) investigated imidazolium-based ionic liquids for liquid-liquid extraction of butanol. Yan and Wang (2013) demonstrated dispersive liquid-liquid microextraction for improved separation of butanol from the fermentation media due to higher butanol solubility in the organic phase. Motghare et al. (2019) evaluated butanol recovery by tetradecyl(trihexyl)phosphonium bis(2,4,4-trimethylpentyl) phosphinate. Teke and Pott (2021) designed a semipartition bioreactor for in situ product extraction and recovery by continuous liquid-liquid extractive fermentation.

9.12.5 Perstraction

This is an extractive fermentation technique with recovery by a membrane contactor providing surface area between two immiscibles for butanol exchange (Kim et al. 2020). Ezeji et al. (2007) reported the complete transfer of butanol from the media to the organic extractant. This technology overcomes the toxicity issues from extraction solvent to cells, phase dispersion, emulsion, and rag layer formation. But membrane fouling is the disadvantage of perstraction (Ezeji et al. 2010).

9.12.6 Development of Innovative Downstream Processes

The hybrid in situ butanol recovery process improves the efficiency of butanol fermentation. Enhancement in butanol separation from fermentation media has been observed with the following two stages or two recovery processes:

- Two-Stage Gas Stripping
- Gas Stripping–Gas Permeation
- Extraction–Gas Stripping

Xue et al. (2016) reported high-efficient product recovery by vapor stripping–vapor permeation (VSVP) process. Cai et al. (2016) evaluated how gas stripping–pervaporation hybrid process integrated with ABE fermentation increased ABE production. Rochón et al. (2020) reported butanol fermentation from sugarcane–sweet sorghum juices combined with gas stripping, followed by pervaporation, helped in improved butanol separation for *Clostridium beijerinckii* DSM 6423. Efficiency was also found to be increased with combination of distillation with effective butanol recovery techniques, such as:

- Vacuum separation–distillation
- Gas stripping–distillation
- Extraction-distillation
- Pervaporation-distillation
- Hybrid separations–distillation

Progress in upstream and downstream stages over the years has resulted in the enhanced recovery of butanol from ABE fermentation by solventogenic clostridia and genetically modified microbes.

9.13 CONCLUSION

Biobutanol has a great potential as an alternative future fuel based on its better fuel properties and less-toxic gas emissions. The global availability and abundance of second- and third-generation butanol feedstocks overcome the limitations of first-generation butanol production. But low butanol yield and butanol inhibition feedback reduce the performance efficiency of ABE fermentation from lignocellulosics. The strains with high-tolerance mechanisms, combined with upstream and downstream processes, are useful for improvement in butanol yield. R&D efforts and strategies to reduce feedstock cost with sustainable and efficient substrates, microbial strains, and improved fermentation and recovery techniques may definitely overcome the hindrance in commercialization by increasing butanol yield in ABE fermentation.

REFERENCES

Abd-Alla, M.H. and El-Enany, A.E. (2012). Production of acetone-butanol-ethanol from spoilage date palm (Phoenix dactylifera L.) fruits by mixed culture of *Clostridium acetobutylicum* and *Bacillus subtilis*. *Biomass Bioenergy*. 2: 172–178.

Alsaker, K.V., Paredes, C.J. and Papoutsakis, E.T. (2005). Design, optimization and validation of genomic DNA microarrays for examining the Clostridium acetobutylicum transcriptome. *Biotechnology and Bioprocess Engineering*. 10: 432–443. Doi:10.1007/BF02989826.

Alsaker, K.V., Spitzer, T.R. and Papoutsakis, E.T. (2004). Transcriptional analysis of spo0A overexpression in Clostridium acetobutylicum and its effect on the cell's response to butanol stress. *Journal of Bacteriology*. 186: 1959–1971. Doi:10.1128/JB.186.7.1959-1971.2004.

Al-Shorgani, N.K.N., Kalil, M.S. and Yusoff, W.M.W. (2012). Biobutanol production from rice bran and de-oiled rice bran by Clostridium saccharoperbutylacetonicum N1–4. *Bioprocess and Biosystems Engineering*. 35: 817–826.

Al-Shorgani, N.K.N., Shukor, H., Abdeshahian, P., Kalil, M.S., Yusoff, W.M.W. and Hamid, A. A. (2018). Enhanced butanol production by optimization of medium parameters using Clostridium acetobutylicum YM1. *Saudi Journal of Biological Sciences*. 25 (7): 1308–1321.

Amiri, H., Karimi, K. and Zilouei, H. (2014). Organosolv pretreatment of rice straw for efficient acetone, butanol, and ethanol production. *Bioresource Technology*. 152: 450–456. Doi:10.1016/j.biortech.2013.11.038.

Ana Maria, Z.A., Plazas, T. L., Ferraz, A.R., Pinto, M.A., Walter, van G., Rubens, M.F. and Sindelia, F. (2021). Co-fermentation of sugarcane bagasse hydrolysate and molasses by Clostridium saccharoperbutylacetonicum: Effect on sugar consumption and butanol production. *Industrial Crops and Products*. 167: 113512. Doi:10.1016/j.indcrop.2021.113512.

Atmanli, A. (2016). Comparative analyses of diesel—waste oil biodiesel and propanol, n-butanol or 1-pentanol blends in a diesel engine. *Fuel*. 176: 209–215. Doi:10.1016/j.fuel.2016.02.076.

Atsumi, S., Hanai, T. and Liao, J.C. (2008). Non-fermentative pathways for synthesis of branched-chain higher alcohols as biofuels. *Nature*. 451: 86–89.

Atsumi, S. and Liao, J.C. (2008). Metabolic engineering for advanced biofuels production from Escherichia coli. *Current Opinion in Bio-technology*. 19: 414–419.

Azambuja, S.P.H. and Goldbeck, R. (2020). Butanol production by Saccharomyces cerevisiae: Perspectives, strategies and challenges. *World Journal of Microbiology and Biotechnology*. 36: 48. https://doi.org/10.1007/s11274-020-02828.

Bellido, C., Infante, C., Coca, M., González-Benito, G., Lucas, S. and García-Cubero, M.T. (2015). Efficient acetone—butanol—ethanol production by Clostridium beijerinckii from sugar beet pulp. *Bioresource Technology*. 190: 332–338. Doi:10.1016/j.biortech.2015.04.082.

Berezina, O.V., Zakharova, N.V., Brandt, A., Yarotsky, S.V., Schwarz, W.H. and Zverlov, V.V. (2010). Reconstructing the Clostridial n-butanol metabolic pathway in Lactobacillus brevis. *Applied Microbiology and Biotechnology*. 87: 635–646.

Birgen, C., Dürre, P., Preisig, H.A. and Wentzel, A. (2019). Butanol production from lignocellulosic biomass: Revisiting fermentation performance indicators with exploratory data analysis. *Biotechnology for Biofuels*. 12: 167. Doi:10.1186/s13068-019-1508-6.

Borden, J.R. and Papoutsakis, E.T. (2007). Dynamics of genomic-library enrichment and identification of solvent tolerance genes for Clostridium acetobutylicum. *Applied and Environmental Microbiology*. 73(9): 3061–3068. Doi:10.1128/AEM.02296-06.

Bowles, L.K. and Ellefson, W.L. 1985. Effects of butanol on Clostridium acetobutylicum. *Applied Environmental Microbiology*. 50: 1165–1170.

Boynton, Z.L., Bennett, G.N. and Rudolph, F.B. (1996). Cloning, sequencing, and expression of genes encoding phospho-transacetylase and acetate kinase from Clostridium aceto-butylicum ATCC 824. *Applied Environmental Microbiology*. 62: 2758–2766.

Bruant, G., Lévesque, M.J., Peter, C., Guiot, S.R. and Masson, L. (2010). Genomic analysis of carbon monoxide utilization and butanol production by Clostridium carboxidivorans strain P7^T. *PLoS One*. 5 (9): e13033. Doi:10.1371/journal.pone.0013033.

Cai, D., Li, P., Luo, Z., Qin, P., Chen, C., Wang, Y., Wang, Z. and Tan, T. (2016). Effect of dilute alkaline pretreatment on the conversion of different parts of corn stalk to fermentable sugars and its application in acetone—butanol—ethanol fermentation. *Bioresource Technology*. 211: 117–124. Doi:10.1016/j.biortech.2016.03.076.

Cai, D., Zhang, T., Zheng, J., Chang, Z., Wang, Z., Qin, P. and Tan, T. (2013). Biobutanol from sweet sorghum bagasse hydrolysate by a hybrid pervaporation process. *Bioresource Technology*. 145: 97–102.

Cannilla, C., Bonura, G., Maisano, S., Frusteri, L., Migliori, M., Giordano, G., Todaro, S. and Frusteri, F. (2020). Zeolite-assisted etherification of glycerol with butanol for biodiesel oxygenated additives production. *Journal of Energy Chemistry*. 48: 136–144. Doi:10.1016/j.jechem.2020.01.002.

Castro, Y.A., Ellis, J.T., Miller, C.D. and Sims, R.C. (2015). Optimization of wastewater microalgae saccharification using dilute acid hydrolysis for acetone, butanol, and ethanol fermentation. *Applied Energy*. 140: 14–19.

Chen, C., Long, S., Li, A., Xiao, G., Wang, L. and Xiao, Z. (2017). Performance comparison of ethanol and butanol production in a continuous and closed-circulating fermentation system with membrane bioreactor. *Preparative Biochemistry & Biotechnology*. 47 (3): 254–260. Doi:10.1080/10826068.2016.1224242.

Chen, T., Xu, F., Zhang, W., Zhou, J., Dong, W., Jiang, Y., Lu, J., Fang, Y., Jiang, M. and Xin, F. (2019). High butanol production from glycerol by using Clostridium sp. strain CT7 integrated with membrane assisted pervaporation. *Bioresource Technology*. 288:121530. Doi:10.1016/j.biortech.2019.121530.

Cheng, H.H., Whang, L.M., Chan, K.C., Chung, M.C., Wu, S.H., Liu, C.P., Tien, S.Y., Chen, S.Y., Chang, J.S. and Lee, W.J. (2015). Biological butanol production from micro-algae based biodiesel residues by Clostridium acetobutylicum. *Bioresource Technology*. 184: 379–385.

Darmayanti, R., Tashiro, Y., Kenji, S., Kenji, S., Ari, S., Bekti, P. and Meta, R. (2020). Biobutanol production using high cell density fermentation in a large extractant. *International Journal of Renewable Energy Development*. 9. Doi:10.14710/ijred.2020.29986.

Darmayanti, R., Tashiro, Y., Noguchi, T., Gao, M., Sakai, K. and Sonomoto, K. (2018). Novel biobutanol fermentation at a large extractant volume ratio using immobilized Clostridium saccharoperbutylacetonicum N1–4. *Journal of Bioscience and Bioengineering*. 126 (6): 750–757. Doi:10.1016/j.jbiosc.2018.06.006.

Díaz, G.V.H. and Tost, G.O. (2018). Economic optimization of in situ extraction of inhibitors in acetone-ethanol-butanol (ABE) fermentation from lignocellulose. *Process Biochemistry*. 70: 1–8. Doi:10.1016/j.procbio.2018.04.014.

Ding, J.C., Xu, G., Han, R. and Ni, Y. (2015). Biobutanol production from corn stover hydrolysate pretreated with recycled ionic liquid by Clostridium saccharobutylicum DSM 13864. *Bioresource Technology*. 199. Doi:10.1016/j.biortech.2015.07.119.

Dolejš, I., Krasňan, V., Stloukal, R., Rosenberg, M. and Rebroš, M. (2014). Butanol production by immobilised Clostridium acetobutylicum in repeated batch, fed-batch, and continuous modes of fermentation. *Bioresource Technology*. 169: 723–730.

Dürre, P. (2007). Biobutanol: An attractive biofuel. *Biotechnology Journal*. 2: 1525–1534.

Dürre, P. (2008). Fermentative butanol production—bulk chemical and biofuel. *Annals of the New York Academy of Sciences*. 1125: 353–362.

Efremenko, E.N., Nikolskaya, A.B., Lyagin, I.V., Senko, O.V., Makhlis, T.A., Stepanov, N. A., Maslova, O.V., Mamedova, F. and Varfolomeev, S.D. (2012). Production of biofuels from pretreated microalgae biomass by anaerobic fermentation with immobilized Clostridium acetobutylicum cells. *Bioresource Technology*. 114: 342–348.

Ellis, J.T., Hengge, N.N., Sims, R.C. and Miller, C.D. (2012). Acetone, butanol, and ethanol production from wastewater algae. *Bioresource Technology*. 111: 491–495.

Ezeji, T., Milne, C., Price, N.D. and Blaschek, H.P. (2010). Achievements and perspectives to overcome the poor solvent resistance in acetone and butanol-producing microorganisms. *Applied Microbiology and Biotechnology*. 85 (6): 1697–1712. Doi:10.1007/s00253-009-2390-0.

Ezeji, T., Qureshi, N. and Blaschek, H.P. (2007). Butanol production from agricultural residues: Impact of degradation products on Clostridium beijerinckii growth and butanol fermentation. *Biotechnology and Bioengineering*. 97: 1460–1469.

Fu, H., Hu, J., Guo, X., Feng, J., Yang, S.T. and Wang, J. (2021). Butanol production from Saccharina japonica hydrolysate by engineered Clostridium tyrobutyricum: The effects of pretreatment method and heat shock protein overexpression. *Bioresource Technology*. 335: 125290. Doi:10.1016/j.biortech.2021.125290.

Gallego-Morales, L.J., Escobar, A., Peñuela, M., Peña, J. and Rios, L. (2015). King Grass: A promising material for the production of second-generation butanol. *Fuel*. 143. Doi:10.1016/j.fuel.2014.11.077.

Gao, K., Boiano, S., Marzocchella, A. and Rehmann, L. (2014). Cellulosic butanol production from alkali-pretreated switchgrass (Panicum virgatum) and phragmites (Phragmites australis). *Bioresource Technology*. 174:176–181.

Gao, X., Zhao, H., Zhang, G., He, K. and Jin, Y. (2012). Genome shuffling of Clostridium acetobutylicum CICC 8012 for improved production of acetone—butanol—ethanol (ABE). *Current Microbiology*. 65 (2):128–132. Doi:10.1007/s00284-012-0134-3.

Gholizadeh, L. (2009). Enhanced butanol production by free and immobilized Clostridium sp. cells using butyric acid as co-substrate. M.Sc. Thesis. School of Engineering, University of Borås, Boras, Sweden. https://www.diva-portal.org/smash/get/diva2:1311624/FULLTEXT01.pdf.

Girbal, L. and Soucaille, P. (1994). Regulation of Clostridium acetobutylicum metabolism as revealed by mixed-substrate steady-state continuous cultures: Role of NADH/NAD ratio and ATP pool. *Journal of Bacteriology*. 176: 6433–6438.

Gottumukkala, L.D., Parameswaran, B., Valappil, S.K., Mathiyazhakan, K., Pandey, A. and Sukumaran, R.K. (2013). Biobutanol production from rice straw by a non-acetone producing *Clostridium sporogenes* BE01. *Bioresource Technology*. 145: 182–187.

Goyal, L., Jalan, N.K. and Khanna, S. (2019). Butanol tolerant bacteria: Isolation and characterization of butanol tolerant Staphylococcus sciuri sp. *Journal of Biotech Research*. 10: 68–77.

Guo, T., Tang, Y., Xi, Y.L., He, A.Y., Sun, B.J., Wu, H., Liang, D.F., Jiang, M. and Ouyang, P.K. (2011). Clostridium beijerinckii mutant obtained by atmospheric pressure glow discharge producing high proportions of butanol and solvent yields. *Biotechnology Letters*. 33: 2379–2383.

Ha, S.H., Mai, N.L. and Koo, Y.M. (2010). Butanol recovery from aqueous solution into ionic liquids by liquid—liquid extraction. *Process Biochemistry*. 45 (12): 1899–1903. Doi:10.1016/j.procbio.2010.03.030.

Han, A.H., Cho, D.H., Kim, Y.H. and Shin, S.J. (2013). Biobutanol production from 2- year-old willow biomass by acid hydrolysis and acetone-butanol-ethanol fermentation. *Energy*. 61: 13–17.

Harris, L.M., Welker, N.E. and Papoutsakis, E.T. (2002). Northern, morphological, and fermentation analysis of spo0A inactivation and overexpression in Clostridium acetobutylicum ATCC 824. *Journal of Bacteriology*. 184: 3586–3597.

Hartmanis, M.G. and Gatenbeck, S. (1984). Intermediary metabolism in Clostridium acetobutylicum: Levels of enzymes involved in the formation of acetate and butyrate. *Applied Environmental Microbiology*. 47: 1277–1283.

Hijosa-Valsero, M., Garita-Cambronero, J., Paniagua-García, A.I. and Díez-Antolínez, R. (2018). Biobutanol production from coffee silverskin. *Microbial Cell Factories*. 17: 154. Doi:10.1186/s12934–018–1002-z.

Hou, X., Nikolaj, F., Irini, A., Wouter, J.J.H. and Anne-Belinda, B. (2017). Butanol fermentation of the brown seaweed Laminaria digitata by Clostridium beijerinckii DSM-6422. *Bioresource Technology*. 238: 16–21. Doi:10.1016/j.biortech.2017.04.035.

Huang, H., Liu, H. and Gan, Y.R. (2010). Genetic modification of critical enzymes and involved genes in butanol biosynthesis from biomass. *Biotechnology Advances*. 28 (5): 651–657. Doi:10.1016/j.biotechadv.2010.05.015.

Isken, S. and de Bont, J.A. (1998). Bacteria tolerant to organic solvents. *Extremophiles*. 2: 229–238.

Jernigan, A., May, M., Potts, T., Rodgers, B., Hestekin, P., May, J. P., McLaughlin, J., Beitle, R.R. and Hestekin, C. (2013). Effects of drying and storage on year-round production of butanol and biodiesel from algal carbohydrates and lipids using algae from water remediation. *Environmental Progress and Sustainable Energy*. 32(4): 1013–1022. doi:10.1002/ep.11852.

Jones, D.T. and Woods, D.R. (1986). Acetone-butanol fermentation revisited. *Microbiological Reviews*. 50: 484–524.

Kanno, M., Katayama, T., Tamaki, H., Mitani, Y., Meng, X. Y., Hori, T., Narihiro, T., Morita, N., Hoshino, T., Yumoto, I., Kimura, N., Hanada, S. and Kamagata, Y. (2013). Isolation of butanol- and isobutanol-tolerant bacteria and physiological characterization of their butanol tolerance. *Applied and Environmental Microbiology*. 79(22): 6998–7005. Doi:10.1128/AEM.02900–13.

Kataoka, N., Tajima, T., Kato, J., Rachadech W. and Vangnai, A. (2011). Development of butanol-tolerant Bacillus subtilis strain GRSW2-B1 as a potential bioproduction host. *AMB Express*. 1: 10.1. Doi:10.1186/2191-0855-1-10.

Khanna, S., Goyal, A. and Moholkar, V.S. (2013). Production of *n*-butanol from biodiesel derived crude glycerol using *Clostridium pasteurianum* immobilized on Amberlite. *Fuel*. 112: 557–561.

Kheyrandish, M., Asadollahi, M., Jeihanipour, A., Doostmohammadi, M., Rismani-Yazdi, H. and Karimi, K. (2015). Direct production of acetone-butanol ethanol from waste starch by free and immobilized Clostridium acetobutylicum. *Fuel*. 142: 129–133.

Kim, J.H., Cook, M., Peeva, L., Yeo, J., Bolton, L.W., Lee, Y.M. and Livingston, A.G. (2020). Low energy intensity production of fuel-grade bio-butanol enabled by membrane-based extraction. *Energy & Environmental Science*. 13: 4862–4871. Doi:10.1039/D0EE02927K.

Knoshaug, E.P. and Zhang, M. (2009). Butanol tolerance in a selection of microorganisms. *Applied Biochemistry and Biotechnology*. 153: 13–20. Doi:10.1007/s12010-008-8460-4.

Komonkiat, I. and Cheirsilp, B. (2013). Felled oil palm trunk as a renewable source for biobutanol production by *Clostridium spp*. *Bioresource Technology*. 146: 200–207.

Kushwaha, D., Srivastava, N., Prasad, D., Mishra, P.K. and Upadhyay, S.N. (2020). Biobutanol production from hydrolysates of cyanobacteria Lyngbya limnetica and Oscillatoria obscura. *Fuel*. 271: 117583. Doi:10.1016/j.fuel.2020.117583.

Lee, J., Jang, Y.S., Han, M.J., Kim, J.Y. and Lee, S.Y. (2016). Deciphering Clostridium tyrobutyricum metabolism based on the whole-genome sequence and proteome analyses. *mBio*. 7 (3): e00743–16. Doi:0.1128/mBio.00743-16.

Li, J., Chen, X., Qi, B., Luo, J., Zhang, Y., Su, Y. and Yinhua, W. (2014). Efficient production of acetone—butanol—ethanol (ABE) from cassava by a fermentation—pervaporation coupled process. *Bioresource Technology*. 169: 251–257. Doi:10.1016/j.biortech.2014.06.102.

Li, J., Zhao, J.B., Zhao, M., Yang, Y.L., Jiang, W.H. and Yang, S. (2010). Screening and characterization of butanol-tolerant micro-organisms. *Letters in Applied Microbiology*. 50: 373–379.Li, L., Ai, H., Zhang, S., Li, S., Liang, Z., Wua, Z., Yang, S. and Wang, J. (2013). Enhanced butanol production by coculture of Clostridium beijerinckii and Clostridium tyrobutyricum. *Bioresource Technology*. 143: 397–404.

Li, L., Xiao, Z., Zhang, Z. and Tan, S. (2004). Pervaporation of acetic acid/water mixtures through carbon molecular sieve-filled PDMS membranes. *Chemical Engineering Journal*. 97(1):83–86. Doi:10.1016/S1385-8947(03)00102-5.

Li, T., Yan, Y. and He, J. (2015). Enhanced direct fermentation of cassava to butanol by Clostridium species strain BOH3 in cofactor-mediated medium. *Biotechnology for Biofuels*. 8: 166. Doi:10.1186/s13068-015-0351-7.

Li, Q., Wu, M., Wen, Z., Jiang, Y., Wang, X., Zhao, Y., Liu, J., Yang, J., Jiang, Y. and Yang, S. (2020). Optimization of n-butanol synthesis in Lactobacillus brevis via the functional expression of thl, hbd, crt and ter. *Journal of Industrial Microbiology and Biotechnology*. 47 (12): 1099–1108. Doi:10.1007/s10295-020-02331-2.

Li, H., Zhang, O., Yu, X., Wei, L. and Wang, Q. (2016). Enhancement of butanol production in Clostridium acetobutylicum SE25 through accelerating phase shift by different phases pH regulation from cassava flour. *Bioresource Technology*. 201: 148–155. Doi:10.1016/j.biortech.2015.11.027.

Li, S.Y., Srivastava, R., Suib, S.L., Li, Y. and Parnas, R.S. (2011). Performance of batch, fed-batch, and continuous A-B-E fermentation with pH-control. *Bioresource Technology*.102 (5): 4241–4250. Doi:10.1016/j.biortech.2010.12.078.

Liu, K., Atiyeh, H.K., Pardo-Planas, O., Ezeji, T.C., Ujor, V., Overton, J.C., Berning, K., Wilkins, M.R. and Tanner, R.S. (2015). Butanol production from hydrothermolysis-pretreated switchgrass: Quantification of inhibitors and detoxification of hydrolyzate. *Bioresource Technology*. 189: 292–301.

Liu, L., Wang, Y., Wang, N., Chen, X., Li, B., Shi, J. and Li, X. (2021). Process optimization of acetone-butanol-ethanol fermentation integrated with pervaporation for enhanced butanol production. *Biochemical Engineering Journal*. 173 (21): 108070. Doi:10.1016/j.bej.2021.108070.

Liu, S., Bischoff, K.M., Leathers, T.D., Qureshi, N., Rich, J.O. and Hughes, S.R. (2012). Adaptation of lactic acid bacteria to butanol. *Biocatalysis and Agricultural Biotechnology*. 1 (1): 57–61. Doi:10.1016/j.bcab.2011.08.008.

Liu, S., Qureshi, N., Bischoff, K. and Darie, C.C. (2021). Proteomic analysis identifies dysregulated proteins in butanol-tolerant gram-positive Lactobacillus mucosae BR0713–33. *ACS Omega*. 6 (5): 4034–4043. Doi:10.1021/acsomega.0c06028.

Lu, C. (2011). Butanol Production from Lignocellulosic Feedstocks by Acetone-Butanol-Ethanol Fermentation with Integrated Product Recovery. Ph.D. Dissertation, Graduate Program in Chemical and Biomolecular Engineering, The Ohio State University, United States of America.

Lu, C., Zhao, J., Yang, S.T. and Wei, D. (2012). Fed-batch fermentation for n-butanol production from cassava bagasse hydrolysate in a fibrous bed bioreactor with continuous gas stripping. *Bioresource Technology*. 104: 380–387.

Madihah, M.S., Ariff, A.B., Sahaid, K.M., Suraini, A.A. and Karim, M.I.A. (2001). Direct fermentation of gelatinized sago starch to acetone—butanol—ethanol by Clostridium acetobutylicum. *World Journal of Microbiology and Biotechnology*. 17: 567–576.

Marchal, R., Blanchet, D. and Vandecasteele, J.P. (1985). Industrial optimization of acetone-butanol fermentation: A study of the utilization of Jerusalem artichokes. *Applied Microbiology and Biotechnology*. 23: 92–98.

Mitchell, W.J. and Tangney, M. (2005). Carbohydrate uptake by the phosphotransferase system and other mechanisms. In: Dürre, P. (ed.), *Handbook on Clostridia*. CRC Press, Boca Raton, pp. 155–175.

Moon, Y., Han, K., Kim, D. and Day, D. (2015). Enhanced production of butanol and isopropanol from sugarcane molasses using Clostridium beijerinckii optinoii. *Biotechnology and Bioprocess Engineering*. 20: 871–877. Doi:10.1007/s12257-015-0323-6.

Motghare, K.A., Wasewar, K.L. and Shende, D.Z. (2019). Separation of butanol using tetradecyl(trihexyl)phosphonium bis(2,4,4-trimethylpentyl)phosphinate, oleyl alcohol, and castor oil. *Journal of Chemical and Engineering*. 64 (12): 5079–5088. Doi:10.1021/acs.jced.9b00211.

Nawab, S., Wang, N., Ma, X. and Huo, Y.X. (2020). Genetic engineering of non-native hosts for 1-butanol production and its challenges: A review. *Microbial Cell Factories*. 19: 79. Doi:10.1186/s12934-020-01337-w.

Ndaba, B., Adeleke, R., Makofane, R., Daramola, M.O. and Moshokoa, M. (2020) Butanol as a drop-in fuel: A perspective on production methods and current status. In: Daramola, M. and Ayeni, A. (eds.), *Valorization of Biomass to Value-Added Commodities. Green Energy and Technology*. Springer, Cham. Doi:10.1007/978-3-030-38032-8_18.

Ni, Y., Xia, Z., Wang, Y. and Sun, Z. (2012). Continuous butanol fermentation from inexpensive sugar-based feedstocks by Clostridium saccharobutylicum DSM 13864. *Bioresource Technology*. 129: 680–685. Doi:10.1016/j.biortech.2012.11.142.

Nielsen, D.R. and Prather, K.J. (2009). In situ product recovery of n-butanol using polymeric resins. *Biotechnology and Bioengineering*. 102: 811–821.

Nielsen, D.R., Yoon, L.E., S.H. Tseng, H.C., Yuan, C. and Prather, K.L.J. (2009). Engineering alternative butanol production platforms in heterologous bacteria. *Metabolic Engineering*. 11: 262–273.

Niglio, S., Marzocchella, A. and Rehmann, L. (2019). Clostridial conversion of corn syrup to Acetone-Butanol-Ethanol (ABE) via batch and fed-batch fermentation. *Heliyon*. 5 (3): e01401. Doi:10.1016/j.heliyon.2019.e01401.

Örs, I., Sarıkoç, S., Atabani, A.E. and Ünalan, S. (2020). Experimental investigation of effects on performance, emissions and combustion parameters of biodiesel—diesel—butanol blends in a direct-injection CI engine. *Biofuels*. 11 (2): 121–134. Doi: 10.1080/17597269.2019.1608682.

Palmqvis, E. and Hahn-Hagerdal, B. (2000). Fermentation of lignocellulosic hydrolysates. II: Inhibitors and mechanisms of inhibition. *Bioresource Technology*. 74: 25–33.

Paniagua-García, A.I., Hijosa-Valsero, M., Díez-Antolínez, R., Sánchez, M.E. and Coca, M. (2018). Enzymatic hydrolysis and detoxification of lignocellulosic biomass are not always necessary for ABE fermentation: The case of Panicum virgatum. *Biomass and Bioenergy*. 116: 131–139. Doi:10.1016/j.biombioe.2018.06.006.

Plaza, P.E., Gallego-Morales, L.J., Peñuela-Vásquez, M., Lucas, S., García-Cubero, M.T. and Coca, M. (2017). Biobutanol production from brewer's spent grain hydrolysates by Clostridium beijerinckii. *Bioresource Technology*. 244 (1): 166–174. Doi:10.1016/j.biortech.2017.07.139.

Potts, T., Du, J., Paul, M., May, P., Beitle, R. and Hestekin, J. (2012). The production of butanol from Jamaica bay macro algae. *Environmental Progress & Sustainable Energy*. 31: 29–36. Doi:10.1002/ep.10606.

Qureshi, N. and Blaschek, H.P. (2000). Butanol production using Clostridium beijerinckii BA101 hyper-butanol producing mutant strain and recovery by pervaporation. *Applied Biochemistry and Biotechnology*. 84–86: 225–235.

Qureshi, N., Lin, X., Liu, S., Saha, B.C., Mariano, A.P., Polaina, J., Ezeji, T.C., Friedl, A., Maddox, I.S., Klasson, K.T., Dien, B.S. and Singh, V. (2020). Global view of biofuel butanol and economics of its production by fermentation from sweet sorghum bagasse, food waste, and yellow top presscake: Application of novel technologies. *Fermentation*. 6: 58. Doi:10.3390/fermentation6020058.

Qureshi, N., Lolas, A. and Blaschek, H.P. (2001). Soy molasses as fermentation substrate for production of butanol using Clostridium beijerinckii BA101. *Journal of Industrial Microbiology and Biotechnology*. 26: 290–295.

Qureshi, N., Saha, B.C., Dien, B., Hector, R.E. and Cotta, M.A. (2010). Production of butanol (a biofuel) from agricultural residues: Part I—Use of barley straw hydrolysate. *Biomass Bioenergy*. 34: 559–565.

Qureshi, N., Schripsema, J., Lienhardt, J. and Blaschek, H.P. (2000). Continuous solvent production by Clostridium beijerinckii BA101 immobilized by adsorption onto brick. *World Journal of Microbiology and Biotechnology*. 16: 377–382. Doi:10.1023/a:1008984509404.

Raganati, F., Procentese, A., Olivieri, G., Russo, M.E., Salatino, P. and Marzocchella, A. (2020). Bio-butanol recovery by adsorption/desorption processes. *Separation and Purification Technology*. 235: 116145. Doi:10.1016/j.seppur.2019.116145.

Rahman, M.M., Rasul, M.G., Hassan, N.M.S., Azad, A.K. and Uddin, M.N. (2017). Effect of small proportion of butanol additive on the performance, emission, and combustion of Australian native first- and second-generation biodiesel in a diesel engine. *Environmental Science and Pollution Research*. 24 (28): 22402–22413. Doi:10.1007/s11356-017-9920-6.

Ranganathan, S. and Maranas, C.D. (2010). Microbial 1-butanol production: Identification of non-native production routes and in silico engineering interventions. *Biotechnology Journal*. 5 (7): 716–725. Doi:10.1002/biot.201000171.

Ravagnani, A., Jennert, K.C., Steiner, E., Grunberg, R., Jefferies, J.R., Wilkinson, S.R., Young, D.I., Tidswell, E.C., Brown, D.P., Youngman, P., Morris, J.G. and Young, M. (2000). Spo0A directly controls the switch from acid to solvent production in solvent-forming clostridia. *Molecular Microbiology*. 37: 1172–1185.

Rochón, E., Cortizo, G., Cabot, M.I., Cubero, M.T.G., Coca, M., Ferrari, M.D. and Lareo, C. (2020). Bioprocess intensification for isopropanol, butanol and ethanol (IBE) production by fermentation from sugarcane and sweet sorghum juices through a gas stripping-pervaporation recovery process. *Fuel*. 281: 118593. Doi:10.1016/j.fuel.2020.118593.

Roffler, S.R., Blanch, H.W. and Wilke, C.R. (1987). In-situ recovery of butanol during fermentation: Part 2. Fed-batch extractive fermentation. *Bioprocess Engineering*. 2: 181–190.

Roussos, A., Misailidis, N., Koulouris, A., Zimbardi, F. and Petrides, D. (2019). A Feasibility study of cellulosic isobutanol production—Process simulation and economic analysis. *Processes*. 7 (10): 667. Doi:10.3390/pr7100667.

Rühl, J., Schmid, A. and Blank, L.M. (2009). Selected pseudomonas putida strains able to grow in the presence of high butanol concentrations. *Applied and Environmental Microbiology*. 75 (13): 4653–4656. Doi:10.1128/AEM.00225-09.

Russmayer, H., Marx, H. and Sauer, M. (2019). Microbial 2-butanol production with Lactobacillus diolivorans. *Biotechnology Biofuels*. 12: 262. Doi:10.1186/s13068-019-1594-5.

Sakuragi, H., Morisaka, H., Kuroda, K. and Ueda, M. (2015). Enhanced butanol production by eukaryotic Saccharomyces cerevisiae engineered to contain an improved pathway. *Bioscience Biotechnology Biochemistry*. 79: 314–320. Doi:10.1080/09168451.2014.972330.

Schadeweg, V. and Boles, E. (2016). Increasing n-butanol production with Saccharomyces cerevisiae by optimizing acetyl-CoA synthesis, NADH levels and trans-2-enoyl-CoA reductase expression. *Biotechnology Biofuels*. 9: 257. Doi:10.1186/s13068-016-0673-0.

Shen, C.R. and Liao, J.C. (2008). Metabolic engineering of Escherichia coli for 1-butanol and 1-propanol production via the keto-acid pathways. *Metabolic Engineering*. 10 (6): 312–320. Doi:10.1016/j.ymben.2008.08.001.

Shen, C.R., Lan, E.I., Dekishima, Y., Baez, A., Cho, K.M. and Liao. J.C. (2011). Driving forces enable high-titer anaerobic 1-butanol synthesis in Escherichia coli. *Applied and Environmental Microbiology*. 77: 2905–2915.

Si, T., Luo, Y., Xiao, H. and Zhao, H. (2014). Utilizing an endogenous pathway for 1-butanol production in Saccharomyces cerevisiae. *Metabolic Engineering*. 22: 60–68. Doi:10.1016/j.ymben.2014.01.002.

Sikkema, J., de Bont J.A. and Poolman, B. (1995). Mechanisms of membrane toxicity of hydrocarbons. *Microbiology Reviews*. 59: 201–222.

Sillers, R., Al-Hinai, M.A. and Papoutsakis, E.T. (2009). Aldehyde—alcohol dehydrogenase and/or thiolase overexpression coupled with CoA transferase downregulation led to higher alcohol titers and selectivity in Clostridium acetobutylicum fermentations. *Biotechnology and Bioengineering*. 102: 38–49.

Stanier, R., Doudororoff, M. and Adelberg, E. (1970). *The Microbial World*, 3rd edition. Prentice Hall, Englewood Cliffs, NJ, p. 186.

Steen, E.J., Chan, R., Prasad, N., Myers, S., Petzold, C.J., Redding, A., Ouellet, M. and Keasling, J.D. (2008). Metabolic engineering of *Saccharomyces cerevisiae* for the production of n-butanol. *Microbial Cell Factories*. 7: 36. Doi:10.1186/1475-2859-7-36.

Survase, S.A., Sklavounos, E., Van, H.A. and Granström, T. (2013). Market refused vegetables as a supplement for improved acetone—butanol—ethanol production by Clostridium acetobutylicum DSM 792. *Industrial Crops and Products*. 45: 349–354.

Taconi, K.A., Venkataramanan K.P. and Johnson, D.T. (2009). Growth and solvent production by Clostridium pasteurianum ATCC® 6013™ utilizing biodiesel derived crude glycerol as the sole carbon source. *Environ Prog Sustainable Energy*. 28: 100–110.

Tashiro, Y., Shinto, H., Hayashi, M., Baba, S.I., Kobayashi, G. and Sonomoto, K. (2007). Novel high-efficient butanol production from butyrate by non-growing Clostridium saccharoperbutylacetonicum N1–4 (ATCC 13564) with methyl viologen. *Journal of Bioscience and Bioengineering*. 104 (3): 238–240.

Teke, G.M. and Pott, R.W.M. (2021). Design and evaluation of a continuous semipartition bioreactor for in situ liquid-liquid extractive fermentation. *Biotechnology and Bioengineering*. 118: 58–71. Doi:10.1002/bit.27550.

Thang, V.H., Kanda, K. and Kobayashi, G. (2010). Production of acetone—butanol—ethanol (ABE) in direct fermentation of cassava by Clostridium saccharoperbutylacetonicum N1–4. *Applied Biochemistry and Biotechnology*. 161: 157–170.

Tomas, C.A., Alsaker, K.V., Bonarius, H.P.J., Hendriksen, W.T., Yang, H., Beamish, J.A., Parades, C.J. and Papoutsakis, E.T. (2003). DNA-array based transcriptional analysis of asporogenous, non-solventogenic Clostridium acetobutylicum strains SKO1 and M5. *Journal of Bacteriology*. 185: 4539–4547. Doi:10.1128/JB.185.15.4539-4547.2003.

Valles, A., Alvarez-Hornos, F.J., Martínez-Soria, V., Marzal, P. and Gabaldon, C. (2020). Comparison of simultaneous saccharification and fermentation and separate hydrolysis and fermentation processes for butanol production from rice straw. *Fuel*. 282: 118831. Doi:10.1016/j.fuel.2020.118831.

Valles, A., Capilla, M., Álvarez-Hornos, F.J., García-Puchol, M., San-Valero, P. and Gabaldón, C. (2021). Optimization of alkali pretreatment to enhance rice straw conversion to butanol. *Biomass and Bioenergy*. 150: 106131. Doi:10.1016/j.biombioe.2021.106131.

van der Wal, H., Sperber, B.L.H.M., Houweling-Tan, B., Bakker, R.R.C., Brandenburg, W. and López-Contreras, A.M. (2013). Production of acetone, butanol, and ethanol from biomass of the green seaweed Ulva lactuca. *Bioresource Technology*. 128: 431–437.

Vane, L.M. (2008). Separation technologies for the recovery and dehydration of alcohols from fermentation broths. *Biofuels, Bioproducts and Biorefining*. 2: 553–588.

Vasconcelos, I., Girbal, L. and Soucaille, P. (1994). Regulation of carbon and electron flow in Clostridium acetobutylicum grown in chemostat culture at neutral pH on mixtures of glucose and glycerol. *Journal of Bacteriology*. 176: 1443–1450.

Vees, C.A., Neuendorf, C.S. and Pflügl, S. (2020). Towards continuous industrial bioprocessing with solventogenic and acetogenic clostridia: Challenges, progress and perspectives. *Springer International Publishing*. Doi:10.1007/s10295-020-02296-2.

Verma, R., Dehury, P., Bharti, A. and Banerjee, T. (2018). Liquid-liquid extraction, COSMO-SAC predictions and process flow sheeting of 1-butanol enhancement using mesitylene and oleyl alcohol. *Journal of Molecular Liquids*. 265: 824–839. Doi:10.1016/j.molliq.2018.06.088.

Virunanon, C., Chantaroopamai, S., Denduangbaripant, J. and Chulalaksananukul, W. (2008). Solventogenic-cellulolytic clostridia from 4-step-screening process in agricultural waste and cow intestinal tract. *Anaerobe*. 14: 109–117.

Wagner, J.L., Lee-Lane, D., Monaghan, M., Sharifzadeh, M. and Hellgardt, K. (2019). Recovery of excreted n-butanol from genetically engineered cyanobacteria cultures: Process modelling to quantify energy and economic costs of different separation technologies. *Algal Research*. 37: 92–102. Doi:10.1016/j.algal.2018.11.008.

Wang, Y., Guo, W., Cheng, C.L., Ho, S.H., Chang, J.S. and Ren, N. (2016). Enhancing bio-butanol production from biomass of Chlorella vulgaris JSC-6 with sequential alkali pretreatment and acid hydrolysis. *Bioresource Technology*. 200: 557–564. Doi:10.1016/j.biortech.2015.10.056.

Wang, Y., Ho, S.H., Cheng, C.L., Nagarajan, D., Guo, W.Q., Lin, C., Li, S., Ren, N. and Chang, J.S. (2017). Nutrients and COD removal of swine wastewater with an isolated microalgal strain Neochloris aquatica CL-M1 accumulating high carbohydrate content used for biobutanol production. *Bioresource Technology*. 242: 7–14. Doi:10.1016/j.biortech.2017.03.122.

Wechgama, K., Laopaiboon, L. and Laopaiboon, P. (2017). Enhancement of batch butanol production from sugarcane molasses using nitrogen supplementation integrated with gas stripping for product recovery. *Industrial Crops and Products*. 95: 216–226. Doi:10.1016/j.indcrop.2016.10.012.

Wyman, C.E., Dale, B.E., Elander, R.T., Holtzapple, M., Ladisch, M.R. and Lee, Y.Y. (2005) Comparative sugar recovery data from laboratory scale application of leading pretreatment technologies to corn stover. *Bioresource Technology*. 96: 2026–2032. Doi:10.1016/j.biortech.2005.01.018.

Xiao, H., Li Z., Jiang Y., Yang Y., Jiang W., Gu Y. and Yang S. (2012). Metabolic engineering of D-xylose pathway in Clostridium beijerinckii to optimize solvent production from D-xylose mother liquid. *Metabolic Engineering*. 14: 569–578.

Xue, C., Wang, Z., Wang, S., Zhang, X., Chen, L., Mu, Y. and Bai, F. (2016). The vital role of citrate buffer in acetone-butanol-ethanol (ABE) fermentation using corn stover and high-efficient product recovery by vapor stripping-vapor permeation (VSVP) process. *Biotechnology for Biofuels and Bioproducts*. 9: 146. Doi:10.1186/s13068-016-0566-2.

Yadav, S., Rawat, G., Tripathi, P. and Saxena, R.K. (2014). A novel approach for biobutanol production by *Clostridium acetobutylicum* using glycerol: A low cost substrate. *Renewable Energy*. 71: 37–42.

Yan, H. and Wang, H. (2013). Recent development and applications of dispersive liquid—liquid micro extraction. *Journal of Chromatography A*. 1295: 1–15. Doi:10.1016/j.chroma.2013.04.053.

Yang, M., Kuittinen, S., Vepsäläinen, J., Zhang, J. and Pappinen, A. (2017). Enhanced acetone butanol-ethanol production from lignocellulosic hydrolysates by using starchy slurry as supplement. *Bioresource Technology*. Doi:10.1016/j.biortech.2017.06.021.

Yoo, M., Croux, C., Meynial-Salles, I. and Soucaille, P. (2016). Elucidation of the roles of adhE1 and adhE2 in the primary metabolism of Clostridium acetobutylicum by combining in-frame gene deletion and a quantitative system-scale approach. *Biotechnology Biofuels*. 9: 92. Doi:10.1186/s13068-016-0507-0.

Yoruklu, H.C., Koroglu, E.O., Demir, A. and Ozkaya, B. (2019). The electromotive-induced regulation of anaerobic fermentation: Electrofermentation. In: Venkata Mohan, S., Varjani, S. and Pandey, A. (eds.), *Biomass, Biofuels and Biochemicals, Microbial Electrochemical Technology*, pp. 739–756. Doi:10.1016/B978-0-444-64052-9.00030-3.

Zautsen, R.R.M., Maugeri-Filho, F., Vaz-Rossell, C.E., Straathof, A.J.J., van der Wielen, L.A.M. and de Bont, J.A.M. (2009). Liquid-liquid extraction of fermentation inhibiting compounds in lignocellulose hydrolysate. *Biotechnology Bioengineering*. 102: 1354–1360.

Zhang, J., Wang, M.Y., Gao, M.T., Fang, X., Yano, S. and Qin, S.L. (2013). Efficient acetone-butanol-ethanol production from corncob with a new pretreatment technology-wet disk milling. *BioEnergy Research*. 6: 35–43.

Zheng, Y.N., Li, L.Z., Xian, M., Ma, Y.J., Yang, J.M., Xu, X. and He, D.Z. (2009). Problems with the microbial production of butanol. *Journal of Industrial Microbiology and Biotechnology*. 36: 1127–1138.

Zhou, Q., Liu, Y. and Yuan, W. (2018). Kinetic modeling of lactic acid and acetic acid effects on butanol fermentation by Clostridium saccharoperbutylacetonicum. *Fuel*. 226: 181–189.

Zhu, H., Li, X., Pan, Y., Liu, G., Wu, H., Jiang, M. and Jin, W. (2020). Fluorinated PDMS membrane with anti-biofouling property for in-situ biobutanol recovery from fermentation-pervaporation coupled process. *Journal of Membrane Science*. 609: 118225. Doi:10.1016/j.memsci.2020.118225.

10 Bioethanol Production from Jute and Mesta Biomass

Prospects and Challenges

A.K. Lavanya, Laxmi Sharma, Gouranga Kar, Pratik Satya, Suman Roy, Srinjoy Ghosh, and Bijan Majumdar
ICAR-Central Research Institute for Jute and Allied Fibers, Barrackpore, Kolkata, India

CONTENTS

10.1 INTRODUCTION

The Anthropocene era has been blessed with ample natural resources to sustain life. However, accelerated population growth on Earth and the human desire to raise standard of living have exhausted the available resources. In the process, the rapidly developing society has also disturbed the ecosystem, and the planet is one the verge of irreplaceable environmental damage. Experts have sought to address the issue using environment-friendly approaches in a sustainable manner. Perhaps this has become one of the most daunting tasks for this civilization. One of such challenges is harvesting plant biomass–derived energy. Transport and industrial sectors

DOI: 10.1201/9781003191247-10

are majorly driven by conventional energy sources. Due to lack of petroleum-based reserves, India is the third largest consumer of crude oil (International Emergency Agency [IEA], 2019) with 80% of its demand met by import (CSO, 2017). It is projected that due to the shrinkage of supply, India's transport sector would be 92% dependent on external sources by 2030 (IEA, 2009). Therefore, attaining self-sufficiency in oil production is the priority keeping in view GHG emissions. Each year, the world emits roughly 50 billion tons of greenhouse gases (GHGs), with the contribution from the transport sector being 16.2% (Ritchie and Max, 2020). Since, India is the world's third largest CO_2 emitter (2.2 giga tons), contributing 12% to the transport sector, it is critical to look for alternative energy sources derived from bio-inexhaustible sources. The United States of America and Brazil are the topmost countries that contribute (84%) to global bioethanol production, mainly using corn and sugarcane as feedstocks, respectively (IEA, 2019). However, in countries like India, where food security is a major issue, dedicated energy crops grown in marginal land is the necessity to rule out the food-versus-fuel-crop issue. Biofuel production in India currently accounts for only 1% of global production (Shinoj *et al.*, 2011). India, China, and Thailand are all set to improve their agricultural biotechnology investments, and they may soon become potential biofuel producers (Gonsalves, 2006; Rajagopal *et al.*, 2007; Scheper, 2008; Huang *et al.*, 2020). The National Biofuel Policy 2018 emphasizes bioethanol generation from lignocelluloses and wastes, considering food, energy, and environmental security. Lignocelluloses are known to comprise half of the global biomass and is a suitable feedstock for bioethanol production. Enriched cell wall is a valuable possession of this category and is a high source of cellulose and hemicelluloses. Hence it would be wise to channelize such abundant resources for energy production. Huge amounts of feedstock are required to meet the expanding demand for biofuel. A number of crops or their residues qualify the criteria for bioethanol feedstocks. However, it is important to mind that their biomass should be higher and input should be lower per unit area. Jute and mesta, known for their fiber, are such sources of biomass. Traditionally, these crops have been cultivated for fiber purposes. The by-product, i.e., the sticks that form around 10% of the total biomass, is generally used as fuel for cooking in rural areas. However, the availability of LPG has now replaced them for cooking purposes. The sticks are now underutilized after fiber extraction. Therefore, it would be wise to mobilize this biomass for bioethanol production in addition to fiber. Although production technologies from these crops are at their infancy, these crops fulfill most of the criteria as energy crops for bioethanol production. The improvement of microbial production technology by focusing on novel strategies for improving hydrolytic enzymes, ethanol fermentation performance, and genetically engineered fermentative, cellulolytic bacteria/fungi that can ferment both hexose and pentose sugars is being studied. To boost microbial conversion efficiency and bioethanol yield, researchers are focusing on bioprospecting new microorganisms with better ethanol yield and hydrolytic enzyme production potential.

Jute and mesta, generally grown in marginal land with low input cost, are tolerant to adverse environmental conditions. The growth rate is very high, and this accounts for higher biomass within 110–120 days. Moreover, the quality of biomass is comparable to woody perennials (Park *et al.*, 2021). Additional benefits of growing jute and mesta include ecosystem services through atmospheric carbon fixation. Jute is known to add 15 tons CO_2 ha^{-1} $year^{1}$ in 120 days, producing 49.7 g m^{-2} day^{-1} biomass (Palit and Meshram, 2008), while kenaf absorbs 21.89 tons CO_2 ha^{-1} $year^{1}$ (Santoso *et al.*, 2015). They play a part in phytoremediation as well; thus, polluted sites are equally favorable for their growth. These crops are thus fully utilized as dedicated energy crops to produce bioethanol at commercial scale in developing nations, including India, to satisfy the tremendous population pressure. It would be worthy to note that the governments of USA and European countries are already adapting mesta as their bioethanol feedstock (Tuck *et al.*, 2006; Webber *et al.*, 2011; Berti *et al.*, 2013). Bioethanol production from mesta biomass has been commercialized in China and Malaysia (Guo *et al.*, 2014; Saba *et al.*, 2015). This will thus ensure to fulfill per capita consumption with added environmental benefits. In this chapter, we will mainly introduce jute and mesta as bioenergy crops, its crop improvement program, as well as strategies for the enhancement of microbial conversion efficiency for clean energy production. Jute plants have carbon dioxide assimilation rate, and it cleans the air by consuming large quantities of carbon dioxide.

10.2 BOTANY, PHYSIOLOGY, AND THE IMPORTANCE OF JUTE AND MESTA CROP

Jute and mesta are vegetable bast fiber plants next to cotton (*Gossypium* spp.) in importance. Jute, commonly known as "golden fiber," is the second most important natural fiber under the family Malvaceae. There are approximately forty wild species of jute, but only two types of jute are cultivated commercially, namely, *Corchorus olitorius* (2n = 2x = 14), the golden jute or tossa jute, originated in Africa, and *C. capsularis* (2n = 2x = 14), the white jute, which is proposed to have evolved in South China (Benor *et al.*, 2012; Kundu *et al.*, 2013). *C. olitorius* is an annual herbaceous plant that can grow up to 3.5–4.5 m in height within a four-month period (Kundu, 1956). *C. capsularis* is slightly shorter and can grow up to 3–4 m. It is Bangladesh's and India's most important bast fiber. India ranks first in area (7.9 lakh ha), and it is mostly grown in West Bengal (74.7% acreage and accounts for around 85% of world jute production). After cotton and jute, mesta ranks third in terms of commercial fiber crops. The genus *Hibiscus* contains over 300 species. Under the family Malvaceae, *Hibiscus cannabinus* (2n = 2x = 36) is also known as kenaf, mesta, Deccan jute, or Guinea hemp, originated in Africa. The origin of *H. sabdariffa* (roselle, mesta, or Bimli jute), an allotetraploid species (2n = 4x = 72), is less defined, with both being two of the most widely grown and economically significant crops. This crop has wider adaptability than jute, is cultivated on marginal soil that isn't appropriate for food crops, and requires less water and fertilizer. Both *H. cannabinus* and *H. sabdariffa* are annual herbaceous crop plants that can grow up to 10 cm per day and may reach a height of 2–5 m. Faster growth rate makes higher CO_2 absorption capacity, making it more robust than other biomass feedstocks (Islam, 2019; Baghban *et al.*, 2020). *H. sabdariffa* is more tolerant to drought and can grow well under low precipitation (<500 mm annual rainfall). However, *H. cannabinus* is more tolerant to waterlogging. For biofuel production, whole crop biomass is more important, so maintaining wider spacing between the plants and lower plant density give better

TABLE 10.1
Area and Production of Jute and Mesta Crops

Crops	Area ('000 hectares)	Production ('000 bales of 180 Kgs.each)	Fibre yield (kg/ha)	Total biomass (t/ha/year)
Jute	685.75	10332.97	2564	10–20
Mesta	56.02	534.06	1556	10–25

Source: Directorate of Economics and Statistics, (2017–18), GOI.

biomass yield and carbohydrate composition (Berti *et al.*, 2013). The area, production, fiber output, and total biomass availability of jute and mesta crops in India are shown in Table 10.1. In comparison to mesta, the jute crop is grown across a larger area, which can generate annual biomass of 10–20 t/ha and 10–25 t/ha from jute and mesta respectively.

Both these crops are extensively distributed around the globe and have high adaptability to diverse climatic and soil conditions and also have low cost of cultivation (Ramesh *et al.*, 2018). The harvestable principal product (fiber) in these fiber crops is less than 20% of the biomass. In kenaf, for example, the dry-fiber-weight-to-green-biomass ratio is roughly 1:20, whereas the dry-fiber-weight-to-dry-biomass ratio is 1:6 (Satya and Maiti, 2013). The biomass can be used either as entire biomass or as waste by-products rather than devouring farm lands committed to providing food and nourishment for sustainable biofuel production.

10.3 STRUCTURAL ORGANIZATION OF JUTE AND MESTA BIOMASS

The chemical makeup of fibrous agricultural biomass can be split into three categories: cellulose, hemicellulose, and lignin. Cellulose is a polymer made up of D-glucose subunits that are linked by β-1,4 glycosidic linkages. Lignin is an amorphous polymer of phenylpropane, mainly made from p-coumaryl alcohol, coniferyl alcohol, and sinapyl alcohol. Hemicellulose is a heteropolymer xylan, glucuronoxylan, arabinoxylan, or glucomannan made up of pentose and hexose sugars linked by β-1,4 glycosidic linkages. Pectin, protein, ash, extractives such as sugars, nitrogenous material, chlorophyll, and waxes are also present in lower amounts (Hendriks and Zeeman, 2009; Azelee *et al.*, 2014). The cellulose strands are packed together to produce cellulose microfibrils. With varied secondary structure and bonding factors, hemicellulose maintains linkage with cellulose and lignin. About 70% of total plant biomass consists of cellulose and hemicelluloses and are closely attached to the lignin via covalent and hydrogenic interactions, which makes them highly robust and resistant to treatment (Edye *et al.*, 2015; Meents *et al.*, 2018). A network of hemicellulose and lignin surrounds the cellulose fibrils. Cross-linking and constituents can differ depending on the plant species, plant age, and growth stage and other conditions (Hu *et al.*, 2012; Tanmoy *et al.*, 2014). Both crops have higher cellulose and hemicellulose content, with cellulose content ranging from 40 to 63% in jute and 37 to 63% in mesta, and hemicellulose content ranging from 18 to 22% and 14 to

24% in jute and mesta, respectively (Kundu *et al.*, 2012; Song *et al.*, 2017). On maturity, jute and mesta contain 390–568 mg/g and 410–520 mg/g glucose, respectively, higher than tree wood. Jute (121–173 mg/g) and mesta (108–142 mg/g) exhibit high xylan concentrations when compared to other complex hemicellulosic components (mannan, arabinan, etc.) and can be converted to biofuels like ethanol and butanol. Lignin has the lowest solubility of the three biomass components and is the most difficult to treat downstream. As a result, a high lignin content increases the cost of producing ethanol. The lignin concentration of jute (12–24%) and mesta (10–21%) is substantially lower than that of wood (Satya and Maiti, 2013); as a result, low lignin content with other additional properties makes them a suitable option for bioethanol production.

10.4 BREEDING AND BIOTECHNOLOGICAL ASPECTS TO IMPROVE THE BIOMASS YIELD WITH REDUCED BIOMASS RECALCITRANCE FOR BIOETHANOL PRODUCTION

The most important plant factors that need to be considered for bioethanol production is the high biomass yield coupled with low lignin content and the arrangement of vascular and epicuticular waxes of the crop for maximum utilization of feedstock (Mosier *et al.*, 2005). Lignin prevents cellulose and hemicellulose from being hydrolyzed into fermentable sugars by restricting enzyme access to them. The possible way to overcome this problem is by selecting cultivars with high biomass yield potential and specified quality traits, boosting plant polysaccharide content, modifying the characteristics and properties of lignocellulose and overall biomass yield. Genetically engineering technique is also employed to evolve crops with more of cellulose and hemicellulose and reduced lignin recalcitrance. The rapidly evolving data from genomics and genetics, as well as more advanced recombinant DNA technology, would give many possibilities in order to boost bioethanol output from these feedstocks. Breeding procedures can be used to bring about changes in the biomass or by genetic modifications in the plant lignin and polysaccharides. The application of breeding strategies in the genomics era appears to be fairly unexpected; hence, traditional breeding techniques pave the way for breeding with the help of a molecular marker to improve biomass quality and yield. Traditional pedigree breeding methods have essentially little potential for developing low-lignin cultivars due to a limited genetic foundation, high sexual incompatibility, and a dearth of appropriate breeding resources. There are different approaches to make these fiber crops suitable for biofuel production, like (a) increasing cell wall polysaccharide content (using functional genomics and mutation approaches) by genetic manipulation to enhance polysaccharides for increased cellulosic ethanol production (Bhatia *et al.*, 2017; Brandon and Scheller, 2020), (b) increasing the overall biomass (using transgenic approach) by identifying possible targets for genetic manipulation to boost biomass (Perveen *et al.*, 2020; Moreno *et al.*, 2020), and (c) modifying features of cellulose and lignin (using RNA interference [RNAi] technology) by diverting plant carbon resources away from lignin production or by downregulating lignin biosynthesis enzymes using antisense oligonucleotides (Bewg *et al.*, 2016; Halpin, 2019).

The plant cell wall biosynthetic pathways are examined to know their linkage with other cell wall components. Modifying targeted genes in jute and kenaf may provide potential biomass for bioethanol production by increasing or decreasing the number of cell wall components or conferring structural alterations in their connections to ease the biomass saccharification for ethanol production (Farrar *et al.*, 2012; Donev *et al.*, 2018; Li *et al.*, 2019; Pazhany *et al.*, 2019). Jute pectin synthesis routes were discovered after identifying genes and their homologues from *C. capsularis* hypocotyl transcriptomes can provide evidence for inhibiting cell wall disintegration and what tactics could be designed to speed biomass deconstruction of lignocellulosic plants (Satya *et al.*, 2018). In one study, Chakraborty *et al.* (2015) examined pathways linked with lignin production in jute fibers and created the first comprehensive bast transcriptome(s) of jute. Transgenic overexpression of enzymes involved in cell wall polysaccharide production can be targeted, which would drastically improve carbohydrate content (Brandon and Scheller, 2020). Advanced techniques like the use of recombinant DNA technology approach may change the composition and ratio of lignin syringyl-to-guaiacyl (S/G) ratio and decrease the recalcitrance of syringyl-rich lignin, (Fu *et al.*, 2011). Another approach is gene editing, CRISPR/CRISPR-associated protein 9 (Cas9), also known as clustered, regularly interspaced short palindromic repeat, which is a relatively novel method for modifying higher plants' genetics. The two most promising strategies for generating low-lignin jute cultivars are genetic engineering and mutagenesis-based reverse genetics (e.g., TILLING) (Sticklen, 2006). Particle bombardment, or *Agrobacterium*-mediated transformation, was tried in jute crop, but due to the lack of an appropriate regeneration technique, it has not been attempted to convert jute protoplasts directly, until now. Recently, in jute, to overcome the regeneration problem, a shoot tip–based stable genetic transformation technique was developed which can be successfully used to introduce foreign genes (Saha *et al.*, 2014; Bhattacharyya *et al.*, 2015). Jute is an appealing model for studying xylan-type bast fiber development and lignification in secondary cell wall because of these characteristics. The recent estimation of *Corchorus* spp. genome size (Benor *et al.*, 2011; Sarkar *et al.*, 2011; Akashi *et al.*, 2012) has prompted a rise in research at affordable prices through genomics or high-throughput, next-generation sequencing (NGS) methods (Chakraborty *et al.*, 2015; Kundu *et al.*, 2015; Zhang *et al.*, 2015a).

Some important jute and mesta varieties developed by ICAR-CRIJAF (Indian Council of Agricultural Research-Central Research Institute for Jute and Allied Fibers) are MT 150 (Nirmal), a mesta variety which can produce higher biomass (256 q/ha green biomass), suitable for bioethanol production, and JBO 1 (Sudhangsu), S 19 (Subala), JRO 2407 (Samapti), JROB 2, JRC 532 (Sashi), and RRPS-27-C-3 (Monalisa), which are jute varieties having higher biomass, higher cellulose, and low lignin content. Recently developed jute variety JROB 2 (Purnendu) can produce 59.1 q/ha green biomass, much higher than the previous varieties. In an ideal production situation, jute produces higher biomass per unit area than mesta, which increases the cultivar's appropriateness for the production of bioethanol (Satya *et al.*, 2013). These biorefining capabilities of jute and kenaf imply a promising bioethanol research area for exploring its inherent potential. Table 10.2 summarizes the various biochemical and biotechnological approaches attempted to improvise the crop biomass and to reduce the recalcitrance in jute and mesta crop, to make them suitable for bioethanol production.

10.5 BIOMASS-TO-BIOETHANOL CONVERSION PROCESS

Bioethanol can be produced from lignocellulosic biomass using one of two methods (biochemical or thermochemical conversion). The process breaks down resistant cell wall structure into lignin, hemicellulose, and cellulose fragments. Each polysaccharide is broken down into sugars before being converted into bioethanol (Mood *et al.*, 2013). The biochemical approaches in the conversion of biomass to bioethanol comprise four unit operations, namely, pretreatment, hydrolysis, fermentation, and distillation (Sharma *et al.*, 2020). Figure 10.1 portrays the process flow diagram for bioethanol production from jute and mesta biomass, in which the biomass sticks are composed of cellulose polymer matrix intertwined with lignin and hemicelluloses. This structure can be disturbed by the pretreatment process, which can disturb the complicated structure and separate the lignin hemicellulose from the cellulose. By using efficient individual or a cocktail of enzymes produced by bacteria/fungus, complex carbohydrates/cellulose/hemicellulose can be converted to fermentable sugars (hexoses and pentoses) through saccharification process, followed by fermentation using efficient strains of yeast/bacteria, which can convert these sugars to ethanol. A few researchers have demonstrated bioethanol production from the cellulose part of kenaf hydrolysates (Guo *et al.*, 2014; Ruan *et al.*, 2011), and this process could be improved by physical or chemical detoxification or delignification (Shah *et al.*, 2019; Azelee *et al.*, 2014).

10.5.1 PRETREATMENT

Pretreatment is the most critical step, and the most expensive, in achieving high bioethanol yield with low-cost capital (Bhutto *et al.*, 2017; Mankar *et al.*, 2021). It

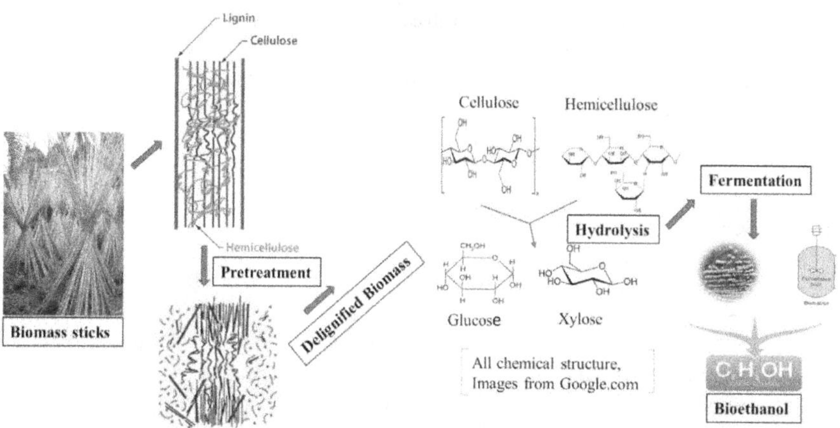

FIGURE 10.1 Process flow diagram of bioethanol production from jute and mesta biomass.

includes physical or mechanical (i.e., biomass size reduction), chemical, and microbiological processes and their combination, based on the chemical composition of raw material being explored (Banerjee *et al.*, 2010). These processes increase the porosity of the substrate, exposing maximum cell surface area to enzymes for successful hydrolysis and sugar recovery (Zhu *et al.*, 2009; Zhu and Pan, 2010). Low quantities of lignin (10–25%) characterize jute and mesta residues, resulting in a less-robust texture. Among the pretreatment procedures, the most promising, cost-effective, less-energy-intensive, chemical-free, and environmentally friendly approach is biological pretreatment, which also generates less inhibitors (Saritha *et al.*, 2012; Wan *et al.*, 2012). The growing interest in using microorganisms (bacteria and fungi, as well as microbial consortia), as they are readily available and can multiply on their own, would be the cost-effective alternative to the highly sought-after physicochemical approaches (Shi *et al.*, 2009). *Oxyporus latemarginatus*, *Rigidoporus vinctus*, *Phanerochaete chrysosporium*, *Coriolus versicolor*, and *Ceriporiopsis subvermispora*, lignin-degrading fungi and actinobacteria, show high levels of lignin peroxidase (LiP) activity in kenaf biomass, which may be explored for biological delignification of these biomasses (Suganya *et al.*, 2007; Mohamed *et al.*, 2013; Brzonova *et al.*, 2014).

TABLE 10.2
Different Biochemical and Biotechnological Approaches Used to Improvise Crop Biomass and Reduce Recalcitrance in Jute and Mesta Crop

Crop	Approaches	Objectives Achieved	Reference
Jute	Mutation (X-ray irradiation)	Observed 50% reduction in lignin and a 50% increase in cellulose at the same time, Lignin-deficient mutant is designated as deficient lignified phloem fibre (dlfp)	Sengupta and Palit (2004)
Kenaf	Microwave-assisted pretreatment processes	High glucose yield	Ooi *et al.* (2011)
Kenaf	Hydrothermal pretreatment under high temperature (160–240°C) and/or high pressure (1.0–3.5 MPa)	Changed pore volume and specific surface area of biomass	Pu *et al.* (2013)
Jute	CCoAOMT (Caffeoyl-CoA 3-O-methyltransferase enzyme) gene	Lowered the lignin content and resulted in increased plant height	Zhang *et al.* (2014)
Jute	EST-SSRs marker located in genes that encode cellulose synthases A (CesAs) and transcription factors (TFs)	Involved in the regulation of secondary wall formation and CesA, a key protein for cellulose biosynthesis	Chakraborty *et al.* (2015); Samanta *et al.* (2015); Zhang *et al.* (2015a)

TABLE 10.2 ***(Continued)***
Different Biochemical and Biotechnological Approaches Used to Improvise Crop Biomass and Reduce Recalcitrance in Jute and Mesta Crop

Crop	Approaches	Objectives Achieved	Reference
Kenaf	De novo assembly of kenaf transcriptome using Illumina sequencing	Gene discovery and marker identification (SNP, SSR); marker-assisted breeding in kenaf; identified cellulose biosynthesis genes	Chen *et al.* (2014); Zhang *et al.* (2015b)
Jute	Artificial miRNA-mediated downregulation of lignin biosynthetic genes (COMT, F5H, C3H and C4H)	F5H-amiRNA and C3H-amiRNA transgenesis for lignin reduction, reduced level of gene expression	Shafrin *et al.* (2015, 2017).
Kenaf	Developed by gamma irradiation	Cultivar "Jangdae," exhibited higher biomass, higher seed yield	Kang *et al.* (2016)
Jute	Lignin biosynthetic genes coumarate 3-hydroxylase (C3H) and ferulate 5-hydroxylase by RNAi technology	Lignin content reduced by 25% with 3–6% increase in cellulose	Shafrin *et al.* (2017)
Jute	CAD (cinnamyl alcohol dehydrogenase) by genomics-assisted molecular techniques	Low-lignin jute fibers	Majumder *et al.* (2020)
Kenaf	Characterization of gene isoforms related to cellulose and lignin biosynthesis	Identified putative enzymes involved in cellulose and lignin biosynthesis using Kyoto encyclopedia of genes and genomes (KEGG) analysis	Lyu *et al.* (2020)
Jute	Identified pectin biosynthesis pathways by transcriptome and genome sequencing data	Low-pectin jute fibers	Satya *et al.* (2021)
Jute	Transgenics (downregulation of lignin biosynthetic genes [COMT, C4H, F5H and C3H])	16–25% reduction in lignin, 3–6% increase in cellulose content, and small enhancement in saccharification	Nath *et al.* (2021)

10.5.2 Integrated Bioprocesses for Saccharification and Fermentation

Saccharification. In the biomass-to-biofuel conversion, the saccharification process, in which the cellulose and hemicellulose portions of the biomass are converted to pentose or hexose fermentable sugars using hydrolytic enzymes, is a very important step, and there is need of making it more efficient and cost-effective. Hydrolases, mainly cellulase (cellulose degrading) and hemicellulases (hemicellulose degrading), are predominantly produced by several species of microorganisms, such as, *Trichoderma, Cellulomonas, Thermonospora, Bacteroides, Aspergillus, Bacillus* and *Clostridium, Ruminococcus, Erwinia, Acetovibrio, Microbispora, Streptomyces, Trichoderma, Penicillium, Fusarium, Phanerochaete, Humicola*, and *Schizophyllum* sp. (Koeck *et al.*, 2014). Use of these enzyme cocktails can increase biomass hydrolysis efficiency (Adav *et al.*, 2012). In the past, Lavanya *et al.* (2020) used cellulase from psychrotolerant *Aspergillus niger* SH3 and β-glucosidase from *Pseudomonas lutea* BG8 to treat pretreated mesta biomass, resulting in 533.25 mg/gds sugar release at saccharification temperature of 40°C, which is 10° less than commercial enzymes, resulting in substantial energy savings.

Fermentation. An ideal microorganism for commercial ethanol production would be one that is able to utilize a wide variety of substrates, has high ethanol productivity, and is resistant to ethanol, temperature, and environmental inhibitors. The following is the conversion reaction for hexose (reaction 1) and pentose sugars (reaction 2):

$$C_6H_{12}O_6 \rightarrow 2C_2H_5OH + 2CO_2 \quad (1)$$

$$3C_5H_{10}O_5 \rightarrow 5C_2H_5OH + 5CO_2 \quad (2)$$

Hexose-metabolizing yeast (*Saccharomyces cerevisiae*) is usually utilized, but that cannot metabolize xylose. The ability of different yeasts and bacteria to convert xylose and other pentoses to ethanol is being studied. Several researchers have investigated microbes capable of fermenting both hexose and pentose sugars to ethanol, e.g., *Escherichia coli, Klebsiella oxytoca*, and *Zymomonas mobilis*, as interesting options (Hahn-Hagerdal *et al.*, 2007). Singh *et al.* (2020) used thermotolerant *Saccharomyces cerevisiae* JRC6 to ferment jute biomass hydrolysate, resulting in 7.55 g/L of ethanol. Integration of different hydrolytic enzymes for hydrolysis and fermentation process is proposed to improve ethanol production, namely:

a. Separate hydrolysis and fermentation (SHF). Two reactors are used to run the reactions. In this process, the liberated cellulose is hydrolyzed, and subsequently, in a separate reactor, ethanol fermentation takes place using efficient yeast strains capable of fermenting various sugars, which simplifies the improvement of specific individual reactors (Kazi *et al.*, 2010).
b. Simultaneous saccharification and fermentation (SSF). This approach integrates cellulose hydrolysis and fermentation simultaneously in a single reactor, using hydrolytic enzymes and fermenting microbes having the same temperature optima. It outperforms SHF because it increases ethanol yields by removing end product inhibition, improves cellulose conversion rates, reduces enzyme loading, is cost-effective, and does not require many reactors (Chandel *et al.*, 2010). The efficient yeast strain *Saccharomyces cerevisiae* LN was used to perform SSF, yielding 4.1 g/L ethanol from mesta biomass (Lavanya *et al.*, 2020) and 7.55 g/L of ethanol from jute biomass using thermotolerant yeast *S. cerevisiae* JRC6 (Singh *et al.*, 2020).

c. Separate hydrolysis and cofermentation (SHCF) and simultaneous saccharification and cofermentation (SSCF). By using a single strain of microbes to confer both saccharification and fermentation in the same reactor, this method improves the process economics and ethanol yield (Kazi *et al.*, 2010; Klein-Marcuschamer *et al.*, 2010). Cofermenting microorganisms achieve simultaneous saccharification of both C5 and C6 sugars in SSCF if the pH and temperature optima of both organisms match well (Kang *et al.*, 2010).
d. Consolidated bioprocessing (CBP). This novel approach uses a single type of microbe to produce all necessary enzymes and ethanol in a single reactor. Because it has no capital or running expenditures for specialized enzyme manufacturing and requires less substrate for enzyme synthesis, CBP is regarded as the pinnacle of biomass-to-bioethanol conversion technology (Klein-Marcuschamer *et al.*, 2010).
e. Simultaneous pretreatment and saccharification (SPS). Using enzymes that are both hydrolytic and oxidizing, a new technique, called the SPS model, is developed. In this process, pretreatment and saccharification are achieved unitedly, making the total procedure simpler and less-expensive. Instead of performing separate pretreatment and hydrolysis operations, the soaked biomasses are blended with a mixture of enzymes, including the surfactant, and the resulting reducing sugars present in the reaction mixture are incubated under ideal fermenting conditions (Dhiman *et al.*, 2015).

10.6 UPSCALING OF BIOETHANOL PRODUCTION THROUGH MICROBIAL CONVERSION

The ability of microorganisms to produce biofuels at a scale that is economical, productive, sustainable, and eco-friendly, a greener approach, needs to be developed at commercial level. The ideal organism for biofuel production should have high substrate consumption and processing capacities, quick and unregulated sugar transport channels, large metabolic fluxes, and good inhibitor and product tolerances, as well as capability to produce only a single fermentation product, i.e., ethanol. Several microorganisms, including genetically modified strains, have been key participants in the production of bio-based compounds from lignocellulosic biomass to date (Su *et al.*, 2015). Recently, several efforts have been made towards improved ethanol production and tolerance ability through engineering *Saccharomyces cerevisiae* (yeast) transcription machinery (Hasunuma *et al.*, 2012; Selim *et al.*, 2018). Extensive study on altering most important yeast (*S. cerevisiae*, *Pichia stipitis*) and bacteria (*Escherichia coli*) have been sparked to make them a potential and effective in digesting both hexose and pentose sugars (Bilal *et al.*, 2018). Engineering biofuel export systems, heat-shock proteins, cell surface, pentose-hexose cofermentation, membrane changes, in situ recovery approaches, and media supplements are all promising ways to engineer strains for biofuel production and tolerance (Ko *et al.*, 2018; Selim *et al.*, 2020). Table 10.3 summarizes the bioconversion of jute and mesta biomass to bioethanol mediated by microorganisms.

TABLE 10.3
Overview of Studies on Microbe-Mediated Biomass-to-Bioethanol Conversion

Crop	Organism/s Used	Objectives Achieved	Reference
Jute	*Bacillus megaterium RB-05*	Cellulase activity	Chowdhury *et al.* (2012)
Kenaf	*Oxyporus latemarginatus*, *Rigidoporus vinctus*, *Phanerochaete chrysosporium*, and *Coriolus versicolor*	High ligninase activity	Mohamed *et al.* (2013)
Jute	*Pleurotus sajor-caju*	High reducing sugar yield by 6%	Mahal *et al.* (2013)
Kenaf	*Trichoderma reesei*	Pretreatment resulted in high xylooligosaccharides production due to its relatively high hemicellulose (xylan) content	Azelee *et al.* (2014)
Kenaf	*Gloeophyllum trabeum*	Enhanced biomass hydrolysis by oxidative lytic polysaccharide monooxygenases (LPMOs) increased hydrolysis rate by 56%	Jung *et al.* (2015)
Kenaf	Enzymatic hydrolysis using Endo-xylanase *(Xyn2)* and arabinofuranosidase *(AnabfA)* from *Pichia pastoris*	High reducing sugar yield (102.66 mg/g of xylan)	Azelee *et al.* (2016)
Kenaf	Recombinant *Escherichia coli*	Substrate utilization is 92.8%, highest xylitol yield of 0.35 g/g based on xylose consumption	Shah *et al.* (2019)
Kenaf	Protein engineering of GH11 xylanase from *Aspergillus fumigatus*	Enhanced catalytic efficiency, increase in sugar yield up to 28%	Damis *et al.* (2019)
Jute	*Saccharomyces cerevisiae* JRC6	Fermentation of jute biomass resulting in 7.55 g/L of ethanol	Singh *et al.* (2020)
Jute	*Mutant UV-8 of Talaromyces verruculosus*	Enzymatic saccharification of cellulose (37%) and xylan (29%) increased	Jain *et al.* (2020)
Jute	*Aspergillus niger*	Enzymatic saccharification (sugar release per unit of biomass) using cellulase enzyme	Nath *et al.* (2021)

10.7 CURRENT ISSUES, CHALLENGES, AND OPPORTUNITIES FOR BIOETHANOL PRODUCTION FROM JUTE AND MESTA BIOMASS

Agricultural biomass from fiber crops is a largely untapped sector for biofuel production. Although having high potential, these crops have largely been neglected so far, indicating a lack of concern and research focus on these biomass as compared to other biofuel crops. Despite the fact that many government and nongovernment researchers are investigating the biofuel potential of bast fiber crops in the United States (Koshel *et al.*, 2010), there is lack of enthusiasm in India to explore these crops for biofuel production. In comparison to other crop biomasses, jute and mesta have many advantages, such as:

1. Nongrain source of biofuel, thus reducing the share of food grain in biofuel production
2. Higher cellulose and hemicellulose, low lignin
3. Higher biomass potential with less input, more adaptable and sustainable in the environment
4. Continuous year-round supply of biomass and long-term benefits (carbon credit, bioremediation, water conservation, etc.)
5. Lower cost of biofuel extraction due to soft, herbaceous biomass
6. Less time of establishment compared to perennial biofuel crops, fits well to regional and local biofuel production system and rotation with other annual bioenergy crops, also benefiting small and marginal farmers
7. Resistant to biotic and abiotic challenges

The abundant vegetative growth of these plants (2.5–3.5 m in height at maturity) shows great opportunities for use in solar energy and CO_2 during photosynthesis (7302.38 thousand tons per year), which strengthens its position in the group of energy plants contributing to renewable energy sources (Islam *et al.*, 2012). These features are also important in connection with global warming.

Crop residue availability as an energy source is mostly determined by the total amount of crop produced, the crop's residue-to-crop ratio (residue coefficient), collection efficiency (which includes storage issues), and the amount used in competing applications. It is considered as an important indicator of biomass waste, which is much higher in fiber crops than other cereal and noncereal crops used as a food source. For example, the residue coefficients of jute and mesta sticks are much higher (1:2.30) compared to other lignocellulosic biofuel sources whose residue coefficients range between 1:0.25 and 1:1.13 (National Productivity Council, 1987). The nonfiber components comprise more than 75% of biomass and are generally wasted or used primarily as a burning fuel or consumed for local household use. The effective availability of jute and kenaf sticks for energy application is about 3.32 mt in India alone, which can efficiently be channeled for biofuel production (Purohit, 2009). The growth rate of jute under a favorable climate, particularly in subtropical, humid climates, is higher than that of kenaf, reaching up to 4.5 m within 90–110 days, which means more biomass can be harvested per unit area (Satya and Maiti, 2013). A number of biofuel firms in the United States have successfully produced biodiesel from whole kenaf biomass. For example, ReadiJet, a jet fuel sold in the United States, is made from kenaf and Camelina as a feedstock (www.ara.com/fuels). The Heilongjiang province of China has set a target to generate cellulosic ethanol from kenaf to produce 28 to 56 million gallons (100,000 to 200,000 t) per year (BiofuelsDigest, 2011). Kenaf has been recognized as a possible biofuel source by the Centre for Jatropha Promotion and Biodiesel in India (Smith-Heisters, 2008).

In terms of sustainability and biodiversity, direct and indirect land use change, feedstock and water availability, and rapid growth of bioethanol production industries in the transportation system have posed several issues (Hess *et al.*, 2016). Given that cultivating these crops as a fuel source on farming land is an unachievable option in Indian context, farmers having agricultural land might earn extra money by selling unutilized agricultural wastes for bioethanol production. However, there is a need for availability of these residues throughout the year. In reality, the amount accessible is determined by harvesting time, storage characteristics, storage facility, and other factors, posing difficulty in its availability throughout the year. Jute and mesta sticks are only accessible for a duration of six months in a year; hence, there is a need for scientific research to increase the periods of availability of these residues by generating varieties which may be grown throughout the year (Tripathi *et al.*, 1998). Efforts should also be made to develop new conversion technologies to improve fiber biomass conversion efficiency to biofuel and to reduce the cost of the chemical conversion process. Kreuger *et al.* (2011) were able to recover 82% hexoses and 46% pentoses from hemp biomass and generated 2,600–3,000 liters of ethanol per hectare of land. Moxley *et al.* (2008) extracted 96% glucose from a cellulosic hemp waste fiber. To make the best use of biomass while reducing waste and emissions, biorefineries play a critical role in addressing the sustainable development of high-value bioproducts (biochemicals, bioenergy) as well as biofuel. Their economic sustainability is dependent on the optimal valorization of hemicelluloses and lignin in addition to cellulose (Ferreira, 2017; Sawatdeenarunat *et al.*, 2018). The important approaches would be genetic engineering, metabolic engineering, and synthetic biology to design microorganisms or plants to make them more efficient, as well as to improve the efficiency of hydrolytic enzymes (cellulolytic, hemicellulolytic), ethanol fermentation performance, creation of genetically engineered fermentative bacteria/fungi to ferment both hexose and pentose sugars, discovery of additional diverse pathways, and isolation of efficient new strain (Ko *et al.*, 2018). The usage of genetically modified organisms (GMOs) is contentious, as their use in large-scale fermentation processes can result to environmental contamination and public health problems (Paoletti *et al.*, 2008). However, for an effective and efficient utilization of these residues, it is necessary to take a comprehensive approach in tackling numerous facets of environmental, social, and economic challenges.

10.8 CONCLUSIONS AND FUTURE PROSPECTS

Diverting cereal grains for biofuel production is inevitably pressurizing the food economy, resulting in the rise of food prices and debates all over the world (Kurowska *et al.*, 2020). With rising energy demands and diminishing energy resources, the use of nonfood fiber crops biomass for biofuel production provides a renewable option. The Indian biofuel strategy focuses primarily on nonfood biomass sources that are grown on deteriorated or barren land that are unsuitable for agriculture, preventing a potential fuel–food security dispute (Anwar *et al.*, 2014). Nonfood energy sources, such as the use of forest tree species for cellulosic ethanol production, pose enormous environmental risk, as they contain high lignin and less cellulose and are less energy-efficient than fiber crop species. Fiber crops like jute and kenaf, which would not disturb the food basket and are more energy-efficient due to their substantially higher photosynthetic rate, will ultimately help to achieve worldwide sustainable development goals as compared to other conventional tree species. These fiber crops are widely adaptable to different parts of the world, can be grown on limiting soil conditions, are less exhaustive in terms of water usage and fertilizers, and also can be grown in fallow lands, thereby not affecting agricultural output. These crops have an inherent ability to produce higher biomass, where input requirement is minimal, and have wider adaptability, are more resistant to biotic and abiotic stresses than other fiber/field crops, and contain high cellulose content and low lignin concentration. They can produce more fuel per unit of farmland used, require less input (chemical and energy) per unit produce, and are environmentally sustainable and can ensure better yield in terms of net energy. However, several researchable issues have to be addressed for increasing efficiency of biofuel production from these crops, such as improving efficiency of extraction process, enzyme processing, pretreatment, and yeast metabolic engineering, reducing cost of biofuel production. But transport of large amounts of biomass feedstock to processing industries, environmental impacts of large-scale cultivation of a single crop on natural habitat, socioeconomic impacts of introduction of fiber crops to new areas, issues related to subsidization of biofuel cultivation, as well as commercial and legal issues related to farm-based industries, need attention. Jute and kenaf will be able to establish themselves as attractive biofuel feedstocks by addressing these fundamental issues and opportunities. Furthermore, the most promising biorefinery should focus on cutting-edge technologies and techniques that may help achieve a more sustainable, more effective exploitation of all biomass fractions while producing bioethanol and other high-value bioproducts at the same time.

REFERENCES

Adav, Sunil S., Anita Ravindran, Esther Sok Hwee Cheow, and Siu Kwan Sze. "Quantitative proteomic analysis of secretome of microbial consortium during saw dust utilization." *Journal of Proteomics* 75, no. 18 (2012): 5590–5603.

Akashi, Ryo, Nurun N. Fancy, Arif M. Tanmoy, and Haseena Khan. "Estimation of genome size of jute (*Corchorus capsularis* (L.) var. CVL-1 using flow cytometry." *Plant Tissue Culture and Biotechnology* 22, no. 1 (2012): 83–86.

Anwar, Zahid, Muhammad Gulfraz, and Muhammad Irshad. "Agro-industrial lignocellulosic biomass a key to unlock the future bio-energy: A brief review." *Journal of Radiation Research and Applied Sciences* 7, no. 2 (2014): 163–173.

Azelee, Nur Izyan Wan, Jamaliah Md Jahim, Ahmad Fauzi Ismail, Siti Fatimah Zaharah Mohamad Fuzi, Roshanida A. Rahman, and Rosli Md Illias. "High xylooligosaccharides (XOS) production from pretreated kenaf stem by enzyme mixture hydrolysis." *Industrial Crops and Products* 81 (2016): 11–19.

Azelee, Nur Izyan Wan, Jamaliah Md Jahim, Amir Rabu, Abdul Munir Abdul Murad, Farah Diba Abu Bakar, and Rosli Md Illias. "Efficient removal of lignin with the maintenance of hemicellulose from kenaf by two-stage pretreatment process." *Carbohydrate Polymers* 99 (2014): 447–453.

Baghban, Mohammad Hajmohammadian, and Reza Mahjoub. "Natural kenaf fiber and LC3 binder for sustainable fiber-reinforced cementitious composite: A review." *Applied Sciences* 10, no. 1 (2020): 357–372.

Banerjee, Saumita, Sandeep Mudliar, Ramkrishna Sen, Balendu Giri, Devanand Satpute, Tapan Chakrabarti, and R. A. Pandey. "Commercializing lignocellulosic bioethanol: Technology bottlenecks and possible remedies." *Biofuels, Bioproducts and Biorefining: Innovation for a Sustainable Economy* 4, no. 1 (2010): 77–93.

Benor, Solomon, Sebsebe Demissew, Karl Hammer, and Frank R. Blattner. "Genetic diversity and relationships in *Corchorus olitorius* (Malvaceae sl) inferred from molecular and morphological data." *Genetic Resources and Crop Evolution* 59, no. 6 (2012): 1125–1146.

Benor, Solomon, Jorg Fuchs, and Frank R. Blattner. "Genome size variation in *Corchorus olitorius* (Malvaceae sl) and its correlation with elevation and phenotypic traits." *Genome* 54, no. 7 (2011): 575–585.

Berti, Marisol T., Srinivas Reddy Kamireddy, and Yun Ji. "Row spacing affects biomass yield and composition of kenaf (*Hibiscus cannabinus* L.) as a lignocellulosic feedstock for bioenergy." *Journal of Sustainable Bioenergy Systems* 3, no. 1 (2013): 68–73.

Bewg, William P., Charleson Poovaiah, Wu Lan, John Ralph, and Heather D. Coleman. "RNAi downregulation of three key lignin genes in sugarcane improves glucose release without reduction in sugar production." *Biotechnology for Biofuels* 9, no. 1 (2016): 1–13.

Bhatia, Rakesh, Joe A. Gallagher, Leonardo D. Gomez, and Maurice Bosch. "Genetic engineering of grass cell wall polysaccharides for biorefining." *Plant Biotechnology Journal* 15, no. 9 (2017): 1071–1092.

Bhattacharyya, Jagannath, Anirban Chakraborty, Souri Roy, Subrata Pradhan, Joy Mitra, Monami Chakraborty, Anulina Manna, Narattam Sikdar, Saikat Chakraborty, and Soumitra Kumar Sen. "Genetic transformation of cultivated jute (*Corchorus capsularis* L.) by particle bombardment using apical meristem tissue and development of stable transgenic plant." *Plant Cell, Tissue and Organ Culture (PCTOC)* 121, no. 2 (2015): 311–324.

Bhutto, Abdul Waheed, Khadija Qureshi, Khanji Harijan, Rashid Abro, Tauqeer Abbas, Aqeel Ahmed Bazmi, Sadia Karim, and Guangren Yu. "Insight into progress in pre-treatment of lignocellulosic biomass." *Energy* 122 (2017): 724–745.

Bilal, Muhammad, Hafiz MN Iqbal, Hongbo Hu, Wei Wang, and Xuehong Zhang. "Metabolic engineering and enzyme-mediated processing: A biotechnological venture towards biofuel production—a review." *Renewable and Sustainable Energy Reviews* 82 (2018): 436–447.

BioFuelsDigest. "Bio-Dynamic, Shanghai Zhongwei Biochemistry JV for China cellulosic ethanol in 2011." (2011). Available at: www.biofuelsdigestbdigest/tag/shan.com/ghai-zhongwei-biochemistry (accessed 11 November 2011).

Brandon, Andrew G., and Henrik V. Scheller. "Engineering of bioenergy crops: Dominant genetic approaches to improve polysaccharide properties and composition in biomass." *Frontiers in Plant Science* 11 (2020): 1–14.

Brzonova, Ivana, Evguenii Kozliak, Alena Kubátová, Michelle Chebeir, Wensheng Qin, Lew Christopher, and Yun Ji. "Kenaf biomass biodecomposition by basidiomycetes and actinobacteria in submerged fermentation for production of carbohydrates and phenolic compounds." *Bioresource Technology* 173 (2014): 352–360.

Chakraborty, Avrajit, Debabrata Sarkar, Pratik Satya, Pran Gobinda Karmakar, and Nagendra Kumar Singh. "Pathways associated with lignin biosynthesis in lignomaniac jute fibers." *Molecular Genetics and Genomics* 290, no. 4 (2015): 1523–1542.

Chandel, Anuj K., G. Chandrasekhar, M. Lakshmi Narasu, and L. Venkateswar Rao. "Simultaneous saccharification and fermentation (SSF) of aqueous ammonia pretreated Saccharum spontaneum (wild sugarcane) for second generation ethanol production." *Sugar Tech* 12, no. 2 (2010): 125–132.

Chen, Peng, Shanmin Ran, Ru Li, Zhipeng Huang, Jinghua Qian, Mingli Yu, and Ruiyang Zhou. "Transcriptome de novo assembly and differentially expressed genes related to cytoplasmic male sterility in kenaf (*Hibiscus cannabinus* L.)." *Molecular Breeding* 34, no. 4 (2014): 1879–1891.

Chowdhury, Sougata Roy, Ratan Kumar Basak, Ramkrishna Sen, and Basudam Adhikari. "Utilization of lignocellulosic natural fiber (jute) components during a microbial polymer production." *Materials Letters* 66, no. 1 (2012): 216–218.

CSO, Indian Enegy Statistics 2017, 2017. https://doi.org/10.1017/CBO9781107415324.004.

Damis, Siti Intan Rosdianah, Abdul Munir Abdul Murad, Farah Diba Abu Bakar, Siti Aishah Rashid, Nardiah Rizwana Jaafar, and Rosli Md Illias. "Protein engineering of GH11 xylanase from Aspergillus fumigatus RT-1 for catalytic efficiency improvement on kenaf biomass hydrolysis." *Enzyme and Microbial Technology* 131 (2019): 109383.

Dhiman, Saurabh Sudha, Jung-Rim Haw, Dayanand Kalyani, Vipin C. Kalia, Yun Chan Kang, and Jung-Kul Lee. "Simultaneous pretreatment and saccharification: Green technology for enhanced sugar yields from biomass using a fungal consortium." *Bioresource Technology* 179 (2015): 50–57.

Directorate of Economics and Statistics, Ministry of Agriculture and Farmers Welfare (2017–18).

Donev, Evgeniy, Madhavi Latha Gandla, Leif J. Jönsson, and Ewa J. Mellerowicz. "Engineering non-cellulosic polysaccharides of wood for the biorefinery." *Frontiers in Plant Science* 9 (2018): 1537.

Edye, Leslie Alan, and William Orlando Sinclair Doherty. "Fractionation of a lignocellulosic material." U.S. Patent 8,999,067, issued April 7, 2015.

Farrar, Kerrie, David N. Bryant, Lesley Turner, Joe A. Gallagher, Ann Thomas, Markku Farrell, Mervyn O. Humphreys, and Iain S. Donnison. "Breeding for bio-ethanol production in Lolium perenne L.: Association of allelic variation with high water-soluble carbohydrate content." *Bioenergy Research* 5, no. 1 (2012): 149–157.

Ferreira, Ana F. "Biorefinery concept." In *Biorefineries*, pp. 1–20. Springer, Cham, 2017.

Fu, Chunxiang, Jonathan R. Mielenz, Xirong Xiao, Yaxin Ge, Choo Y. Hamilton, Miguel Rodriguez, Fang Chen et al. "Genetic manipulation of lignin reduces recalcitrance and improves ethanol production from switchgrass." *Proceedings of the National Academy of Sciences* 108, no. 9 (2011): 3803–3808.

Gonsalves, Joseph B. *An Assessment of the Biofuels Industry in India*. Unctad, Geneva, 2006.

Guo, Fenfen, Wan Sun, Xuezhi Li, Jian Zhao, and Yinbo Qu. "Pretreatment of ramie and kenaf stalk for bioethanol production." Sheng wu gong cheng xue bao= *Chinese journal of Biotechnology* 30, no. 5 (2014): 774–783.

Hahn-Hagerdal, Barbel, Kaisa Karhumaa, César Fonseca, Isabel Spencer-Martins, and Marie F. Gorwa-Grauslund. "Towards industrial pentose-fermenting yeast strains." *Applied Microbiology and Biotechnology* 74, no. 5 (2007): 937–953.

Halpin, Claire. "Lignin engineering to improve saccharification and digestibility in grasses." *Current Opinion in Biotechnology* 56 (2019): 223–229.

Hasunuma, Tomohisa, and Akihiko Kondo. "Development of yeast cell factories for consolidated bioprocessing of lignocellulose to bioethanol through cell surface engineering." *Biotechnology Advances* 30, no. 6 (2012): 1207–1218.

Hendriks, A. T. W. M., and G. Zeeman. "Pretreatments to enhance the digestibility of lignocellulosic biomass." *Bioresource Technology* 100, no. 1 (2009): 10–18.

Hess, Tim M., J. Sumberg, T. Biggs, Matei Georgescu, David Haro-Monteagudo, G. Jewitt, M. Ozdogan et al. "A sweet deal? Sugarcane, water and agricultural transformation in Sub-Saharan Africa." *Global Environmental Change* 39 (2016): 181–194.

Hu, Fan, and Art Ragauskas. "Pretreatment and lignocellulosic chemistry." *Bioenergy Research* 5, no. 4 (2012): 1043–1066.

Huang, Jiangfeng, Muhammad Tahir Khan, Danilo Perecin, Suani T. Coelho, and Muqing Zhang. "Sugarcane for bioethanol production: Potential of bagasse in Chinese perspective." *Renewable and Sustainable Energy Reviews* 133 (2020): 110296.

IEA (International Energy Agency), International Energy Outlook 2009. Paris, France: International Energy Agency, 2009.

IEA. "How competitive is biofuel production in Brazil and the United States?" (2019). Available at: www.iea.org/articles/how-competitive-is-biofuel-production-in-brazil-and-the-united-states.

Islam, Mohammad Mahbubul. "Varietal advances of jute, kenaf and mesta crops in Bangladesh: A review." *International journal of Bioorganic Chemistry* 4, no. 1 (2019): 24–41.

Islam, Mohammad Shahidul, and Sheikh Kamal Ahmed. "The impacts of jute on environment: An analytical review of Bangladesh." *Journal of Environment and Earth Science* 2, no. 5 (2012): 24–31.

Jain, Lavika, Akhilesh Kumar Kurmi, Avnish Kumar, Anand Narani, Thallada Bhaskar, and Deepti Agrawal. "Exploring the flexibility of cellulase cocktail obtained from mutant UV-8 of Talaromyces verruculosus IIPC 324 in depolymerising multiple agro-industrial lignocellulosic feedstocks." *International Journal of Biological Macromolecules* 154 (2020): 538–544.

Jung, Sera, Younho Song, Ho Myeong Kim, and Hyeun-Jong Bae. "Enhanced lignocellulosic biomass hydrolysis by oxidative lytic polysaccharide monooxygenases (LPMOs) GH61 from *Gloeophyllum trabeum*." *Enzyme and Microbial Technology* 77 (2015): 38–45.

Kang, Li, Wei Wang, and Yoon Y. Lee. "Bioconversion of kraft paper mill sludges to ethanol by SSF and SSCF." *Applied Biochemistry and Biotechnology* 161, no. 1 (2010): 53–66.

Kang, SiYong, SoonJae Kwon, SangWook Jeong, JinBaek Kim, SangHoon Kim, and JaiHyunk Ryu. "An improved kenaf cultivar 'Jangdae' with seed harvesting in Korea." *Korean Journal of Breeding Science* 48, no. 3 (2016): 349–354.

Kazi, Feroz Kabir, Joshua A. Fortman, Robert P. Anex, David D. Hsu, Andy Aden, Abhijit Dutta, and Geetha Kothandaraman. "Techno-economic comparison of process technologies for biochemical ethanol production from corn stover." *Fuel* 89 (2010): S20–S28.

Klein-Marcuschamer, Daniel, Piotr Oleskowicz-Popiel, Blake A. Simmons, and Harvey W. Blanch. "Technoeconomic analysis of biofuels: A wiki-based platform for lignocellulosic biorefineries." *Biomass and Bioenergy* 34, no. 12 (2010): 1914–1921.

Ko, Ja Kyong, Je Hyeong Jung, Fredy Altpeter, Baskaran Kannan, Ha Eun Kim, Kyoung Heon Kim, Hal S. Alper, Youngsoon Um, and Sun-Mi Lee. "Largely enhanced bioethanol production through the combined use of lignin-modified sugarcane and xylose fermenting yeast strain." *Bioresource Technology* 256 (2018): 312–320.

Koeck, Daniela E., Alexander Pechtl, Vladimir V. Zverlov, and Wolfgang H. Schwarz. "Genomics of cellulolytic bacteria." *Current Opinion in Biotechnology* 29 (2014): 171–183.

Koshel, Patricia, and Kathleen McAllister. "Expanding biofuel production and the transition to advanced biofuels: Lessons for sustainability from the upper Midwest: Summary of a workshop." *The National Academies Press* (2010): 1–179.

Kreuger, Emma, Bálint Sipos, Guido Zacchi, Sven-Erik Svensson, and Lovisa Björnsson. "Bioconversion of industrial hemp to ethanol and methane: The benefits of steam pretreatment and co-production." *Bioresource Technology* 102, no. 3 (2011): 3457–3465.

Kundu, Avijit, Avrajit Chakraborty, Nur Alam Mandal, Debajeet Das, Pran Gobinda Karmakar, Nagendra Kumar Singh, and Debabrata Sarkar. "A restriction-site-associated DNA (RAD) linkage map, comparative genomics and identification of QTL for histological fiber content coincident with those for retted bast fiber yield and its major components in jute (*Corchorus olitorius* L., Malvaceae sl)." *Molecular Breeding* 35, no. 1 (2015): 1–17.

Kundu, Avijit, Debabrata Sarkar, Nur Alam Mandal, Mohit Kumar Sinha, and Bikash Sinha Mahapatra. "A secondary phloic (bast) fiber-shy (bfs) mutant of dark jute (*Corchorus olitorius* L.) develops lignified fiber cells but is defective in cambial activity." *Plant Growth Regulation* 67, no. 1 (2012): 45–55.

Kundu, Avijit, Niladri Topdar, Debabrata Sarkar, Mohit K. Sinha, Amrita Ghosh, Sumana Banerjee, Moumita Das, Harindra S. Balyan, B. S. Mahapatra, and Puspendra K. Gupta. "Origins of white (*Corchorus capsularis* L.) and dark (*C. olitorius* L.) jute: A reevaluation based on nuclear and chloroplast microsatellites." *Journal of Plant Biochemistry and Biotechnology* 22, no. 4 (2013): 372–381.

Kundu, B. C. "Jute—World's foremost bast fiber. I. Botany, agronomy, diseases and pests." *Economic Botany* 10, no. 2 (1956): 103–133.

Kurowska, Krystyna, Renata Marks-Bielska, Stanisław Bielski, Hubert Kryszk, and Algirdas Jasinskas. "Food security in the context of liquid biofuels production." *Energies* 13, no. 23 (2020): 6247.

Lavanya, A. K., Anamika Sharma, Shashi Bhushan Choudhary, Hariom Kumar Sharma, Pawan Kumar Singh Nain, Surender Singh, and Lata Nain. "Mesta (*Hibiscus* spp.)—a potential feedstock for bioethanol production." *Energy Sources, Part A: Recovery, Utilization, and Environmental Effects* 42, no. 21 (2020): 2664–2677.

Li, Mi, Chang Geun Yoo, Yunqiao Pu, Ajaya K. Biswal, Allison K. Tolbert, Debra Mohnen, and Arthur J. Ragauskas. "Downregulation of pectin biosynthesis gene GAUT4 leads to reduced ferulate and lignin-carbohydrate cross-linking in switchgrass." *Communications Biology* 2, no. 1 (2019): 1–11.

Lyu, Jae Il, Hong-Il Choi, Jaihyunk Ryu, Soon-Jae Kwon, Yeong Deuk Jo, Min Jeong Hong, Jin-Baek Kim, Joon-Woo Ahn, and Si-Yong Kang. "Transcriptome analysis and identification of genes related to biosynthesis of anthocyanins and kaempferitrin in kenaf (*Hibiscus cannabinus* L.)." *Journal of Plant Biology* 63, no. 1 (2020): 51–62.

Mahal, Zinat, Tanzir Ahmed, Md Hossain, Abdullah Al Mahin, Harun or Rashid, and Tabassum Mumtaz. "Use of *Pleurotus sajor-caju* in upgrading green jute plants and jute sticks as ruminant feed." *Journal of BioScience and Biotechnology* 2, no. 2 (2013). 101-107.

Majumder, Shuvobrata, Prosanta Saha, Karabi Datta, and Swapan K. Datta. "Fiber crop, jute improvement by using genomics and genetic engineering." In *Advancement in Crop Improvement Techniques*, pp. 363–383. Woodhead Publishing with Elsevier Inc, Amsterdam, Netherlands, 2020.

Mankar, Akshay R., Ashish Pandey, Arindam Modak, and K. K. Pant. "Pre-treatment of lignocellulosic biomass: A review on recent advances." *Bioresource Technology* (2021): 125235.

Meents, Miranda J., Yoichiro Watanabe, and A. Lacey Samuels. "The cell biology of secondary cell wall biosynthesis." *Annals of Botany* 121, no. 6 (2018): 1107–1125.

Mohamed, R., M. T. Lim, and R. Halis. "Biodegrading ability and enzymatic activities of some white rot fungi on kenaf (*Hibiscus cannabinus*)." *Sains Malaysiana* 42, no. 10 (2013): 1365–1370.

Mood, Sohrab Haghighi, Amir Hossein Golfeshan, Meisam Tabatabaei, Gholamreza Salehi Jouzani, Gholam Hassan Najafi, Mehdi Gholami, and Mehdi Ardjmand. "Lignocellulosic biomass to bioethanol, a comprehensive review with a focus on pretreatment." *Renewable and Sustainable Energy Reviews* 27 (2013): 77–93.

Moreno, Juan C., Jianing Mi, Shreya Agrawal, Stella Kössler, Veronika Tureckova, Danuse Tarkowska, Wolfram Thiele, Salim Al-Babili, Ralph Bock, and Mark Aurel Schottler. "Expression of a carotenogenic gene allows faster biomass production by redesigning plant architecture and improving photosynthetic efficiency in tobacco." *The Plant Journal* 103, no. 6 (2020): 1967–1984.

Mosier, Nathan, Charles Wyman, Bruce Dale, Richard Elander, Y. Y. Lee, Mark Holtzapple, and Michael Ladisch. "Features of promising technologies for pretreatment of lignocellulosic biomass." *Bioresource Technology* 96, no. 6 (2005): 673–686.

Moxley, Geoffrey, Zhiguang Zhu, and Y-H. Percival Zhang. "Efficient sugar release by the cellulose solvent-based lignocellulose fractionation technology and enzymatic cellulose hydrolysis." *Journal of Agricultural and Food Chemistry* 56, no. 17 (2008): 7885–7890.

Nath, Mousumi, Farhana Tasnim Chowdhury, Shabbir Ahmed, Avizit Das, Mohammad Riazul Islam, and Haseena Khan. "Value addition to jute: Assessing the effect of artificial reduction of lignin on jute diversification." *Heliyon* 7, no. 3 (2021): 1–10.

NPC. *Report on Improvement of Agricultural Residues and Agro-Industrial by-Products Utilization*. National Productivity Council (NPC), New Delhi, 1987.

Ooi, Beng Guat, Ashley L. Rambo, and Miguel A. Hurtado. "Overcoming the recalcitrance for the conversion of kenaf pulp to glucose via microwave-assisted pre-treatment processes." *International Journal of Molecular Sciences* 12, no. 3 (2011): 1451–1463.

Palit, P., and J. H. Meshram. "Physiology of jute yield and quality." In *Jute and Allied Fiber Updates,* pp. 112–124. Central Research Institute for Jute and Allied Fibers, Kolkata, 2008.

Paoletti, Claudia, Eric Flamm, William Yan, Sue Meek, Suzy Renckens, Marc Fellous, and Harry Kuiper. "GMO risk assessment around the world: Some examples." *Trends in Food Science and Technology* 19 (2008): S70–S78.

Park, Heeyoung, Sang Un Park, Byeong-Kwan Jang, Jeong Jae Lee, and Yong Suk Chung. "Germplasm evaluation of Kenaf (*Hibiscus cannabinus*) for alternative biomass for cellulosic ethanol production." *GCB Bioenergy* 13, no. 1 (2021): 201–210.

Pazhany, Adhini S., and Robert J. Henry. "Genetic modification of biomass to alter lignin content and structure." *Industrial and Engineering Chemistry Research* 58, no. 35 (2019): 16190–16203.

Perveen, Shahnaz, Mingnan Qu, Faming Chen, Jemaa Essemine, Naveed Khan, Ming-Ju Amy Lyu, Tiangen Chang, Qingfeng Song, Gen-Yun Chen, and Xin-Guang Zhu. "Overexpression of maize transcription factor mEmBP-1 increases photosynthesis, biomass, and yield in rice." *Journal of Experimental Botany* 71, no. 16 (2020): 4944–4957.

Pu, Yunqiao, Fan Hu, Fang Huang, Brian H. Davison, and Arthur J. Ragauskas. "Assessing the molecular structure basis for biomass recalcitrance during dilute acid and hydrothermal pretreatments." *Biotechnology for Biofuels* 6, no. 1 (2013): 1–13.

Purohit, Pallav. "Economic potential of biomass gasification projects under clean development mechanism in India." *Journal of Cleaner Production* 17, no. 2 (2009): 181–193.

Rajagopal, Deepak, Steven E. Sexton, David Roland-Holst, and David Zilberman. "Challenge of biofuel: Filling the tank without emptying the stomach?" *Environmental Research Letters* 2, no. 4 (2007): 1–9.

Ramesh, P., B. Durga Prasad, and K. L. Narayana. "Characterization of kenaf fiber and its composites: A review." *Journal of Reinforced Plastics and Composites* 37, no. 11 (2018): 731–737.

Ritchie, Hannah, and Max Roser. "CO_2 and greenhouse gas emissions." *Our World in Data* (2020). https://ourworldindata.org/co2-and-other-greenhouse-gas-emissions.

Ruan, Qicheng, Jianmin Qi, Kaihui Hu, Pingping Fang, Haihong Lin, Jiantang Xu, Aifen Tao, Guolong Lin, and Lifu Yi. "Effects of microbial pretreatment of kenaf stalk by the white-rot fungus *Pleurotus sajor-caju* on bioconversion of fuel ethanol production." Sheng wu gong cheng xue bao= *Chinese Journal of Biotechnology* 27, no. 10 (2011): 1464–1471.

Saba, N., M. Jawaid, K. R. Hakeem, M. T. Paridah, A. Khalina, and O. Y. Alothman. "Potential of bioenergy production from industrial kenaf (*Hibiscus cannabinus* L.) based on Malaysian perspective." *Renewable and Sustainable Energy Reviews* 42 (2015): 446–459.

Saha, Prosanta, Karabi Datta, Shuvobrata Majumder, Chirabrata Sarkar, Shyamsundar P. China, Sailendra N. Sarkar, Debabrata Sarkar, and Swapan K. Datta. "*Agrobacterium* mediated genetic transformation of commercial jute cultivar *Corchorus capsularis* cv. JRC 321 using shoot tip explants." *Plant Cell, Tissue and Organ Culture (PCTOC)* 118, no. 2 (2014): 313–326.

Samanta, Pradipta, Sanjoy Sadhukhan, and Asitava Basu. "Identification of differentially expressed transcripts associated with bast fiber development in *Corchorus capsularis* by suppression subtractive hybridization." *Planta* 241, no. 2 (2015): 371–385.

Santoso, B., A. H. Jamil, and M. Machfud. "Kenaf (*Hibiscus cannabinus* L.) benefits in carbon dioxide (CO_2) sequestration." *Perspektif* 14, no. 2 (2015): 125–133.

Sarkar, Debabrata, Avijit Kundu, Anindita Saha, Nur Alam Mondal, Mohit Kumar Sinha, and Bikash Sinha Mahapatra. "First nuclear DNA amounts in diploid (2 n= 2 x= 14) *Corchorus* spp. by flow cytometry: Genome sizes in the cultivated jute species (*C. capsularis* L. and *C. olitorius* L.) are~ 300% smaller than the reported estimate of 1100–1350 Mb." *Caryologia* 64, no. 2 (2011): 147–153.

Saritha, M., and Anju Lata Arora. "Biological pretreatment of lignocellulosic substrates for enhanced delignification and enzymatic digestibility." *Indian Journal of Microbiology* 52, no. 2) (2012): 122–130. Doi: 10.1007/s12088-011-0199-x.

Satya, P., A. Chakraborty, D. Sarkar, M. Karan, D. Das, N. A. Mandal, D. Saha, S. Datta, S. Ray, C. S. Kar, and P. G. Karmakar. "Transcriptome profling uncovers β-galactosidases of diverse domain classes infuencing hypocotyl development in jute (*Corchorus capsularis* L.)." *Phytochemistry* 156 (2018): 20–32.

Satya, Pratik, and Ratikanta Maiti. "14 Bast and leaf fiber crops: Kenaf, hemp, jute, agave, etc." *Biofuel Crops: Production, Physiology and Genetics* (2013): 292–311.

Satya, Pratik, Debabrata Sarkar, Joshitha Vijayan, Soham Ray, Deb Prasad Ray, Nur Alam Mandal, Suman Roy et al. "Pectin biosynthesis pathways are adapted to higher rhamnogalacturonan formation in lignocellulosic jute (*Corchorus* spp.)." *Plant Growth Regulation* 93, no. 1 (2021): 131–147.

Sawatdeenarunat, C., H. Nam, S. Adhikari, S. Sung, and S. K. Khanal. "Decentralized biorefinery for lignocellulosic biomass: Integrating anaerobic digestion with thermochemical conversion." *Bioresource Technology* 250 (2018): 140–147.

Scheper, T. Food biotechnology. *Advances in Biochemical Engineering/Biotechnology*, pp. 1–269. Springer, 2008.

Selim, Khaled A., Saadia M. Easa, and Ahmed I. El-Diwany. "The xylose metabolizing yeast *Spathaspora passalidarum* is a promising genetic treasure for improving bioethanol production." *Fermentation* 6, no. 1 (2020): 33–45.

Selim, Khaled A., Dina E. El-Ghwas, Saadia M. Easa, and Mohamed I. Abdelwahab Hassan. "Bioethanol a microbial biofuel metabolite; new insights of yeasts metabolic engineering." *Fermentation* 4, no. 1 (2018): 16–43.

Sengupta, Gargi, and P. Palit. "Characterization of a lignified secondary phloem fiber-deficient mutant of jute (*Corchorus capsularis*)." *Annals of Botany* 93, no. 2 (2004): 211–220.

Shafrin, Farhana, Ahlan Sabah Ferdous, Suprovath Kumar Sarkar, Rajib Ahmed, Kawsar Hossain, Mrinmoy Sarker, Jorge Rencoret et al. "Modification of monolignol biosynthetic pathway in jute: Different gene, different consequence." *Scientific Reports* 7, no. 1 (2017): 1–12.

Shafrin, Farhana, Sudhanshu Sekhar Das, Neeti Sanan-Mishra, and Haseena Khan. "Artificial miRNA-mediated down-regulation of two monolignoid biosynthetic genes (C3H and F5H) cause reduction in lignin content in jute." *Plant Molecular Biology* 89, no. 4 (2015): 511–527.

Shah, Siti Syazwani Mohd, Abdullah Amru Indera Luthfi, Kheng Oon Low, Shuhaida Harun, Shareena Fairuz Abdul Manaf, Rosli Md Illias, and Jamaliah Md Jahim. "Preparation of kenaf stem hemicellulosic hydrolysate and its fermentability in microbial production of xylitol by *Escherichia coli* BL21." *Scientific Reports* 9, no. 1 (2019): 1–13.

Sharma, Bhawna, Christian Larroche, and Claude-Gilles Dussap. "Comprehensive assessment of 2G bioethanol production." *Bioresource Technology* (2020): 123630.

Shi, Jian, Ratna R. Sharma-Shivappa, Mari Chinn, and Noura Howell. "Effect of microbial pretreatment on enzymatic hydrolysis and fermentation of cotton stalks for ethanol production." *Biomass and Bioenergy* 33, no. 1 (2009): 88–96.

Shinoj, P., S. S. Raju, and P. K. Joshi. "India's biofuels production programme: Need for prioritizing the alternative options." *Indian Journal of Agricultural Sciences* 81, no. 5 (2011): 391–397.

Singh, Jyoti, Abha Sharma, Pushpendra Sharma, Surender Singh, Debarup Das, Gautam Chawla, Atul Singha, and Lata Nain. "Valorization of jute (*Corchorus* sp.) biomass for bioethanol production." *Biomass Conversion and Biorefinery* (2020): 1–12.

Smith-Heisters, Skaidra. *Illegally Green: Environmental Costs of Hemp Prohibition*. Reason Foundation, Los Angeles, 2008.

Song, Yan, Guangting Han, and Wei Jiang. "Comparison of the performance of kenaf fiber using different reagents presoak combined with steam explosion treatment." *The Journal of the Textile Institute* 108, no. 10 (2017): 1762–1767.

Sticklen, Mariam. "Plant genetic engineering to improve biomass characteristics for biofuels." *Current Opinion in Biotechnology* 17, no. 3 (2006): 315–319.

Su, Buli, Mianbin Wu, Zhe Zhang, Jianping Lin, and Lirong Yang. "Efficient production of xylitol from hemicellulosic hydrolysate using engineered *Escherichia coli*." *Metabolic Engineering* 31 (2015): 112–122.

Suganya, D. S., S. Pradeep, J. Jayapriya, and S. Subramanian. "Bio-softening of mature coconut husk for facile coir recovery." *Indian Journal of Microbiology* 47, no. 2 (2007): 164–166.

Tanmoy, A. M., M. A. Alum, M. S. Islam, T. Farzana, and H. Khan. "Jute (*Corchorus olitorius* var. O-72) stem lignin: Variation in content with age." *Bangladesh Journal of Botany* 43, no. 3 (2014): 309–314.

Tripathi, Arun K., P. V. R. Iyer, Tara Chandra Kandpal, and K. K. Singh. "Assessment of availability and costs of some agricultural residues used as feedstocks for biomass gasification and briquetting in India." *Energy Conversion and Management* 39, no. 15 (1998): 1611–1618.

Tuck, Gill, Margaret J. Glendining, Pete Smith, Jo I. House, and Martin Wattenbach. "The potential distribution of bioenergy crops in Europe under present and future climate." *Biomass and Bioenergy* 30, no. 3 (2006): 183–197.

Wan, Caixia, and Yebo Li. "Fungal pretreatment of lignocellulosic biomass." *Biotechnology Advances* 30, no. 6 (2012): 1447–1457.

Webber III, C. L., V. K. Robert, and R. E. Bledsoe. "United States kenaf (*Hibiscus cannabinus* L.) cultivar review." In Webber, C. L. III and Liu, A. (eds.) *Plant Fibers as Renewable Feedstocks for Biofuel and Biobased Products*, pp. 117–126. CCG International, St Paul, Minnesota, 2011.

Zhang, Gaoyang, Yujia Zhang, Jiantang Xu, Xiaoping Niu, Jianmin Qi, Aifen Tao, Liwu Zhang, Pingping Fang, LiHui Lin, and Jianguang Su. "The CCoAOMT1 gene from jute (*Corchorus capsularis* L.) is involved in lignin biosynthesis in *Arabidopsis thaliana*." *Gene* 546, no. 2 (2014): 398–402.

Zhang, Liwu, Ray Ming, Jisen Zhang, Aifen Tao, Pingping Fang, and Jianmin Qi. "De novo transcriptome sequence and identification of major bast-related genes involved in cellulose biosynthesis in jute (*Corchorus capsularis* L.)." *BMC Genomics* 16, no. 1 (2015a): 1–13.

Zhang, Liwu, Xuebei Wan, Jiantang Xu, Lihui Lin, and Jianmin Qi. "De novo assembly of kenaf (*Hibiscus cannabinus*) transcriptome using Illumina sequencing for gene discovery and marker identification." *Molecular Breeding* 35, no. 10 (2015b): 1–11.

Zhu, J. Y., and X. J. Pan. "Woody biomass pretreatment for cellulosic ethanol production: Technology and energy consumption evaluation." *Bioresource Technology* 101, no. 13 (2010): 4992–5002.

Zhu, J. Y., G. S. Wang, X. J. Pan, and Roland Gleisner. "Specific surface to evaluate the efficiencies of milling and pretreatment of wood for enzymatic saccharification." *Chemical Engineering Science* 64, no. 3 (2009): 474–485.

11 Exploring the Potential of Cyanobacterial Biomass for Bioethanol Production

Nirmal Renuka[ad], *Sachitra Kumar Ratha*[b], *Virthie Bhola*[a], *Kokila, V.*[c], *Lata Nain*[c], *Faizal Bux*[a], *and Radha Prasanna*[c]

[a]Institute for Water and Wastewater Technology, Durban University of Technology, Durban, South Africa-4001

[b]Phycology Laboratory, CSIR-National Botanical Research Institute, Lucknow, India-226001

[c]Division of Microbiology, ICAR-Indian Agricultural Research Institute, New Delhi, India-110012

[d]Algal Biotechnology Laboratory, Department of Botany, Central University of Punjab, Bathinda, India-151401

CONTENTS

DOI: 10.1201/9781003191247-11

11.1 INTRODUCTION

With the energy crisis and rapid dwindling of petroleum reserves, exploration of renewable sources is gaining importance, particularly third-generation biofuels (Chisti et al. 2013; Holmatov et al. 2021; María et al. 2021). Additionally, the problems associated with the use of food crops–based biomass and their poor degradation efficiency, combined with high processing costs of lignocelluloses, emphasize the need to look for alternative feedstocks (Harun et al. 2014; Devi et al. 2021). Researchers are looking towards developing innovations in deciphering and modulating technical pathways for bioethanol production, which are directly photosynthesis-derived rather than biomass-based (Ashokkumar et al. 2019; Deb et al. 2021; Dave et al. 2021; Rawat et al. 2021). Photosynthetic biomass, including algae and aquatic plants, represents a promising feedstock for production of biogas, bioethanol, and biodiesel. Among photosynthetic organisms, besides eukaryotic microalgae, cyanobacteria are potential candidates and amenable to genetic manipulation, as the plasticity of their metabolism permits comprehensive and in-depth analyses of the intracellular components and energy fluxes (Lasry et al. 2019; Huang et al. 2021). The ability to generate organic carbon metabolites by utilizing solar energy and carbon dioxide through the Calvin cycle makes them ideal miniature factories to metabolically convert organic carbon metabolites to multiple products, which, through introduction of relevant gene(s) and assembling an ethanol-producing pathway, can lead to ethanol as major end product (Deng and Coleman 1999; Wang et al. 2020). The creation of this novel pathway led to ethanol yields of 230 mg/L. Dexter and Fu (2009) were able to further increase that to 460 mg/L, and Gao et al. (2012) increased this further to 5.5 g/L in *Synechocystis* sp. PCC 6803. Ethanol productivity from direct conversion process to ethanol using glycogen-rich *Arthrospira platensis* in combination with amylase-expressing strain of *Saccharomyces cerevisiae* (MT8–1dGS) was found to be 6.5 g/L (ethanol productivity of 1.08 g/L/d), as shown by Aikawa et al. (2013). Additionally, low yields of different molecules, such as isoprene, sucrose, hydrogen, β-caryophyllene, butanol, isobutyraldehyde, isobutanol, fatty alcohols, and fatty acids, have also been recorded in cyanobacteria modified using tools of genetic or metabolic engineering (Tan et al. 2011; Yao et al. 2014). Ethanol production under dark metabolism is also reported; however, the ethanol yields are too low for it to be exploited at a commercial scale.

Ethanol production under photoautotrophic mode using a model cyanobacterium, e.g., *Synechocystis* PCC6803, is an attractive technology. This technology involves the use of metabolic engineering tools using a two-step process, involving, firstly, using the cloned heterologous pyruvate decarboxylase to convert pyruvate to acetaldehyde, and its conversion to ethanol using the cloned alcohol dehydrogenase gene, taken from either homologous or heterologous sources (Pembroke et al. 2019). Theoretical calculations illustrate that photosynthetic organisms can show a productivity of ethanol, ca. 5280 gal/ac/yr, vis-à-vis annual yield of ethanol being 727 and 321 gal/ac/yr from corn and sugarcane, with switchgrass and corn stover contributing 330–810 and 290–580 gal/ac/yr (Sanderson 2006). However, in the last decade, focus has moved to a consolidated bioprocessing strategy involving one single biological system, i.e., integrating photosynthetic biomass production with microbial conversion, leading to ethanol production using a bacterium/fungus which converts carbohydrates to ethanol.

11.2 CYANOBACTERIAL BIOMASS AS FEEDSTOCK FOR BIOETHANOL PRODUCTION

Among microalgae, cyanobacteria are a preferred source for use as biomass for ethanol production, as they possess certain advantages over eukaryotic microalgae. Despite having a cell envelope resembling gram-negative bacteria, the peptidoglycan layer in the cell wall shows more similarities with that of gram-positive bacteria, in cyanobacteria (Hoiczyk and Hansel 2000; Domozych 2011), which is easily degradable by lysozyme. On the other hand, most microalgae exhibit complex cell walls made up of complex polysaccharides and proteoglycans (Domozych 2011). The type of storage carbohydrate in the biomass is most important for it to be used as a fermentation substrate by microorganisms such as yeast/fungi, and cyanobacteria have glycogen (Allen 1984; Ball and Morell 2003), while most eukaryotic green algae store as starch or β-glucans, as recorded in green/red algae and diatoms/brown algae, respectively (Ball et al. 2011).

Looking at the structural features of these storage polymers, in glycogen and starch, essentially α-1,4-glucans with α-1, 6-branching is observed, which are water-insoluble and much larger (0.1–100 μm), while glycogen particles are small (0.04–0.05 μm) and water-soluble. Depending on the strain and growth conditions, cyanobacteria normally accumulate glycogen from 10 to 50% of their biomass content, and this polysaccharide serves as substrate for biofuel fermentation (John et al. 2011). As a fermentation feedstock, glycogen is preferred over starch as glycogen mobilization is less energy-intensive than *in vitro* starch mobilization, which involves heating and/or enzymatic treatments (Mamo et al. 2013).

Towards increasing the efficiency of alcohol fermentation using cyanobacterial biomass, the first step involves identifying a pretreatment method which is effective. Enzymatic hydrolysis with the efficient mix of enzymes hydrolyzing a complex of carbohydrates has been the focus of research. The main carbohydrate substrate for fermentation in cyanobacteria is glycogen, and its mobilization is comparatively easier than that of starch. In recent years, the whole-cell material from glycogen-enriched cyanobacteria and starch-enriched green microalgae (Harun and Danquah 2011; Harun et al. 2011) represents useful feedstocks for bioethanol production by yeast fermentation. Physical treatments such as drying, heating, and acid and base treatment, or enzymatic or chemical treatments of the biomass, have been explored as a means of releasing the monomeric hexoses. The ethanol productivity of whole-cell biomass from microalgae (Harun et al. 2011) or cyanobacteria (Aikawa et al. 2013) has focused on unicellular cyanobacteria; they have been the choice of feedstock, particularly *Synechocystis* and *Synechococcus* species, as their genomes have sequenced, making them more amenable to modulation using biochemical or genetic engineering approaches. In cyanobacteria, glycogen is stored as a main energy compound in chloroplast, whose accumulation can be improved by subjecting to stress conditions, such as imbalanced nutrients ratio, nutrient limitation, and salt stress in the cultivation medium (González-Fernández and Ballesteros 2012).

The marine cyanobacterium *Synechococcus* sp. PCC 7002 (previously known as *Agmenellum quadruplicatum* PR-6) can be a promising biomass feedstock for anaerobic fermentation in combination with the yeast *Saccharomyces cerevisiae*. The accumulation of glycogen and cyanophycin by the cyanobacterium in the form of reserves of carbon and nitrogen, under N-limitation in the growth medium, is advantageous. Several coordinated and complex physiological adaptations are involved, which result in increased C:N ratio of the biomass, greater carbohydrate content (mostly glycogen), along with a degradation of the light-harvesting phycobilisome (PBS) antenna proteins (Luque and Forchhammer 2008; Beck et al. 2012). Interestingly, photosynthesis and growth were not completely affected under N-limitation (Sauer et al. 2001).

Investigations undertaken in this context have led to the development of a number of mutants which do not store photosynthate in the form of glycogen or polyhydroxybutyrate, such as those having a deletion of *glg*C (glucose-1-phosphate adenylyl transferase) (Xu et al. 2012). The *glg*C mutant of *Synechocystis* sp. PCC 6803 is able to grow photoautotrophically in continuous light, with no decrease in growth rate, but exhibit unusual physiology under mixotrophic growth, light/dark cycles, high light intensities, and nitrogen deprivation than wild type (WT) (Carrieri et al. 2012). Accumulation of polyhydroxy butyrate is also observed in some cyanobacteria (Beck et al. 2012). Huang et al. (2016) targeted the rewiring of metabolic networks in cyanobacteria by using CRISPR interference (CRISPRi), by suppressing the transcript levels of genes essential for the accumulation of glycogen (*glg*C) and the conversion of succinate to fumarate (*sdh*A and *sdh*B) in *Synechococcus elongatus* (PCC 7942). More importantly, cyanobacteria can be efficient as cell factories for ethanol production. The stoichiometric energy yield for ethanol was found comparable with other potential fuel metabolites, paving the way for focused research on utilizing these feedstocks for bioethanol production (Kamarainen et al. 2012).

Another line of research involves exploring ethanol productivity using cell material of cyanobacteria after hydrolysis and enzymatic treatment, in combination with other types of feedstock for ethanol production. A high concentration of this carbohydrate-enriched *Synechococcus* biomass, as fermentation feedstock, after treatment with only enzymes, was promising, not only as fermentation feedstock, but also as a source of nutrients to enhance the productivity of yeast fermentations using various biomass feedstock (Möllers et al. 2014). Microwave-based heating of biomass has shown to be effective in the breakdown of polysaccharides in the cyanobacterial biomass of *Arthrospira platensis*, in combination with yeast fermentation to yield 0.97 g/L of alcohol (Aikawa et al. 2013); this was higher than conventional treatments.

11.3 PRETREATMENT AND SACCHARIFICATION

Pretreatment is an expensive but essential procedure in bioethanol production. Cyanobacterial biomass after biomass recovery is subjected to cell wall rupture (pretreatment) for polysaccharide release and saccharification (Astolfi et al. 2020; Maia et al. 2020). Efficient pretreatment of the cyanobacterial biomass is one of the foremost requirements for the commercial-scale production of bioethanol. This step also results in a greater surface area, which improves bioavailability or breakdown of the substrate to enable the hydrolysis of the biomass using microbial enzymes more effectively. In essence, during pretreatment, complex carbohydrates (often entrapped within the cell wall) or starch/glycogen polymers (intracellularly) are broken down into monomers, such as glucose. Thereafter, the resultant simple and fermentable sugars can be easily metabolized by microorganisms to produce bioethanol *via* fermentation (Table 11.1).

TABLE 11.1
Different Reports on Bioethanol Production from Cyanobacterial Biomass Using Different Pretreatment and Fermentation Procedures

Organism	Carbohydrate Content (% in Dry Cell Weight)	Biomass Concentration (g/L in Dry Cell Weight)	Type of Pretreatment	Pretreatment Condition	Fermentation	Productivity (g Ethanol/ L/d) and/or Yield (g Ethanol/g Biomass or g Ethanol/L)	Ethanol Yield on Sugar Consumption (%)	Yield of Treatment (%)	Yield of Fermentation (%)	Reference
Spirulina platensis LEB-52	50–60 (% DCW)	-	Physical and enzymatic methods	Freezing/thawing Thereafter, saccharification by addition of α-amylase and amylo-glucosidase	-	-	-	-	-	Rempel et al. 2018
Arthrospira platensis		-	Thermohydrolysis	Microwave heating (150°C)	Fermentation using industrial strain of *Saccharomyces cerevisiae* (As4)	0.97 g/L	-	-	-	Nowicka et al. 2020
Microcystis aeruginosa (FACHB 915)	50.0 ± 0.3 (% DCW)	30	Acid hydrolysis and autoclaving	1% H_2SO_4	Fermentation using industrial strain of *Saccharomyces cerevisiae* (CICC33068)	0.22 ± 0.02 g ethanol/g biomass	-	-	84.4 ± 7.4	Huang et al. 2021

(*Continued*)

TABLE 11.1 *(Continued)*
Different Reports on Bioethanol Production from Cyanobacterial Biomass Using Different Pretreatment and Fermentation Procedures

Organism	Carbohydrate Content (% in Dry Cell Weight)	Biomass Concentration (g/L in Dry Cell Weight)	Type of Pretreatment	Pretreatment Condition	Fermentation	Productivity (g Ethanol/L/d) and/or Yield (g Ethanol/g Biomass or g Ethanol/L)	Ethanol Yield on Sugar Consumption (%)	Yield of Treatment (%)	Yield of Fermentation (%)	Reference
Antrosphira platensis		12–13	Acid hydrolysis	HNO_3 (0.5 N) H_2SO_4 (0.5 N)	Fermentation using industrial strain of *Saccharomyces cerevisiae* MV 92081 (Martin Vialatte)	16.32% ± 0.90% (g ethanol/g biomass) 16.27% ± 0.97% (g ethanol/g biomass)	-	80	56	Markou et al. 2013
Synechococcus PCC 7002		100	Enzymatic hydrolysis	Lysozyme and α-glucanases	Fermentation by industrial strain of *Saccharomyces cerevisiae*	0.27 g ethanol/g biomass; 20 g ethanol/L/d	Conversion of 90% glucose in the biomass into ethanol	80	86	Möllers et al. 2014
Arthrospira platensis	40.02 ± 0.47 (% DCW)	-	Enzymatic hydrolysis	α-amylase enzyme for 2 h (90°C) followed by Amyloglucosydase Units of glucoamylase enzyme (60°C) for 2 h	Fermentation using metabolically engineered *E. coli* MSo4	1.4 g ethanol/L/h	92% conversion yield of the glucose content in the hydrolysate	-	-	Werlang et al. 2020

Arthrospira (Spirulina) platensis	70.9 ± 1.6 (% DCW) (SOT medium containing 3 mM $NaNO_3$)	-	None	None	Consolidated bioprocess (direct fermentation by yeast)	1.08 g ethanol/L/d	86%	-	-	Aikawa et al. 2013
Arthrospira platensis	90 g/L of glycogen	-	None	None	Consolidated bioprocess (Direct fermentation by yeast (*Saccharomyces cerevisiae*) with the addition of lysozyme and calcium chloride)	48 g ethanol/L; 1.0 g ethanol/L/h; 0.32 g ethanol/g biomass	-	-	-	Aikawa et al. 2018

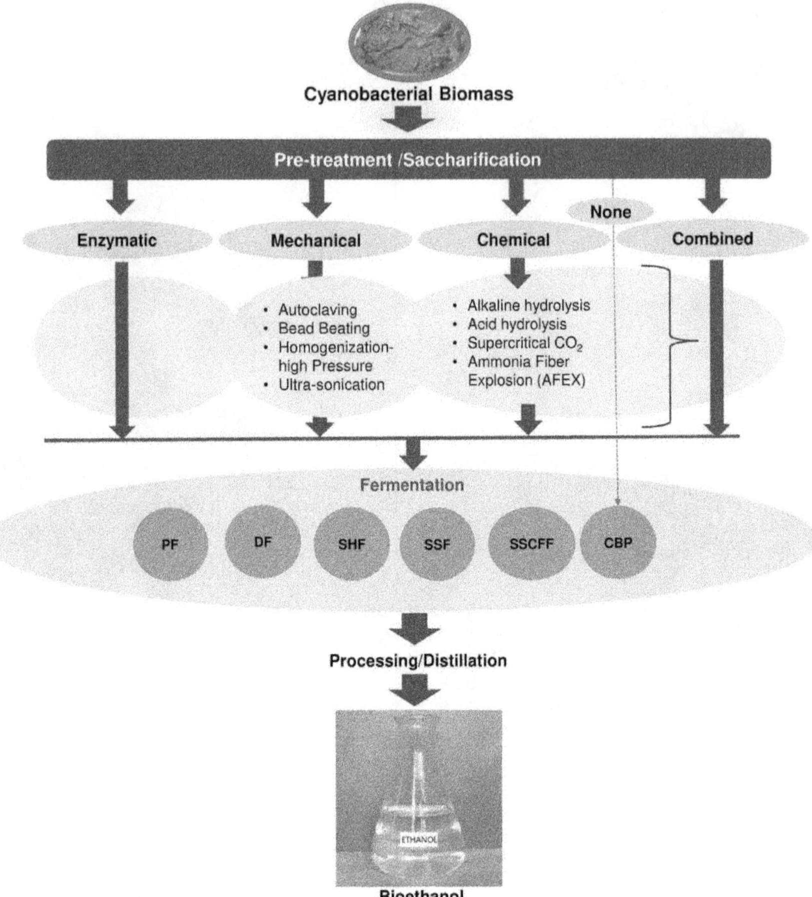

FIGURE 11.1 Schematic presentation of different pretreatment, saccharification, and fermentation processes involved in bioethanol production using cyanobacterial biomass. PF: photofermentation; DF: dark fermentation; SHF: separate hydrolysis and fermentation; SSF: simultaneous saccharification and fermentation; SSCF: simultaneous saccharification and cofermentation; CBP: consolidated bioprocessing.

Numerous pretreatment methods exist and have been employed, which include mechanical, chemical, and enzymatic pretreatments (Figure 11.1). The ultimate goal of these methods is to disrupt cyanobacterial cell walls in order to release more carbohydrates and break down into simple monosaccharide units (Hernandez et al. 2015; Phwan et al. 2018). The competence of pretreatment methods largely depends upon their ease of implementation, cost, energy efficiency, amount of fermentable sugars, and degradation capacity.

11.3.1 Mechanical Techniques

Mechanical pretreatments, such as grinding and milling, are generally used for the disruption of cell membrane, increasing cell surface area and making an easy saccharification of biomass (Smachetti et al. 2019). Mechanical forces that are applied to disrupt cell walls can be divided into the following categories: liquid-shear forces (high-pressure homogenization), solid-shear forces (bead beating), and wave transmission (ultrasonication) (Phwan et al. 2018; Maia et al. 2020; Nowicka et al. 2020). A major advantage of utilizing this physical technique is that most of the functionality of the cellular materials is preserved. Furthermore, chemical contamination of the biomass is prevented. However, it is important to note that when compared to chemical methods, this form of pretreatment is often more energy-intensive, with higher associated costs.

High-pressure homogenization (HPH) is an effective pretreatment method that is dependent on controllable parameters (e.g,. intensity of the pressure applied and the number of cycles). The idea behind HPH is to apply pressure drop–induced shear stress on the cell suspension, leading to the stationary valve surface receiving an accelerated cellular jet of intense pressure. Bead beating, or bead milling, is a technique that uses beads in motion to apply shear stress on biomass to disrupt cell walls. This method would require cylindrical shaking vessels and appropriate glass/steel beads. Cyanobacterial cell walls are disintegrated when they are continuously ground against the rough surface of the beads. Bead size is a significant factor, as the rate of cell disruption would be faster if there was a higher volume of beads to cell suspension. Even though this is a fairly rapid and simple technique, it is often an unattractive pretreatment option, owing to the large amounts of heat generated and the high energy usage. Ultrasonication is a mechanical cavitation process that uses acoustic waves at high frequency. This produces jet streams in the targeted biomass medium, resulting in tiny cavitation bubbles around the cells, and cell rupture occurs when the bubbles collapse and emit shock waves (Al hattab and Abdel 2015).

11.3.2 Chemical Methods

Chemical pretreatment is easier and more cost-effective in comparison to enzymatic pretreatment. If stored correctly, chemicals have higher durability and are easier to handle, making them less trouble than enzymes. Acid hydrolysis uses common acids, such as sulfuric acid (H_2SO_4) or hydrochloric acid (HCl), to convert feedstock into fermentable sugars (Phwan et al. 2018). Higher acid concentrations could improve the sugar extraction yield. This, however, is a trade-off, as a strong acid leads to the complex substrate getting completely degraded, resulting in fermentable sugars getting lost (Miranda et al. 2012). Moreover, severe acid pretreatment often leads to the formation of inhibitors, like furfural (from pentoses) or hydroxymethyl furfural (from hexoses). Formation of these inhibitors could have adverse effects on the fermentation process. A neutralization step would therefore be required prior to the fermentation process, causing an increase in overall production costs (Phwan et al. 2018). During acid hydrolysis, the bonds which connect polysaccharide chains are broken. The initial step involves the destruction of these hydrogen bonds. This, in turn, leads to the rupture of the polysaccharide chains, causing them to transition into a complete amorphous state. Following the destruction of the bonds, the polysaccharide is extremely vulnerable to hydrolysis. The acid then acts as a catalyzer and cleaves polysaccharide by hydrolyzing the glycosidic bonds. The final step entails addition of water at a moderate temperature, which leads to complete and rapid hydrolysis of the hydrolysate into monosaccharide (Jambo et al. 2016).

Markou et al. (2013) investigated the bioethanol production potential of the cyanobacterium *Arthrospira platensis*. Four acids were employed for saccharification of the carbohydrate-rich biomass. During this study, varying concentrations (2.5 N, 1 N, 0.5 N and 0.25 N respectively) of H_2SO_4, HNO_3, HCl, and H_3PO_4 were tested, and saccharification was done under different temperatures (40°C, 60°C, 80°C, and 100°C, respectively). From the results obtained, these researchers were able to conclude that higher reducing sugar yields (%, g reducing sugars/g total sugars) were obtained at higher acid concentrations. Higher rates were also noted when the temperature was increased at lower acid concentrations. It was further noted that the subsequent bioethanol yield (after fermentation) was significantly dependent on the acid concentration used during the saccharification step. Hydrolysates that were produced using HNO_3 (0.5 N) and H_2SO_4 (0.5 N) exhibited the highest ethanol yields of 16.32% ± 0.90% (g ethanol/g biomass) and 16.27% ± 0.97% (g Ethanol/g Biomass), respectively (Markou et al. 2013).

Similarly, alkali solutions can also be utilized as a chemical pretreatment option. Selection and concentration of the alkaline reagent as well as temperature would be significant factors for the success of this method (Maia et al. 2020). The supercritical fluid (CO_2) method can be used as a pretreatment for bioethanol production (Phwan et al. 2018). This process is based on the interaction among fluid properties, such as viscosity, density, diffusivity, and surface-tension, and on operating conditions, such as biomass concentration, temperature, and pressure. However, it was observed that supercritical CO_2 technique was mostly used for lipid extraction and not for carbohydrate extractions pertaining to the bioethanol production (Phwan et al. 2018). One of the rare chemical forms of pretreatment include ammonia fiber explosion (AFEX), where biomass is subjected to the liquid anhydrous ammonia treatment for a limited time period at a specific temperature and usually under high pressure. Thereafter, liquid ammonia is vaporized to permit its retrieval and/or recycling. Pretreatment exploiting of the AFEX protocol can be beneficial as no by-product inhibitors like furans are formed. Also, ammonia is relatively cheaper than H_2SO_4, making it a commodity chemical that is more economically viable in comparison to acid hydrolysis. One downside to this method is the recovery of ammonia following pretreatment. The recovery step could increase operational and capital costs. More research is, however, required on this form of pretreatment for bioethanol production, as most studies have been focused on the use of this technique for lipid extraction instead.

11.3.3 Enzymatic Hydrolysis

When compared to microalgae, cyanobacterial biomass is easily degradable by enzymatic hydrolysis techniques. This is because their cell walls contain a peptidoglycan layer which is less complex as opposed to microalgal cell walls, which contain complex polysaccharide and proteoglycan structures (Ismail et al. 2020). Furthermore, the storage carbohydrate within cyanobacterial cell walls is glycogen. Glycogen is not only water-soluble but also smaller in particle size as compared to starch, therefore requiring less energy for its hydrolysis (Möllers et al. 2014).

In recent years, the use of enzymatic hydrolysis as a pretreatment for bioethanol production has proved more promising as compared to acid/alkali hydrolysis methods (Jambo et al. 2016; Shokrkar et al. 2017; Phwan et al. 2018; Maia et al. 2020). Even though the quantities of selected enzymes required for effective saccharification using enzymatic hydrolysis are generally high, this process utilizes less energy, making this technique very attractive. Mild enzymatic conditions for hydrolysis can also help achieve a higher glucose yield, with no production of toxic by-products such as formic acid, hydroxymethyl furfural, furfural, levulinic acid, and other phenolic compounds (Shokrkar et al. 2017). If toxic compounds were produced during the pretreatment step, they could possibly affect the enzymatic reactions involved in the fermentation process. Effective enzymatic hydrolysis and biomass digestibility are dependent on many factors, such as the type, concentration, and efficiency of the enzymes used, sugar-release patterns, duration of treatment, pH, temperature, as well as the combination of the substrate (Al hattab and Abdel 2015; Phwan et al. 2018). Enzymes commonly employed for cell wall degradation include alkaline protease, sanilase, cellulase, papain, lysozyme, and neutral protease (Al hattab and Abdel 2015). In a study by Möllers et al. (2014), cyanobacterial cells (*Synechococcus* sp.) were harvested and thereafter subjected to enzymatic hydrolysis with lysozyme and 2-α-glucanases. No further treatment was required, and the resulting hydrolysate was thereafter fermented using *S. cerevisiae* into ethanol. Approximately 90% conversion of glucose into bioethanol through fermentation of the cyanobacterial biomass hydrolysate (about 20 g of ethanol/L/d) was achieved. It was also interesting to note that no additional nutrients were required for the enzyme and fermentation treatments, and all processes were conducted in the residual growth medium of the cyanobacterium (Möllers et al. 2014).

11.3.4 Combined Pretreatment Approaches

Among different pretreatment and saccharification procedures, chemical methods seem to be the most popular; however, they produce inhibitors which can often have an adverse effect on the fermentation process. The use of enzymes has certain inherent advantages, but high costs associated with specific enzymes for the hydrolysis process would hinder large-scale commercial applications. For effective mechanical/physical pretreatment, high amounts of energy and additional treatment procedures are often required. Some studies also employed the combined pretreatment, i.e., the use of more than one pretreatment procedures for efficient carbohydrate recovery (Khan et al. 2017; Rempel et al. 2018). Rempel et al. (2018) reported that the physical pretreatment of *Spirulina platensis* biomass using freeze-and-thaw procedure, followed by gelatinization, proved promising for the saccharification using free and immobilized enzymatic hydrolysis. They could achieve hydrolysis efficiency of 99% using free amylolytic enzyme (1% v/v). According to Khan et al. (2017), pretreatment with calcium oxide prior to enzymatic treatment led to a twofold increase in the monomeric sugars in *Microcystis aeruginosa* biomass. However, owing to these aforementioned challenges and high cost associated with pretreatment procedures, recently a combined system approach or single-step conversion is gaining attention and could be more valuable, where pretreatment and saccharification are being carried out in the same system for bioethanol production; this is discussed in the following section (Section 4, Subsection 4.3).

11.4 PATHWAYS FOR BIOETHANOL PRODUCTION IN CYANOBACTERIA

Three possible routes are envisaged for cyanobacterial biomass processing towards bioethanol production, *viz*., firstly, the traditional one, wherein hydrolysis, followed by fermentation of biomass, is undertaken; the second method involves the production of hydrogen (H_2), acids, and ethanol under dark conditions by employing the metabolic pathways in which photosynthesis is redirected, while the third route is photofermentation, involving the use of cyanobacteria which have been engineered to redirect the routine biochemical pathways (Dexter et al. 2015; Hernandez et al. 2015; de Farias and Bertucco 2016). The production of ethanol is enabled by sequestering CO_2 as a carbon source under photoautotrophic growth conditions, involving the reactions forming a part of glycolysis, pentose phosphate pathway, citric acid cycle, and Calvin cycle by fermentation (Paulo et al. 2011). The conversion of the biomass into products using fermentation process is higher due to the higher efficiency of the enzymes in yeasts, such as *Saccharomyces*, or bacterial genera, such as *Zymomonas*, commonly involved in ethanolic fermentation. The foremost advantage of the fermentation from the photosynthetic microorganisms is to produce ATP, which is necessary for driving energy requiring metabolic processes (Ohta et al. 1987; Ueno et al. 1998). The algal species (microalgae and cyanobacteria) mostly contain common carbohydrates, such as starch, glycogen, cyanophycean starch, and cellulose, which can be used for the production of bioethanol. Glycogen (energy storage compound which is a glucose polymer) has higher solubility in water and possesses shorter polymer chains, which is among several beneficial characteristics, including easy hydrolysis; this makes cyanobacteria feasible for the production of bioethanol (Möllers et al. 2014).

11.4.1 Bioethanol Production by Dark Fermentation

The process whereby organic substrates are converted into biohydrogen/bioethanol is referred to as dark fermentation. This process utilizes fermentative and hydrolytic microorganisms to hydrolyze complex organic polymers into monomers. Thereafter, the monomers are transformed into a mixture of low molecular weight organic acids and alcohols (mainly ethanol and acetic acid). The accumulation of carbohydrates (glucose, xylose, mannose, galactose, and arabinose) as a result of photosynthesis in the cyanobacterial cells favors the production of ethanol, making it a possible feedstock for bioethanol production through fermentative metabolism when switching growth from light to dark conditions (Ueno et al. 1998). Dark fermentation converts organic substrates into biohydrogen (Heyer and Krumbein 1991; Hirano et al. 1997); this involves hydrolysis of the starch/glycogen reserves to simpler sugars by amylase activity, followed by conversion to pyruvate *via* glycolysis (Catalanotti et al. 2013). The use of dark fermentation for hydrogen production is usually not favored, as acids and alcohols remain after the overall process, which represents approximately

80–90% of the initial chemical oxygen demand (de Farias Silva and Bertucco 2016). Yields as low as 1–2 mol H_2/mol of glucose have been reported under optimal experimental conditions, rendering this an ineffective process for biohydrogen production (Ueno et al. 1998). Dark fermentation has also been investigated for ethanol production from cyanobacteria that possess high concentrations of carbohydrate. Cyanobacteria accumulate carbohydrates within their cells during photosynthesis. If the growth mode were thereafter switched to dark conditions, the cyanobacterial cells would be forced to utilize their carbohydrate and lipid reserves and synthesize ethanol through direct fermentative metabolism. Furthermore, the inclusion of certain additives has been reported to increase ethanol productivity. However, the efficiency of bioethanol production can be hampered due to the interference of light and oxygen (de Farias Silva and Bertucco 2016). Cyanobacterial species, *viz.*, *Oscillatoria limosa*, *O. limnetica*, *Gleocapsa alpicola*, *Cyanothece* sp., *Spirulina* sp., and *Synechococcus* sp., in the absence of light, can expel ethanol through the cell wall, using intracellular processes, after hydrolysis of complex organic polymers into monomers, and their conversion into a mixture of organic acids and alcohols (acetic acid and ethanol) (Ueno et al. 1998; Deng and Coleman 1999).

The Embden-Meyerhof-Parnas and pentose phosphate pathway are metabolic pathways that are responsible for the degradation of intracellular polysaccharides into pyruvate by using two enzymes, *viz.*, pyruvate decarboxylase and alcohol dehydrogenase (Jambo et al. 2016; Phwan et al. 2018). During dark fermentation, cellular polysaccharides are converted to pyruvate by generating the energy-rich molecules ATP and NADH, which must be reoxidized to sustain fermentation (Ben-Amotz 1975). Pyruvate acts as a main intermediate compound which serves as a substrate for acetyl coenzyme A (acetyl-CoA) and is converted into different end products and produces ATP through conversion to acetate and, finally, to ethanol in *Chlamydomonas* (Hemschemeier and Happe 2005). The alcohol/acetaldehyde dehydrogenase (ADHI), having putative dual functions, converts acetyl-CoA to acetaldehyde or ethanol using three enzymes, *viz.*, pyruvate fumarate lyase (PFL1), pyruvate ferredoxin oxidoreductase (PFOR), and pyruvate dehydrogenase (PDH) complex. The NADH generated by PDH needs to be reoxidized for the maintenance of redox balance. The enzyme alcohol/aldehyde dehydrogenase 1 (ADH1) converts acetyl-CoA produced by PFL1 and PFOR reactions to ethanol. Alternatively, ethanol production through direct decarboxylation of pyruvate to CO_2 and acetaldehyde through the activity of pyruvate decarboxylase (PDC3) is also documented. Two molecules of NADH are oxidized during the conversion of acetyl-CoA to ethanol by ADH1, while during PDC3 pathway, only a single NADH molecule is oxidized (Catalanotti et al. 2013).

11.4.2 Bioethanol Production by Photofermentation (The Photanol Route)

The process of converting sunlight into products of fermentation is called "photofermentation" (Photanol), which is a highly efficient metabolic pathway carried out by genetically modified cyanobacteria. This involves a single step wherein bioethanol is generated from CO_2 and water using light as the energy source (Lakatos et al. 2019); however, the key factors of each stage (photosynthesis and fermentation), along with the metabolic needs of the cyanobacteria, determine the efficiency of the overall process. This method would require the use of genetically modified microorganisms (de Farias Silva and Bertucco 2016). Molecular engineering procedures have led to the successful genetic modification of certain cyanobacterial species (examples include *Synechocystis* sp. PCC 6803, *Synechococcus elongatus* sp. PCC 7992, and *Anabaena* sp. PCC 7120 and PCC 7002) for use as fermentative organisms. These modifications introduce specific fermentation cassettes into the cyanobacterial species. *Synechocystis* sp. PCC 6803 is one of the best characterized cyanobacteria and the first photosynthetic organism to have its genome sequenced (Savakis and Hellingwerf 2015). *Thermosynechococcus* also represents a cyanobacterial species that is naturally transformable (Rosgaard et al. 2012). The unicellular cyanobacterium *Synechococcus* is a freshwater organism which is an excellent workhorse, as it can tolerate the insertion of foreign DNA into the chromosome facilitated through homologous recombination at specific active sites, or through transformation and replication using shuttle vectors between *Escherichia coli* and cyanobacteria.

In recent years, photofermentation has gained much interest. Photofermentation is not just an efficient ethanol production process but also one that can be applied to produce numerous naturally occurring products by glycolysis-based fermentation. Bioethanol production by photofermentation can be briefly summarized into two stages: photosynthesis and fermentation. During photosynthesis, inorganic carbon is fixed as phosphoglycerate by Calvin cycle and then converted into pyruvate by the action of pyruvate decarboxylase (PDC) and alcohol dehydrogenase (ADH) enzymes. The final step entails the production of ethanol (Equation 11.1).

$$CO_2 \underset{\text{Calvin cycle}}{\rightarrow} GAP \underset{\text{Intermediate (PEP)}}{\rightarrow} Pyruvate \overset{\text{PDC}}{\rightarrow} Acetaldehyde \overset{\text{ADH}}{\rightarrow} Ethanol \tag{11.1}$$

Where, GAP is phosphoglycerate; PEP, phosphoenolpyruvate; PDC, pyruvate decarboxylase; ADH, alcohol dehydrogenase (de Farias Silva and Bertucco 2016).

The genetic modification is aimed at increasing the activities of PDC and ADH towards maximizing the production of ethanol efficiently, as observed in *Synechococcus* sp. PCC 7942, which then yielded 0.025 mg/L ethanol after six days (Deng and Coleman 1999). Similarly, Chow and coworkers (2015) were able to genetically engineer the production of bacterial cellulose by *S. elongatus* from 28% to 35% (DW), with a concomitant increase in productivity by fourfold (144–564 mg/L/d). After fixation of inorganic carbon by Calvin cycle, the reducing power of photosynthesis is used to form phosphoglycerate (GAP), which is converted into an intermediate compound called phosphoenolpyruvate (PEP), then to pyruvate, and finally to acetaldehyde and ethanol by enzymes pyruvate decarboxylase (PDC) and alcohol dehydrogenase (ADH), respectively. Glycerol was detected (at a concentration of 3 to 4 g/L) as the major by-product in *Synechococcus* hydrolysate, when it was fermented (Sluiter et al. 2008). Acetate and lactate were also detected at low concentrations (< 1 g/L), and the compounds released by lysed cells can be a nutrient source for the yeast during the fermentation process (Möllers et al. 2014). Figure 11.2 illustrates the pathways for ethanol production using cyanobacteria, with biochemical/genetic interventions indicated.

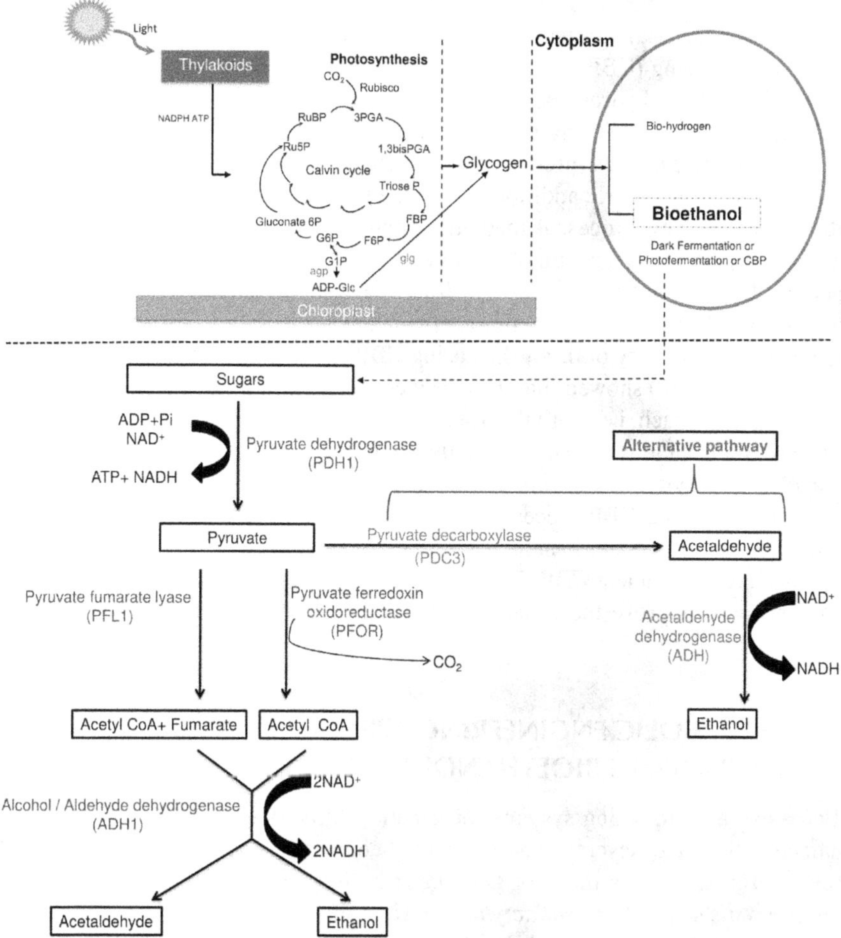

FIGURE 11.2 Pathways of ethanol production using wild and genetically engineered strains of cyanobacteria. NADPH = nicotinamide adenine dinucleotide phosphate; ATP = adenosine triphosphate; CO_2 = carbon dioxide; Rubisco = Ribulose-1,5-bisphosphate carboxylase-oxygenase; 3PGA = 3-phosphoglycerate; 1,3 bisPGA = 1,3-bisphosphoglyceric acid; Triose P = triose phosphate; Ru5P = ribose-5-phosphate; RuBP = ribulose 1,5-bisphosphate; FBP = fructose-bisphosphate; F6P = fructose 6-phosphate; G6P = glucose 6-phosphate; gluconate 6P = gluconate 6-phosphate; glg = glycogen operon; G1P = glucose 1-phosphate; agp = agp gene; ADP Glc = adenosine diphosphate glucose; ADP + Pi = adenosine triphosphate; NAD^+ = nicotinamide adenine dinucleotide; NADH = nicotinamide adenine dinucleotide + hydrogen; Acetyl CoA = acetyl coenzyme A; CBP = consolidated bioprocessing.

11.4.3 Consolidated Bioprocessing (CBP) for Ethanol Production

A higher process efficiency can be achieved at lower operational costs with less contamination and low inhibitory effects by an integration process called consolidated bioprocessing or simultaneous saccharification and fermentation (SSF). This process, however, is still in the infancy stage, and research pertaining to its use for cyanobacterial species is required (Olguin-Maciel et al. 2020). The consolidated bioprocessing (CBP) process does not need multiple steps involving pretreatment or enzymatic hydrolysis, as required for conventional bioethanol production (Figure 11.1). The underlying principle of CBP involves combining the production of saccharolytic enzymes, pretreatment, saccharification, and fermentation steps in a single reactor by orderly addition so that the sugars released are rapidly converted into bioethanol. This process is mediated through the use of a single microorganism and/or a microbial consortium. Aikawa et al. (2013) investigated a single-step production of bioethanol from a carbohydrate-rich cyanobacterium, *Arthrospira platensis*, using a lysozyme and recombinant amylase expressing yeast. They achieved a bioethanol productivity of 0.14 g/L/h using CBP processing of *A. platensis* biomass. Aikawa et al. (2018) showed that the direct conversion of *A. platensis* biomass using a combined approach, i.e., with the use of lysozyme (a recombinant amylase-expressing yeast) and calcium chloride, not only improved ethanol yield but also improved ethanol productivity. Bioethanol yield and productivity of 0.32 g/g and 0.14 g/L/h were achieved using CBP processing of *A. platensis* respectively. However, to date, there is no report of a single natural microorganism that possesses all the characteristics desired for efficient CBP. Therefore, future research should focus on microbial engineering to improve the feasibility of this technique (Jambo et al. 2016; Phwan et al. 2018).

11.5 METABOLIC ENGINEERING APPROACHES FOR IMPROVING BIOETHANOL PRODUCTION

Microalgae are promising systems for genetic engineering, as they possess unique characteristics, and several genomes have been fully sequenced and available at the GeneBank, such as those of *Dunaliella salina*, *Chlorella*, *Synechococcus* sp., *Synechocystis* sp., and *Phaeodactylum* sp. How such systems, with their potential to utilize sunlight and CO_2 to produce ethanol as a biofuel, at yields, compare with other production systems of industrially important microorganisms has received a lot of attention (Luan et al. 2015; Testa et al. 2018; Wang et al. 2020). Genetic engineering offers powerful tools to diversify the utilization of fixed carbon from cellular biomass to products; however, in cyanobacteria, the fundamental understanding underlying the complex physiological responses at the molecular level involved in carbon fluxes is meager. Most of the research has focused towards diverting the pathways related to stored carbon reserves to value addition, i.e., industrially useful compounds. These may involve intensifying the native production or involve genetic engineering by heterologous expression through the introduction of one or more foreign gene(s) or pathways (Ducat et al. 2011; Ducat and Silver 2012; Gao et al. 2012). The details of molecular tools/technologies used in the model organisms are given in Table 11.2.

TABLE 11.2
Molecular Tools and Technologies Used in Engineering Cyanobacteria for Improving Ethanol Production

Cyanobacterium	Tools/Technologies	References
Synechocystis sp. PCC6803	Genome-wide analysis of differentially expressed small RNAs	Hu et al. 2017
	Global acetylome analyses	Mo et al. 2015
	sRNAs deep sequencing	Xu et al. 2014
	Homologous recombination	Savakis and Hellingwerf 2015
	Site-directed mutagenesis	Mo et al. 2017
	Transformation	Deng and Coleman 1999
	Knocking out competing pathways involved in PHB production or storage	van der Woude et al. 2014 Lopes da Silva et al. 2018
	Transformation and metabolomics approach	Zhang et al. 2015
Synechococcus elongatus PCC 7942	Homologous recombination	Savakis and Hellingwerf 2015
	CRISPRi	Huang et al. 2016
	Introduction of heterologous genes	Deng et al. 1999 Dexter and Fu 2009

11.5.1 Engineering Cyanobacteria to Produce High Yields of Ethanol

Several efforts have been made to introduce gene from heterologous hosts, and the transformed cyanobacterium was able to exhibit a mix of three cellular functions, *viz.*, oxygenic photosynthesis, C-fixation, and ethanol production within the single organism (Deng et al. 1999; Dexter and Fu 2009). US biotechnology companies such as Algenol and Joule Unlimited improvised further and filed patents involving manipulation of these systems to further improve yields to 0.552 g/L/d from 0.07 g/L/d using cyanobacterial strains. However, problems related to potential evaporative losses and nonaxenized cultures leading to contamination problems led to setbacks in commercialization. This stimulated research on strain improvement and engineering strains through metabolic interventions based molecular tools. Manipulating the carbon accumulation through transporters involved or carbon flux within the cell or directly the enzymes involved in the Calvin-Benson-Bassham cycle has been envisaged, but with limited success; however, this can also be important for improving cyanobacterial growth and production of metabolites (Liang et al. 2018).

A lot of work on knocking out competing pathways has been undertaken with respect to PHB, as some mutants deficient in their storage can be engineered to produce lactate (Van der Woude et al. 2014; Borirak et al. 2015). In terms of metabolic networks, gene knockouts are important for the maximization of ethanol production, particularly in organisms undergoing photoautotrophic growth, because engineered microorganisms are able to use both photosynthesis and fermentation. It has been shown that malic enzyme and, from TCA cycle, the citrate synthase are the two enzymes whose deletions do not influence biomass yield under photoautotrophic growth (Hong and Lee 2007). Under photoautotrophic conditions, knockout of NAD(P)H dehydrogenase (*ndhF1*) using ammonium as N-source can enhance ethanol production. Another strategy involves employing N-starvation along with the combinatorial deletions of selected reactions of the glycogen synthesis and the poly(3-hydroxybutyrate) (PHB) synthesis pathways, which leads to enhanced ethanol production rate in *Synechocystis* sp. PCC 6803, leading to specific rate of 240 mg/g (dry cell weight)/d, while in a high-cell-density culture (OD_{730} = 50), ethanol production rates of 1.08 and 2.01 g/L/d were recorded under light conditions of 40 and 80 µmol/m^2/s, respectively (Namakoshi et al. 2016); interestingly, deleting only the PHB synthesis pathway was not effective. Another intervention involves diverting the flux away from storage; in this context, increased production of ethanol was recorded with direct insertion of the ethanol cassette into the *pha* genes (da Silva et al. 2018). Analyses of large libraries of mutants in unicellular organisms sometimes can be tedious; Abalde-Cela et al. (2015) developed a novel microfluidics strategy to rapidly screen mutants having higher ethanol production levels in microdroplets.

Among the important reactions targeted in cyanobacteria is the diversion of pyruvate through metabolic engineering into useful industrial products, including ethanol. A key enzyme catalyzing the decarboxylation of pyruvate to acetaldehyde in alcohol fermentation is pyruvate decarboxylase (PDC, EC 4.1.1.1), present in fungi, plants, and yeasts, but absent in humans and rare and found only in very few bacterial species. The PDC *of Zymomonas mobilis* has been extensively explored in ethanol production; however, the spotlight has shifted to a number of other bacterial PDCs, which have pH optimum or lower Km, such as *Gluconacetobacter diazotrophicus*, *Zymobacter* species, etc. The *Zymomonas mobilis* PDC enzyme with an optimum pH of 6.0 is a homotetramer of 240 kDa that has been utilized the most; however, as a majority of cyanobacteria have optimal pH in the range of 7.5–8.0, exploring other PDCs is in progress. Preference is always for organisms with a lower Km, which can increase the flux from pyruvate and catalyze the coupling of acetaldehyde with ADH better.

Pretreatment and separation of carbohydrate components from cyanobacterial biomass is one of the major challenges in bioethanol production. This has led to an urge for the development of approaches for conversion to bioethanols that are simpler, economical, less-energy-intensive, and scalable. One of the useful approaches involve the metabolic engineering of cyanobacteria for the accumulation of simple sugars, such as sucrose, which can be easily taken by microorganisms for fermentation to bioethanol. Some of the genetically modified cyanobacterial strains are reported to accumulate sucrose at even higher levels than that of sugar-rich crops, such as sugarcaneC:\Users\91971\Downloads\15064-4254-FullBook.docx - Ref_348_FILE150644254011. Smachetti et al. (2019) developed a sucrose-accumulating mutant strain of *Anabaena* sp. PCC 7120 *via* genetic modification (cloning native spsB gene in pRL1404 plasmid). The sucrose accumulation in cyanobacterium was further induced *via* salt stress (120 mM NaCl). They reported that a simple drying and milling of sucrose-rich biomass, followed by hot aqueous extraction of carbohydrates, was useful for the further fermentation of carbohydrates to bioethanol (0.51 g/g of sucrose) using *S. cerevisiae*.

As a prelude to the use of seawater for growing cyanobacteria, overexpression studies on the pyruvate decarboxylase (*pdc*) from *Zymomonas mobilis* and the native alcohol dehydrogenase (*adh*A) in the ethanol-producing strain of *Synechocystis* sp. PCC 6803 were investigated under different salt concentrations (Pade et al. 2017). The negative impact on ethanol production could be attributed to intracellular carbon limitation caused by drain of carbon from central metabolism to ethanol, which could be partially recovered by overexpression of glycogen-degrading enzyme (GlgP). However, Carrieri et al. (2010) demonstrated a one-hundred-fold higher ethanol production in marine cyanobacterium *Synechocystis* sp. PCC7002, grown in high-salt medium, illustrating that the compensatory mechanisms for carbon limitation may vary from organism to organism, as also reported by other researchers.

11.5.2 Problems and Prospects for Improving Yields

One of the striking highlights of cyanobacterial research in this context has been microevolution, or genetic changes in a strain growing under high light in laboratory conditions, which can be detected through genome sequencing. This has led to substrains emerging from the original strains and naming from various laboratories all over the world which have interesting variations of biotechnological interest, such as high/low growth rates/doubling times, low transformation rates, buoyancy, etc. The slow growth of these photoautotrophic microbes is attributed to their polyploidy—e.g., the different strains of *Synechocystis* contain 12–52 copies of the chromosome, which is a major issue with cloning interventions. Often, growth and biomass production need to be decoupled to obtain greater yields; in this context, tunable promoters or riboswitches to be introduced along with gene cassettes are being evaluated. A number of promoters-hybrid-promoter or strong synthetic systems besides the native constitutive promoters are available presently, which need to be evaluated. Another important aspect which needs further research is regarding identifying and characterizing the neutral sites or integration sites for the stability and functionality of the insert, particularly for integrative vectors used in cloning (Pinto et al. 2015), besides aspects related to the polyploidy nature hindering segregation and stability of vectors. The need for complex synthetic machinery for vitamins, cofactors, nucleotides, etc., additionally, may not be an economically viable option towards developing them as cell factories (Pembroke et al. 2019).

Studies have also been undertaken to improve stress tolerance in cyanobacteria under outdoor conditions for improved biomass, carbohydrate, and bioethanol productivity (Su et al. 2017; Zhu et al. 2017). Su et al. (2017) showed that the genetic manipulation and overexpression of stress genes (*hsp*A, osmotin) in *Synechococcus elongatus* led to thirty- to forty-fold increase in stress tolerance under outdoor cultivation in seawater. The problem of contaminants in outdoor nonsterilized conditions can significantly reduce ethanol yields. Zhu et al. (2017) were able to completely eliminate an ethanol-consuming contaminant *Pannonibacter phragmitetus* by maintaining alkaline conditions (pH 11.0) in the cultivation system for *Synechocystis* sp. PCC 6803, by adopting a bicarbonate-based integrated carbon capture system. However, ethanol yield decreased; this needs further optimization. Recently, Velmurugan et al.

(2020) revealed that cocultivation of two genetically engineered *Synechocystis* sp. PCC 6803 (engineered for *pdc-adh* overexpression and *glgC-phaA* knockout) in a photobioreactor (20 L bioreactor with 15 L capacity) led to significant increase in bioethanol production. The metabolites released by *glgC-phaA* knockout strain were utilized by *pdc-adh* overexpressing strain to produce bioethanol. Besides biochemical modulation through several approaches, such as N-limitation, etc., there is now a need to supplement such strategies with interventions at the genetic level, particularly by exploring the transcriptome or proteome to decipher the mechanisms underlying ethanol hyperaccumulation. Furthermore, the challenges associated with microbial intolerance to high ethanol yields need to be addressed fully. High ethanol content in medium may affect the proton motive force and increase the permeability of cell membrane in cyanobacteria, which can lead to leakage of metabolites from cells. Therefore, the identification of membrane transport proteins for ethanol tolerance has been of great scientific interest to improve cyanobacterial bioethanol production (Zhang et al. 2015; Vidal 2017). Additionally, ethanol tolerance also needs to be determined if higher yields are recorded in engineered systems (Lopes et al. 2018). Some engineered cyanobacteria, such as strains of *Synechocystis* (UL 004 and UL 030), which has been developed for the direct conversion to bioethanol, are reported to tolerate up to 15 g/L of ethanol concentration (da Silva et al. 2018). The determination and identification of ethanol-tolerant strains is crucial for developing a process with high yields of bioethanol.

11.6 STATUS AND PROSPECTS OF TECHNOECONOMIC AND LIFE CYCLE ASSESSMENT OF BIOETHANOL PRODUCTION

Regardless of the technological advances, commercial feasibility of bioethanol production *via* cyanobacterial feedstock is much dependent upon the economics of the technology. To attract the commercial marketplace, it is crucial that the total cost of the final product is competitive enough to the low cost of fossil fuels. The main modules for efficacious commercial-scale production of bioethanol include the cost of raw material (biomass production), pretreatment, production of enzymes used in fermentation, ethanol productivity, and downstream recovery efficiency (Chandel et al. 2010). Different advances in the engineering of metabolism and the use of innovative approaches, such as CBP, have enhanced the possibility of cost-effective bioethanol production using cyanobacteria. However, one of the foremost shortcomings of cyanobacterial fermentation and bioethanol production is the requirement of additional separation step to recover low concentration of bioethanol and convert it into a drop-in fuel with high purity. A critical aspect is the ethanol recovery efficiency through cyanobacterial production, which cannot be done by distillation as done for yeasts, as the diluted stream would involve extremely high costs. Alternatively, several methods have been advocated, such as vacuum or gas stripping, pervaporation or membrane permeation, or solvent extraction. Adsorption as well as various hybrid processes have been developed (Lopes et al. 2019; da Silva et al. 2020). Life cycle assessment (LCA) of cyanobacterial bioethanol production has shown that the downstream purification of ethanol significantly contributes to energy consumption and carbon footprints. LCA studies have also revealed that the environmental benefits in terms of CO_2 sequestration are generally lower than with other biofuels (Luo et al. 2010; Nilsson et al. 2020). However, these environmental benefits can be improved if the technology is coupled with the use of heat and CO_2 from waste sources (Arora et al. 2020; Quiroz-Arita et al. 2017; Nilsson et al. 2020). Additionally, bioethanol production requires additional pretreatment and hydrolysis step before fermentation, which significantly contributes to the cost of production (Lopes et al. 2019). Therefore, recently, the use of CBP method for the direct production of bioethanol by cyanobacteria is gaining interest, which primarily involves the use of genetically engineered cyanobacteria. Lopes et al. (2019) performed a techno-economic assessment by considering capital (CAPEX) and operating expenditures (OPEX) of direct production of fuel-grade bioethanol using genetically modified *Synechocystis* sp. in a unilayer horizontal tubular photobioreactor with microfiltration unit. They reported that the process was not feasible under studied conditions for bioethanol production, while bioethanol production integrated with the recovery of value-added coproducts zeaxanthin and phycocyanin and CHP (combined heat and power cogeneration) was found to be economically feasible with payback period of ten years. Recently, da Silva et al. (2020) performed a comprehensive technoecomic assessment of process optimization in different scenarios of *Synechocystis*-based biorefinery. They revealed that the biorefinery concept may be useful to certain extent for improving the economic feasibility of cyanobacterial bioethanol production. They revealed that the pretreatment and purification steps are the most crucial steps and sometimes may cost more than the revenue gained, depending upon the method used. The use of membrane filtration for harvesting, ball mill for cell rupturing, and conventional solvent extraction with pervaporation assisted in the reduction of cost as compared to the use of dilation with pervaporation. However, it was observed that the operational cost was higher than the revenue from bioethanol, emphasizing the need for exploration of alternative process development for improving economic feasibility (Deb et al. 2019; Rempel et al. 2019; da Silva et al. 2020; Chou et al. 2021; Tsolcha et al. 2021). Therefore, bioethanol production using cyanobacterial feedstocks still needs improvements and interventions in the process development aspects, in particular, for making the technology economically viable (Chou et al. 2021; Deb et al. 2021). Future developments can focus on improving biomass and ethanol productivity, minimizing the use of chemical fertilizers through cultivation in wastewater, improving wastewater/stress tolerance in strains, minimizing resource input in pretreatment and downstream purification of bioethanol, and searching for strategies for enhanced environmental benefits.

11.7 CONCLUSIONS

Cyanobacterial biomass of wild and genetically modified strains has been proved a promising feedstock for bioethanol production at laboratory and industrial scale. From an industrial standpoint, for efficient ethanol production using cyanobacteria, once the metabolic and genetic engineering challenges are overcome, there is a need to undertake overall life cycle analyses. Capital and processing costs, as well as development of suitable energy-efficient photobioreactors, are among the major parameters to be considered besides biosafety and containment considerations if genetically engineered strains are used. The challenges associated with the additional cost for pretreatment, bioethanol recovery efficiency in the downstream, and high cost of bioethanol recovery procedures need to be addressed. Future studies must be made on the techno-economic aspects and quest for approaches, such as coupling bioethanol production with the coproduction of high- and low-value-added products; using waste resources, such as flue gas and wastewater, for biomass production; identifying the tolerant strains and/or genetic engineering for improving ethanol and/or wastewater tolerance. This will help in improving the cost-effectiveness of the technology and is needed to further prove the potential for industrial-scale implementation.

11.8 ACKNOWLEDGMENTS

Authors Kokila V., L. Nain, and R. Prasanna are thankful to the ICAR-Indian Agricultural Research Institute, New Delhi, India. Authors N. Renuka, V. Bhola, and F. Bux are thankful to the Durban University of Technology and NRF-SARChI (UID: 84166) of South Africa for the research facilities and funding, respectively. Author S.K. Ratha is thankful to the Director, CSIR-National Botanical Research Institute, Lucknow, India, for providing the research facilities. Author N. Renuka is also thankful to the Central University of Punjab, Bathinda, India.

REFERENCES

Abalde-Cela, S., Gould, A., Liu, X., Kazamia, E., Smith, A. G. and C. Abell. 2015. High-throughput detection of ethanol-producing cyanobacteria in a microdroplet platform. *Journal of the Royal Society Interface* 12(106): 216.

Aikawa, S., Inokuma, K., Wakai, S., et al. 2018. Direct and highly productive conversion of cyanobacteria *Arthrospira platensis* to ethanol with $CaCl_2$ addition. *Biotechnology for Biofuels* 11(1): 1–9.

Aikawa, S., Joseph, A., Yamada, R., et al. 2013. Direct conversion of *Spirulina* to ethanol without pretreatment or enzymatic hydrolysis processes. *Energy and Environmental Science* 6: 1844–1849.

Al hattab, M. and G. Abdel. 2015. Microalgae oil extraction pre-treatment methods: Critical review and comparative analysis. *Journal of Fundamentals of Renewable Energy and Applications* 5: 4.

Allen, M. M. 1984. Cyanobacterial cell inclusions. *Annual Review of Microbiology* 38: 1–25.

Arora, P., Chance, R. R., Fishbeck, T., et al. 2020. Lifecycle greenhouse gas emissions for an ethanol production process based on genetically modified cyanobacteria: CO_2 sourcing options. *Biofuels, Bioproducts and Biorefining* 14(6): 1324–1334.

Ashokkumar, V., Chen, W. H., Ngamcharussrivichai, C., Agila, E. and F. N. Ani. 2019. Potential of sustainable bioenergy production from Synechocystis sp. cultivated in wastewater at large scale—a low cost biorefinery approach. *Energy Conversion and Management* 186: 188–199.

Astolfi, A. L., Rempel, A., Cavanhi, V. A. F., Alves, M., Deamici, K. M., Colla, L. M. and J.A.V. Costa. 2020. Simultaneous saccharification and fermentation of *Spirulina* sp. and corn starch for the production of bioethanol and obtaining biopeptides with high antioxidant activity. *Bioresource Technology* 301: 122698.

Ball, S. G. and M. K. Morell. 2003. From bacterial glycogen to starch: Understanding the biogenesis of the plant starch granule. *Annual Review of Plant Biology* 54: 207–233.

Ball, S. G., Colleoni, C., Cenci, U., Raj, J. N. and C. Tirtiaux. 2011. The evolution of glycogen and starch metabolism in eukaryotes gives molecular clues to understand the establishment of plastid endosymbiosis. *Journal of Experimental Botany* 62: 1775–1801.

Beck, C., Knoop, H., Axmann, I. M. and R. Steuer. 2012. The diversity of cyanobacterial metabolism: Genome analysis of multiple phototrophic microorganisms. *BMC Genomics* 13: 56.

Ben-Amotz, A. 1975. Adaptation of the unicellular alga *Dunaliella parva* to a saline environment. *Journal of Phycology* 11: 50–55.

Borirak, O., de Koning, L. J., van der Woude, A. D., et al. 2015. Quantitative proteomics analysis of an ethanol and a lactate-producing mutant strain of *Synechocystis* sp. PCC6803. *Biotechnology for Biofuels* 8: 111.

Carrieri, D., Momot, D., Brasg, I. A., et al. 2010. Boosting auto fermentation rates and product yields with sodium stress cycling: Application to production of renewable fuels by cyanobacteria. *Applied and Environmental Microbiology* 76: 6455–6462.

Carrieri, D., Paddock, T., Maness, P. C., Seibert, M. and J. Yu. 2012. Photo-catalytic conversion of carbon dioxide to organic acids by a recombinant cyanobacterium incapable of glycogen storage. *Energy and Environmental Science* 5(11): 9457–9461.

Catalanotti, C., Yang, W., Posewitz, M. C. and A. R. Grossman. 2013. Fermentation metabolism and its evolution in algae. *Frontiers in Plant Science* 4: 150.

Chandel, A. K., Singh, O. V., Chandrasekhar, G., Rao, L. V. and M. L. Narasu. 2010. Key drivers influencing the commercialization of ethanol-based biorefineries. *Journal of Commercial Biotechnology* 16(3): 239–257.

Chisti, Y. 2013. Constraints to commercialization of algal fuels. *Journal of Biotechnology* 167: 201–214

Chou, H. H., Su, H. Y., Chow, T. J., Lee, T. M., Cheng, W. H., Chang, J. S. and H. J. Chen. 2021. Engineering cyanobacteria with enhanced growth in simulated flue gases for high-yield bioethanol production. *Biochemical Engineering Journal* 165: 107823.

Chow, T., Su, H., Tsai, T., Chou, H., Lee, T. and J. Chang. 2015. Using recombinant cyanobacterium (*Synechococcus elongatus*) with increased carbohydrate productivity as feedstock for bioethanol production via separate hydrolysis and fermentation process. *Bioresource Technology* 184: 133–141.

da Silva, A. F., Brazinha, C., Costa, L. and N. S. Caetano. 2020. Techno-economic assessment of a *Synechocystis* based biorefinery through process optimization. *Energy Reports* 6: 509–514.

da Silva, T. L., Passarinho, P. C., Galriça, R., et al. 2018. Evaluation of the ethanol tolerance for wild and mutant *Synechocystis* strains by flow cytometry. *Biotechnology Reports* 17: 137–147.

de Farias Silva, C. E. and A. Bertucco. 2016. Bioethanol from microalgae and cyanobacteria: A review and technological outlook. *Process Biochemistry* 51(11): 1833–1842.

Dave, N., Varadavenkatesan, T., Selvaraj, R. and R. Vinayagam. 2021. Modelling of fermentative bioethanol production from indigenous *Ulva prolifera* biomass by Saccharomyces cerevisiae NFCCI1248 using an integrated ANN-GA approach. *Science of The Total Environment* 791: 148429.

Deb, D., Mallick, N. and P. B. S. Bhadori. 2019. Analytical studies on carbohydrates of two cyanobacterial species for enhanced bioethanol production along with poly-β-hydroxybutyrate, C-phycocyanin, sodium copper chlorophyllin, and exopolysaccharides as co-products. *Journal of Cleaner Production* 221: 695–709.

Deb, D., Mallick, N. and P. B. S. Bhadoria. 2021. Engineering culture medium for enhanced carbohydrate accumulation in *Anabaena variabilis* to stimulate production of bioethanol and other high-value co-products under cyanobacterial refinery approach. *Renewable Energy* 163: 1786–1801.

Deng, M-D. and J. R. Coleman. 1999. Ethanol synthesis by genetic engineering in cyanobacteria. *Applied and Environmental Microbiology* 65: 523–528.

Devi, A., Niazi, A., Ramteke, M. and S. Upadhyayula. 2021. Techno-economic analysis of ethanol production from lignocellulosic biomass—a comparison of fermentation, thermo catalytic, and chemocatalytic technologies. *Bioprocess and Biosystems Engineering* 44(6): 1093–1107.

Dexter, J., Armshaw, P., Sheahan, C. and J. Pembroke. 2015. The state of autotrophic ethanol production in cyanobacteria. *Journal of Applied Microbiology* 119: 11–24.

Dexter, J. and P. Fu. 2009. Metabolic engineering of cyanobacteria for ethanol production. *Energy and Environmental Science* 2(8): 857–864.

Domozych, D. S. 2011. *Algal cell walls.* eLS Chichester: John Wiley and Sons.

Ducat, D. and P. Silver. 2012. Improving carbon fixation pathways. *Current Opinion in Chemical Biology* 16: 337–344. doi: 10.1016/j.cbpa.2012.05.002

Ducat, D. C., Way, J. C. and P. A. Silver. 2011. Engineering cyanobacteria to generate high-value products. *Trends in Biotechnology* 29: 95–103.

Gao, Z., Zhao, H., Li, Z., Tan, X. and X. Lu. 2012. Photosynthetic production of ethanol from carbon dioxide in genetically engineered cyanobacteria. *Energy and Environmental Science* 5: 9857–9865.

González-Fernández, C. and M. Ballesteros. 2012. Linking microalgae and cyanobacteria culture conditions and key-enzymes for carbohydrate accumulation. *Biotechnology Advances* 30(6): 1655–1661.

Harun, R. and M. K. Danquah. 2011. Influence of acid pre-treatment on microalgal biomass for bioethanol production. *Process Biochemistry* 46: 304–309.

Harun, R., Danquah, M. K. and G. M. Forde. 2010. Microalgal biomass as a fermentation feedstock for bioethanol production. *Journal of Chemical Technology and Biotechnology* 85: 199–203.

Harun, R., Jason, W., Cherrington, T. and M. K. Danquah. 2011. Exploring alkaline pre-treatment of microalgal biomass for bioethanol production. *Applied Energy* 88: 3464–3467.

Harun, R., Yip, J. W., Thiruvenkadam, S., Ghani, W. A., Cherrington, T. and M. K. Danquah. 2014. Algal biomass conversion to bioethanol—a step-by-step assessment. *Biotechnology Journal* 9(1): 73–86.

Hernandez, D., Riano, B., Coca, M. and M. Garcia-Gonzalez. 2015. Saccharification of carbohydrates in microalgal biomass by physical, chemical and enzymatic pre-treatments as a previous step for bioethanol production. *Chemical Engineering Journal* 262: 939–945.

Heyer, H. and W. E. Krumbein. 1991. Excretion of fermentation products in dark and anaerobically incubated cyanobacteria. *Archives of Microbiology* 155: 284–287.

Hemschemeier, A. and T. Happe. 2005. The exceptional photo-fermentative hydrogen metabolism of the green alga *Chlamydomonas reinhardtii*. *Biochemical Society Transcations* 33: 39–41.

Hirano, A., Ueda, R., Hirayama, S. and Y. Ogushi. 1997. CO_2 fixation and ethanol production with microalgal photosynthesis and intracellular anaerobic fermentation. *Energy* 22: 137–142.

Hoiczyk, E. and A. Hansel. 2000. Cyanobacterial cell walls: News from an unusual prokaryotic envelope. *Journal of Bacteriology* 182: 1191–1199.

Holmatov, B., Schyns, J. F., Krol, M. S., Gerbens-Leenes, P. W. and A. Y. Hoekstra. 2021. Can crop residues provide fuel for future transport? Limited global residue bioethanol potentials and large associated land, water and carbon footprints. *Renewable and Sustainable Energy Reviews* 149: 111417.

Hong, S. J. and C. G. Lee. 2007. Evaluation of central metabolism based on a genomic database of *Synechocystis* PCC6803. *Biotechnology and Bioprocess Engineering* 12(2): 165–173.

Hu, J., Li, T., Xu, W., Zhan, J., Chen, H., He, C., et al. 2017. Small antisense RNA RblR positively regulates RuBisCo in *Synechocystis* sp. PCC 6803. *Frontiers in Microbiology* 8: 231. https://doi.org/10.3389/fmicb.2017.00231

Huang, C. H., Shen, C. R., Li, H., Sung, L. Y., Wu, M. Y. and Y. C. Hu. 2016. CRISPR interference (CRISPRi) for gene regulation and succinate production in cyanobacterium *S. elongatus* PCC 7942. *Microbial Cell Factories* 15: 196.

Huang, Y., Chen, X., Liu, S., Lu, J., Shen, Y., Li, L., Peng, L., Hong, J., Zhang, Q. and I. Ostrovsky. 2021. Converting of nuisance cyanobacterial biomass to feedstock for bioethanol production by regulation of intracellular carbon flow: Killing two birds with one stone. *Renewable and Sustainable Energy Reviews* 149: 111364.

Ismail, M. M., Ismail, G. A. and M. M. El-Sheekh. 2020. Potential assessment of some micro- and macroalgal species for bioethanol and biodiesel production. *Energy Sources, Part A: Recovery, Utilization, and Environmental Effects*: 1–17. https://doi.org/10.1080/15567036.2020.1758853.

Jambo, S. A., Abdulla, R., Mohd Azhar, S. H., Marbawi, H., Gansau, J. A. and P. Ravindra. 2016. A review on third generation bioethanol feedstock. *Renewable and Sustainable Energy Reviews* 65: 756–769.

John, R. P., Anisha, G., Nampoothiri, K. M. and A. Pandey. 2011. Micro and macroalgal biomass: A renewable source for bioethanol. *Bioresource Technology* 102: 186–193.13.

Kamarainen, J., Knoop, H., Stanford, N. J., et al. 2012. Physiological tolerance and stoichiometric potential of Cyanobacteria for hydrocarbon fuel production. *Journal of Biotechnology* 162: 67–74.

Khan, M. I., Lee, M. G., Shin, J. H. and J. D. Kim. 2017. Pretreatment optimization of the biomass of Microcystis aeruginosa for efficient bioethanol production. *Amb Express* 7(1): 1–9.

Lakatos, G. E., Ranglová, K., Manoel, J. C., Grivalský, T., Kopecký, J. and J. Masojídek. 2019. Bioethanol production from microalgae polysaccharides. *Folia Microbiologica* 64(5): 627–644.

Lasry Testa, R., Delpino, C., Estrada, V. and S. M. Diaz. 2019. In silico strategies to couple production of bioethanol with growth in cyanobacteria. *Biotechnology and Bioengineering* 116(8): 2061–2073.

Liang, F., Englund, E., Lindberg, P. and P. Lindblad. 2018. Engineered cyanobacteria with enhanced growth show increased ethanol production and higher biofuel to biomass ratio. *Metabolic Engineering* 46: 51–59.

Lopes da Silva, T., Passarinho, P. C., Galriça, R., et al. 2018. Evaluation of the ethanol tolerance for wild and mutant *Synechocystis* strains by flow cytometry. *Biotechnology Reports* 17: 137–147.

Lopes, T. F., Cabanas, C., Silva, A., et al. 2019. Process simulation and techno-economic assessment for direct production of advanced bioethanol using a genetically modified *Synechocystis* sp. *Bioresource Technology Reports* 6: 113–122.

Luan, G., Qi, Y., Wang, M., Li, Z., Duan, Y., Tan, X. and X. Lu. 2015. Combinatory strategy for characterizing and understanding the ethanol synthesis pathway in cyanobacteria cell factories. *Biotechnology for Biofuels* 8(1): 1–12.

Luo, D., Hu, Z., Choi, D. G., Thomas, V. M., Realff, M. J. and R. R. Chance. 2010. Life cycle energy and greenhouse gas emissions for an ethanol production process based on blue-green algae. *Environmental Science and Technology* 44(22): 8670–8677.

Luque, I. and K. Forchhammer. 2008. Nitrogen assimilation and C/N balance sensing. In *The Cyanobacteria: Molecular Biology, Genomics and Evolution*. Ed. Herrero, A. and E. Flores, 335–382. Norfolk: Caister Academic Press.

Maia, J. L. D., Cardoso, J. S., Mastrantonio, D., et al. 2020. Microalgae starch: A promising raw material for the bioethanol production. *International Journal of Biological Macromolecules* 165: 2739–2749.

Mamo, G., Faryar, R. and E. N. Karlsson. 2013. Microbial glycoside hydrolases for biomass utilization in biofuels applications. In *Biofuel Technologies*. Ed. Gupta V. K. and M. G. Tuohy, 171–188. Heidelberg: Springer.

María, A. D., Edwin, O. S., Patrick, O. U., et al. 2021. A review on cyanobacteria cultivation for carbohydrate-based biofuels: Cultivation aspects, polysaccharides accumulation strategies, and biofuels production scenarios. *Science of The Total Environment* 148636. https://doi.org/10.1016/j.scitotenv.2021.148636.

Markou, G., Angelidaki, I., Nerantzis, E. and D. Georgakakis. 2013. Bioethanol production by carbohydrate-enriched biomass of *Arthrospira* (*Spirulina*) *platensis*. *Energies* 6: 3937–3950.

Miranda, M., Vega-Gálvez, A., Martinez, E., López, J., Rodríguez, M. J., Henríquez, K., Fuentes, F. 2012. Genetic diversity and comparison of physicochemical and nutritional characteristics of six quinoa (*Chenopodium quinoa* willd.) genotypes cultivated in Chile. *Ciencia y tecnología de alimentos* 32(4): 835–843.

Mo, H., Xie, X., Zhu, T. and Lu, X. 2017. Effects of global transcription factor NtcA on photosynthetic production of ethylene in recombinant *Synechocystis* sp. PCC 6803. *Biotechnology for Biofuels* 10: 145.

Möllers, K. B., Cannella, D., Jørgensen, H. and N. U. Frigaard. 2014. Cyanobacterial biomass as carbohydrate and nutrient feedstock for bioethanol production by yeast fermentation. *Biotechnology for Biofuels* 7: 1–11.

Namakoshi, K., Nakajima, T., Yoshikawa, K., Toya, Y. and H. Shimizu. 2016. Combinatorial deletions of glgC and phaCE enhance ethanol production in *Synechocystis* sp. PCC 6803. *Journal of Biotechnology* 239: 13–19.

Nilsson, A., Shabestary, K., Brandão, M. and E. P. Hudson. 2020. Environmental impacts and limitations of third-generation biobutanol: Life cycle assessment of n-butanol produced by genetically engineered cyanobacteria. *Journal of Industrial Ecology* 24(1): 205–216.

Nowicka, A., Zielinski, M. and M. Debowski. 2020. Microwave support of the alcoholic fermentation process of cyanobacteria *Arthrospira platensis*. *Environmental Science and Pollution Research* 27: 118–124.

Ohta, S., Miyamoto, K. and Y. Miura 1987. Hydrogen evolution as a consumption mode of reducing equivalents in green algal fermentation. *Plant Physiology* 83: 1022–1026.

Olguin-Maciel, E., Singh, A., Chable-Villacis, R., Tapia-Tussell, R. and H. A. Ruiz. 2020. Consolidated bioprocessing, an innovative strategy towards sustainability for biofuels production from crop residues: An overview. *Agronomy* 10(11): 1834.

Pade, N., Mikkat, S. and Hageman, M. 2017. Ethanol, glycogen and glucosylglycerol represent competing carbon pools in ethanol-producing cells of *Synechocystis* sp. PCC 6803 under high-salt conditions. *Microbiology (Reading)* 63: 300–307.

Paulo, C., Di Maggio, J., Estrada, V. and M. S. Diaz. 2011. Optimizing cyanobacteria metabolic network for ethanol production. *Computer Aided Chemical Engineering* 29: 1366–1370.

Pembroke, J. T., Armshaw, P. and M. P. Ryan. 2019. Metabolic engineering of the model photoautotrophic cyanobacterium *synechocystis* for ethanol production: Optimization strategies and challenges. In *Fuel Ethanol Production from Sugarcane*. Ed. Basso, T. P. and L. C. Basso, 199–219. London, UK: Intech Open.

Phwan, C. K., Ong, H. C., Chen, W-H., Ling, T. C., Ng, E. P. and P. L. Show. 2018. Overview: Comparison of pretreatment technologies and fermentation processes of bioethanol from microalgae. *Energy Conversion and Management* 173: 81–94.

Pinto, F., Pacheco, C. C., Oliveira, P., et al. 2015. Improving a *Synechocystis*-based photoautotrophic chassis through systematic genome mapping and validation of neutral sites. *DNA Research* 22(6): 425–437.

Quiroz-Arita, C., Sheehan, J. J. and T. H. Bradley. 2017. Life cycle net energy and greenhouse gas emissions of photosynthetic cyanobacterial biorefineries: Challenges for industrial production of biofuels. *Algal Research* 26: 445–452.

Rawat, J., Gupta, P. K., Pandit, S., et al. 2021. Current perspectives on integrated approaches to enhance lipid accumulation in microalgae. *3 Biotech* 11: 303.

Rempel, A., de Souza Sossella, F., Margarites, A. C., et al. 2019. Bioethanol from *Spirulina platensis* biomass and the use of residuals to produce biomethane: An energy efficient approach. *Bioresource Technology* 288: 121588.

Rempel, A., Machado, T., Treichel, H., Colla, E., Margarites, A. C. and L. M. Colla. 2018. Saccharification of *Spirulina platensis* biomass using free and immobilized amylolytic enzymes. *Bioresource Technology* 263: 163–171.

Rosgaard, L., A. J. de Porcellinis adn, J. H. Jacobsen, N. U. Frigaard, Y. Sakuragi. 2012. Bioengineering of carbon fixation, biofuels, and biochemcials in cyanobacteria and plants. *Journal of Biotechnology* 162: 134–147.

Sanderson, M., Adler, P., Boateng, A., Casler, M. D. and G. Sarath. 2006. Switchgrass as a biofuels feedstock in the USA. *Canadian Journal of Plant Science* 86 (Special Issue): 1315–1325.

Sauer, J., Schreiber, U., Schmid, R., Völker, U. and K. Forchhammer. 2001. Nitrogen starvation-induced chlorosis in *Synechococcus* PCC 7942. Low-level photosynthesis as a mechanism of long-term survival. *Plant Physiology* 126: 233–243.

Savakis, P. and K. J. Hellingwerf. 2015. Engineering cyanobacteria for direct biofuel production from CO_2. *Current Opinion in Biotechnology* 33: 8–14.

Shokrkar, H., Ebrahimi, S. and M. Zamani. 2017. Bioethanol production from acidic and enzymatic hydrolysates of mixed microalgae culture. *Fuel* 200: 380–386.

Sluiter A. D., Hames, B. R., Ruiz, R. O., et al. 2008. Determination of structural carbohydrates and lignin in biomass. National Renewable Energy Laboratory Technical Report NREL/TP-510–42618. www.nrel.gov/docs/gen/fy11/42618.pdf.

Smachetti, M. E. S., Cenci, M. P., Salerno, G. L. and L. Curatti. 2019. Ethanol and protein production from minimally processed biomass of a genetically-modified cyanobacterium over-accumulating sucrose. *Bioresource Technology Reports* 5: 230–237.

Su, H. Y., Chou, H. H., Chow, T. J., et al. 2017. Improvement of outdoor culture efficiency of cyanobacteria by over-expression of stress tolerance genes and its implication as bio-refinery feedstock. *Bioresource Technology* 244: 1294–1303.

Tan, X. M., Yao, L., Gao, Q. Q., Wang, W. H., Qi, F. X. and X. F. Lu. 2011. Photosynthesis driven conversion of carbon dioxide to fatty alcohols and hydrocarbons in cyanobacteria. *Metabolic Engineering* 13: 169–176.

Testa, F., O. Boiral and F. Iraldo, F. 2018. Internalization of environmental practices and institutional complexity: Can stakeholders pressures encourage greenwashing? *Journal of Business Ethics* 147: 287–307. https://doi.org/10.1007/s10551-015-2960-2

Tsolcha, O. N., Patrinou, V., Economou, C. N., Dourou, M., Aggelis, G. and A. G. Tekerlekopoulou. 2021. Utilization of biomass derived from cyanobacteria-based agro-industrial wastewater treatment and raisin residue extract for bioethanol production. *Water* 13(4): 486.

Ueno, Y., Kurano, N. and S. Miyachi. 1998. Ethanol production by dark fermentation in the marine green alga, *Chlorococcum littorale*. *Journal of Fermentation and Bioengineering* 86: 38–43.

van der Woude, A. D, Angermayr, S. A., Veetil, V. P., Osnato, A. and K. J. Hellingwerf. 2014. Carbon sink removal: Increased photosynthetic production of lactic acid by *Synechocystis* sp PCC6803 in a glycogen storage mutant. *Journal of Biotechnology* 184: 100–102.

Velmurugan, R. and A. Incharoensakdi. 2020. Co-cultivation of two engineered strains of *Synechocystis* sp. PCC 6803 results in improved bioethanol production. *Renewable Energy* 146: 1124–1133.

Vidal, R. 2017. Alcohol dehydrogenase AdhA plays a role in ethanol tolerance in model cyanobacterium *Synechocystis* sp. PCC 6803. *Applied Microbiology and Biotechnology* 101(8): 3473–3482.

Wang, M., Luan, G. and X. Lu. 2020. Engineering ethanol production in a marine cyanobacterium *Synechococcus* sp. PCC7002 through simultaneously removing glycogen synthesis genes and introducing ethanologenic cassettes. *Journal of Biotechnology* 317: 1–4.

Werlang, E. B., Julich, J., Muller, M. V., de Farias Neves, F., Sierra-Ibarra, E., Martinez, A. and R. D. C. D. S. Schneider. 2020. Bioethanol from hydrolyzed *Spirulina* (*Arthrospira platensis*) biomass using ethanologenic bacteria. *Bioresources and Bioprocessing* 7(1): 1–9.

Xu, Y., Tiago Guerra, L., Li, Z., Ludwig, M., Charles Dismukes, G., and Bryant, D.A., 2012. Altered carbohydrate metabolism in glycogen synthase mutants of Synechococcus sp. strain PCC 7002: Cell factories for soluble sugars. *Metabolic Engineering* 16 C, 56–67.

Xu, W., Chen, H., He, C. L. and Q. Wang. 2014. Deep sequencing-based identification of small regulatory RNAs in *Synechocystis* sp PCC 6803. *Plos One* 9: e92711.

Yao, L., Qi, F., Tan, X. and X. Lu. 2014. Improved production of fatty alcohols in cyanobacteria by metabolic engineering. *Biotechnology for Biofuels* 7: 94.

Zhang, Y., Niu, X., Shi, M., Pei, G., Zhang, X., Chen, L. and W. Zhang. 2015. Identification of a transporter Slr0982 involved in ethanol tolerance in cyanobacterium *Synechocystis* sp. PCC 6803. *Frontiers in Microbiology* 6: 487.

Zhu, Z., Luan, G., Tan, X., Zhang, H. and X. Lu. 2017. Rescuing ethanol photosynthetic production of cyanobacteria in non-sterilized outdoor cultivations with a bicarbonate-based pH-rising strategy. *Biotechnology for Biofuels* 10(1): 1–11.

12 Recent Trends in Application of Feather Keratin Hydrolysate Produced through Bioconversion of Poultry Feather Waste

Pintubala Kshetri[a], Subhra Saikat Roy[a], Thangjam Surchandra Singh[a], Pangambam Langamba[a], K. Tamreihao[a], Susheel Kumar Sharma[b], and Ayekpam Bimolini Devi[c]

[a]ICAR Research Complex for NEH Region, Manipur Centre, Imphal, India

[b]ICAR-Indian Agricultural Research Institute, New Delhi, India

[c]Mizoram University, Aizawl, Mizoram 796004, India

CONTENTS

DOI: 10.1201/9781003191247-12

12.1 INTRODUCTION

To meet the global demand for meat, the poultry industry has become one of the fastest-growing and more profitable agricultural sectors all over the world. Consequently, the generation of feather waste as a by-product of poultry processing plants is also increasing at an alarming rate (Zaghloul et al. 2011). Hence, feather waste is posing a serious challenge for solid waste management in a country. Moreover, feathers are very recalcitrant to degradation in nature, taking longer time to decompose as compared to other biodegradable waste. Over 90% of the weight of feathers is constituted by a pure protein known as keratin. The recalcitrant nature of feathers is due to the presence of keratin, a cysteine-rich, highly cross-linked fibrous and insoluble protein (Gupta and Ramnani 2006). The most abundant amino acids present in keratin are cysteine, glycine, alanine, serine, and valine, while lysine, methionine, and tryptophan are present in lesser proportions. Other than feather, keratin is also an integral part of the skin, scales, hair, wool, hooves, and nails of animals (Wang et al. 2016). Structurally, keratin can be grouped into two: hard keratins (β-keratin) and soft keratins (α-keratin) (Gupta and Ramnani 2006). The keratin present in feathers is β-keratin, so named because they are rich in stacked β-pleated sheets. The β-pleated sheets of feather keratin are stabilized by cross-linked disulfide bridges (Qiu et al. 2020). Hence, feathers are not easily degraded due to their recalcitrant nature by commonly reported proteases (Meyers et al. 2008). Many efforts have been made to degrade feathers and utilize it in various fields.

Conventionally, feathers are degraded by various chemical or physical methods. These methods involve the use of harmful chemicals and large energy expenses which compromise the environment (Park and Son 2009). Nevertheless, in the year 1990, Shih and his coworkers have shown that feathers could be degraded by microbial action (Williams et al. 1990). They have isolated and characterized a nonpathogenic feather-degrading bacterium, *Bacillus licheniformis* PWD1. Since then, researchers have focused on isolation, identification, and characterization of feather-degrading microbes and their mechanism of action. From their collective research findings, it was shown that the mechanism of keratin degradation is a complex process involving a synergistic action, sulfitolysis/reduction (to cleave the disulfide bond) and proteolysis (to cleave the peptide bond). The enzyme responsible for the degradation of keratin is known as keratinase (Sharma and Devi 2018). So far, many keratinolytic microbes belonging to various genera and keratinase enzymes with different catalytic activity have been reported. Microbial or enzymatic hydrolysis of feather keratin is an eco-friendly approach for the valorization of feather waste into value-added products (Haq et al. 2020). Microbial or enzymatic hydrolyzed feather has a wide range of applicability in many industrial sectors, which include agricultural, leather, pharmaceutical, cosmeceutical, etc. Further exploration of microbial or enzymatic hydrolyzed feathers for applications in other important sectors is continuing among many researchers.

Keeping this in view, the chapter focuses on the importance of valorization of feather waste, the role of keratinolytic bacteria and their keratinase enzymes in bioconversion of feather waste, as well as potential applications of feather hydrolysate (FH).

12.2 IMPORTANCE OF VALORIZATION OF POULTRY FEATHER WASTE BIOMASS

Protein is an important part of the diet for both humans and animals, and daily protein intake in appropriate doses plays a key role in the normal growth and functioning of cells (Carbone and Pasiakos 2019). The United Nations (UN) projected that the world's population will increase from its current 8.0 billion to 9.8 billion in 2050 (UN World Population 2017 revision). The key challenges that are going to be faced by the world's population in the coming decades will be energy, food, water crisis, environmental and climatic changes, and so on. Among these challenges, the production of adequate, protein-rich food to feed this anticipated rise in population is also noteworthy. The main source of dietary protein is plants and animals. While rearing animals also requires protein-rich feeds, there is always a competition between animals and humans in terms of food consumption (Schader et al. 2015). This dictates the search for an alternative source for feedstuffs. One such promising area is the valorization of waste (plant/animal origin) into food ingredients. In this context, valorization of keratinous waste, especially feather waste, has become an emerging area in the past three decades. Since the poultry industry is one of the largest industries in production of protein-rich meat and poultry products have become an important source of dietary protein consumed globally, valorization of feather waste into animal feed, fertilizer, and other valuable products has been initiated by many researchers across the globe. Hydrothermal or chemical treatments are commonly used conventional methods for preparing feather meal. These methods are known to be associated with many drawbacks, such as requirement of large energy inputs, deterioration in the quality of feather meal, etc. (Onifade et al. 1998). Microbial/enzymatic degradation of poultry feather wastes is a preferable method, and several reports have revealed the biotechnological potential of this approach (Kshetri and Ningthoujam 2016). Increasing demand for animal products for human consumption will lead to further demand for animal feed for poultry production (Thornton 2010). The incorporation of feather meal into feed ingredients is expected to help in alleviating the competition between humans and animals for the dietary protein source. Moreover, feathers can also be converted into other valuable products for application in other industrial sectors. Hence, valorization of feather waste has become an indispensable part for generating a circular economy targeting zero waste.

12.3 MICROBIAL DEGRADATION OF FEATHER

Even though feather waste is hard to degrade, they are biodegradable in nature (Onifade et al. 1998). Biodegradation of feathers in nature is performed by microbial actions by secreting keratinolytic enzymes—the keratinases (Gupta and Ramnani 2006). However, the time taken for natural microbial degradation of feather waste is comparatively longer as compared to degradation of other proteinaceous waste. In this section, a brief review of the keratinolytic bacteria and their keratinases is highlighted.

12.3.1 Keratinolytic Bacteria

A plethora of feather-degrading microorganisms has been reported since Williams et al. (1990) first reported feather-degrading bacteria *Bacillus licheniformis* PWD1. It was a serendipitous discovery; when their group was working on anaerobic digestion of organic waste into biogas (methane), a hen accidentally fell into the digester. When they opened the digester, to their surprise, there was no trace of the hen, including feather, and from that digester *Bacillus licheniformis* PWD1 was isolated (Shih 2012). Before their discovery, keratinolytic microorganisms were notorious as a dermatophyte (mostly pathogenic fungi) (Gradisar et al. 2005). For the past three decades, a diverse group of keratinolytic microorganisms has been isolated and identified from different environments. Most of the reported feather-degrading bacteria were isolated from feather waste dumping sites, poultry farms, or directly from feathers, since these habitats have more chance to harbor keratinolytic bacteria. However, some bacteria have been reported from unusual habitats. For instance, Bach et al. (2011) have identified twenty feather-degrading bacteria isolated from the Brazilian Amazon forest and Atlantic forest; most of the isolates belong to genus *Bacillus*, followed by *Serratia*, *Aeromonas*, and *Chryseobacterium*. Similarly, Saarela et al. (2017) isolated 122 keratinolytic bacteria from a bird's nest. Partial sequencing of the 16S rRNA gene had revealed that the keratinolytic isolates belong to genera *Pseudomonas* and *Stenotrophomonas*, *Bacillus*, *Exiguobacterium*, *Paenibacillus*, *Rummeliibacillus*, and *Sporosarcina*. Daroit et al. (2009) reported a keratinolytic bacterium, *Bacillus* sp. P45, from fish intestine. Most of the keratinolytic bacteria reported in literature belong to the genus *Bacillus*, and a majority of them were found to be neutrophilic or mesophilic bacteria. However, some thermophilic and alkaliphilic keratinolytic bacteria belonging to genera *Fervidobacterium* (Friedrich and Antranikian 1996; Nam et al. 2002), *Thermoanaerobacter* (Riessen and Antranikian 2001), *Bacillus* (Gessesse et al. 2003), and *Nesterenkonia* (Bakhtiar et al. 2005) have been reported. In India, the research team working at ICAR Research Complex for NEH Region, Manipur Centre, Imphal, has also isolated more than forty feather-degrading bacterial strains from different habitats of Northeast Indian Himalayan Region (data not published). Of these, seven strains belong to the genus *Streptomyces*, and the other two strains to *Chryseobacterium* and *Bacillus* genera, respectively (Kshetri et al. 2018, 2019a, 2020). Some of the keratinolytic bacteria reported since the past three decades are summarized in Table 12.1.

TABLE 12.1
Overview of Keratinolytic Bacteria Isolated from Different Habitats for the Past Three Decades

Year	Habitat	Reference
1990–2000		
Bacillus licheniformis PWD1	Poultry waste digester	Williams et al. 1990;
Burkholderia	Feather waste	Riffel and Brandelli 2006

TABLE 12.1 *(Continued)*
Overview of Keratinolytic Bacteria Isolated from Different Habitats for the Past Three Decades

Year	Habitat	Reference
Chryseobacterium sp. Kr6, Kr9	Feather waste	Riffel et al. 2003; Riffel and Brandelli 2006
Fervidobacterium pennavorans	Hot spring	Friedrich and Antranikian 1996,
Microbacterium sp.	Feather waste	Riffel and Brandelli 2006
Pseudomonas	Feather waste	Riffel and Brandelli 2006
Streptomyces graminofaciens	Feather waste	Szabo et al. 2000
Streptomyces thermoviolaceus strain SD8	Lake	Chitte et al. 1999
Vibrio sp. strain kr2	Poultry farm	Sangali and Brandelli 2000
2001–2010		
Amycolatopsis keratiniphila	Marsh soil	Al-Musallam et al. 2003
Bacillus cereus	Keratinous waste	Laba and Rodziewicz 2010
Bacillus cereus 1268	Feather waste	Mazotto et al. 2010
Bacillus cereus KB043	Decomposing feather	Nagal and Jain 2010
Bacillus cereus LAU 08	Poultry farm	Lateef et al. 2010
Bacillus licheniformis	Partially degraded chicken feather	Rozs et al. 2001
Bacillus licheniformis 1269	Feather waste	Mazotto et al. 2010
Bacillus licheniformis RG1	-	Ramani and Gupta 2004
Bacillus licheniformis RPk	River sediments	Fakhfakh et al. 2009
Bacillus megaterium F7–1	Poultry waste	Park and Son 2009
Bacillus megaterium SN1	Feather waste dumping site	Agrahari and Wadhwa 2010
Bacillus polymyxa	Soil	Laba and Rodziewicz 2010
Bacillus pseudofirmus	Alkaline soda lake	Gessesse et al. 2003
Bacillus pumilus A1	Wastewater effluent released from a slaughterhouse	Fakhfakh et al. 2010
Bacillus pumilus FH9	Feather waste dumping site	El-Refai et al. 2005
Bacillus pumilus SN3	Feather waste dumping site	Agrahari and Wadhwa 2010
Bacillus sp. FK 46	Soil	Suntornsuk and Suntornsuk 2003
Bacillus sp. P45	Intestine of fish (*Piaractus mesopotamicus*)	Daroit et al. 2009
Bacillus sp. JB99	Sugarcane molasses	Kainoor and Naik 2010
Bacillus subtilis	Forest soil	Jeong et al. 2010a
Bacillus subtilis AMR	Feather waste	Mazotto et al. 2010
Bacillus subtilis S14	Soil	Macedo et al. 2005
Bacillus thuringiensis SN2	Feather waste dumping site	Agrahari and Wadhwa 2010
Fervidobacterium islandicum AW-1	Geothermal hot stream	Nam et al. 2002
Kocuria rosea	Soil	Bernal et al. 2006
Nesterenkonia sp.	Alkaline soda lake	Gessesse et al. 2003
Pseudomonas aeruginosa KS-1	-	Sharma and Gupta 2010
Stenotrophomonas maltophilia	Poultry plant	Cao et al. 2009
Stenotrophomonas maltophilia	Rhizospheric soil	Jeong et al. 2010b
Streptomyces fradiae Var S-221	-	Cheng et al. 2010
Streptomyces sp. MS-2	-	Mabrouk 2008
Streptomyces sp. strain AB1	Soil	Jaouadi et al. 2010

(Continued)

TABLE 12.1 ***(Continued)***
Overview of Keratinolytic Bacteria Isolated from Different Habitats for the Past Three Decades

Year	Habitat	Reference
Thermoanaerobacter keratinophilus	Geothermal hot spring	Riessen and Antranikian 2001
Xanthomonas sp. P5	Rhizospheric soil	Jeong et al. 2010c
2011–2021		
Azotobacter chroococcum B4	-	Mamangkey et al. 2020
Bacillus altitudinis GVC11	Slaughterhouse waste dump yard	Kumar et al. 2011
Bacillus cereus	Poultry farm waste	Lakshmi et al. 2013
Bacillus cereus	Mushroom farm	Ahmadpour et al. 2016
Bacillus haynessi	Poultry farms	Balakrishnan and Padmanabhan 2019
Bacillus licheniformis	Soil from oil refinery	Alahyaribeik et al. 2020
Bacillus pumilus	Soil from oil refinery	Alahyaribeik et al. 2020
Bacillus sp. 8A6	-	Huang et al. 2020
Bacillus sp. CSK2	Agro-waste dump site	Nnolim et al. 2020
Bacillus subtilis 1271	Feather waste	Mazotto et al. 2011
Bacillus subtilis	Poultry farm waste	Lakshmi et al. 2013
Bacillus subtilis	Feather waste dumping site	Sekar et al. 2016
Chryseobacterium sp. RBT	Feather waste dumping site	Gurav and Jadhav 2013b
Geobacillus stearothermophilus	Soil from oil refinery	Alahyaribeik et al. 2020
Kocuria rhizophila p3	Live poultry birds	Laba et al. 2018
Meiothermus taiwanensis WR-220	Hot spring	Wu et al. 2017
Raoultella electrica	Feather waste dumping site	Prasanna and Moorthy 2020
Rhodococcus erythropolis	Soil from oil refinery	Alahyaribeik et al. 2020
Stenotrophomonas maltophilia BBE11–1	-	Fang et al. 2013
Streptomyces sampsonii GS 1322	-	Jain et al. 2016
Streptomyces sp.	Feather dumping site	Li et al. 2020
Streptomyces sp. IF 5	Poultry farm	Ramakrishnan et al. 2011

12.3.2 Microbial Keratinases

The ability of feather degradation by certain microbes is due to the secretion of a special class of protease known as keratinases. Bacterial keratinases (EC3.4.21/24/99.11) belong to serine or metalloproteases (Gupta and Ramnani 2006). The remarkable characteristics of keratinases are their high affinity towards hard-to-degrade proteins and broad substrate specificities. Conventional proteases, such as trypsin, pepsin, and chymotrypsin, cannot degrade keratin. However, bacterial keratinases can degrade both keratin and other common proteinaceous substrates, such as casein, BSA, gelatin, etc. The defining criteria for a protease to be assigned as keratinase is the K:C ratio (keratin: casein), i.e., a keratinase should have a K:C ratio above 0.5 (Gupta et al. 2013). The applications of keratinases in food, cosmetic, detergent, leather, and pharmaceutical industries are well documented in various literature. In addition, keratinase also finds application in emerging sectors, such as decontamination of prion proteins, production of bioplastics, biomaterials, bioenergy, etc. Thus, keratinases are suitably known as "modern proteases" (Cao et al. 2009). There are several reports on the purification and identification of keratinases from a diverse bacterial species. However, all the commercially available keratinases were obtained from native *Bacillus licheniformis* strain, and some of them are also produced by recombinant technology. The keratinase from *B. licheniformis* is marketed in different trade names with a varied range of applications in feed, food, pharmaceutical, cosmeceutical, and waste treatment industries. Table 12.2 shows that till now, more than ten keratinase-based products have been commercialized successfully in the market.

The promising application of bacterial keratinases in various industrial sectors has motivated researchers to isolate and identify microbial strains that produce novel keratinases having efficient keratinolytic activity under extreme environmental conditions (high pH, temperature, salt concentration, etc.). Many thermostable or alkaline keratinases have been reported in the literature. For instance, Wu et al. (2017) had isolated heat-stable keratinase from thermophilic *Meiothermus taiwanensis* WR-220, and they further produced the keratinase as recombinant keratinase in *E. coli*. The recombinant keratinase showed optimal activity at 65°C and pH 10. Heat- and alkali-stable keratinases have considerable advantages in industrial use where high pH, heating, and drying are involved during processing. Recently, an alkaline, salt-stable keratinase (KER-102) has also been reported from *Bacillus* sp.

TABLE 12.2
Some of Commercially Available Keratinase-Based Products (Sharma and Devi 2018)

Trade Name	Source	Manufacturer
Keratopeel® PB	*Bacillus licheniformis*	Proteos biotech
BioGuard Plus	Proprietary blend of microorganisms, including keratinase producer	RuShay Inc.
Cibenza DP100	*Bacillus licheniformis* PWD-1	Novus International
FEED-0001	Native *Bacillus licheniformis*	Creative Enzymes, USA
Keratoclean® HYDRA PB	*Bacillus licheniformis*	Proteos biotech
Keratoclean® PB	*Bacillus licheniformis*	Proteos biotech
Keratoclean® Sensitive PB	*Bacillus licheniformis*	Proteos biotech
NATE-0853	Recombinant	Creative Enzymes, USA
Prionzyme ™	*Bacillus licheniformis*	Genencor International and Health Protection Agency, UK
PURE100	Recombinant (original source: *Bacillus licheniformis* PWD-1)	Proteos biotech
Valkerase®	*Bacillus licheniformis*	Bioresource International Inc (BRI), North Carolina
Versazyme®	*Bacillus licheniformis*	Bioresource International Inc (BRI), North Carolina

RCM-SSR-102 which showed optimal activity at pH 10 and could tolerate up to 10% NaCl concentration (Kshetri et al. 2020).

12.4 APPLICATIONS OF FH

Bioconversion of feather waste into keratin hydrolysate can be performed by either using whole microbial cells or by enzymatic hydrolysis using keratinase enzymes. There are pros and cons of these two methods. For example, several studies have revealed that the whole-cell microbe has a higher keratin-degrading efficiency than the purified keratinase (Li 2019). However, during fermentation, the microbe utilizes the nutrients and essential amino acids, decreasing the nutritional value of feather hydrolysate for being used as an animal feed ingredient, so enzymatic hydrolysis is more preferable over microbial degradation (Tiwary and Gupta 2012). On the contrary, some researchers have reported that incorporation of microbial biomass in feather meal improves the nutritional value of feather meal (Onifade et al. 1998). Considering these contradictory observations, further in-depth research on keratin hydrolysis by both microbial and enzymatic methods is highly demanded. Nevertheless, feather hydrolysate produced either by microbial or enzymatic hydrolysis has many biotechnological applications. Previously, FH was commonly used as a feather meal for animal feed supplement and slow-releasing nitrogen (N) fertilizers. However, the use of feather hydrolysate is not confined to feed and fertilizer industries; rather, its application is expanded in new industrial sectors, as shown in Figure 12.1. The recent trends in the application of feather hydrolysates in various industrial sectors are discussed in the following section.

12.4.1 Feed Industries

The feed industry is one of the largest and most important agro-industrial sectors where hydrolyzed feather waste is being utilized. The potential of feather meal for use as animal feed supplement has been reported by Binkley and Vasak as early as the year 1950. Later, many researchers also analyzed the nutritional quality of feather

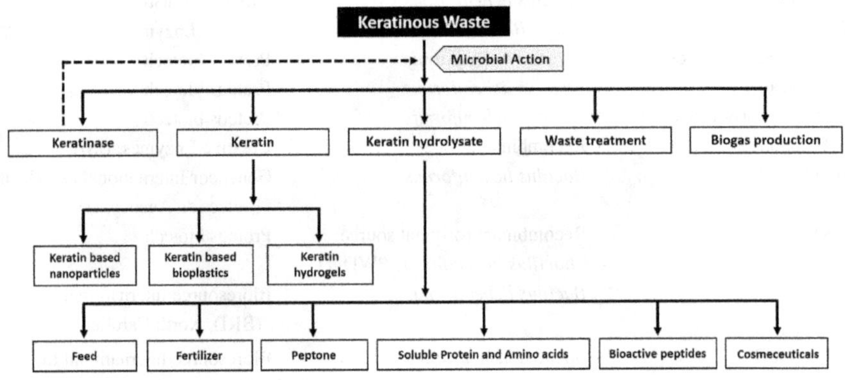

FIGURE 12.1 Potential application of microbial hydrolyzed feather.

meal and studied the effect of feather meal on the growth performance of nonruminant animals. However, the use of feather meal as animal feed supplement had been questioned during the fifties and sixties due to its poor digestibility and deficiencies in essential amino acids, such as methionine, lysine, histidine, and tryptophan. This was because of the poor processing method used at that time. Conventionally, feather meal was prepared from poultry feathers by partially grinding them under elevated heat and pressure. These methods not only destroyed some essential amino acids but also produced nonnutritive amino acids, such as lysinoalanine and lanthionine. However, a significant amount of research on the improvement of feather meal was carried out by several workers. These include the incorporation of keratinase enzymes in feed and microbial/enzyme hydrolysis of feather waste (Onifade et al. 1998). The current methods of feather meal processing, viz., enzymatic or microbial hydrolysis, increased nutritional quality and digestibility. Since protein is the most limiting ingredient in preparation of animal feed, feather meal is considered as an economically viable alternative protein source for livestock feed. The current global feather meal market size (US$ 359.5 million) is projected to increase twofold (US$ 820 million) by 2025 with a CAGR of around 8.6%. Other than animal feed, feather meal has also become a better alternative protein source for aquaculture feeds. Bishop et al. (1995) had observed that the growth performance of *Oreochromis niloticus* fry was not compromised when 66% of fish meal and meat meal present in the total diet was replaced with feather meal. Similarly, Campos et al. (2017) also showed that inclusion of up to 12.5% HF (76% fishmeal replacement) had not affected the feed intake, growth, immune response, or EPA and DHA levels in the muscle of European seabass fish (*Dicentrarchus labrax*).

12.4.2 Fertilizer Industries

Feather hydrolysate is one of the best slow-releasing organic nitrogenous fertilizer which provides a constant nitrogen supply to plants. It contains around 10–12% nitrogen, which is considered as the highest nitrogen content among the commonly used organic fertilizers. FH can be applied as both solid and liquid form. Application of FH not only supplies nutrients to the plant but also attracts other rhizospheric soil microorganisms, which lead to increase in soil fertility. FH prepared either by microbial or hydrothermal hydrolysis can be used as N fertilizer. Choi and Nelson (1996) compared the performance of microbial-hydrolyzed and steam-hydrolyzed feather as N fertilizer. They observed that steam hydrolysis resulted in the quick release of N, whereas microbial hydrolysis increased the capacity of slow release of N. The slow N release is an attractive property due to a lesser chance of N leaching as compared to fast-releasing inorganic nitrogenous fertilizer, like urea. The FH produced by keratinolytic bacteria has been shown to be a good organic fertilizer for a wide variety of crops, such as rice (Tamreihao et al. 2017), mung bean (Bhange et al. 2016), banana (Gurav and Jadhav. 2013a), lettuce (Sobucki et al. 2019), chickpea (Paul et al. 2013), wheat (Jain et al. 2016; Genc and Atici 2019), etc. Moreover, during microbial fermentation of feathers, some other metabolites are also produced, which directly or indirectly enhance plant growth. For instance, the production of amino acid tryptophan from feather hydrolysis stimulates the production of plant growth hormone Indole-3-acetic acid (IAA) (Bhange et al. 2016). Microbial-hydrolyzed feather has also shown antifungal activity against some phytopathogens (Gousterova et al. 2011). Hence, the application of multifaceted bacteria having feather-degrading, plant-growth-promoting, or biocontrol activity as a biofertilizer is a promising approach to organic agriculture. Many of such multifaceted bacterial strains have been reported recently (Jeong et al. 2010a, 2010b, 2010c; Tamreihao et al. 2017.) Seven *Streptomyces* spp. exhibiting keratinolytic, plant-growth-promoting (phosphate solubilization; IAA, ammonia, and siderophore production), and biocontrol activity have also been reported by Kshetri et al. (2018). They reported the enhancement of in vitro germination of garden pea seeds by feather hydrolysate produced by *Streptomyces* sp. RCM-SSR-6.

12.4.3 Cosmetic Industries

Keratin hydrolysate (KH) has been used as an ingredient in all types of cosmetic products of skin and hair, such as conditioners, shampoo and serum, lotions, nail polish, mascaras, and eye makeup agents. The skincare property of keratin hydrolysate includes formation of a protective film, strong moisturizing capacity, and increased elasticity of the skin. The small peptides present in KH can penetrate the skin and hair follicles (Mokrejš et al. 2017). Moreover, these small peptides have a high buffering capability, which, in turn, stabilizes the pH of cosmetics. KH from hair, wool, and feather have been produced by alkali, acid, and chemical hydrolysis using reducing agents. Microbial and enzymatic hydrolysis of keratin into KH has also been reported. For instance, Villa et al. (2013) reported the production of feather keratin hydrolysate employing *Bacillus subtilis* AMR, where they observed that application of hydrolysate in hair follicles increased the hydration, brightness, and softness of hair fibers. The antioxidant property of KH has been demonstrated by many workers (Callegaro et al. 2018; Fakhfakh et al. 2013). In addition to antioxidant activity, the antityrosinase activity of feather KH prepared with enzymatic hydrolysis has also been reported by Kshetri et al. (2020). The antioxidant and antityrosinase properties of feather keratin hydrolysate will further help in the value addition of KH-based cosmetic products. Mokrejš et al. (2017) also demonstrated that the application of KH prepared with a combination of enzyme and alkali hydrolysis decreases transepidermal water loss from skin.

12.4.4 Pharmaceutical Sector

Hydrolysis of feathers either by microbial action or enzymatic process releases bioactive peptides. Bioactive peptides are short-chain amino acids (MW$\leq$ 20 kDa) that have health-beneficial biological activity (Kshetri et al. 2019b). Peptides produced from feather hydrolysis have shown various biological activities, such as antioxidant, angiotensin-I-converting enzyme inhibitory, antityrosinase, abd dipeptidyl peptidase-IV-inhibitory activities (Callegaro et al. 2018). Wan et al. (2016) have identified an antioxidant peptide, SALCRPCG, from feather hydrolysate prepared by fermentation using *Bacillus subtilis* S1-4. Similarly, Fontoura et al. (2019) also reported an antioxidant peptide, LPGPILSSFPQ, obtained through submerged cultivation of feathers with *Chryseobacterium* sp. kr6. Antioxidant compounds have wide applicability in the pharmaceutical sector. Hence, feathers could be a potential source for antioxidant peptides for use in pharmaceutical industries. Moreover, Paul et al. (2015) isolated and characterized a peptide (Molecular Weight =4.66 kDa) from the fermentation of feather with *Paenibacillus woosongensis*; the peptide exhibited antimicrobial activity against multiple-antibiotic-resistant (MAR) *Staphylococcus aureus*. This showed that a diversity of bioactive peptides having different amino acid sequences can be generated from hydrolysis of feathers by using different strains of keratinolytic bacteria and enzymes. These bioactive peptides can be directly or indirectly (using the information of amino acid sequence) exploited for being used in the formulation of new/novel peptide-based drugs.

12.4.5 Bioenergy Sector

The application of feather waste into the production of biogas is an emerging and very important area since it has not only recycled the feather waste into energy but also will help in the mitigation of environmental pollution. This sector has the capacity to recycle a huge amount of feather waste generated from the poultry industry. Since feathers are rich in protein content, they represent an excellent raw material for biogas production. Bioconversion of feather waste into biogas requires two steps; at first, the feather is hydrolyzed by microbial or enzymatic action, and then the partially hydrolyzed feather is again anaerobically digested to produce biogas (Forgács et al. 2011). Bálint et al. (2005) demonstrated the two-step fermentation system to produce biogas from feather waste. First, in the fermentation system, the feather waste is aerobically hydrolyzed using keratinolytic bacterial strain *Bacillus licheniformis* KK1. FH was further fermented anaerobically using an anaerobic, hyperthermophilic archaebacteria, *Thermococcus litoralis*. The anaerobic digestion of chicken feather waste pretreated with a recombinant keratinolytic *Bacillus megaterium* was able to produce 0.35 Nm3/kg dry feathers of methane gas (Forgács et al. 2011). Similarly, enzymatic pretreatment of feathers resulted in the production of 0.37 ± 0.16 Nm3/kg V of methane (Me´zes and Tama´s 2015).

12.4.6 Keratin-Based Biomaterial

Keratin extracted from wool, hair, and feathers is also used in the preparation of biomaterials, such as biocomposites, biopolymers, and bioplastics. Keratin can also be cast as sponges, films, and hydrogels. As these biomaterials are biocompatible, they can be used in various biomedical applications. Moreover, because of their biodegradability and high mechanical strength, keratin-based bioplastics have a promising potential for being used as biodegradable packaging material. Keratin present in feathers is small (about 10 KDa) and uniform in size, having self-assembling properties (Sharma and Gupta 2016). Keratin extracted from chicken feathers is utilized in the preparation of fabricated keratin films and used as an agent in controlled drug delivery systems (Poole et al. 2009; Yin et al. 2013). Extraction of keratin from feather by microbial or enzymatic hydrolysis is not recommended for the preparation of keratin-based biomaterials as the enzymatic or microbial hydrolysis cannot extract the intact keratin, and this approach results in the degradation of some parts of the keratin protein (Shavandi et al. 2017). However, microbial or enzymatic methods for keratin extraction have advantages, especially from an environmental perspective, since it does not use harmful chemicals and consumes less energy. Implementation of these two methods for extraction of keratin in combination with other methods will reduce the use of harmful chemicals. For example, Eslahi et al. (2013) extracted keratin from wool and feather employing enzymatic (savinase) hydrolysis, followed by treatment with a reducing agent (hydrogen sulfite), and the molecular weight of keratin was reported in the range of 11–28 kD. The finding was in agreement with the reported size of the keratin extracted from feathers with other nonenzymatic methods (Yin et al. 2013; Alahyaribeik and Ullah 2020). Hence, further studies on optimization and improvement of these methods for efficient extraction of keratin are highly demanded.

12.4.7 Other Uses

KH has been used in filling-cum-retaining operation in leather processing (Karthikeyan et al. 2007). The use of KH during the tanning process enhances chromium uptake by the leather, which in turn minimizes chromium contamination. Another emerging area for the application of keratin is the preparation of keratin nanoparticles (Perotto et al. 2019). Xu et al. (2014) reported the preparation of highly water-soluble feather keratin nanoparticles of 70 nm that can be a potential drug delivery vehicle in treatment of livestock diseases. In another study, Mousavia et al. (2018) synthesized feather keratin nanoparticles (about 42 nm) having the potential to remove the Cu from contaminated water. Preliminary investigation by Jiang and his coworkers (2008) demonstrated that feather keratin hydrolysate (FKH) can be used as an effective and cheap ingredient in the formulation of wood-adhesive resins. The investigators observed that 30% of phenol content in the resin can be replaced by FKH. FKH can also be employed as a source of peptone for use in microbiological and cell culture media preparation (Taskin et al. 2016; Orak et al. 2018). Peptone is one of the most commonly used substrates for microbial growth and metabolite production. Replacement of expensive peptones with low-cost feather peptone can help in the production of cost-effective microbial metabolites.

12.5 CHALLENGES IN THE VALORIZATION OF FEATHER WASTE AT INDUSTRIAL SCALE

For the past few decades, research on the valorization of feather waste has made significant advancement, and their applications are also expanded in different industrial sectors. However, some challenges hinder the efficient utilization of these protein-rich wastes. Some of the major challenges are discussed in the following passages:

- The most important challenge is to get the best-quality feather by-products. As chickens are exposed to many drugs (antibiotics, fungicides, sex hormones), pathogenic microbes, and other potentially harmful chemicals at the time of rearing in poultry farms, there is a high chance of contamination of feathers with these harmful chemicals. Exposing feathers to these harmful chemicals leads to poor quality of the end product. The sterilization or sanitization of the feather waste in bulk is a big challenge for commercial production. If the information on the source of feathers and rearing practice in the poultry farms can be tracked, the potentially harmful effect may be reduced substantially. Hence, farmers' awareness regarding good agricultural practice is highly encouraged.

- Another challenge is the scaling-up of the biodegradation process in fermenters for industrial-scale production. Most of the feather degradation studies reported in literature are in laboratory-scale shake flask conditions. There are meager reports on the feather-degradation process studied in laboratory-scale fermenters or pilot fermenters (Zaghloul et al. 2011). The degradation process may be significantly different due to differences in approach techniques. Hence, optimization studies on microbial degradation of feather waste in laboratory-scale fermenters instead of shake flask will help in scaling-up of feather hydrolysate production in large-scale industrial fermenters.
- Another constraint which limits the bioconversion of feather waste into feather hydrolysate is the less substrate tolerance of the feather-degrading bacteria. Most of the reported feather-degrading bacteria can degrade low feather substrate concentration ($\leq 2\%$). High substrate tolerance is preferable for higher-end product recovery and ease in downstream processing. Therefore, identification of microbial strains having high substrate tolerance either by exploring efficient degraders from various habitats, improving degradation efficiency of microbial strains using recombinant DNA, and/or use of microbial consortia for effective feather degradation rather than single bacteria are the need of the hour.

12.6 FUTURE PERSPECTIVES

There will be a huge accumulation of feather waste in the future due to an increase in the demand, production, and consumption of poultry products, adversely affecting the environment and health. Moreover, in countries where landfill space is limited, poultry-processing industries spend huge money for the disposal of feather waste. Exploration of feather as an inexpensive raw material for the production of a value-added product with varied applications in many industrial sectors will help in the mitigation of environmental issues and alleviate the economy of resource-poor farmers. Even though feather keratin–based biomaterials have a potential for biomedical application, in most of the reported literature, the extraction of keratin involved the use of many harmful chemicals. Efforts have been made by many researchers to develop eco-friendly methods for the extraction of keratin. One such approach is microbial or enzymatic hydrolysis, but still, a scope exists for further investigation on this process so that degradation efficiency and yield can be improved up to the desired level. In addition, more investigations on feather degradation in pilot-scale fermenters are highly needed to ease the problems faced during large-scale industrial production. In coming decades, feather waste is expected to be considered no more as a waste but as a treasure trove. Hence, further in-depth research on improvement in microbial KH production, characterization of new keratinases, and other novel applications of feather waste is an emerging area for researchers in the coming future.

12.7 CONCLUSION

From thirty years of research on microbial hydrolysis of feathers, it has been established that feathers are a potential source of many industrially useful products. Moreover, it is demonstrated that microbial or enzymatic hydrolysis of feathers is an eco-friendly alternative approach to chemical-based and physical methods which utilize either harmful chemicals or high energy. Feather keratin–based products have a wide range of applications, including agricultural, pharmaceutical, cosmeceutical, leather, and many other industrial sectors. Poultry feather waste, which once was notorious as an environmental pollutant, has now become a valuable raw material for a wide array of applications. In-depth research on the exploration of feather waste application in new areas will further augment the value of feather. In brief, valorization of feather waste has become one of the integral parts of "zero waste circular economy" models, contributing towards achieving sustainable development goals.

12.8 ACKNOWLEDGMENTS

Support from Department of Biotechnology, MoST, Govt. of India under the DBT-BioCARe Scheme (BT/PR30381/BIC/101/1069/2018) is gratefully acknowledged.

REFERENCES

Agrahari, S., Wadhwa, N. 2010. Degradation of a chicken feather, a poultry waste product by keratinolytic bacteria isolated from dumping site at Ghazipur poultry processing plant. *Int J Poult Sci*, *9*: 482–489.

Ahmadpour, F., Yakhchali, B., Musavi, M.S. 2016. Isolation and identification of a keratinolytic Bacillus cereus and optimization of keratinase production. *JABR*, 3: 507–512.

Alahyaribeik, S., Sharifi, S.D., Tabandeh, F., Honarbakhsh, S., Ghazanfari, S. 2020. Bioconversion of chicken feather wastes by keratinolytic bacteria. *Process Saf Environ*, 135: 171–178.

Alahyaribeik, S., Ullah, A. 2020. Methods of keratin extraction from poultry feathers and their effects on antioxidant activity of extracted keratin. *Int J Biol Macromol*, 148: 449–456.

Al-Musallam, A., A., Al-Zarban, S.S., Fasasi, A., Kroppenstedt, R.M., Stackebrandt, E. 2003 *Amycolatopsis keratiniphilia* sp.nov., a novel keratinolytic soil actinomycete from Kuwait. *Int J Syst Evol Micr*, 53: 871–874.

Bach, E., Cannavan, F. S., Duarte, F. R. S., Taffarel, J. A. S., Tsai, S. M., Brandelli, A. 2011. Characterization of feather-degrading bacteria from Brazilian soils. *Int Biodeter Biodegr*, 65: 102–107.

Bakhtiar, S., Estiveira, R.J., Hatti-Kaul, R. 2005. Substrate specificity of alkaline protease from alkaliphilic feather-degrading *Nesterenkonia sp*. AL 20, *Enzyme Microb Tech*, 37: 534–540.

Balakrishnan, D., Padmanabhan, N.S. 2019. Identification and phylogenetic analysis of keratinase producing bacteria snp1 from poultry field. *IJBB*, 15: 39–51.

Bálint, B., Bagi, Z., Tóth, A., Rákhely, G., Perei, K., Kovács, K.L. 2005. Utilization of keratin-containing biowaste to produce biohydrogen. *Appl Microbiol Biotechnol*, 69:404–410.

Bernal, C., Cairó, J., Coello, N. 2006. Purification and characterization of a novel exocellular keratinase from Kocuria rosea. *Enzyme Microb Tech*, 38: 49–54.

Bhange, K., Chaturvedi, V., Bhatt, R. 2016. Ameliorating effects of chicken feathers in plant growth promotion activity by a keratinolytic strain of *Bacillus subtilis* PF1. *Bioresour Bioprocess*, 3: 13.

Binkley, C.H., Vasak, O.R. 1950. Production of a friable meal from feathers. AIC-274, Bureau Agricultural and Ind. Chem. Agr. Res. Administration, U.S.D.A.

Bishop, C.D., Angus, R.A., Watts, S.A. 1995. The use of feather meal as a replacement for fish meal in the diet of *Oreochromis niloticus* fry. *Bioresource Technol*, 54: 291–295.

Callegaro, K., Welter, N., Daroit, D.J., 2018. Feathers as bioresource: Microbial conversion into bioactive protein hydrolysates. *Process Biochem*, 75: 1–9.

Campos, I., Matos, E., Marques, A., Valente, L.M.P. 2017. Hydrolyzed feather meal as a partial fishmeal replacement in diets for European seabass (*Dicentrarchus labrax*) juveniles. *Aquaculture*, 476: 152–159.

Cao, Z.J., Zhang, Q., Wei, D.K., Chen, J.W., Zhang, X.Q., Zhou, M.H. 2009. Characterization of a novel *Stenotrophomonas* isolate with high keratinase activity and purification of the enzyme. *J Ind Microbiol Biot*, 36: 181–188.

Carbone, J.W., Pasiakos, S.M. 2019. Dietary protein and muscle mass: Translating science to application and health benefit. *Nutrients*, *11*: 1136.

Cheng, X., Huang, L., Tu, E.R., Li, K.T. 2010. Medium optimization for the feather-degradation by *Streptomyces fradiae* var S-221 using the response surface methodology. *Biodegradation*, 21: 117–122.

Chitte, R.R., Nalawade, V.K., Dey, S. 1999. Keratinolytic activity from the broth of a feather-degrading thermophilic *Streptomyces thermoviolaceus* strain SD8. *Lett Appl Microbiol*, 28: 131–136.

Choi, J.M., Nelson, P.V. 1996. Developing a slow-release nitrogen fertilizer from organic sources: II. Using poultry feathers. *J Amer Soc Hort Sci*. 121: 634–638.

Daroit, D.J., Corrêa, A.P.F., Brandelli, A. 2009. Keratinolytic potential of a novel Bacillus sp. P45 isolated from the Amazon basin fish *Piaractus mesopotamicus*. *Int Biodeter Biodeg*, 63: 358–363.

El-Refai, H.A., AbdelNaby, M.A., Gaballa, A., El-Araby, M.H., Abdel Fattah, A.F. 2005. Improvement of the newly isolated *Bacillus pumilus* FH9 keratinolytic activity. *Process Biochem*, 40: 2325–2332.

Eslahi, N., Dadashian, F. Nejad, N.H. 2013. An investigation on keratin extraction from wool and feather waste by enzymatic hydrolysis. *Prep Biochem Biotech*, 43(7): 624–648.

Fakhfakh, N., Hmidet, N., Haddar, A., Kanoun, S., Nasari, M. 2010. A novel serine metallokeratinase from a newly isolated *Bacillus pumilus* A1 grown on chicken feather meal:Biochemical and molecular characterization. *Appl Biochem Biotech*, 162: 329–344.

Fakhfakh, N., Kanoun, S., Manni, L., Nasri, M. 2009. Production and biochemical and molecular characterization of a keratinolytic serine protease from chicken feather-degrading *Bacillus licheniformis* RPk. *Can J Microbiol*, 55: 427–436.

Fakhfakh, N., Ktari, N., Siala, R., Nasri, M., 2013. Wool-waste valorization: Production of protein hydrolysate with high antioxidative potential by fermentation with a new keratinolytic bacterium, *Bacillus pumilus* A1. *J Appl Microbiol*. 115: 424–433.

Fang., Zhang, J., Liu, B., Du, G., Chen, J. 2013. Biodegradation of wool waste and keratinase production in scale-up fermenter with different strategies by *Stenotrophomonas maltophilia* BBE11–1. *Bioresource Technol*, 140: 286–291.

Fontoura, R., Daroit, D.J., Corrêa, A.P.F., Moresco, K.S., Santi, L., Beys-da-Silva, W.O., Yates, J.R. 3rd, Moreira, J.C.F., Brandelli, A. 2019. Characterization of a novel antioxidant peptide from feather keratin hydrolysates. *New Biotechnol*, 49: 71–76.

Forgács, G., Alinezhad, S., Mirabdollah, A., Feuk-Lagerstedt, E., Horváth, I.S. 2011. Biological treatment of chicken feather waste for improved biogas production. *J Envirn Sci*, 23: 1747–1753.

Friedrich, A.B., Antranikian, G. 1996. Keratin degradation by *Fervidobacterium pennavorans*, a novel thermophilic anaerobic species of the order thermotogales. *Appl Environ Microb*, 62: 2875–2882.

Genc, E., Atici, O. 2019. Chicken feather protein hydrolysate as a biostimulant improves the growth of wheat seedlings by affecting biochemical and physiological parameters. *Turk J Bot*, 43: 67–79.

Gessesse, A., Hatti-Kaul R., Gashe, B.A., Mattiasson, B. 2003. Novel alkaline proteases from alkaliphilic bacteria grown on chicken feather. *Enzyme Microb Tech*, 32: 519–524.

Gousterova, A., Nustorova, M., Paskaleva, D., Naydenov, M., Neshev, G., Vasileva-Tonkova, E. 2011. Assessment of feather hydrolysate from thermophilic Actinomycetes for soil amendment and biological control application. *Int J Environ Res*, 5: 1065–1070.

Gradisar, H., Friedrich, J., Krizaj, I., Jerala, R. 2005. Similarities and specificities of fungal keratinolytic proteases: Comparison of keratinases of Paecilomyces marquandii and Doratomyces tipiteses to some known proteases. *Appl Environ Microb*, 71: 3420–3426.

Gupta, R., Ramnani, P. 2006. Microbial keratinases and their prospective applications: An overview. *Appl Microbiol Biot*, 70: 21–33.

Gupta, R., Sharma, R., Beg, Q.K. 2013. Revisiting microbial keratinases: Next generation proteases for sustainable biotechnology. *Crit Rev Biotechnol*, 33: 216–228.

Gurav, R.G., Jadhav, J.P. 2013a. A novel source of biofertilizer from feather biomass for banana cultivation. *Environ Sci Pollut Res Int*, 20: 4532–4539.

Gurav, R.G., Jadhav, J.P. 2013b. Biodegradation of keratinous waste by *Chryseobacterium* sp. RBT isolated from soil contaminated with poultry waste. *J Basic Microbiol*, 53: 128–135.

Haq, I.U., Akram, F., Jabbar, Z. 2020. Keratinolytic enzyme-mediated biodegradation of recalcitrant poultry feathers waste by newly isolated *Bacillus* sp. NKSP-7 under submerged fermentation. *Folia Microbiol,* 65: 823–834.

Huang, Y., Łężyk, M., Herbst, F.A., Busk, P.K., Lange, L. 2020. Novel keratinolytic enzymes, discovered from a talented and efficient bacterial keratin degrader. *Sci Rep*, 10: 10033.

Jain, R., Jain, A., Rawat, N., Nair, M., Gumashta, R. 2016. Feather hydrolysate from *Streptomyces sampsonii* GS 1322: A potential low cost soil amendment. *J Biosci Bioeng,* 121: 672–677.

Jaouadi, B., Abdelmale, k B., Fodil, D., Ferradji, F.Z., Rekik, H., Zaraî, N., Bejar, S. 2010. Purification and characterization of a thermostable keratinolytic serine alkaline proteinase from *Streptomyces* sp. Strain AB1 with high stability in organic solvents. *Bioresour Technol*, 101:8361–8369.

Jeong, J.H., Jeon, Y.D., Lee, O.M., Kim, J.D., Lee, N.R., Park, G.T., Son, H.J. 2010a Characterization of a multifunctional feather-degrading *Bacillus subtilis* isolated from forest soil. *Biodegradation*, 21: 1029–1040.

Jeong, J.H., Lee, O.M., Jeon, Y.D., Kim, J.D., Lee, N.R., Lee, C.Y., Son, H.J. 2010b Production of keratinolytic enzyme by a newly isolated feather-degrading *Stenotrophomona smaltophilia* that produces plant growth-promoting activity. *Process Biochem*, 45: 1738–1745.

Jeong, J.H., Park, K.H., Oh, D.J., Hwang, D.Y., Kim, H.S., Lee, C.Y., Son, H.J. 2010c. Keratinolytic enzyme-mediated biodegradation of recalcitrant feather by a newly isolated *Xanthomonas* sp. P5. *Polym Degrad Stab*, 95: 1969–1977.

Jiang, Z., Qin, D., Hse, C.Y, Kuo, M., Luo, Z., Wang, G., Yan, Y. 2008. Preliminary study on chicken feather protein-based wood adhesives. *J Wood Chem Technol*, 28: 240–246.

Kainoor, P.S., Naik, G.R. 2010. Production and characterization of feather degrading keratinase from *Bacillus* sp. JB 99, *Indian J Biotechnol*, 9: 384–390.

Karthikeyan, R., Balaji, S., Sehgal, P.K. 2007. Industrial applications of keratins – A review. *Journal of Scientific & Industrial Research*, 66: 710–715.

Kshetri, P. and Ningthoujam, D.S. 2016. Keratinolytic activities of alkaliphilic *Bacillus sp.* MBRL 575 from a novel habitat, limestone deposit site in Manipur, India. *Springerplus*, 5: 1–16.

Kshetri, P., Roy, S.S., Chanu, S.B., Singh, T.S., Tamreihao, K., Sharma, S.K., Ansari, M.A., Prakash, N. 2020. Valorization of chicken feather waste into bioactive keratin hydrolysate by a newly purified keratinase from *Bacillus* sp. RCM-SSR-102. *J Environ Manage*, 273: 111195.

Kshetri, P., Roy, S.S., Sharma, S.K., Singh, T.S., Ansari, M.A., Prakash, N., Ngachan, S.V. 2019a. Transforming chicken feather waste into feather protein hydrolysate using a newly isolated multifaceted keratinolytic bacterium *Chryseobacterium sediminis* RCM-SSR-7. *Waste Biomass Valori*, 10: 1–11.

Kshetri, P., Roy, S.S., Sharma, S.K., Singh, T.S., Ansari, M.A., Sailo, B., Singh, S., Prakash, N. 2018. Feather degrading, phytostimulating, and biocontrol potential of native actinobacteria from North Eastern Indian Himalayan Region. *J Basic Microb*, 58: 730–738.

Kshetri, P., Singh, T.S., Roy, S.S. 2019b. Bioactive peptides and their therapeutic potential as antimicrobial drugs. In. *Frontiers in Anti-Infective Agents*. Ed. Tamreihao, K., Mukherjee, S. and Ningthoujam, D.S., 67–85. Bentham Science.

Kumar, E.V., Srijana, M., Chaitanya, K., Reddy, Y.H.K., Reddy, G. 2011. Biodegradation of poultry feathers by a novel isolate *Bacillus altitudinis* GVC11. *Indian J Biotechnol*. 10: 502–507.

Laba, W., Rodziewicz, A. 2010. Keratinolytic potential of feather-degrading *Bacillus polymyxa* and *Bacillus cereus*. *Pol J Environ Stud*, 19: 371–378.

Łaba, W., Żarowska, B., Chorążyk, D., Pudło, A., Piegza, M., Kancelista, A., Kopeć, W. 2018. New keratinolytic bacteria in valorization of chicken feather waste. *AMB Express*. 8: 9.

Lakshmi, P.J., Chitturi, C.M.K., Lakshmi, V.V. 2013. Efficient degradation of feather by keratinase producing *Bacillus* sp. *Int J Microbial*: 608321.

Lateef, A., Oloke, J.K., Gueguim Kana, E.B., Sobowale, B.O., Ajao, S.O., Bello, B.Y. 2010. Keratinolytic activities of a new feather-degrading isolate of *Bacillus cereus* LAU 08 isolated from Nigerian soil. *Int Biodeter Biodeg*, 64: 162–165.

Li, Q. 2019. Progress in microbial degradation of feather waste. *Front Microbiol*, 10: 2717.

Li, Z.W., Liang, S., Ke, Y., Deng, J.J., Zhang, M.S., Lu, D.L., Li, Z., Luo, X.C. 2020. The feather degradation mechanisms of a new *Streptomyces* sp. Isolate SCUT-3. *Commun Biol*, 3: 191.

Mabrouk, M.E.M. 2008. Feather degradation by a new keratinolytic Streptomyces sp. MS-2. *World J Microb Biot*, 24: 2331–2338.

Macedo, A.J., Beys da Silva, W.O., Gava, R., Driemeier, D., Henriques, J.A.P., Termignoni, C. 2005. Novel keratinase from *Bacillus subtilis* S14 exhibiting remarkable deharing capabilities. *Appl Environ Microbiol*, 71: 594–596.

Mamangkey, J., Suryanto, D., Munir, E., Mustopa, A.Z. 2020. Keratinase activity of a newly keratinolytic bacteria, *Azotobacter chroococcum* B4. *J Pure Appl Microbiol*, 14: 1203–1211.

Mazotto, A.M., Cedrola, S.M.L., Lins, U., Rosado, A.S., Silva, K.T., Chaves, J.Q., Rabinovitch, L., Zingali, R.B., Vermelho, A.B. 2010. Keratinolytic activity of *Bacillus subtilis* AMR using human hair. *Lett Appl Microbiol*, 50: 89–96.

Mazotto, A.M., de Melo, A.C., Macrae, A., Rosado, A.S., Peixoto, R., Cedrola, S.M., Couri, S., Zingali, R.B., Villa, A.L., Rabinovitch, L., Chaves, J.Q., Vermelho, A.B. 2011. Biodegradation of feather waste by extracellular keratinases and gelatinases from *Bacillus* spp. *World J Microb Biot*, 27: 1355–1365.

Me´zes, L., Tama´s, J. 2015. Feather waste recycling for biogas production. *Waste Biomass Valori*, 6: 899–911.

Meyers, M.A., Chen, P.Y., Lin, A.Y.M., Seki, Y. 2008. Biological materials: Structure and mechanical properties. *Prog Mater Sci*, 53: 1–206.

Mokrejš, P., Huťťa, M., Pavlačková, J., Egner, P. 2017. Preparation of Keratin hydrolysate from chicken feathers and its application in cosmetics. *Jove -J Vis Exp*. 129: 56254.

Mousavia, S.Z., Manteghiana, M., Shojaosadati, S.A., Pahlavanzadeh, H. 2018. Keratin nanoparticles: Synthesis and application for Cu(II) removal. *Adv Environ Technol*, 2: 83–93.

Nagal, S., Jain, P.C. 2010. Feather degradation by strains of Bacillus isolated from decomposing feathers. *Braz J Microbiol*, 41: 196–200.

Nam, G.W., Lee, D.W., Lee, H.S., Lee, N.J., Kim, B.C., Choe, E.A., Hwang, J.K., Suhartono M.T., Pyun, Y.R. 2002. Native-feather degradation by *Fervidobacterium islandicum* AW-1, a newly isolated keratinase-producing thermophilic anaerobe. *Arch Microbiol*, 178: 538–547.

Nnolim, N.E., Okoh, A.I., Nwodo, U.U. 2020. *Bacillus* sp. FPF-1 Produced keratinase with high potential for chicken feather degradation. *Molecules*, 25: 1505.

Onifade, A.A., Al-Sane, N.A., Al-Musallam, A.A., Al-Zarban, S. 1998. A review: Potentials for biotechnological applications of keratin-degrading microorganisms and their enzymes for nutritional improvement of feathers and other keratins as livestock feed resources, *Bioresource Technol*, 66: 1–11.

Orak, T., Caglar, O., Ortucu, S., Ozkan, H., Taskin, M., 2018. Chicken feather peptone: A new alternative nitrogen source for pigment production by *Monascus purpureus*. *J Biotechnol*, 271: 56–62.

Park, G.T., Son, H.J. 2009. Keratinolytic activity of *Bacillus megaterium* F7–1, a feather-degrading Mesophilic bacterium. *Microbiol Res*, 164: 478–485.

Paul, T., Halder, S.K., Das, A., Bera, S., Maity, C., Mandal, A., Das, P.S., Mohapatra, P.K.D., Pati, B.R., Mondal, K.C. 2013. Exploitation of chicken feather waste as a plant growth promoting agent using keratinase producing novel isolate *Paenibacillus woosongensis* TKB2. *Biocatal Agric Biotechnol*, 2: 50–57.

Paul, T., Mandal, A., Mandal, S.M., Ghosh, K., Mandal, A.K., Halder, S.K., Das, A., Maji, S.K., Kati, A., Mohapatra, P.K.D., Pati, B.R., Monad, K.C. 2015. Enzymatic hydrolyzed feather peptide, a welcoming drug for multiple-antibiotic-resistant *Staphylococcus aureus:* Structural analysis and characterization. *Appl Biochem Biotech*, 175: 3371–3386.

Perotto, G., Sandri, G., Pignatelli, C., Milanesi, G., Athanassiou, A. 2019. Water-based synthesis of keratin micro- and nanoparticles with tunable mucoadhesive properties for drug delivery. *J Mater Chem B*, 7: 4385–4392.

Poole, A.J., Church, J.S., Huson, M.G. 2009. Environmentally sustainable fibers from regenerated protein. *Biomacromolecules*, 10: 1–8.

Prasanna, I., Moorthy, K. 2020. Isolation, identification and characterization of feather degrading *Raoultella electrica* from poultry waste containing soil. *Gorteria J*, 33: 342–346.

Qiu, J., Wilkens, C., Barrett, K., Meyer, A.S. 2020. Microbial enzymes catalyzing keratin degradation: Classification, structure, function. *Biotechnology Advances*, 44: 107607.

Ramakrishnan, J., Balakrishnan, H., Raja, S.T., Sundararamakrishnan, N., Renganathan, S., Radha, V.N. 2011. Formulation of economical microbial feed using degraded chicken feathers by a novel *Streptomyces* sp: Mitigation of environmental pollution. *Braz J Microbiol*, 42: 825–834.

Ramani, P., Gupta, R. 2004. Optimization of medium composition for keratinase production on feather by *Bacillus licheniformis* RG1 using stastical methods involving response surface methodology. *Biotechnol Appl Biochem*, 40: 191–196.

Riessen, S., Antranikian, G. 2001. Isolation of *Thermoanaerobacter keratinophilus* sp. Nov., a novel thermophilic, anaerobic bacterium with keratinolytic activity. *Extremophiles*, 5: 399–408.

Riffel, A., Brandelli, A. 2006. Keratinolytic bacteria isolated from feather waste. *Braz J Microbiol*, 37: 395–399.

Riffel, A., Lucas, F., Heeb, P., Brandelli, A. 2003. Characterization of a new keratinolytic bacterium that completely degrades native feather keratin. *Arch Microbiol*, 179: 258–265.

Rozs, M., Manczinger, L., Vágvölgyi, C., Kevei, F., Hochkoeppler, A., Rodríguez, A.G.V. 2001. Fermentation characteristics and secretion of proteases of a new keratinolytic strain of *Bacillus licheniformis*. *Biotechnol Lett*, 23: 1925–1929.

Saarela, M., Berlin, M., Nygren, H., Lahtinen, P., Honkapää, K., Lantto, R., Maukonen, J. 2017. Characterization of feather-degrading bacterial populations from birds' nests—Potential strains for biomass production for animal feed. *Int Biodeter Biodegr*, 123: 262–268.

Sangali, S., Brandelli, A. 2000. Feather keratin hydrolysis by a *Vibrio sp*. Strain kr2. *J Appl Microbiol*, 89: 735–743.

Schader, C., Muller, A., Scialabba, N., Hecht, J., Isensee, A., Erb, K. H., Smith, P., Makkar, H. P., Klocke, P., Leiber, F., Schwegler, P., Stolze, M., Niggli, U. 2015. Impacts of feeding less food-competing feedstuffs to livestock on global food system sustainability. *J R Soc Interface*, 12: 20150891.

Sekar, V., Kannan, M., Ganesan, R., Dheeba, B., Sivakumar, N., Kannan, K. 2016. Isolation and screening of keratinolytic bacteria from feather dumping soil in and around Cuddalore and Villupuram, Tamil Nadu. *Proc Natl Acad Sci, India, Sect B Biol Sci*, 86: 567–575.

Sharma, R., Devi, S. 2018. Versatility and commercial status of microbial keratinases: A review. *Rev Environ Sci Biotechnol,* 17: 19–45.

Sharma, R., Gupta, R. 2010. Thermostable thiol activated keratinase from *Pseudomonas aeruginosa* KS-1 for prospective application in prion decontamination. *Res J Microbiol*, 5: 954–965.

Sharma, S., Gupta, A. 2016. Sustainable management of keratin waste biomass: Applications and future perspectives. *Braz Arch Biol Techn*, *59*, April 29. https://doi.org/10.1590/1678-4324-2016150684.

Shavandi, A., Silva, T.H., Bekhit, A.A., Bekhit, A.E.A. 2017. Keratin: Dissolution, extraction and biomedical application. *Biomater Sci*, 22: 1699–1735.

Shih, J. 2012. From biogas energy, biotechnology to new agriculture. *World Poultry Sci J*, *68*: 409–416.

Sobucki, L., Ramos, R.F., Gubiani, E., Brunetto, G., Kaiser, D.R., Dariot D.J. 2019. Feather hydrolysate as a promising nitrogen-rich fertilizer for greenhouse lettuce cultivation. *Int J Recycl Org Waste Agricult,* 8: 493–499.

Suntornsuk, W., Suntornsuk, L. 2003. Feather degradation by *Bacillus sp*. FK 46 in submerged cultivation, *Bioresource Technol*, 86: 239–243.

Szabo, I., Benedek, A., Sazabo, I.M., Barabas, G. 2000. Feather degradation with a thermotolerant *Streptomyces graminofaciens* strain. *World J Microb Biot*, 16: 253–255.

Tamreihao, K., Devi, L.J., Khunjamayum, R., Mukherjee, S., Ashem, R.S., Ningthoujam, D.S. 2017. Biofertilizing potential of feather hydrolysate produced by indigenous keratinolytic *Amycolatopsis* sp. MBRL 40 for rice cultivation under field conditions. *Biocatal Agric Biotechnol*, 10: 317–320.

Taskin, M., Unver, Y., Firat, A., Ortucu, S., Yildiz, M. 2016. Sheep wool protein hydrolysate: A new peptone source for microorganisms. *J Chem Technol Biotechnol*, 91: 1675–1680.

Thornton, P.K. 2010. Livestock production: Recent trends, future prospects. *Philos T R Soc B*, 365: 2853–2867.

Tiwary, E., Gupta, R. 2012. Rapid conversion of chicken feather to feather meal using dimeric keratinase from *Bacillus licheniformis* ER-15. *J Bioprocess Biotech*, 2: 123.

United Nations, Department of economics and social welfare. 2017 World Population Prospects: The 2017 revision. www.un.org/development/desa/en/news/population/world-population-prospects-2017.html (accessed April 10, 2021).

Villa, A.L.V., Aragão, M.R.S., dos Santos, E.P., Mazotto, A.M., Zingali, R.B., Souza E.P., Vermelho, A.B. 2013. Feather keratin hydrolysates obtained from microbial keratinases: Effect on hair fiber. *BMC Biotechnol,* 13: 15.

Wan, M.Y., Dong, G., Yang, B.Q., Feng, H. 2016. Identification and characterization of a novel antioxidant peptide from feather keratin hydrolysate. *Biotechnol Lett*, 38: 643–649.

Wang, B., Yang, W., Mckittrick, J., Meyers, M.A. 2016. Keratin: Structure, mechanical properties, occurrence in biological organisms, and efforts at bioinspiration. *Prog Mater Sci*, 76: 229–318.

Williams, C.M., Richter, C.S., Mackenzie, J.M., Shih, J.C. 1990. Isolation, identification, and characterization of a feather-degrading bacterium. *Appl Environ Microbiol*, 56: 1509–15015.

Wu, W.L., Chen, M.Y., Tu, I.F., Lin, Y.C., EswarKumar, N., Chen, M.Y., Ho, M.C., Wu, S.H. 2017. The discovery of novel heat-stable keratinases from *Meiothermus taiwanensis* WR-220 and other extremophiles. *Sci Rep*, 7: 4658.

Xu, H., Shi, Z., Reddy, N., Yang, Y., 2014. Intrinsically water-stable keratin nanoparticles and them in vivo biodistribution for targeted delivery. *J Agr Food Chem*, 62: 9145–9150.

Yin, X.-C., Li, F.-Y., He, Y.-F., Wang, Y., Wang, R.-M. 2013. Study on effective extraction of chicken feather keratins and their films for controlling drug release. *Biomater Sci-UK, 1*: 528.

Zaghloul, T.I., Embaby, A.M., Elmahdy A.R. 2011. Biodegradation of chicken feathers waste directed by Bacillus subtilis recombinant cells: Scaling up in a laboratory scale fermentor. *Bioresource Technol*, 102: 2387–2393.

13 Biological Pathways for the Production of Levulinic Acid from Lignocellulosic Resources

Laura G. Covinich and María Cristina Area
IMAM, UNaM, CONICET, FCEQYN, Programa de Celulosa y Papel (PROCYP), Misiones, Argentina, Félix de Azara 1552, Posadas, Argentina

CONTENTS

Lignocellulosic agro-industrial wastes are abundantly available renewable sources of fermentable sugars and phenolic compounds. The agro-industrial waste generated is sold at prices less than USD 10/t up to USD 113/t. Meanwhile, forest industrial wastes are generated mainly by the timber and the pulp and paper industries. The costs of various forest industrial wastes depend on the region and species but could vary between USD 7/t and USD 123/t (Clauser et al. 2021). Both kinds of residues are the raw material for several products. Studies developed in the last years have identified high-value chemical platforms or building blocks from lignocellulosic biomass (Werpy and Petersen 2004). Based on this evaluation, a list of twelve promising building blocks was determined, where most of these products were reaffirmed recently (Bozell and Petersen 2010a; Choi et al. 2015). One of the auspicious products is levulinic acid (LA). LA is a high-valued-added product, and estimations state that its market will grow up to 15.4% of CAGR (compound annual growth rate) in the following years, with estimated revenue of more than 57 million USD (Clauser et al. 2021).

DOI: 10.1201/9781003191247-13

Lignocellulosic biomass is a source of high commercial value and is a bio-based chemical of great industrial importance (Kamm and Kamm 2004) produced by biological and biochemical methods. Generally, the biotransformation process to the desired end product is a challenging task (Schmidt, Castiglione, and Kourist 2018) that can be affected by several factors, such as the specificity of the enzyme (D. Zhu et al. 2006), temperature, and composition of the medium (Elgharbawy et al. 2016), pH (Cronk et al. 2001), available nutrients, generated inhibitors and their cross-reactivity (Rudroff et al. 2018), end product inhibition (Gregg and Saddler 1996), substrate main features and concentration (Olsson et al. 2014), enzyme-substrate-product interactions (Gan, Allen, and Taylor 2003), and sustained growth reaction (Clomburg et al. 2019), among others.

In addition, there are some challenges to address that will turn in the bioprocesses development and implementation, rendering it more efficient. For example, getting greater efficiency than chemical processes and finding new, highly competitive synthetic routes (for high-volume and low-cost products) and selective processes (for low-volume and highly complex products) (Rudroff et al. 2018).

Among bio-based chemical compounds, keto acids (or oxoacids) are known to have been widely applied in the pharmaceutical industry (Yang Song et al. 2016), cosmetic industry, food industry (Spaepen et al. 2007), health system (Porter et al. 2000), etc. Traditionally, chemical-based production of keto acid relies on the use of dehydration and decarboxylation (Porter et al. 2000), hydrolysis (Nshimiyimana, Liu, and Du 2019; Morone, Apte, and Pandey 2015), hydantoin (Luo et al. 2021), oxidation (Zamora et al. 2006), double carbonylation (Das and Bhanage 2020), etc.

The biotechnological industrial production of bio-based keto acids has been discovered and explored using microorganisms as cell factories (Luo et al. 2021). These "cell factories" are highly effective thanks to the development of metabolic engineering (see Figure 13.1) (Otero and Nielsen 2010) to control the biosynthesis pathways (Yang Song et al. 2017; Jin and Stephanopoulos 2007). The former implies the optimization of strain development (Na, Kim, and Lee 2010), fermentation processes (K. H. Lee et al. 2007), recovery and purification processes (Park et al. 2008) through combining the existing pathways, the engineering of the existing pathways, and *de novo* pathways design (Prather and Martin 2008). These strategies allow the development of genetic and metabolic cell factories for rapid cell growth, high productivity of the entire system, and optimal flows towards producing the desired product without generating unwanted by-products (J. W. Lee et al. 2012).

There are a group of microorganisms that were metabolically engineered to produce keto acids. Among the bacteria mostly used to produce keto acids are *Escherichia coli* (for pyruvic acid) (Hossain et al. 2016), α-Ketoisovaleric acid (Li et al. 2017), levulinic acid (Cheong, Clomburg, and Gonzalez 2016), *Pseudomonas aeruginosa* (for 2-Keto-D-gluconic acid) (Chia, Van Nguyen, and Choi 2008), *Proteus vulgaris* (for phenylpyruvic acid) (Coban et al. 2016), *Pseudomonas fluorescens* (for ketoglutaric acid) (Otto, Yovkova, and Barth 2011), *Corynebacterium glutamicum* (for α-ketoisovaleric acid) (Buchholz et al. 2013), 5-aminolevulinic acid (Ramzi et al. 2015), α-Ketoisocaproic acid (Bückle-Vallant et al. 2014), ketoglutaric acids (Brüsseler et al. 2019), and *Streptomyces cinnamonensis* (for α-ketoisovaleric acid) (Pospíšil, Kopecký, and Přikrylová 1998), among others. Among yeast mostly

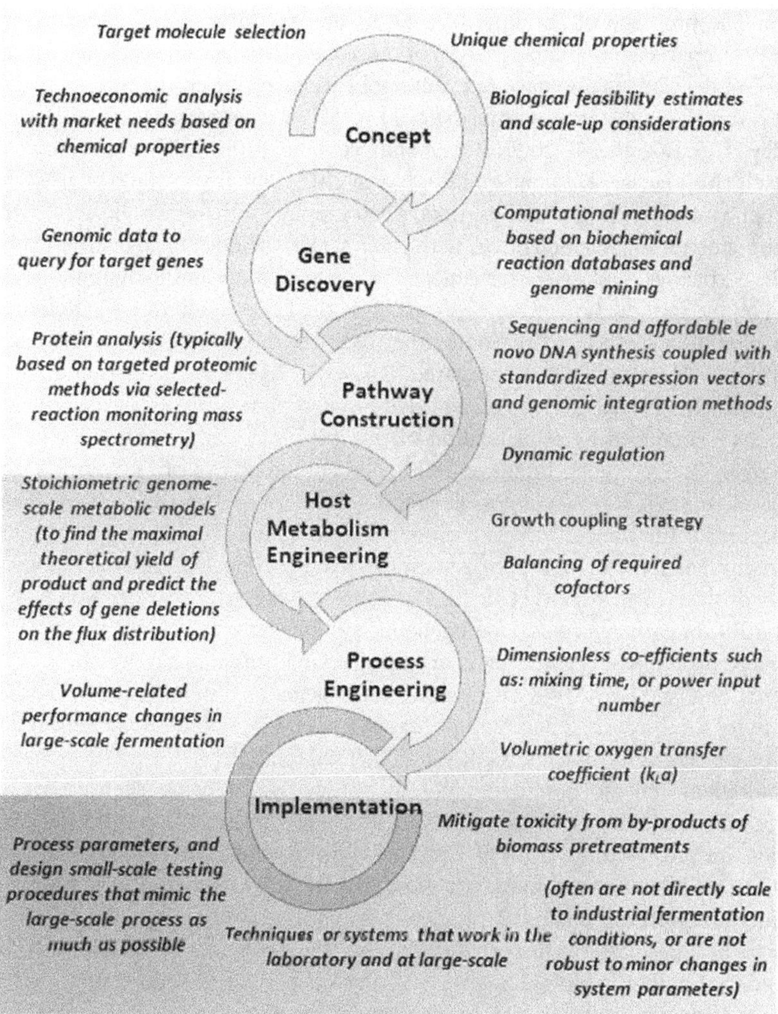

FIGURE 13.1 The process of metabolic engineering for controlling biosynthesis pathway of chemicals.

Source: Adapted from Chubukov et al. 2018.

used to produce keto acids are *Candida glabrata* (for pyruvic acid) (S. Yang et al. 2014), *Saccharomyces cerevisiae* (for pyruvic acid) (Van Maris et al. 2004), and *Yarrowia lipolytica* (for ketoglutaric acid) (Guo et al. 2014), among others.

13.1 BIOLOGICAL PATHWAYS INVOLVED IN OBTAINING LA

Among keto acids, levulinic acid (LA) is the simplest and leading γ-oxocarboxylic acid (or 4-oxopentanoic acid of the molecular formula $C_5H_8O_3$) (Klingler and Ebertz

2000). The presence of carbonyl ($C = O$, as an acid ketone) and carboxyl groups ($COOH$) imparts a remarkable pattern of reactivity that makes possible the formation of several derivatives that have relevant applications in varied fields, thus making it a very versatile substance within the field of green chemistry (Morone, Apte, and Pandey 2015; Bozell et al. 2000; Bozell and Petersen 2010b). LA is a specialty chemical itself (Mukherjee, Dumont, and Raghavan 2015) which also could be used in varied applications, such as coatings, inks, photography (with diphenolic acid prepared by the condensation reaction of LA with phenol) (Kricheldorf, Hobzova, and Schwarz 2003), herbicides, and drugs for cancer treatment (with 5-aminolevulinic acid prepared by the introduction of an $-NH_2$ group at the C5-position of LA) (Rebeiz et al. 1984; Mahmoudi et al. 2019). Some other uses are in solvents, fuel additives, food additives (with γ-valerolactone prepared by LA hydrogenation) (Mellmer et al. 2014; Dutta et al. 2019), energy-rich fuel additives, foam-comprising materials (with levulinate esters produced by esterification of the levulinic acid, among other methods) (Badgujar, Badgujar, and Bhanage 2020), and commodity plastics (with LA polymerization through Ugi-4-component reaction) (Hartweg and Becer 2018; Hartweg and Becer 2016).

Traditionally, the most widely used processes for producing LA are homogeneous acid hydrolysis (Yan, Jarvis, et al. 2015) and heterogeneous acid hydrolysis (Covinich et al. 2019) of carbohydrates. Homogeneous systems have excellent catalytic activity and thermal stability, because there are no diffusion limitations (Girisuta et al. 2013; Nhien, Long, and Lee 2016). Meanwhile, heterogeneous systems present an excellent selectivity for the desired end product, catalyst recovery and reuse, and minimal volume of generated wastewater (Gallo et al. 2013; Abdullah et al. 2017).

Biological and biochemical routes to produce LA rely on chemical methods and remain a challenge (Morone, Apte, and Pandey 2015; Luo et al. 2021). The studies are mostly limited to "hybrid" processes, where biological processes are used for biomass pretreatment, whereas hydrolysis or isomerization, followed by traditional catalyzed chemical processes (such as dehydration and hydration), is used to produce LA from monosaccharides (Alipour and Omidvarborna 2017). Considering only the biological and biochemical pathways, some authors reported the simultaneous production of polyhydroxyalkanoates and LA through fermentation, using hemicellulose hydrolysates as a carbon source (Pinto-Ibieta et al. 2020). However, they do not explain the reaction mechanism or the enzymes involved in the process.

In this regard, a recent study (Cheong, Clomburg, and Gonzalez 2016) demonstrated the application of fermentative processes carried out with metabolically engineered *Escherichia coli*[1] with succinyl-CoA as primer and acetyl-CoA as extender unit, which uses glycerol as a carbon source through an orthogonal iterative pathway based on nondecarboxylative Claisen condensation reactions (and later β-reductions).

Referred to previously, a US 2019 Patent (Zanghellini 2019) claims to convert a carbon source (C6 or C5) into pyruvate, followed by an aldol addition reactions and subsequent reaction, such as reduction, oxidation, dehydration, group transfer, hydrolysis, and (or) lactonization, to prepare the desired C5 product, LA in this case (Figure 13.2). Additionally, alternative paths and cyclical processes arise

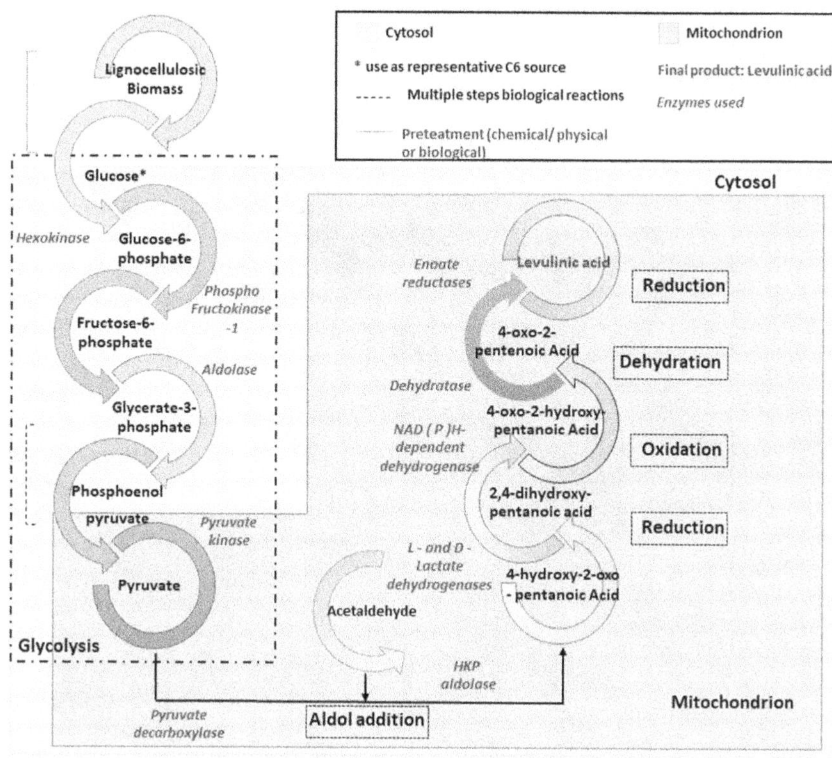

FIGURE 13.2 Metabolic route for the conversion of a C6 source from lignocellulosic feedstock to pyruvate and its subsequent biochemical conversion to levulinic acid.

Source: Adapted from Zanghellini (2019).

from the main one that allows obtaining, for example, levulinate esters or lactones (Zanghellini 2019).

A key point of converting lignocellulosic biomass into LA through biochemical and biological methods is pyruvate obtained from the used carbon source. For example, using a C5 source like xylose, arabinose, etc. implies following a route for pyruvate production different from that of glycolysis (see glycolysis path in Figure 13.2). Thus, fermentation of a mixed carbon source (C6 and C5) means selecting microbial strains that can simultaneously exploit C6 and C5 of lignocellulosic biomass as a feedstock (Hu et al. 2011; Hasona et al. 2004; da Conceição Gomes et al. 2019) to obtain pyruvate. This needs a deep understanding of the intracellular incorporation of C6 and C5 sources by the selected microorganism since it uses different catabolic routes for carbon fluxes to process the monosaccharide desired mixture (Wushensky et al. 2018). Possible metabolic pathways followed by the C5 sources from lignocellulosic feedstock to reach pyruvate are shown in Figure 11.3.

For example, when using metabolically engineered *Saccharomyces cerevisiae* or *Pichia stipitis*, which can simultaneously ferment C5 and C6 monosaccharides,

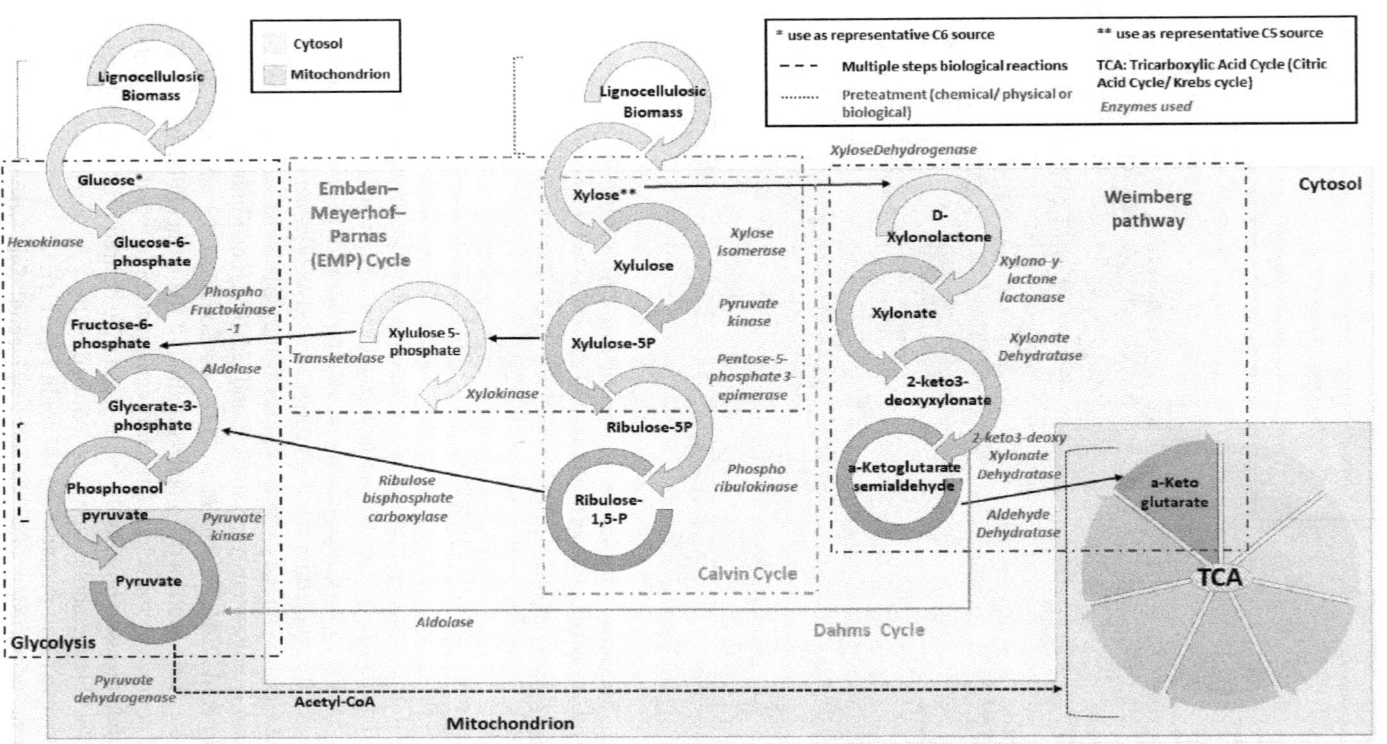

FIGURE 13.3 Metabolic routes followed for the conversion of a C5 source from lignocellulosic feedstock to pyruvate.

Source: Adapted from Nduko et al. (2013; Gonzalez et al. (2002; Wushensky et al. (2018; Wasserstrom et al. (2018; Bator et al. (2020).

the direct fermentation of levulinic acid could be accomplished (Zanghellini 2019). Usually, xylose is not used with *Saccharomyces cerevisiae*. However, metabolically engineered *Saccharomyces cerevisiae* can be used in xylose/glucose medium as a carbon source (Sandström et al. 2014) due to its high osmolarity and low pH tolerance (Borgström et al. 2019). Also, *Clostridium acetobutylicum* (Gheshlaghi et al. 2009) and *Scheffersomyces stipites* (Ma et al. 2013) can metabolize several carbon sources. The metabolically engineered yeast *Kluyveromyces marxianus*[2,3] has a broad substrate spectrum (Fonseca et al. 2008) and thermotolerant properties (Anderson, McNeil, and Watson 1986). So it could be used for saccharification of biomass with C6 and C5 components (Zhang et al. 2017; Zhang et al. 2013). The metabolically engineered bacteria *Pseudomonas putida* could also be used when working with biomass with C6 and C for simultaneous fermentation (Bator et al. 2020).

Several authors demonstrated the application of successful simultaneous fermentative processes of glucose and xylose (as examples of C6 and C5, respectively) carried out with metabolically engineered *Hansenula polymorpha*[4] (Ishchuk et al. 2008). *Escherichia coli* could also metabolize a carbon source different from glucose, like lactose, arabinose, and xylose, among others (Ammar, Wang, and Rao 2018). For example, *Escherichia coli*[5,6] (Nduko et al. 2013) is also used with glucose/glycerol medium as a carbon source (Cheong, Clomburg, and Gonzalez 2018). Thus, metabolic engineering allows fine-tuning yeast or bacteria not usually used for a specific objective, based on the *de novo* design of the desired pathways. So enzymes and targeted final products can be designed through appropriate termination pathways (S. Kim, Clomburg, and Gonzalez 2015), accomplishing higher final yield, rate, and titer (Cheong, Clomburg, and Gonzalez 2018).

Table 13.1 shows some of the enzymes involved in the pathways of Figures 13.2–13.4 that are of interest from metabolically engineered yeast or bacteria for biological processes applied to C6 and C5 from biomass (alone or combined).

13.2 LEVULINIC ACID DERIVATIVES THROUGHOUT BIOLOGICAL PATHWAYS

As aforementioned, 5-aminolevulinic (5-ALA) acid belongs to the keto acid group, also known as 5-amino-4-oxopentanoic acid, one of the most relevant levulinic acid derivatives (Mellmer et al. 2015). Chemical-based methods to produce 5-ALA are based on the selective introduction of the $-NH_2$ group on the LA C5 (Pileidis and Titirici 2016).

On the other hand, biological routes for 5-ALA production can be accomplished from lignocellulosic biomass (Nishikawa and Murooka 2001), divided into two main groups, known as C4 and C5 pathways (Morone, Apte, and Pandey 2015; Luo et al. 2021). The C4 path corresponds to the direct condensation of succinyl-CoA from tricarboxylic acid cycle (TCA) and glycine (see Figure 13.4) throughout 5-aminolevulinate synthase (Hunter and Ferreira 2009). Contrarily, the C5 path involves several enzymatic steps, following the a-ketoglutarate (from TCA cycle)/glutamate/glutamate-1 semialdehyde/5 ALA sequence (Luo et al. 2021) (see Figure 13.4).

5-ALA is generally obtained from *Rhodobacter sphaeroides*, *Clostridium thernloaceticlun*, and *Chlorella spp.* (Nishikawa and Murooka 2001). The C4 pathway to

TABLE 13.1
Enzymes from Metabolically Engineered Yeast or Bacteria for Biological Processes Applied to C6 and (or) C5 from Biomass for Several End Products

Enzyme	Main Function	Ref	Yeast* or Bacteria**	Strain	Carbon Source	Ref
Pyruvate decarboxylase (Pdc)	Regulation of pyruvate dehydrogenase (PDH)	(Rapala-Kozik 2011)	*Kluyveromyces marxianus* *	KCTC 17555	Glucose/glycerol	(Choo et al. 2018)
Pyruvate decarboxylase (Pdc)	Regulation of pyruvate dehydrogenase (PDH)	(Rapala-Kozik 2011)	*Saccharomyces cerevisiae* *	GG 570	Glucose	(Steensmati, Dijkent, and Pronkt 1996)
Pyruvate dehydrogenase (PDH)	Conversion of pyruvate to acetyl coenzyme A (CoA)	(Milne 2013)	*Escherichia coli* **	MG1655	Glucose	(M. Yang and Zhang 2017)
Phosphofrutokinase-1	Phosphorylation of fructose-6-phosphate	(Uyeda and Kurooka 1970)	*Clostridium acetobutylicum* **	-	Glucose	(Gheshlaghi et al. 2009)
D-xylose isomerase	Isomerization of Xylose into Xylose	(Jeffries 1983)	*Escherichia coli* **	MJ133K-1	Xylose/glucose	(Gao, Gao, and Dong 2017)
α-ketoglutarate dehydrogenase	Oxidative decarboxylation of α-ketoglutarate to produce succinyl-CoA	(N.V.Bhagavan and Chung-EunHa 2015)	*Saccharomyces cerevisiae* *	IMX1401	Glucose	(Baldi et al. 2019)
β-glucosidase	Capable of hydrolyzing β-glycosidic linkages	(Mario Pinto 1999)	*Pseudomonas putida* **	W619	Xylose/glucose/ Arabinose/ mannose	(Davis et al. 2013)
Xylose dehydrogenase	Xylose is oxidized to Xylonolactone	(WEIMBERG 1961)	*Caulobacter crescentus* **	-	Xylose	(Stephens et al. 2007)
NAD(P)H- dependent dehydrogenase	Oxidation of the OH at C4 position 2,4-dihydroxy-pentanoic acid	(Zanghellini 2019)	*Scheffersomyces stipites* *	CBS 6054	Xylose	(Ma et al. 2013)

Glutamate-1-semialdehyde-2,1-aminomutase	Isomerization of glutamate-1-semialdehyde to 5-aminolevulinate	(Yingxian Song et al. 2016)	*Escherichia coli* **	DALA	Glucose	(Kang et al. 2011)
Transketolase	Catalyzes the reversible transfer of a ketol group	(Il, Ki, and Kim 2005)	*Corynebacterium glutamicum* **	YL1 pNOG2	Glucose	(Y. Lee, Cho, and Woo 2020)
5-Aminolevulinate synthase	Condensation of glycine with succinyl-coenzyme A from Krebs cycle	(Hunter and Ferreira 2009)	*Escherichia coli* **	Rosetta(DE3)	Glucose/fructose, xylose/mannose/ arabinose/ galactose/lactose	(Lin, Fu, and Cen 2009)
Enoate reductases	Asymmetric bioreduction of activated C=C bonds from 4-oxo-2- pentenoic acid to levulinic acid	(Stuermer et al. 2007; Zanghellini 2019)	*Escherichia coli* **	JM109	Glucose	(Kawasaki et al. 2018)
Xylose dehydrogenase	Catalyzes the NADP+-oxidation of xylose to xylono-1,5-lactone	(Toivari et al. 2012)	*Saccharomyces cerevisiae* *	H158 pXks	Xylose/glucose/ mannose/ galactose	(Johansson et al. 2001)
Glutamate dehydrogenase	Reversibly converts α-ketoglutarate to glutamate	(Plaitakis et al. 2017)	*Escherichia coli* **	Transetta GTR/ GBP	Glucose	(Aiguo and Meizhi 2019)
Succinyl-CoA synthetase	Substrate level Phosphorylation generates succinate in Krebs cycle	(Bridger 1974)	*Escherichia coli* **	MG1655	Glucose	(Veit, Polen, and Wendisch 2007)
Pentose-5-phosphate 3-epimerase	Catalyzes the interconversion of ribulose-5-phosphate and xylulose-5-phosphate	(Jelakovic et al. 2003)	*Escherichia coli* **	K-12	Several carbon sources	(Bridger 1974)
Endoglucanase	Capable of hydrolyzing β-glycosidic linkages	(Rees, Chan, and Kim 1993)	*Aspergillus niger (fungus)*	38	Glucose	(Jecu 2000)
Endoglucanase	Capable of hydrolyzing β-glycosidic linkages	(Rees, Chan, and Kim 1993)	*Acidothermus cellulolyticus***	-	Several carbon sources (rice)	(Oraby et al. 2007)

(*Continued*)

TABLE 13.1 *(Continued)*
Enzymes from Metabolically Engineered Yeast or Bacteria for Biological Processes Applied to C6 and (or) C5 from Biomass for Several End Products

Enzyme	Main Function	Ref	Yeast* or Bacteria**	Strain	Carbon Source	Ref
Aconitate hydratase	Stereospecific isomerization of citrate to isocitrate	(Chen and Russo 2012)	*Yarrowia lipolytica**	H222-S4	Glycerol/glucose/sucrose	(Holz et al. 2009)
Pyruvate kinase	Transfer of a phosphate from phosphoenolpyruvate to ADP, yielding of pyruvate	(F.J.Kayne and N.C.Price 1973)	*Escherichia coli* **	VH34	Glucose	(Meza et al. 2012)
Malate dehydrogenase	Oxidation of malate to oxaloacetate	(Grotjohann, Huang, and Kowallik 2001)	*Escherichia coli* **	SBS550MG-cms243	Glucose	(F. Zhu, San, and Bennett 2020)
Fumarase	Reversible hydration/dehydration of fumarate to malate	(Grotjohann, Huang, and Kowallik 2001)	*Saccharomyces cerevisiae* *	77	Glucose/galactose	(Polakis, Bartley, and Meek 1965)

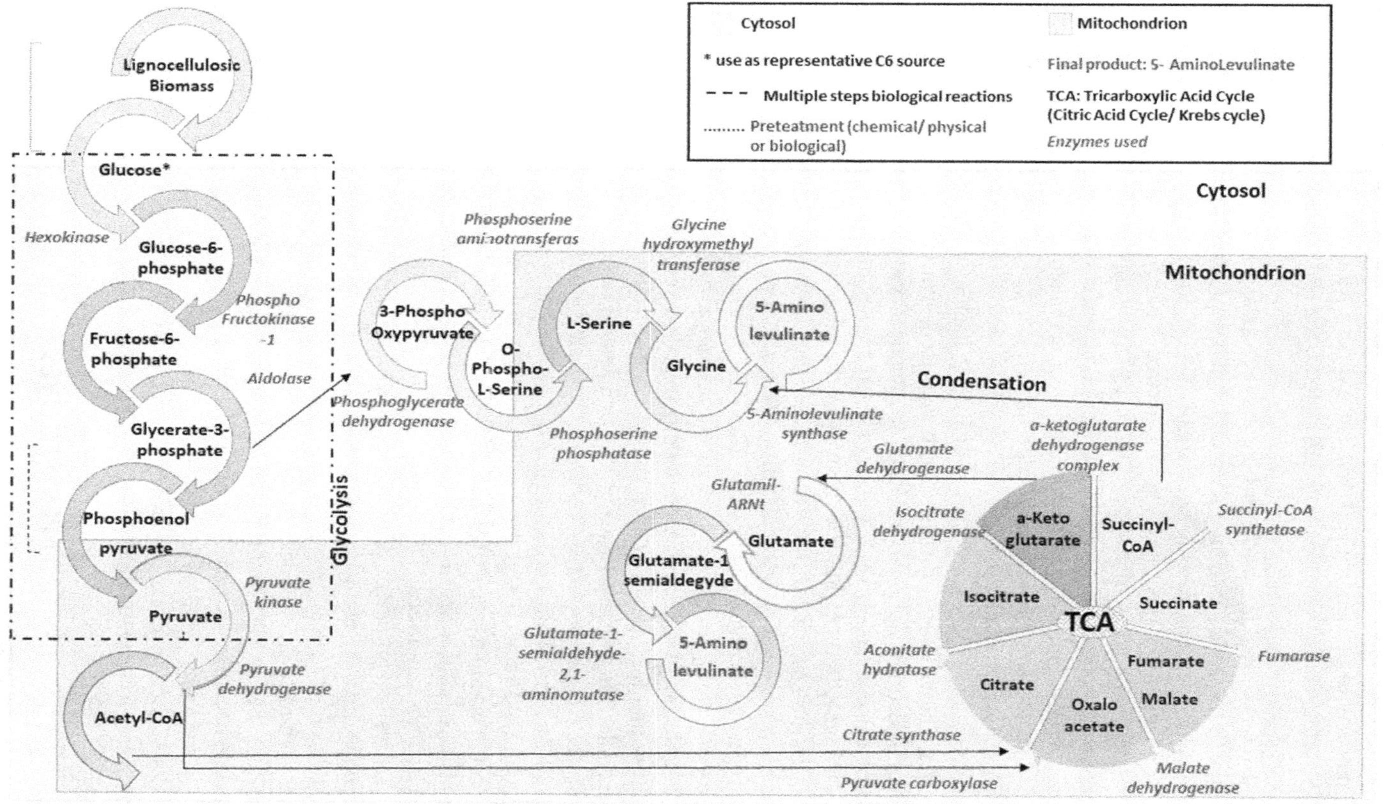

FIGURE 13.4 C4 and C5 biochemical routes to produce 5-aminolevulinic acid from biomass.

Source: Adapted from Yi et al. (2012; Peterson and Young (1968; Luo et al. (2021; S. Y. Lee et al. (2019; P. Yang et al. (2016; D. Jahn and Heinz (2008).

produce 5-ALA was carried out with *Agrobacterium radiobacter*[7,8] (Lin, Fu, and Cen 2009; Qin et al. 2006) or *Corynebacterium glutamicum*[9] (Feng et al. 2016).

The C5 pathway to produce 5-ALA was carried out with *Corynebacterium glutamicum*[10,11] (Ramzi et al. 2015) (Yu et al. 2015) and *Chlorella regularis*[12] (Ano et al. 1999).

LA can also serve itself as the sole carbon source and successfully be metabolized (Richards et al. 2018; Habe, Sato, and Kirimura 2020) to produce some value-added end products. However, considering particularly 5-ALA from LA as a sole carbon source throughout biological pathways has the drawback of LA potentially inhibiting the 5-aminolevulinic acid synthase (see Figure 13.4) by suppressing its generation and (or) enzymatic activity of glycine-succinyl-CoA condensation in C4 pathways (Habe, Sato, and Kirimura 2020). To solve this, an organism such as *Agmenellum quadruplicum*, *Anacystis marina*, and *Chlorella vulgaris* can secrete 5-ALA extracellularly (Harutyunyan et al. 2018).

Throughout the biological path, LA can also serve itself as the sole carbon source and can suffer, among others, a reduction to produce 4-hydroxyvalerate (Gorenflo et al. 2001) or a decarboxylation to produce 2-butanone (Šivec et al. 2019).

The biological conversion of LA is initialized by a CoA thioesterase (Habe, Sato, and Kirimura 2020), yielding levulinyl-CoA by a ligation process (Novackova et al. 2019). In the case of levulinyl-CoA production from LA, several strains of *Pseudomonas citronellolis* (Habe et al. 2019), *Cupriavidus necator* (Jaremko and Yu 2011), or *Ralstonia eutropha* (Chung et al. 2001) can be used.

4-hydroxyvalerate is produced through the activation of LA by the coenzyme A (CoA), followed by a reduction from a dehydrogenase reductase (Sathesh-Prabu and Lee 2019), and then the 4-hydroxyvalerate is extracellularly secreted because of an intracellular thioesterase (Martin, Wu, and Jones Prather 2010). Besides, the 4-hydroxyvalerate could be further lactonized to yield gamma-valerolactone (Yan, Yang et al. 2015). For 4-hydroxyvalerate production, several strains of *Pseudomonas putida* (Martin and Prather 2009), *Escherichia coli* (D. Kim et al. 2019), *Alcaligenes faecalis* (Yeon, Park, and Yoo 2013), and *Ralstonia eutropha* (Gorenflo et al. 2001) proved to be successful when using LA as a carbon source. *Pseudomonas putida* KT2440 (Martínez-García et al. 2014), one of the most used strains, is mainly characterized by its excellent properties of high tolerance to oxidative stress and organic solvents, metabolic diversity features, and minimal formation of unwanted by-products (Nikel and de Lorenzo 2018).

LA is converted to 2-butanone through its conversion to 3-ketovaleryl-CoA (a transferase) and acetyl-CoA (Mehrer et al. 2019) and subsequent decarboxylation by an acetoacetate decarboxylase (Min et al. 2013). For 2-butanone production, several strains of *Clostridium acetobutylicum* (Min et al. 2013), *Escherichia coli* (Richards et al. 2018), and *Pseudomonas putida* (Mehrer et al. 2019) proved to be successful when using LA as a carbon source.

13.3 CONCLUSIONS AND PERSPECTIVES

This book chapter deals with the biotechnological processes currently available for levulinic acid and its derivatives production.

The biotransformation process of the desired end product from lignocellulosic biomass is a challenging task.

The analyzed studies show that the key to converting lignocellulosic biomass into LA through biochemical and biological methods is pyruvate obtained from used carbon source. Thus, fermentation of a mixed carbon source (C6 and C5) requires selecting microbial strains that can simultaneously use C6 and C5 from lignocellulosic biomass as a feedstock using different catabolic routes for carbon fluxes.

In this regard, the development of metabolic engineering pathways for levulinic acid is definitively in the early stage and, at the moment of writing, cannot meet the requirements for its industrial production. Therefore, there is a need to bridge the gap between the well-developed chemical conversion processes and the underdeveloped biological conversion processes for levulinic acid.

NOTES

1 Strain: JST06, genotype: JC01 ΔyciA:: FRT ΔybgC:: FRT Δydil:: FRT ΔtesA:: FRT Δ fadM:: FRT ΔtesB:: FRT; overexpressing: cationic amino acid transporters-1 (CAT-1) as activation, Beta-ketoadipyl-CoA thiolase (PaaJ), β-ketoadipate: succinyl-coenzyme A transferase (PcaIJ) as ACT.

2 Strain: YZB058, genotype: YZB057, pZJ061, ScGAL2-N376F, ScURA3, overexpressing: xylose-specific transporter (ScGAL2-N376F).

3 Strain: YZB014, genotype: YZB001, Kmura3: pPSXRPTUM1.

4 Strain: NCYC 495 leu1–1, overexpressing: pyruvate decarboxylase isozyme 1 (PDC1), GAPDH as a promoter.

5 Strain: JW0885, genotype: pct, phaC1 Ps(ST/QK), phaAB, overexpressing: polyhydroxyalkanoate (synthase PhaC1Ps(ST/QK).

6 Strain: MG1655(DE3), genotype: MG1655 with the DE3 prophage integrated, overexpressing: 2-hydroxyacid dehydrogenase, T7 as a promoter.

7 Strain: ATCC4718, overexpressing: *Escherichia coli* Rosetta(DE3)/pET28a(+)-hemA.

8 Strain: zju-0121, overexpressing: E. coli BL21 (DE3).

9 Strain: ALA1, genotype: Wild-type ATCC 13032 harboring pXA, overexpressing: ppc.

10 Strain: ATCC 13032, genotype: wild-type.

11 Strain: SEAL, genotype: $hemA^{M}$(S. arizona) and hemL (E. coli), overexpressing: hemAM (S. arizona) and hemL (E. coli).

12 Strain: YA-603,45.

REFERENCES

Abdullah, Sharifah Hanis Yasmin Sayid, Nur Hanis Mohamad Hanapi, Azman Azid, Roslan Umar, Hafizan Juahir, Helena Khatoon, and Azizah Endut. 2017. "A Review of Biomass-Derived Heterogeneous Catalyst for a Sustainable Biodiesel Production." *Renewable and Sustainable Energy Reviews* 70 (July 2016): 1040–51. Doi:10.1016/j.rser.2016.12.008.

Aiguo, Zhao, and Zhai Meizhi. 2019. "Production of 5-Aminolevulinic Acid from Glutamate by Overexpressing HemA1 and Pgr7 from Arabidopsis Thaliana in Escherichia Coli." *World Journal of Microbiology and Biotechnology* 35 (11). Springer Netherlands: 1–9. Doi:10.1007/s11274-019-2750-6.

Alipour, Siamak, and Hamid Omidvarborna. 2017. "Enzymatic and Catalytic Hybrid Method for Levulinic Acid Synthesis from Biomass Sugars." *Journal of Cleaner Production* 143. Elsevier Ltd: 490–96. Doi:10.1016/j.jclepro.2016.12.086.

Ammar, Ehab M., Xiaoyi Wang, and Christopher V. Rao. 2018. "Regulation of Metabolism in Escherichia Coli during Growth on Mixtures of the Non-Glucose Sugars: Arabinose, Lactose, and Xylose." *Scientific Reports* 8 (1). Springer US: 1–11. Doi:10.1038/s41598-017-18704-0.

Anderson, P. J., K. McNeil, and K. Watson. 1986. "High-Efficiency Carbohydrate Fermentation to Ethanol at Temperatures above 40°C by Kuyveromyces Marxianus Var. Marxianus Isolated from Sugar Mills." *Applied and Environmental Microbiology* 51 (6): 1314–20. Doi:10.1128/aem.51.6.1314-1320.1986.

Ano, Akihiko, Hitoshi Funahashi, Katsumi Nakao, and Yoshinori Nishizawa. 1999. "Effect of Glycine on 5-Aminolevulinic Acid Biosynthesis in Heterotrophic Culture of Chlorella Regularis YA-603." *Journal of Bioscience and Bioengineering* 88 (1): 57–60. Doi:10.1016/S1389-1723(99)80176-5.

Badgujar, Kirtikumar C., Vivek C. Badgujar, and Bhalchandra M. Bhanage. 2020. "A Review on Catalytic Synthesis of Energy Rich Fuel Additive Levulinate Compounds from Biomass Derived Levulinic Acid." *Fuel Processing Technology* 197 (May 2019). Elsevier: 106213. Doi:10.1016/j.fuproc.2019.106213.

Baldi, Nicolò, James C. Dykstra, Marijke A. H. Luttik, Martin Pabst, Liang Wu, Kirsten R. Benjamin, André Vente, Jack T. Pronk, and Robert Mans. 2019. "Functional Expression of a Bacterial α-Ketoglutarate Dehydrogenase in the Cytosol of Saccharomyces Cerevisiae." *Metabolic Engineering* 56 (September). Elsevier Inc.: 190–97. Doi:10.1016/j.ymben.2019.10.001.

Bator, Isabel, Andreas Wittgens, Frank Rosenau, Till Tiso, and Lars M. Blank. 2020. "Comparison of Three Xylose Pathways in Pseudomonas Putida KT2440 for the Synthesis of Valuable Products." *Frontiers in Bioengineering and Biotechnology* 7 (January): 1–18. Doi:10.3389/fbioe.2019.00480.

Bhagavan, N. V., and Chung-EunHa. 2015. "Carbohydrate Metabolism I: Glycolysis and the Tricarboxylic Acid CycleCarbohydrate Metabolism. Chapter 12." In *Essentials of Medical Biochemistry*, 165–85. Doi:10.1016/b978-0-12-416687-5.00012-9.

Borgström, Celina, Lisa Wasserstrom, Henrik Almqvist, Kristina Broberg, Bianca Klein, Stephan Noack, Gunnar Lidén, and Marie F. Gorwa-Grauslund. 2019. "Identification of Modifications Procuring Growth on Xylose in Recombinant Saccharomyces Cerevisiae Strains Carrying the Weimberg Pathway." *Metabolic Engineering* 55 (May). Elsevier Inc.: 1–11. Doi:10.1016/j.ymben.2019.05.010.

Bozell, Joseph J., L. Moens, D. C. Elliott, Y. Wang, G. G. Neuenscwander, S. W. Fitzpatrick, R. J. Bilski, and J. L. Jarnefeld. 2000. "Production of Levulinic Acid and Use as a Platform Chemical for Derived Products." *Resources, Conservation and Recycling* 28 (3–4): 227–39. Doi:10.1016/S0921-3449(99)00047-6.

Bozell, Joseph J., and Gene R. Petersen. 2010a. "Technology Development for the Production of Biobased Products from Biorefinery Carbohydrates—the US Department of Energy's 'Top 10' Revisited." *Green Chemistry* 12 (4): 539. Doi:10.1039/b922014c.

———. 2010b. "Technology Development for the Production of Biobased Products from Biorefinery Carbohydrates—The US Department of Energy's 'Top 10' Revisited." *Green Chemistry* 12 (4): 539–54. Doi:10.1039/b922014c.

Bridger, William A. 1974. "Succinyl- CoA Synthetase. Chapter 18." In *The Enzymes*, 581–606. Academic Press.

Brüsseler, Christian, Anja Späth, Sascha Sokolowsky, and Jan Marienhagen. 2019. "Alone at Last!—Heterologous Expression of a Single Gene Is Sufficient for Establishing the Five-Step Weimberg Pathway in Corynebacterium Glutamicum." *Metabolic Engineering Communications* 9 (March). Doi:10.1016/j.mec.2019.e00090.

Buchholz, Jens, Andreas Schwentner, Britta Brunnenkan, Christina Gabris, Simon Grimm, Robert Gerstmeir, Ralf Takors, Bernhard J. Eikmanns, and Bastian Blombacha. 2013.

"Platform Engineering of Corynebacterium Glutamicum with Reduced Pyruvate Dehydrogenase Complex Activity for Improved Production of L-Lysine, l-Valine, and 2-Ketoisovalerate." *Applied and Environmental Microbiology* 79 (18): 5566–75. Doi:10.1128/AEM.01741-13.

Bückle-Vallant, Verena, Felix S. Krause, Sonja Messerschmidt, and Bernhard J. Eikmanns. 2014. "Metabolic Engineering of Corynebacterium Glutamicum for 2-Ketoisocaproate Production." *Applied Microbiology and Biotechnology* 98 (1): 297–311. Doi:10.1007/s00253-013-5310-2.

Chen, Jin Qiang, and Jose Russo. 2012. "Dysregulation of Glucose Transport, Glycolysis, TCA Cycle and Glutaminolysis by Oncogenes and Tumor Suppressors in Cancer Cells." *Biochimica et Biophysica Acta—Reviews on Cancer* 1826 (2). Elsevier B.V.: 370–84. Doi:10.1016/j.bbcan.2012.06.004.

Cheong, Seokjung, James M. Clomburg, and Ramon Gonzalez. 2016. "Energy-and Carbon-Efficient Synthesis of Functionalized Small Molecules in Bacteria Using Non-Decarboxylative Claisen Condensation Reactions." *Nature Biotechnology* 34 (5). Nature Publishing Group: 556–61. Doi:10.1038/nbt.3505.

———. 2018. "A Synthetic Pathway for the Production of 2-Hydroxyisovaleric Acid in Escherichia Coli." *Journal of Industrial Microbiology and Biotechnology* 45 (7). Springer Berlin Heidelberg: 579–88. Doi:10.1007/s10295-018-2005-9.

Chia, Mei, Thi Bich Van Nguyen, and Won Jae Choi. 2008. "DO-Stat Fed-Batch Production of 2-Keto-d-Gluconic Acid from Cassava Using Immobilized Pseudomonas Aeruginosa." *Applied Microbiology and Biotechnology* 78 (5): 759–65. Doi:10.1007/s00253-008-1374-9.

Choi, Sol, Chan Woo Song, Jae Ho Shin, and Sang Yup Lee. 2015. "Biorefineries for the Production of Top Building Block Chemicals and Their Derivatives." *Metabolic Engineering* 28. Elsevier: 223–39. Doi:10.1016/j.ymben.2014.12.007.

Choo, Jin Ho, Changpyo Han, Dong Wook Lee, Gyu Hun Sim, Hye Yun Moon, Jae Young Kim, Ji Yoon Song, and Hyun Ah Kang. 2018. "Molecular and Functional Characterization of Two Pyruvate Decarboxylase Genes, PDC1 and PDC5, in the Thermotolerant Yeast Kluyveromyces Marxianus." *Applied Microbiology and Biotechnology* 102 (8). Applied Microbiology and Biotechnology: 3723–37. Doi:10.1007/s00253-018-8862-3.

Chubukov, Victor, Aindrila Mukhopadhyay, Christopher J. Petzold, Jay D. Keasling, and Héctor García Martín. 2018. "Synthetic and Systems Biology for Microbial Production of Commodity Chemicals." *Npj Systems Biology and Applications* 2 (December 2015): 1–11. Doi:10.1038/npjsba.2016.9.

Chung, Sun Ho, Gang Guk Choi, Hyung Woo Kim, and Young Ha Rhee. 2001. "Effect of Levulinic Acid on the Production of Poly(3-Hydroxybutyrate-Co-3-Hydroxyvalerate) by Ralstonia Eutropha KHB-8862." *Journal of Microbiology.*

Clauser, Nicolás M., Fernando E. Felissia, María C. Area, and María E. Vallejos. 2021. "A Framework for the Design and Analysis of Integrated Multi-Product Biorefineries from Agricultural and Forestry Wastes." *Renewable and Sustainable Energy Reviews.* Elsevier Ltd. Doi:10.1016/j.rser.2020.110687.

Clomburg, James M., Shuai Qian, Zaigao Tan, Seokjung Cheong, and Ramon Gonzalez. 2019. "The Isoprenoid Alcohol Pathway, a Synthetic Route for Isoprenoid Biosynthesis." *Proceedings of the National Academy of Sciences of the United States of America* 116 (26): 12810–15. Doi:10.1073/pnas.1821004116.

Coban, Hasan B., Ali Demirci, Paul H. Patterson, and Ryan J. Elias. 2016. "Enhanced Phenylpyruvic Acid Production with Proteus Vulgaris in Fed-Batch and Continuous Fermentation." *Preparative Biochemistry and Biotechnology* 46 (2): 157–60. Doi:10.1080/10826068.2014.995813.

Covinich, Laura G., Nicolás M. Clauser, Fernando E. Felissia, María E. Vallejos, and María C. Area. 2019. "The Challenge of Converting Biomass Polysaccharides into Levulinic Acid through Heterogeneous Catalytic Processes." *Biofuels, Bioproducts and Biorefining*, 1–29. Doi:10.1002/bbb.2062.

Cronk, Jeff D., James A. Endrizzi, Michelle R. Cronk, Jason W. O'neill, and Kam Y.J. Zhang. 2001. "Crystal Structure of E. Coli β-Carbonic Anhydrase, an Enzyme with an Unusual PH-Dependent Activity." *Protein Science* 10 (5): 911–22. Doi:10.1110/ps.46301.

da Conceição Gomes, Absai, Maria Isabel Rodrigues, Douglas de França Passos, Aline Machado de Castro, Lidia Maria Mello Santa Anna, and Nei Pereira. 2019. "Acetone—butanol—ethanol Fermentation from Sugarcane Bagasse Hydrolysates: Utilization of C5 and C6 Sugars." *Electronic Journal of Biotechnology* 42. Elsevier España, S.L.U.: 16–22. Doi:10.1016/j.ejbt.2019.10.004.

Das, Debarati, and Bhalchandra M. Bhanage. 2020. *Double Carbonylation Reactions: Overview and Recent Advances. Advanced Synthesis and Catalysis*. Vol. 362. Doi:10.1002/adsc.202000245.

Davis, Reeta, Rashmi Kataria, Federico Cerrone, Trevor Woods, Shane Kenny, Anthonia O'Donovan, Maciej Guzik, et al. 2013. "Conversion of Grass Biomass into Fermentable Sugars and Its Utilization for Medium Chain Length Polyhydroxyalkanoate (Mcl-PHA) Production by Pseudomonas Strains." *Bioresource Technology* 150. Elsevier Ltd: 202–9. Doi:10.1016/j.biortech.2013.10.001.

Dutta, Shanta, Iris K.M. Yu, Daniel C.W. Tsang, Yun Hau Ng, Yong Sik Ok, James Sherwood, and James H. Clark. 2019. "Green Synthesis of Gamma-Valerolactone (GVL) through Hydrogenation of Biomass-Derived Levulinic Acid Using Non-Noble Metal Catalysts: A Critical Review." *Chemical Engineering Journal* 372 (April). Elsevier: 992–1006. Doi:10.1016/j.cej.2019.04.199.

Elgharbawy, Amal A., Md Zahangir Alam, Muhammad Moniruzzaman, and Masahiro Goto. 2016. "Ionic Liquid Pretreatment as Emerging Approaches for Enhanced Enzymatic Hydrolysis of Lignocellulosic Biomass." *Biochemical Engineering Journal* 109. Elsevier B.V.: 252–67. Doi:10.1016/j.bej.2016.01.021.

Feng, Lili, Ya Zhang, Jing Fu, Yufeng Mao, Tao Chen, Xueming Zhao, and Zhiwen Wang. 2016. "Metabolic Engineering of Corynebacterium Glutamicum for Efficient Production of 5-Aminolevulinic Acid." *Biotechnology and Bioengineering* 113 (6): 1284–93. Doi:10.1002/bit.25886.

Fonseca, Gustavo Graciano, Elmar Heinzle, Christoph Wittmann, and Andreas K. Gombert. 2008. "The Yeast Kluyveromyces Marxianus and Its Biotechnological Potential." *Applied Microbiology and Biotechnology* 79 (3): 339–54. Doi:10.1007/s00253-008-1458-6.

Gallo, Jean Marcel R., David Martin Alonso, Max A. Mellmer, and James A. Dumesic. 2013. "Production and Upgrading of 5-Hydroxymethylfurfural Using Heterogeneous Catalysts and Biomass-Derived Solvents." *Green Chemistry* 15 (1): 85–90. Doi:10.1039/c2gc36536g.

Gan, Q., S. J. Allen, and G. Taylor. 2003. "Kinetic Dynamics in Heterogeneous Enzymatic Hydrolysis of Cellulose: An Overview, an Experimental Study and Mathematical Modelling." *Process Biochemistry* 38 (7): 1003–18. Doi:10.1016/S0032-9592(02)00220-0.

Gao, Haijun, Yu Gao, and Runan Dong. 2017. "Enhanced Biosynthesis of 3,4-Dihydroxybutyric Acid by Engineered Escherichia Coli in a Dual-Substrate System." *Bioresource Technology* 245. Elsevier Ltd: 794–800. Doi:10.1016/j.biortech.2017.09.017.

Gheshlaghi, R., J. M. Scharer, M. Moo-Young, and C. P. Chou. 2009. "Metabolic Pathways of Clostridia for Producing Butanol." *Biotechnology Advances* 27 (6). Elsevier B.V.: 764–81. Doi:10.1016/j.biotechadv.2009.06.002.

Girisuta, B., K. Dussan, D. Haverty, J. J. Leahy, and M. H.B. Hayes. 2013. "A Kinetic Study of Acid Catalysed Hydrolysis of Sugar Cane Bagasse to Levulinic Acid." *Chemical Engineering Journal* 217:61–70. Doi:10.1016/j.cej.2012.11.094.

Gonzalez, Ramon, Han Tao, K. T. Shanmugam, S. W. York, and L. O. Ingram. 2002. "Global Gene Expression Differences Associated with Changes in Glycolytic Flux and Growth Rate in Escherichia Coli during the Fermentation of Glucose and Xylose." *Biotechnology Progress* 18 (1): 6–20. Doi:10.1021/bp010121i.

Gorenflo, V., G. Schmack, R. Vogel, and A. Steinbüchel. 2001. "Development of a Process for the Biotechnological Large-Scale Production of 4-Hydroxyvalerate-Containing Polyesters and Characterization of Their Physical and Mechanical Properties." *Biomacromolecules* 2 (1): 45–57. Doi:10.1021/bm0000992.

Gregg, David J., and John N. Saddler. 1996. "Factors Affecting Cellulose Hydrolysis and the Potential of Enzyme Recycle to Enhance the Efficiency of an Integrated Wood to Ethanol Process." *Biotechnology and Bioengineering* 51 (4): 375–83. Doi:10.1002/(SICI)1097-0290(19960820)51:4<375::AID-BIT1>3.0.CO;2-F.

Grotjohann, Norbert, Yi Huang, and Wolfgang Kowallik. 2001. "Tricarboxylic Acid Cycle Enzymes of the Ectomycorrhizal Basidiomycete, Suillus Bovinus." *Zeitschrift Fur Naturforschung—Section C Journal of Biosciences* 56 (5–6): 334–42. Doi:10.1515/znc-2001-5-603.

Guo, Hongwei, Catherine Madzak, Guocheng Du, Jingwen Zhou, and Jian Chen. 2014. "Effects of Pyruvate Dehydrogenase Subunits Overexpression on the α-Ketoglutarate Production in Yarrowia Lipolytica WSH-Z06." *Applied Microbiology and Biotechnology* 98 (16): 7003–12. Doi:10.1007/s00253-014-5745-0.

Habe, Hiroshi, Hideaki Koike, Yuya Sato, Yosuke Iimura, Tomoyuki Hori, Manabu Kanno, Nobutada Kimura, and Kohtaro Kirimura. 2019. "Identification and Characterization of Levulinyl-CoA Synthetase from Pseudomonas Citronellolis, Which Differs Phylogenetically from LvaE of Pseudomonas Putida." *AMB Express* 9 (1). Springer Berlin Heidelberg. Doi:10.1186/s13568-019-0853-y.

Habe, Hiroshi, Yuya Sato, and Kohtaro Kirimura. 2020. "Microbial and Enzymatic Conversion of Levulinic Acid, an Alternative Building Block to Fermentable Sugars from Cellulosic Biomass." *Applied Microbiology and Biotechnology* 104 (18). Applied Microbiology and Biotechnology: 7767–75. Doi:10.1007/s00253-020-10813-7.

Hartweg, Manuel, and C. Remzi Becer. 2016. "Direct Polymerization of Levulinic Acid via Ugi Multicomponent Reaction." *Green Chemistry* 18 (11). Royal Society of Chemistry: 3272–77. Doi:10.1039/c6gc00372a.

———. 2018. "Levulinic Acid as Sustainable Feedstock in Polymer Chemistry." In *Green Polymer Chemistry: New Products, Processes, and Applications*, Vol. 1310, 331–38. Washington, DC: ACS Symposium Series. American Chemical Society. Doi:10.1021/bk-2018-1310.ch020.

Harutyunyan, Baghish, Mario Novak, Mladen Pavlečić, Vigen Goginyan, Ruzanna Hovhannesyan, Inna Melkumyan, and Božidar Šantek. 2018. "Influence of Glycine, Succinate, Levulinic Acid and Glutamate Concentrations on Growth of Purple Non-Sulfur Photosynthetic Bacteria and 5-Aminolevulinic Acid Production." *International Journal of Innovative Research in Science, Engineering and Technology* 7 (4): 3839–46. Doi:10.15680/ijirset.2018.0704094.

Hasona, Adnan, Youngnyun Kim, F. G. Healy, L. O. Ingram, and K. T. Shanmugam. 2004. "Pyruvate Formate Lyase and Acetate Kinase Are Essential for Anaerobic Growth of Escherichia Coli on Xylose." *Journal of Bacteriology* 186 (22): 7593–7600. Doi:10.1128/JB.186.22.7593-7600.2004.

Holz, Martina, André Förster, Stephan Mauersberger, and Gerold Barth. 2009. "Aconitase Overexpression Changes the Product Ratio of Citric Acid Production by Yarrowia Lipolytica." *Applied Microbiology and Biotechnology* 81 (6): 1087–96. Doi:10.1007/s00253-008-1725-6.

Hossain, Gazi Sakir, Hyun Dong Shin, Jianghua Li, Guocheng Du, Jian Chen, and Long Liu. 2016. "Transporter Engineering and Enzyme Evolution for Pyruvate Production from d/l-Alanine with a Whole-Cell Biocatalyst Expressing l-Amino Acid Deaminase from Proteus Mirabilis." *RSC Advances* 6 (86). Royal Society of Chemistry: 82676–84. Doi:10.1039/c6ra16507a.

Hu, Cuimin, Siguo Wu, Qian Wang, Guojie Jin, Hongwei Shen, and Zongbao K. Zhao. 2011. "Simultaneous Utilization of Glucose and Xylose for Lipid Production by Trichosporon Cutaneum." *Biotechnology for Biofuels* 4. Doi:10.1186/1754-6834-4-25.

Hunter, G. A., and Gloria C. Ferreira. 2009. "5-Aminolevulinate Synthase: Catalysis of the First Step of Heme Biosynthesis." *Cellular and Molecular Biology* 55 (1): 102–10. Doi:10.1170/T844.

Il, Lae Jung, Heon Phyo Ki, and In Gyu Kim. 2005. "RpoS-Mediated Growth-Dependent Expression of the Escherichia Coli Tkt Genes Encoding Transketolases Isoenzymes." *Current Microbiology* 50 (6): 314–18. Doi:10.1007/s00284-005-4501-1.

Ishchuk, Olena P., Andriy Y. Voronovsky, Oleh V. Stasyk, Galina Z. Gayda, Mykhailo V. Gonchar, Charles A. Abbas, and Andriy A. Sibirny. 2008. "Overexpression of Pyruvate Decarboxylase in the Yeast Hansenula Polymorpha Results in Increased Ethanol Yield in High-Temperature Fermentation of Xylose." *FEMS Yeast Research* 8 (7): 1164–74. Doi:10.1111/j.1567-1364.2008.00429.x.

Jahn, D., and D. W. Heinz. 2008. "Biosynthesis of 5-Aminolevulinic Acid. Chapter 2." In *Tetrapyrroles. Molecular Biology Intelligence Unit*, 29–42. New York: SpringerLink.

Jaremko, Matt, and Jian Yu. 2011. "The Initial Metabolic Conversion of Levulinic Acid in Cupriavidus Necator." *Journal of Biotechnology* 155 (3). Elsevier B.V.: 293–98. Doi:10.1016/j.jbiotec.2011.07.027.

Jecu, Luiza. 2000. "Solid State Fermentation of Agricultural Wastes for Endoglucanase Production." *Industrial Crops and Products* 11 (1): 1–5. Doi:10.1016/S0926-6690(99)00022-9.

Jeffries, T. W. 1983. "Utilization of Xylose by Bacteria, Yeasts, and Fungi." In *Advances in Biochemical Engineering/Biotechnology*, Vol. 27, 1–32. Doi:10.1007/bfb0009101.

Jelakovic, Stefan, Stanislav Kopriva, Karl Heinz Süss, and Georg E. Schulz. 2003. "Structure and Catalytic Mechanism of the Cytosolic D-Ribulose-5-Phosphate 3-Epimerase from Rice." *Journal of Molecular Biology* 326 (1): 127–35. Doi:10.1016/S0022-2836(02)01374-8.

Jin, Yong Su, and Gregory Stephanopoulos. 2007. "Multi-Dimensional Gene Target Search for Improving Lycopene Biosynthesis in Escherichia Coli." *Metabolic Engineering* 9 (4): 337–47. Doi:10.1016/j.ymben.2007.03.003.

Johansson, Björn, Camilla Christensson, Timothy Hobley, and Bärbel Hahn-Hägerdal. 2001. "Xylulokinase Overexpression in Two Strains of Saccharomyces Cerevisiae Also Expressing Xylose Reductase and Xylitol Dehydrogenase and Its Effect on Fermentation of Xylose and Lignocellulosic Hydrolysate." *Applied and Environmental Microbiology* 67 (9): 4249–55. Doi:10.1128/AEM.67.9.4249-4255.2001.

Kamm, B., and M. Kamm. 2004. "Principles of Biorefineries." *Applied Microbiology and Biotechnology* 64 (2): 137–45. Doi:10.1007/s00253-003-1537-7.

Kang, Zhen, Yang Wang, Pengfei Gu, Qian Wang, and Qingsheng Qi. 2011. "Engineering Escherichia Coli for Efficient Production of 5-Aminolevulinic Acid from Glucose." *Metabolic Engineering* 13 (5). Elsevier: 492–98. Doi:10.1016/j.ymben.2011.05.003.

Kawasaki, Yukie, Nag Aniruddha, Hajime Minakawa, Shunsuke Masuo, Tatsuo Kaneko, and Naoki Takaya. 2018. "Novel Polycondensed Biopolyamide Generated from Biomass-Derived 4-Aminohydrocinnamic Acid." *Applied Microbiology and Biotechnology* 102 (2). Applied Microbiology and Biotechnology: 631–39. Doi:10.1007/s00253-017-8617-6.

Kayne, F. J., and N. C. Price. 1973. "Amino Acid Effector Binding to Rabbit Muscle Pyruvate Kinase'." *Archives of Biochemistry and Biophysics* 159 (1): 292–96. Doi:10.1016/S1874-6047(08)60071-2.

Kim, Doyun, Chandran Sathesh-Prabu, Young Jooyeon, and Sung Kuk Lee. 2019. "High-Level Production of 4-Hydroxyvalerate from Levulinic Acid via Whole-Cell Biotransformation

Decoupled from Cell Metabolism." Research-article. *Journal of Agricultural and Food Chemistry* 67 (38). American Chemical Society: 10678–84. Doi:10.1021/acs.jafc.9b04304.

Kim, Seohyoung, James M. Clomburg, and Ramon Gonzalez. 2015. "Synthesis of Medium-Chain Length (C6—C10) Fuels and Chemicals via β-Oxidation Reversal in Escherichia Coli." *Journal of Industrial Microbiology and Biotechnology* 42 (3): 465–75. Doi:10.1007/s10295-015-1589-6.

Klingler, Franz Dietrich, and Wolfgang Ebertz. 2000. "Oxocarboxylic Acids." *Ullmann's Encyclopedia of Industrial Chemistry*. Doi:10.1002/14356007.a18_313.

Kricheldorf, Hans R., Radka Hobzova, and Gert Schwarz. 2003. "Cyclic Hyperbranched Polyesters Derived from 4,4-Bis(4′ -Hydroxyphenyl) Valeric Acid." *Polymer* 44 (24): 7361–68. Doi:10.1016/j.polymer.2003.09.041.

Lee, Jeong Wook, Dokyun Na, Jong Myoung Park, Joungmin Lee, Sol Choi, and Sang Yup Lee. 2012. "Systems Metabolic Engineering of Microorganisms for Natural and Non-Natural Chemicals." *Nature Chemical Biology* 8 (6). Nature Publishing Group: 536–46. Doi:10.1038/nchembio.970.

Lee, Kwang Ho, Jin Hwan Park, Tae Yong Kim, Hyun Uk Kim, and Sang Yup Lee. 2007. "Systems Metabolic Engineering of Escherichia Coli for L-Threonine Production." *Molecular Systems Biology* 3 (149). Doi:10.1038/msb4100196.

Lee, Sang Yup, Hyun Uk Kim, Tong Un Chae, Jae Sung Cho, Je Woong Kim, Jae Ho Shin, Dong In Kim, Yoo Sung Ko, Woo Dae Jang, and Yu Sin Jang. 2019. "A Comprehensive Metabolic Map for Production of Bio-Based Chemicals." *Nature Catalysis* 2 (1). Springer US: 18–33. Doi:10.1038/s41929-018-0212-4.

Lee, Yoojin, Hye Jeong Cho, and Han Min Woo. 2020. "A Hybrid Embden—Meyerhof—Parnas Pathway Provides a Synthetic Link between Sugar and Phosphate Metabolism." *BioRxiv*, 0–2. Doi:10.1101/2020.06.11.147082.

Li, Ruoxi, Hossain Gazi Sakir, Jianghua Li, Hyun Dong Shin, Guocheng Du, Jian Chen, and Long Liu. 2017. "Rational Molecular Engineering of L-Amino Acid Deaminase for Production of α-Ketoisovaleric Acid from l-Valine by: Escherichia Coli." *RSC Advances* 7 (11). Royal Society of Chemistry: 6615–21. Doi:10.1039/c6ra26972a.

Lin, Jianping, Weiqi Fu, and Peilin Cen. 2009. "Characterization of 5-Aminolevulinate Synthase from Agrobacterium Radiobacter, Screening New Inhibitors for 5-Aminolevulinate Dehydratase from Escherichia Coli and Their Potential Use for High 5-Aminolevulinate Production." *Bioresource Technology* 100 (7). Elsevier Ltd: 2293–97. Doi:10.1016/j.biortech.2008.11.008.

Luo, Zhengshan, Shiqin Yu, Weizhu Zeng, and Jingwen Zhou. 2021. "Comparative Analysis of the Chemical and Biochemical Synthesis of Keto Acids." *Biotechnology Advances* 47 (January). Elsevier Inc.: 107706. Doi:10.1016/j.biotechadv.2021.107706.

Ma, Menggen, Xu Wang, Xiaoping Zhang, and Xianxian Zhao. 2013. "Alcohol Dehydrogenases from Scheffersomyces Stipitis Involved in the Detoxification of Aldehyde Inhibitors Derived from Lignocellulosic Biomass Conversion." *Applied Microbiology and Biotechnology* 97 (18): 8411–25. Doi:10.1007/s00253-013-5110-8.

Mahmoudi, K., K. L. Garvey, A. Bouras, G. Cramer, H. Stepp, J. G. Jesu Raj, D. Bozec, T. M. Busch, and C. G. Hadjipanayis. 2019. "5-Aminolevulinic Acid Photodynamic Therapy for the Treatment of High-Grade Gliomas." *Journal of Neuro-Oncology* 141 (3). Springer US: 595–607. Doi:10.1007/s11060-019-03103-4.

Mario Pinto, B. 1999. "The World of Carbohydrates and Associated Natural Products. Chapter 3.01." In *Comprehensive Natural Products Chemistry*, 1–11. Elsevier. Doi:10.1016/b978-0-08-091283-7.00160-0.

Martin, Collin H., and Kristala L. Jones Prather. 2009. "High-Titer Production of Monomeric Hydroxyvalerates from Levulinic Acid in Pseudomonas Putida." *Journal of Biotechnology* 139 (1): 61–67. Doi:10.1016/j.jbiotec.2008.09.002.

Martin, Collin H., Danyi Wu, and Kristala L. Jones Prather. 2010. "Integrated Bioprocessing for the PH-Dependent Production of 4-Valerolactone from Levulinate in Pseudomonas Putida KT2440." *Applied and Environmental Microbiology* 76 (2): 417–24. Doi:10.1128/AEM.01769-09.

Martínez-García, Esteban, Pablo I Nikel, Tomás Aparicio, and Víctor De Lorenzo. 2014. "Pseudomonas 2.0: Genetic Upgrading of P. Putida KT2440 as an Enhanced Host for Heterologous Gene Expression." *Microbial Cell Factories* 13 (159): 1–15.

Mehrer, Christopher R., Jacqueline M. Rand, Matthew R. Incha, Taylor B. Cook, Benginur Demir, Ali Hussain Motagamwala, Daniel Kim, James A. Dumesic, and Brian F. Pfleger. 2019. "Growth-Coupled Bioconversion of Levulinic Acid to Butanone." *Metabolic Engineering* 55 (June): 92–101. Doi:10.1016/j.ymben.2019.06.003.

Mellmer, Max A., Jean Marcel R. Gallo, David Martin Alonso, and James A. Dumesic. 2015. "Selective Production of Levulinic Acid from Furfuryl Alcohol in THF Solvent Systems over H-ZSM-5." *ACS Catalysis* 5 (6): 3354–59. Doi:10.1021/acscatal.5b00274.

Mellmer, Max A., Canan Sener, Jean Marcel R. Gallo, Jeremy S. Luterbacher, David Martin Alonso, and James A. Dumesic. 2014. "Solvent Effects in Acid-Catalyzed Biomass Conversion Reactions." *Angewandte Chemie—International Edition* 53 (44): 11872–75. Doi:10.1002/anie.201408359.

Meza, Eugenio, Judith Becker, Francisco Bolivar, Guillermo Gosset, and Christoph Wittmann. 2012. "Consequences of Phosphoenolpyruvate:Sugar Phosphotranferase System and Pyruvate Kinase Isozymes Inactivation in Central Carbon Metabolism Flux Distribution in Escherichia Coli." *Microbial Cell Factories* 11: 1–13. Doi:10.1186/1475-2859-11-127.

Milne, J. L. S. 2013. *Structure and Regulation of Pyruvate Dehydrogenases. Encyclopedia of Biological Chemistry*. 2nd ed. Published by Elsevier Inc. doi:10.1016/B978-0-12-378630-2.00079-7.

Min, Kyoungseon, Seil Kim, Taewoo Yum, Yunje Kim, Byoung In Sang, and Youngsoon Um. 2013. "Conversion of Levulinic Acid to 2-Butanone by Acetoacetate Decarboxylase from Clostridium Acetobutylicum." *Applied Microbiology and Biotechnology* 97 (12): 5627–34. Doi:10.1007/s00253-013-4879-9.

Morone, Amruta, Mayura Apte, and R. A. Pandey. 2015. "Levulinic Acid Production from Renewable Waste Resources: Bottlenecks, Potential Remedies, Advancements and Applications." *Renewable and Sustainable Energy Reviews* 51. Elsevier: 548–65. Doi:10.1016Zj/rser.2015.06.032.

Mukherjee, Agneev, Marie Josée Dumont, and Vijaya Raghavan. 2015. "Review: Sustainable Production of Hydroxymethylfurfural and Levulinic Acid: Challenges and Opportunities." *Biomass and Bioenergy* 72: 143–83. Doi:10.1016/j.biombioe.2014.11.007.

Na, Dokyun, Tae Yong Kim, and Sang Yup Lee. 2010. "Construction and Optimization of Synthetic Pathways in Metabolic Engineering." *Current Opinion in Microbiology* 13 (3). Elsevier Ltd: 363–70. Doi:10.1016/j.mib.2010.02.004.

Nduko, John Masani, Ken'ichiro Matsumoto, Toshihiko Ooi, and Seiichi Taguchi. 2013. "Effectiveness of Xylose Utilization for High Yield Production of Lactate-Enriched P(Lactate-Co-3-Hydroxybutyrate) Using a Lactate-Overproducing Strain of Escherichia Coli and an Evolved Lactate-Polymerizing Enzyme." *Metabolic Engineering* 15 (1). Elsevier: 159–66. Doi:10.1016/j.ymben.2012.11.007.

Nhien, Le Cao, Nguyen Van Duc Long, and Moonyong Lee. 2016. "Design and Optimization of the Levulinic Acid Recovery Process from Lignocellulosic Biomass." *Chemical Engineering Research and Design* 107. Institution of Chemical Engineers: 126–36. Doi:10.1016/j.cherd.2015.09.013.

Nikel, Pablo I., and Víctor de Lorenzo. 2018. "Pseudomonas Putida as a Functional Chassis for Industrial Biocatalysis: From Native Biochemistry to Trans-Metabolism." *Metabolic Engineering* 50 (May). Elsevier Inc.: 142–55. Doi:10.1016/j.ymben.2018.05.005.

Nishikawa, Seiji, and Yoshikatsu Murooka. 2001. "5-Aminolevulinic Acid: Production by Fermentation, and Agricultural and Biomedical Applications." *Biotechnology and Genetic Engineering Reviews* 18 (March 2014): 149–70. Doi:10.1080/02648725.2001.10648012.

Novackova, Ivana, Dan Kucera, Jaromir Porizka, Iva Pernicova, Petr Sedlacek, Martin Koller, Adriana Kovalcik, and Stanislav Obruca. 2019. "Adaptation of Cupriavidus Necator to Levulinic Acid for Enhanced Production of P(3HB-Co-3HV) Copolyesters." *Biochemical Engineering Journal* 151 (August). Elsevier: 107350. Doi:10.1016/j.bej.2019.107350.

Nshimiyimana, Project, Long Liu, and Guocheng Du. 2019. "Engineering of L-Amino Acid Deaminases for the Production of α-Keto Acids from L-Amino Acids." *Bioengineered* 10 (1). Taylor & Francis: 43–51. Doi:10.1080/21655979.2019.1595990.

Olsson, Carina, Alexander Idström, Lars Nordstierna, and Gunnar Westman. 2014. "Influence of Water on Swelling and Dissolution of Cellulose in 1-Ethyl-3-Methylimidazolium Acetate." *Carbohydrate Polymers* 99: 438–46. Doi:10.1016/j.carbpol.2013.08.042.

Oraby, Hesham, Balan Venkatesh, Bruce Dale, Rashid Ahmad, Callista Ransom, James Oehmke, and Mariam Sticklen. 2007. "Enhanced Conversion of Plant Biomass into Glucose Using Transgenic Rice-Produced Endoglucanase for Cellulosic Ethanol." *Transgenic Research* 16 (6): 739–49. Doi:10.1007/s11248-006-9064-9.

Otero, José Manuel, and Jens Nielsen. 2010. "Industrial Systems Biology." *Biotechnology and Bioengineering* 105 (3): 439–60. Doi:10.1002/bit.22592.

Otto, Christina, Venelina Yovkova, and Gerold Barth. 2011. "Overproduction and Secretion of α-Ketoglutaric Acid by Microorganisms." *Applied Microbiology and Biotechnology* 92 (4): 689–95. Doi:10.1007/s00253-011-3597-4.

Park, Jin Hwan, Sang Yup Lee, Tae Yong Kim, and Hyun Uk Kim. 2008. "Application of Systems Biology for Bioprocess Development." *Trends in Biotechnology* 26 (8): 404–12. Doi:10.1016/j.tibtech.2008.05.001.

Peterson, John I., and Donald S. Young. 1968. "Evaluation of the Hexokinase/Glucose-6-Phosphate Dehydrogenase Method of Determination of Glucose in Urine." *Biochemistry, Analytical* 316: 301–16.

Pileidis, Filoklis D., and Maria Magdalena Titirici. 2016. "Levulinic Acid Biorefineries: New Challenges for Efficient Utilization of Biomass." *ChemSusChem* 9 (6): 562–82. Doi:10.1002/cssc.201501405.

Pinto-Ibieta, F., M. Cea, F. Cabrera, M. Abanto, F. E. Felissia, M. C. Area, and G. Ciudad. 2020. "Strategy for Biological Co-Production of Levulinic Acid and Polyhydroxyalkanoates by Using Mixed Microbial Cultures Fed with Synthetic Hemicellulose Hydrolysate." *Bioresource Technology* 309 (April). Elsevier: 123323. Doi:10.1016/j.biortech.2020.123323.

Plaitakis, Andreas, Ester Kalef-Ezra, Dimitra Kotzamani, Ioannis Zaganas, and Cleanthe Spanaki. 2017. "The Glutamate Dehydrogenase Pathway and Its Roles in Cell and Tissue Biology in Health and Disease." *Biology* 6 (1): 1–26. Doi:10.3390/biology6010011.

Polakis, E. S., W. Bartley, and G. A. Meek. 1965. "Changes in the Enzyme Activities of Saccharomyces Cerevisiae during the Aerobic Growth of Yeast on Different Carbon Sources." *Biochemical Journal* 97 (1): 298–302. Doi:10.1042/bj0970298.

Porter, Emilie A., Xifang Wang, Hee-seung Lee, and Bernard Weisblum. 2000. "Non-Haemolytic β-Amino-Acid Oligomers." *Nature* 404 (April): 2000.

Pospíšil, S., J. Kopecký, and V. Přikrylová. 1998. "Derepression and Altered Feedback Regulation of Valine Biosynthetic Pathway in Analogue-Resistant Mutants of Streptomyces Cinnamonensis Resulting in 2-Ketoisovalerate Excretion." *Journal of Applied Microbiology*. Doi:10.1046/j.1365-2672.1998.00459.x.

Prather, Kristala L. Jones, and Collin H. Martin. 2008. "De Novo Biosynthetic Pathways: Rational Design of Microbial Chemical Factories." *Current Opinion in Biotechnology* 19 (5): 468–74. Doi:10.1016/j.copbio.2008.07.009.

Qin, Gang, Jianping Lin, Xiaoxia Liu, and Peilin Cen. 2006. "Effects of Medium Composition on Production of 5-Aminolevulinic Acid by Recombinant Escherichia Coli." *Journal of Bioscience and Bioengineering* 102 (4): 316–22. Doi:10.1263/jbb.102.316.

Ramzi, Ahmad Bazli, Jeong Eun Hyeon, Seung Wook Kim, Chulhwan Park, and Sung Ok Han. 2015. "5-Aminolevulinic Acid Production in Engineered Corynebacterium Glutamicum via C5 Biosynthesis Pathway." *Enzyme and Microbial Technology* 81. Elsevier Inc.: 1–7. Doi:10.1016/j.enzmictec.2015.07.004.

Rapala-Kozik, Maria. 2011. *Vitamin b 1 (Thiamine). A Cofactor for Enzymes Involved in the Main Metabolic Pathways and an Environmental Stress Protectant. Advances in Botanical Research.* 1st ed. Vol. 58. Elsevier Ltd. Doi:10.1016/B978-0-12-386479-6.00004-4.

Rebeiz, C. A., A. Montazer-Zouhoor, H. J. Hopen, and S. M. Wu. 1984. "Photodynamic Herbicides: 1. Concept and Phenomenology." *Enzyme and Microbial Technology* 6 (9): 390–96. Doi:10.1016/0141-0229(84)90012-7.

Rees, Douglas C., Michael K. Chan, and Jongsun Kim. 1993. "Structure and Function of Endoglucanase V." *Advances in Inorganic Chemistry* 40 I: 89–98. doi:10.1016/S0898-8838(08)60182-8.

Richards et al. 2018. "A Metabolic Pathway for Catabolizing Levulinic Acid in Bacteria." *Physiology & Behavior* 176 (5): 139–48. doi:10.1038/s41564-017-0028-z.A.

Rudroff, Florian, Marko D. Mihovilovic, Harald Gröger, Radka Snajdrova, Hans Iding, and Uwe T. Bornscheuer. 2018. "Opportunities and Challenges for Combining Chemo- and Biocatalysis." *Nature Catalysis* 1 (1). Springer US: 12–22. doi:10.1038/s41929-017-0010-4.

Sandström, Anders G., Henrik Almqvist, Diogo Portugal-Nunes, Dário Neves, Gunnar Lidén, and Marie F. Gorwa-Grauslund. 2014. "Saccharomyces Cerevisiae: A Potential Host for Carboxylic Acid Production from Lignocellulosic Feedstock?" *Applied Microbiology and Biotechnology* 98 (17): 7299–318. doi:10.1007/s00253-014-5866-5.

Sathesh-Prabu, Chandran, and Sung Kuk Lee. 2019. "Engineering the Lva Operon and Optimization of Culture Conditions for Enhanced Production of 4-Hydroxyvalerate from Levulinic Acid in Pseudomonas Putida KT2440." Research-article. *Journal of Agricultural and Food Chemistry* 67 (9). American Chemical Society: 2540–46. doi:10.1021/acs.jafc.8b06884.

Schmidt, Sandy, Kathrin Castiglione, and Robert Kourist. 2018. "Overcoming the Incompatibility Challenge in Chemoenzymatic and Multi-Catalytic Cascade Reactions." *Chemistry—A European Journal* 24 (8): 1755–68. doi:10.1002/chem.201703353.

Šivec, Rok, Miha Grilc, Matej Huš, and Blaž Likozar. 2019. "Multiscale Modeling of (Hemi) Cellulose Hydrolysis and Cascade Hydrotreatment of 5-Hydroxymethylfurfural, Furfural, and Levulinic Acid." *Industrial and Engineering Chemistry Research* 58 (35): 16018–32. doi:10.1021/acs.iecr.9b00898.

Song, Yang, Jianghua Li, Hyun Dong Shin, Long Liu, Guocheng Du, and Jian Chen. 2016. "Biotechnological Production of Alpha-Keto Acids: Current Status and Perspectives." *Bioresource Technology* 219: 716–24. doi:10.1016/j.biortech.2016.08.015.

———. 2017. "Tuning the Transcription and Translation of L-Amino Acid Deaminase in Escherichia Coli Improves α-Ketoisocaproate Production from L-Leucine." *PLoS One* 12 (6): 1–14. doi:10.1371/journal.pone.0179229.

Song, Yingxian, Hua Pu, Tian Jiang, Lixin Zhang, and Min Ouyang. 2016. "Crystal Structure of Glutamate-1-Semialdehyde-2,1-Aminomutase from Arabidopsis Thaliana." *Acta Crystallographica Section:F Structural Biology Communications* 72: 448–56. doi:10.1107/S2053230X16007263.

Spaepen, Stijn, Wim Versées, Dörte Gocke, Martina Pohl, Jan Steyaert, and Jos Vanderleyden. 2007. "Characterization of Phenylpyruvate Decarboxylase, Involved in Auxin Production of Azospirillum Brasilense." *Journal of Bacteriology* 189 (21): 7626–33. doi:10.1128/JB.00830-07.

Steensmati, H. Y. D. E., Johannes P. V. A. N. Dijkent, and Jack T. Pronkt. 1996. "Pyruvate Decarboxylase: An Indispensable Enzyme for Growth of Saccharomyces Cerevisiae on Glucose." *Utilization of Xylose by Bacteria, Yeasts, and Fungi* 12: 247–57.

Stephens, Craig, Beat Christen, Thomas Fuchs, Vidyodhaya Sundaram, Kelly Watanabe, and Urs Jenal. 2007. "Genetic Analysis of a Novel Pathway for D-Xylose Metabolism in Caulobacter Crescentus." *Journal of Bacteriology* 189 (5): 2181–85. doi:10.1128/JB.01438-06.

Stuermer, Rainer, Bernhard Hauer, Melanie Hall, and Kurt Faber. 2007. "Asymmetric Bioreduction of Activated C=C Bonds Using Enoate Reductases from the Old Yellow Enzyme Family." *Current Opinion in Chemical Biology* 11 (2): 203–13. doi:10.1016/j.cbpa.2007.02.025.

Toivari, Mervi, Yvonne Nygård, Esa Pekka Kumpula, Maija Leena Vehkomäki, Mojca Benčina, Mari Valkonen, Hannu Maaheimo, et al. 2012. "Metabolic Engineering of Saccharomyces Cerevisiae for Bioconversion of D-Xylose to d-Xylonate." *Metabolic Engineering* 14 (4): 427–36. doi:10.1016/j.ymben.2012.03.002.

Uyeda, K., and S. Kurooka. 1970. "Crystallization and Properties of Phosphofructokinase from Clostridium Pasteurianum." *Journal of Biological Chemistry* 245 (13). Â© 1970 ASBMB. Currently published by Elsevier Inc; originally published by American Society for Biochemistry and Molecular Biology.: 3315–24. doi:10.1016/s0021-9258(18)62997-7.

Van Maris, Antonius J. A., Jan Maarten A. Geertman, Alexander Vermeulen, Matthijs K. Groothuizen, Aaron A. Winkler, Matthew D. W. Piper, Johannes P. Van Dijken, and Jack T. Pronk. 2004. "Directed Evolution of Pyruvate Decarboxylase-Negative Saccharomyces Cerevisiae, Yielding a C2-Independent, Glucose-Tolerant, and Pyruvate-Hyperproducing Yeast." *Applied and Environmental Microbiology* 70 (1): 159–66. doi:10.1128/AEM.70.1.159-166.2004.

Veit, Andrea, Tino Polen, and Volker F. Wendisch. 2007. "Global Gene Expression Analysis of Glucose Overflow Metabolism in Escherichia Coli and Reduction of Aerobic Acetate Formation." *Applied Microbiology and Biotechnology* 74 (2): 406–21. doi:10.1007/s00253-006-0680-3.

Wasserstrom, Lisa, Diogo Portugal-Nunes, Henrik Almqvist, Anders G. Sandström, Gunnar Lidén, and Marie F. Gorwa-Grauslund. 2018. "Exploring D-Xylose Oxidation in Saccharomyces Cerevisiae through the Weimberg Pathway." *AMB Express* 8 (1). Springer Berlin Heidelberg. doi:10.1186/s13568-018-0564-9.

WEIMBERG, R. 1961. "Pentose Oxidation by Pseudomonas Fragi." *The Journal of Biological Chemistry* 236 (3). Â© 1961 ASBMB. Currently published by Elsevier Inc; originally published by American Society for Biochemistry and Molecular Biology.: 629–35. doi:10.1016/s0021-9258(18)64279-6.

Werpy, T., and G. Petersen. 2004. "Top Value Added Chemicals from Biomass Volume I—Results of Screening for Potential Candidates from Sugars and Synthesis Gas Top Value Added Chemicals from Biomass Volume I: Results of Screening for Potential Candidates." *Other Information: PBD: 1 Aug 2004*, Medium: ED; Size: 76 pp. pages. doi:10.2172/15008859.

Wushensky, Julie A., Tracy Youngster, Caroll M. Mendonca, and Ludmilla Aristilde. 2018. "Flux Connections between Gluconate Pathway, Glycolysis, and Pentose-Phosphate Pathway during Carbohydrate Metabolism in Bacillus Megaterium QM B1551." *Frontiers in Microbiology* 9 (November): 1–13. doi:10.3389/fmicb.2018.02789.

Yan, Kai, Cody Jarvis, Jing Gu, and Yong Yan. 2015. "Production and Catalytic Transformation of Levulinic Acid: A Platform for Speciality Chemicals and Fuels." *Renewable and Sustainable Energy Reviews* 51. Elsevier: 986–97. doi:10.1016/j.rser.2015.07.021.

Yan, Kai, Yiyi Yang, Jiajue Chai, and Yiran Lu. 2015. "Catalytic Reactions of Gamma-Valerolactone: A Platform to Fuels and Value-Added Chemicals." *Applied Catalysis B: Environmental* 179. Elsevier B.V.: 292–304. doi:10.1016/j.apcatb.2015.04.030.

Yang, Maohua, and Xiang Zhang. 2017. "Construction of Pyruvate Producing Strain with Intact Pyruvate Dehydrogenase and Genome-Wide Transcription Analysis." *World Journal of Microbiology and Biotechnology* 33 (3). Springer Netherlands: 1–9. doi:10.1007/s11274-016-2202-5.

Yang, Peng, Wenjing Liu, Xuelian Cheng, Jing Wang, Qian Wang, and Qingsheng Qi. 2016. "A New Strategy for Production of 5-Aminolevulinic Acid in Recombinant Corynebacterium Glutamicum with High Yield." *Applied and Environmental Microbiology* 82 (9): 2709–17. doi:10.1128/AEM.00224-16.

Yang, Songxin, Xiulai Chen, Nan Xu, Liming Liu, and Jian Chen. 2014. "Urea Enhances Cell Growth and Pyruvate Production in Torulopsis Glabrata." *Biotechnology Progress* 30 (1): 19–27. doi:10.1002/btpr.1817.

Yeon, Young Joo, Hyung Yeon Park, and Young Je Yoo. 2013. "Enzymatic Reduction of Levulinic Acid by Engineering the Substrate Specificity of 3-Hydroxybutyrate Dehydrogenase." *Bioresource Technology* 134. Elsevier Ltd: 377–80. doi:10.1016/j.biortech.2013.01.078.

Yi, Wen, Peter M. Clark, Daniel E. Mason, Marie C. Keenan, Collin Hill, William A. Goddard, Eric C. Peters, Edward M. Driggers, and Linda C. Hsieh-Wilson. 2012. "Phosphofructokinase 1 Glycosylation Regulates Cell Growth and Metabolism." *Science* 337 (6097): 975–80. doi:10.1126/science.1222278.

Yu, Xiaoli, Haiying Jin, Wenjing Liu, Qian Wang, and Qingsheng Qi. 2015. "Engineering Corynebacterium Glutamicum to Produce 5-Aminolevulinic Acid from Glucose." *Microbial Cell Factories* 14 (1). BioMed Central: 1–10. doi:10.1186/s12934-015-0364-8.

Zamora, Rosario, José L. Navarro, Emerenciana Gallardo, and Francisco J. Hidalgo. 2006. "Chemical Conversion of α-Amino Acids into α-Keto Acids by 4,5-Epoxy-2-Decenal." *Journal of Agricultural and Food Chemistry* 54 (16): 6101–5. doi:10.1021/jf061239n.

Zanghellini, A. L. 2019. Fermentation route for the production of levulinic acid, levulinate esters and valerolactone and derivatives thereof. U.S. Patent No 10,246,727, issued 2019.

Zhang, Biao, Lulu Li, Jia Zhang, Xiaolian Gao, Dongmei Wang, and Jiong Hong. 2013. "Improving Ethanol and Xylitol Fermentation at Elevated Temperature through Substitution of Xylose Reductase in Kluyveromyces Marxianus." *Journal of Industrial Microbiology and Biotechnology* 40 (3–4): 305–16. doi:10.1007/s10295-013-1230-5.

Zhang, Biao, Yelin Zhu, Jia Zhang, Dongmei Wang, Lianhong Sun, and Jiong Hong. 2017. "Engineered Kluyveromyces Marxianus for Pyruvate Production at Elevated Temperature with Simultaneous Consumption of Xylose and Glucose." *Bioresource Technology* 224. Elsevier Ltd: 553–62. doi:10.1016/j.biortech.2016.11.110.

Zhu, Dunming, Yan Yang, John D. Buynak, and Ling Hua. 2006. "Stereoselective Ketone Reduction by a Carbonyl Reductase from Sporobolomyces Salmonicolor. Substrate Specificity, Enantioselectivity and Enzyme-Substrate Docking Studies." *Organic and Biomolecular Chemistry* 4 (14): 2690–95. doi:10.1039/b606001c.

Zhu, Fayin, Ka Yiu San, and George N. Bennett. 2020. "Metabolic Engineering of Escherichia Coli for Malate Production with a Temperature Sensitive Malate Dehydrogenase." *Biochemical Engineering Journal* 164 (August). Elsevier: 107762. doi:10.1016/j.bej.2020.107762.

14 Pernicious *Parthenium* Weed

An Insight into Its Biogenic Control and Transformation to Organic Fertilizer

Praveen Jain[a], *Anuj Kumar Chandel*[b*], *Akhilesh Kumar Singh*[c*], *and Sashi Sonkar*[d]

[a]Department of Botany, Government Chandulal Chandrakar Arts and Science PG college Patan, Durg Chhattisgarh, India

[b]Department of Biotechnology, Engineering School of Lorena (EEL), University of São Paulo, Lorena, Brazil

[c]Department of Biotechnology, School of Life Sciences, Mahatma Gandhi Central University, Motihari, East Champaran, Bihar, India

[d]Department of Botany, Bankim Sardar College, Tangrakhali, South 24 Parganas, West Bengal, India

CONTENTS

DOI: 10.1201/9781003191247-14

14.1 INTRODUCTION

The *Parthenium hysterophorus* L. belongs to the Asteraceae family and is henceforth known as *Parthenium*. It is a well-known flowering, erect, hazardous weed with many branches. Owing to the nonexistence of usual opponents/attackers/combatants, it grows violently in a varied range of environments with the ability to colonize huge areas of land, including the tropical as well as subtropical realm (Tanveer et al. 2015). For instance, it has a tendency to be aggressively fast growing as well as thriving in the agricultural fields of paddy, wheat, and so on, owing to its wide range of adaptability towards various ecological conditions compared to other weeds (Ghosh et. al. 2011). This weed is differentiated through its bulky seed bank, fast seed germination, as well as speedy growth rate, along with extremely viable seed-making capability. Such characteristics enable it to be a forbidding colonizer of parks, etc. (Lawes and Grice 2010; Nigatu et al. 2010; Qureshi et al. 2014). It badly influences the growth as well as development, together with yield of crops, through competing for nutrients. In addition, *P. hysterophorus* L. undergoes flowering as well as seeding throughout the year (Bhowmick 2000). This weed is not only allergic but also unpleasant in taste for grazers. The consumption of this weed by cattle leads to losses owing to severe health issues/threats, including contamination of milk as well as meat (Tudor et al. 1982). In human beings, *Parthenium* weed and its pollen result in health issues like asthma, etc. (Patel 2011). Overall, invasion of this weed causes loss of biodiversity, nutrient loss, as well as other types of environmental imbalances (Singh and Beck 2006; Akter and Zuberi 2009; Sema Patel 2011; Bhateria 2015). It also possesses various allelopathic compounds, especially parthenin, hysterin, ambrosin, as well as flavonoids. These compounds inhibit/toxify other flora/plants and therefore hinder/inhibit their development/growth (Wubneh 2019; Patel 2011). All earlier efforts/approaches to destroy/get rid of/control this weed have been unsuccessful, owing to its higher reproductive potential, hardihood, competitiveness, as well as environmental adaptability (Sema Patel 2011; Manoj 2014). Thus, millions of tons of biomass of this weed remain unexploited. This weed also increases global warming as the remains as well as the dead *Parthenium* weeds disintegrate in the presence or absence of oxygen, liberating carbon dioxide and methane (Abbasi and Abbasi 2010; Abbasi et al. 2012a, 2012b). Henceforth, there is a need to explore strategies through which this weed may be used cost-effectively by repeated harvesting and, therefore, regulating its invasiveness. In view of this, biogenic approaches to control *Parthenium* provide the utmost long-term option for weed control over physicochemical approaches. The chemical approaches for regulation of this weed depend on the exploitation of chemical herbicides, thereby causing ecological issues. Likewise, physical approaches of uprooting as well as ploughing are not only labor-intensive but also comparatively ineffectual. On the other hand, the burning strategy of *Parthenium* weed is found to be associated with detrimental impacts on the ecosystem. In addition, because *Parthenium* weed is richer in nutrient contents, it can therefore also be exploited as a source of organic fertilizer/manure via the composting process. The exploitation of such organic fertilizer/manure in organic farming depicts several advantages in terms of enhancing food quality/security, enriching soil structure, minimizing ecological stresses, as well as improving soil biodiversity. The present chapter provides an overview on viable biogenic control/management approaches, like insects that attack and feed on *Parthenium* weeds, fungal diseases of the *Parthenium* weed plants, rhizospheric fungi– and soil bacteria–mediated decomposition/control/management of *Parthenium*, along with composting strategies for transforming this weed into eco-friendly fertilizers. In addition, this chapter also highlights the compositional characteristics of compost/manures with their influence on different crops.

14.2 BIOGENIC CONTROL/MANAGEMENT OF *PARTHENIUM* WEED

The following feasible biogenic approaches are available (Table 14.1; Figure 14.1) for the control/management of *Parthenium*.

TABLE 14.1
Comparative Account on Insect-, Microbiota-, Sawdust-, Composting-, and Vermicomposting-Mediated Control/Management of *Parthenium* Weed

S. No.	Strategy/Method	Parameters/Results	References
(1)	Insect-mediated digestion	Characterization of insect excreta as a result of feeding on *Parthenium* for manure formation is lacking. Extensive research is needed in the same direction.	No scientific literature available
(2)	Microbiota- and sawdust-mediated microbial digestion	The temperature rose from 29 to 56°C on the 45th day and reached 30°C on the 90th day. The increase in pH from 6.8 to 7.4 on the 60th day produced a shift in microflora from 34.68106 to 21.73107 on the 30th day, 43.91105 on the 60th day, and 36.75106 on the 90th day. The cellulose content diminished from 12.05 to 6.84%. However, within 90 days, the lignin content declined from 11.43 to 6.12%.	Jelin and Dhanarajan (2013)
(3)	Green manure + NPK	It shows good soil health and records significantly higher crop yield over the regular chemical fertilizer-based practice of rice cultivation. Both the initial and ultimate nitrogen availability in the soil increased significantly in the second year compared to the control.	Dolai et al. (2019)
(4)	Compost	*Parthenium* compost can suppress both manure-originated pathogens as well as plant disease–causing microorganisms. The plant pathogen suppression was caused by bio-factors, as inhibition reduced after sterilization, but the decrease in fecal microorganisms might be caused by alkaloids found in *Parthenium* compost.	Rai and Suthar (2020)
(5)	Vermicomposting	Decreases in C:N ratio from 23.86 to 9.09.	Hussain et al. (2016b)

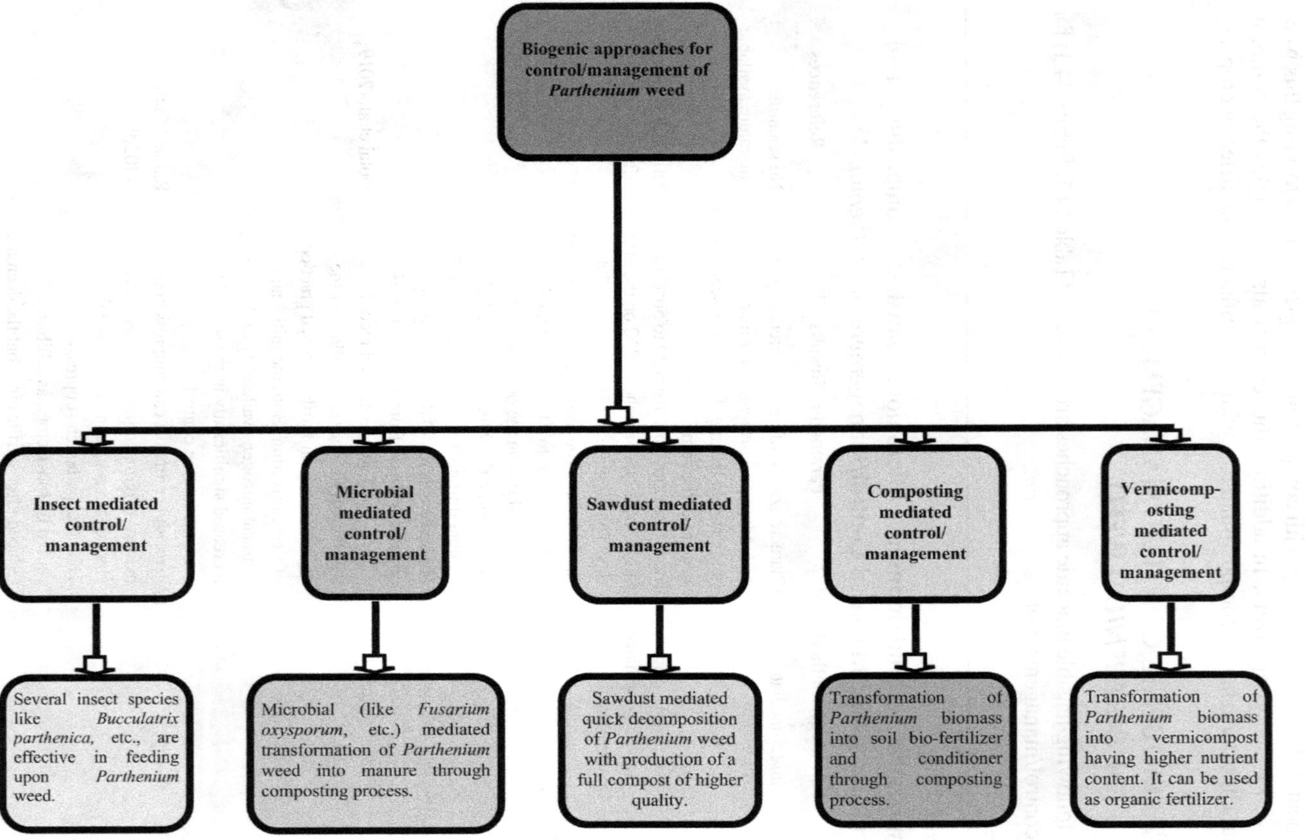

FIGURE 14.1 Summary of viable biogenic control/management of hazardous *Parthenium* weed.

14.2.1 Insect-Mediated Control/Management

Several insect species have been found to be massively effective in feeding different portions of the *Parthenium* weed while not attacking crops in typical field conditions. For instance, various insects like *Platophalonidia mystica* (stem boring moth), *Carmenta ithacae* (stem boring moth), *Conotrachelus albocinereus*, *Epiblema sternuana* (stem galling moth), *Zygogramma bicolorata* (leaf beetle), and so on have been reported to be exploited towards *Parthenium* biocontrol (Saini et al. 2014). Nevertheless, substantial research as well as data concerning insect-mediated feeding, along with characterization of insect excreta and manure formation, is lacking.

14.2.2 Microbiota-Mediated Control/Management

Microbes, like fungal and bacterial species, can be exploited for the control/management of *Parthenium* weed. The fungal species utilized for the control/management of *Parthenium* weed that have been isolated from the rhizosphere (and of endophytic origin) include *Alternaria alternata*, *Fusarium oxysporum*, *Colletotrichum capsici*, etc. (Idrees and Javaid 2008; Kelaniyangoda et al. 2008; Bekeko et al. 2012; Gnanavel 2013). The bacterial species accountable for the composting process have not been well studied, although soil bacteria are the most responsible. A substantial number of viable bacteria, fungus, *Azotobacter*, Actinomycetes, and phosphate-solubilizing bacteria (PSB) were found to be 3.66×10^6, 6.67×10^4, 2.33×10^6, 7.67×10^5, and 2.67×10^6, correspondingly, in per gram of compost (Kishor et al. 2010a). The fate of manure-based pathogens as well as the antibacterial capabilities of ready compost has received little attention from earlier researchers. However, research shows that pathogenic microorganisms are inactivated in composting processes primarily by temperature-time phenomena as well as antagonistic action of advantageous microorganisms which exist in the compost (Chen et al. 2013; Ballardo et al. 2020). Most recently, Wang et al. (2020) discovered 629 possible human/animal bacterial pathogens in compost samples collected from various areas of China. Such microbes should be systematically examined to ensure compost quality as well as food safety, as compost is a significant constituent of nutrient delivery in agricultural plants/cropping systems.

14.2.3 Sawdust-Mediated Control/Management

Jelin and Dhanarajan (2013) discovered a superior decomposition technique by combining sawdust and *Parthenium* biomass in their experimental study. The comparative investigation of sawdust deterioration was executed for ninety days at a constant temperature. On the other hand, the combination of *Parthenium* and sawdust exhibited a higher degradation rate within forty-five days. The findings revealed that the combination of *Parthenium* and sawdust in composting takes a very short time to degrade/decompose to yield a mature compost of high quality. The combination of *Parthenium* and sawdust will accelerate the deterioration process more than either *Parthenium* or sawdust alone. The compost's odor gradually diminishes from either a fishy or rotten odor in the onset to an acceptable level, and then to a nice, earthy fragrance towards the end of the degradation process. The cellulose (g mL^{-1}) content of *Parthenium* manure was observed to be lowered from 34.42 (control) to 16.54 and 13.21 on the forty-fifth and ninetieth days, correspondingly. Also, the combination retains nitrogen more stably in the compost even after ninety days of composting since only *Parthenium* compost retains nitrogen from 1.72 (first day) to 0.9 mg on the ninetieth day, whereas the combination with sawdust retains nitrogen at 1.62 mg from 2.12 mg g^{-1} of compost after ninety days of composting.

14.2.4 Composting-Mediated Control/Management

Composting is a natural bioprocess that involves the disintegration of organic garbage/waste into a stable end product known as compost. It can be exploited in different agricultural uses. Composting may be an attractive and feasible option for converting *Parthenium* biomass into a beneficial product, which may be utilized as a soil enricher and conditioner. The biomass of *Parthenium* weed was composted quickly. In comparison to farmyard manure, the composition of *Parthenium* compost is double in terms of macronutrients, such as NPK, as well as micronutrients, such as Cu, Mn, Zn, and Fe (Channappagoudar et al. 2007). Organic acids produced in the course of composting aid towards the release of insoluble K as well as enhance P and K absorption (Murthy et al. 2010). Compost also includes a high concentration of enzymes, antibiotics, a significant number of beneficial bacteria, such as *Azotobacter*, phosphate solubilizers (PSB), etc. (Kishor et al. 2010b). An investigation on various crops treated with *Parthenium* manure shows an improvement in crop growth and production. Biradar et al. (2006) employed the *Parthenium* manure generated in pits with microbial inoculum alone at 5 t ha^{-1}, which resulted in a 16.1% increase in rice yield. The *Parthenium* compost also promoted growth of chili and sorghum (Channappagoudar et al. 2007), *Vigna radiata* and *Triticum* sp. (Khaket et al. 2012), and *Arachis hypogaea* (Rajiv et al. 2013). Likewise, the 3% green manure treatment produced the maximum root and shoot biomass in maize, which was substantially higher over the control and equal to the NPK fertilizer treatments (Javaid and Shah 2008). The quantities of chemical fertilizer were reduced by the application of weed biomass required for crop growth by ~25% in potatoes (Saravanane et al. 2008). The potential of *Parthenium* compost was demonstrated by Kishor et al. (2010a) in supporting organic farming methods, wherein *Parthenium* compost was mixed with urea and applied to *Azotobacter chroococcum*–pretreated wheat seeds grown in pots under controlled conditions. It was also revealed that *Parthenium* compost, when exploited with chemical fertilizer in appropriate ratio, led to greater crop yields with improved nutrient availability in consequent crop years. Further, Javaid and Shah (2010) stated that the growth of wheat plants has also been found to improve once treated with *Parthenium* green manure. Also, green manure from *Parthenium* supports the development of maize (Suryawanshi 2011). The addition of *Parthenium* leaf manure to rice crops under submerged conditions resulted in the improvement of plant height, greater grain yield, and straw production with no weed emergence (Saravanane et al. 2012).

14.2.5 Vermicomposting-Mediated Control/Management

Vermicomposting offers an excellent platform for the control/management of *Parthenium* weed. In this context, investigations were carried out by researchers. For instance, the application of *Parthenium* vermicompost (PV) in ladies' finger plants results in a positive impact on growth parameters, such as seed germination, flowering, as well as decrease in the incidence of diseases due to pests and pathogens (Hussain et al. 2017). A substantial decrease towards the leaf spot disease, fruit borers, etc. was also reported when grown in agricultural land supplemented with diverse concentrations of PV ranging from 2.5–5 t ha^{-1}. With the application of 5 t ha^{-1} of PV, the maximum carotenoid and chlorophyll content was calculated to be 0.90 ± 0.07 and 1.43 ± 0.09 mg g^{-1}, respectively, in leaves of ladies' finger. Plants cultivated in vermicompost-amended soils also showed a rise in protein and glucose contents. Hussain et al. (2016b) stated the C:N ratio of *Parthenium* falls sharply from 23.86 to 9.09 due to vermicomposting, which signifies its conversion to a N_2-enriched fertilizer. The C:N ratio of less than 20 reveals substantial stabilization with a stage of appropriate maturity, whereas ≤15 is desirable towards the agronomical application of compost (Ndegwa et al. 2000; Deka et al. 2011). The decrease in the C:N ratio is primarily propelled via the reduction of carbon substances of *Parthenium*, owing to microorganisms-mediated respiration. Certain organic nitrogen has been perhaps mineralized through earthworms with the addition of mucus as well as excreta in the form of vermicompost. The UV-visible, FTIR spectroscopy, thermogravimetric, differential thermal analysis, GS-MS analysis, and SEM were used to investigate the transition of *Parthenium* into vermicompost. In the course of vermicomposting, a substantial reduction was observed in mineralization of organic content together with the breakdown of complicated aromatic compounds like lignin/polyphenols to monomeric carbohydrate as well as lipid substances. The vermicompost exhibited substantial disintegration as well as biological oxidation, including molecular rearrangement of chemical constituents over the original substrate. Further, scanning electron microscope micrographs of vermicompost depict extremely broken-down substances. "Parthenin" that has been primarily accountable towards *Parthenium* toxicity as well as allelopathy was shown to be destroyed during vermicomposting, which converts such potentially toxic weed to an eco-friendly organic fertilizer. The results imply that some additional invasives well-known for their negative allelopathy as well as noxiousness can also generate vermicompost having plant-friendly characteristics (Hussain et al. 2016b). When applied at concentrations ranging between 0.75 and 4%, PV promoted greater growth in green gram with respect to shoot length (SL), root length (RL), shoot dry weight (SDW), as well as root dry weight (RDW) over control. Cucumber seedlings grew faster in 0.75–20% PV with respect to RL, SDW, as well as RDW, whereas SL was found to increase up to 4% under PV treatment (Hussain et al. 2016a). Hussain et al. (2016a, 2016b) found that PV includes a variety of fatty acids, nitrogenous compounds, etc. that improve the soil's microbial biomass carbon. Apart from these characteristics, PV-like, manure-based vermicomposts have been shown to improve soil physical qualities. Vermicompost produced by earthworm (*Eisenia fetida*) activity on *P. hysterophorus* has been exploited towards the germination as well as quick growth of *Vigna radiata*, *Abelmoschus esculentus*, as well as *Cucumis sativus*. It was found that the vermicompost lacked the allelopathy as well as soil/plant/animal noxiousness linked to *Parthenium*. Vermicomposting was discovered to be more effective over composting. For instance, Vyankatrao (2017) carried out an investigation on the decomposition of *Parthenium* weed by Bangalore pit compost and NADEP compost approaches as well as vermicomposting by worm bin method and pit method. Furthermore, the N, P, K, Ca, and C:N ratios of the composts and vermicomposts were determined. Both compost and vermicompost exhibit a significant rise in N, P, K, and Ca, but a reduction in C:N ratio. When compared to pit vermicompost and NADEP compost, Bangalore pit compost as well as worm bin vermicompost provide more nutrients. The poisonous weed *Parthenium* utilized in this experiment is thus transformed into useful resources, such as compost and vermicompost with greater nutrient concentrations that may be used as an agricultural amendment, i.e., organic fertilizer.

14.2.6 Nutrient Analysis and Characteristics of *Parthenium* Compost

The nutrient analysis of *Parthenium* compost revealed that it is rich in nutrients. This is supported by the fact that *Parthenium* has been found to be rich in N_2 (2.54%), P (0.44%), K (1.23%), Zn (13.9 ppm), Mn (161.2 ppm), Fe (528.3 ppm), as well as Cu (9.0 ppm) (Javaid and Shah 2008). The *Parthenium* compost depicts the electrical conductivity of 1 dS m^{-1} and a pH value of 7.8. The concentrations of macronutrients such as total N, P, K, and S were 1.58%, 0.33%, 1.645, and 0.29%, correspondingly, while micronutrients Fe, Mn, Zn, and Cu were 7829, 306, 116, and 66 ppm, correspondingly (Kishor et al. 2010a).

14.3 CONCLUSION

Biogenic approaches, like insect, microbiota, sawdust, composting, and vermicomposting, offer control/management of pernicious *Parthenium* weed. Such approaches also facilitate the transformation of *Parthenium* into eco-friendly organic fertilizer for agricultural uses. Although compost is rich in nutrients, the addition of sawdust to *Parthenium* composting results in improved nutrient-retention capacity. Further, along with conventional chemical fertilizers, the addition of compost from *Parthenium* biomass resulted in higher productivity with enhanced nutrient availability in subsequent crops. Moreover, soil porosity together with water-retention capacity were found to increase. However, the antipathogenic activity of compost is an area that has been opened up and studied a little bit. The vermicomposting method has been found to significantly reduce the level of toxic compounds such as "parthenin," apart from faster rate of the composting process. It also proved to be more beneficial in terms of nutrients content, manageability, year-round production, and scope for commercial vermiwash production that can be stored for longer durations.

Furthermore, no scientific literature is available so far on the characterization of insect excreta produced after feeding on *Parthenium* weed for manure formation. Therefore, exhaustive research is needed in the same direction as such research will not only be helpful for the control/management of *Parthenium* weed but also in paving a way towards the exploitation of insect excreta (produced after feeding on *Parthenium* weed) as source of manure for agricultural uses.

ACKNOWLEDGEMENT

A.K. Chandel gratefully acknowledges Conselho Nacional de Desenvolvimento Científico e Tecnológico-CNPq, Brazil for scientific productivity program (Process number: 309214/2021-1).

REFERENCES

Abbasi, T. and Abbasi, S.A. (2010). Biomass energy and the environmental impacts associated with its production and utilization. *Renewable and Sustainable Energy Reviews* 14: 919–937.

Abbasi, T., Tauseef, S.M. and Abbasi, S.A. (2012a). Biogas Energy (New York, NY: Springer), pp. 25–34.

Abbasi, T., Tauseef, S.M. and Abbasi, S.A. (2012b). Anaerobic digestion for global warming control and energy generation: An overview. *Renewable and Sustainable Energy Reviews* 16: 3228–3242.

Akter, A. and Zuberi, M.I. (2009). Invasive alien species in Northern Bangladesh: Identification, inventory and impacts. *International Journal of Biodiversity Conservation* 15: 129–134.

Ballardo, C., Vargas-García, M.C., Sánchez, A., Barrena, R. and Artola, A. (2020). Adding value to home compost: Biopesticide properties through *Bacillus thuringiensis* inoculation. *Waste Management* 106: 32–43.

Bekeko, Z., Hussien, T. and Tessema, T. (2012). Distribution, incidence, severity and effect of the rust (*Puccinia abrupta* var. *partheniicola*) on *Parthenium hysterophorus* L. in Western Hararghe Zone, Ethiopia. *African Journal of Plant Science* 6: 337–345.

Bhateria, R. (2015). *Parthenium hysterophorus* L.: Harmful and beneficial aspects—A review. *Nature Environment and Pollution Technology* 14: 463–474.

Bhowmick, M.K. (2000). *Parthenium hysterophorus* L.: A prolific weed of West Bengal and its management. *Science and Culture* 66: 277–278.

Biradar, D.P., Shivkumar, K.S., Prakash, S.S. and Pujar, B.T. (2006). Bionutrient potentiality of *Parthenium hysterophorus* and its utility as green manure in rice ecosystem. *Karnataka Journal of Agricultural Sciences* 19: 256–263.

Channappagoudar, B.B., Biradar, N.R., Patil, J.B. and Gasimani, C.A.A. (2007). Utilization of weed biomass as an organic source in sorghum. *Karnataka Journal of Agricultural Sciences* 20: 245–248.

Chen, Z., Diao, J.S., Dharmasena, M., Ionita, C., Jiang, X.P. and Rieck, J. (2013). Thermal inactivation of desiccation-adapted *Salmonella* spp. in aged chicken litter. *Applied and Environmental Microbiology Journal* 79: 7013–7020.

Deka, H., Deka, S., Baruah, C.K., Das, J., Hoque, S. and Sanna, N.S. (2011). Vermicomposting of distillation waste of citronella plant (*Cymbopogon winterianus* Jowitt.) employing *Eudrilus eugeniae*. *Bioresource Technology* 102: 6944–6950.

Dolai, A.K., Bhowmick, M.K., Ghosh, P. and Ghosh, R.K. (2019). Utilization of congress grass (*Parthenium hysterophorus* L.) for soil fertility enhancement and improved productivity of potential crop sequences in West Bengal. *Journal of Pharmacognosy and Phytochemistry* 8: 2241–2245.

Ghosh, R.K., Mallick, S., Bera, S., Barman, S., Jana, P.K., Barui, K. and Bhowmik, M.K. (2011). Ecosafe management of invasive weeds in Gangetic Inceptisol of India. Proceedings of 23rd Asian-Pacific Weed Science Society (APWSS) International Conference Organized by APWSS and IWSS at Sebel Cairns in North Queensland, Australia 25–30 September with focus on the theme "Weed Management in a Changing World", p. 35.

Gnanavel, I. (2013). *Parthenium hysetrophorus* L.- a major threat to natural and agro-ecosystems in India. *Science International* 1: 124–131.

Hussain, N., Abbasi, T. and Abbasi, S.A. (2016a). Vermicomposting transforms allelopathic *Parthenium* into a benign organic fertilizer. *Journal of Environmental Management* 180: 180–189.

Hussain, N., Abbasi, T. and Abbasi, S.A. (2016b). Transformation of a highly pernicious and toxic weed Parthenium into an eco-friendly organic fertilizer by vermicomposting. *International Journal of Environmental Studies* 73: 731–745.

Hussain, N., Abbasi, T. and Abbasi, S.A. (2017). Detoxification of Parthenium (*Parthenium hysterophorus*) and its metamorphosis into an organic fertilizer and biopesticide. *Bioresources and Bioprocessing* 4: 1–9.

Idrees, H. and Javaid, A. (2008). Screening of some pathogenic fungi for their herbicidal potential against Parthenium weed. *Pakistan Journal of Phytopathology* 20: 150–155.

Javaid, A. and Shah, M.B.M. (2008). Use of *Parthenium* weed as green manure for maize and Mungbean production. *Philippine Agricultural Scientist* 91: 478–482.

Javaid, A. and Shah, M.B.M. (2010). Growth and yield response of wheat to EM (effective microorganisms) and *Parthenium* green manure. *African Journal of Biotechnology* 9: 3373–3381.

Jelin, J. and Dhanarajan, M.S. (2013). Comparative physicochemical analysis of degrading Parthenium (*Parthenium hysterophorus*) and saw dust by a new approach to accelerate the composting rate. *International Journal of Chemical, Environmental & Biological Sciences* 1: 535–537.

Kelaniyangoda, D.B. and Ekanayake, H.M.R.K. (2008). *Puccinia melampodii* Diet and Holow. as a biological control agent of *Parthenium hysterophorus. Journal of Food and Agriculture* 1: 13–19.

Khaket, T.P., Singh, M., Dhanda, S., Singh, T. and Singh, J. (2012). Biochemical characterization of consortium compost of toxic weeds *Parthenium hysterophorus* and *Eichhornia crassipe*. *Bioresource Technology* 123: 360–365.

Kishor, P., Ghosh, A.K., Singh, S. and Maurya, B.R. (2010b). Potential use of *Parthenium* (*Parthenium hysterophorus* L.) in agriculture. *Asian Journal of Agricultural Research* 4: 220–225.

Kishor, P., Maurya, B.R. and Ghosh, A.K. (2010a). Use of uprooted *Parthenium* before flowering as compost: A way to reduce its hazards worldwide. *International Journal of Soil Science* 5: 73–81.

Lawes, R.A. and Grice, A.C. (2010). War of the weeds: Competition hierarchies in invasive species. *Austral Ecology* 35: 871–878.

Manoj, E.M. (2014). Invasive plants a threat to wildlife. *The Hindu*, 22 September.

Murthy, R.K., Raveendra, H.R. and Manjunatha, R.T.B. (2010). Effect of *Chromolaena* and *Parthenium* as green manure and their compost on yield, uptake and nutrient use efficiency on typic paleustalf. *European Journal of Biological Sciences* 4: 41–45.

Ndegwa, P.M. and Thompson, S.A. (2000). Effects of C-to-N ratio on vermicomposting of biosolids. *Bioresource Technology* 75: 7–12.

Nigatu, L., Hassen, A., Sharma, J. and Adkins, S.W. (2010). Impact of Parthenium hysterophorus on grazing land communities in north-eastern Ethiopia. *Weed Biology and Management* 10: 143–152.

Patel, S. (2011). Harmful and beneficial aspects of *Parthenium hysterophorus*: An update. *3 Biotech* 1: 1–9.

Qureshi, H., Arshad, M. and Bibi, Y. (2014). Invasive flora of Pakistan: A critical analysis. *International Journal of Bioscience* 4: 407–424.

Rai, R. and Suthar, S. (2020). Composting of toxic weed *Parthenium hysterophorus*: Nutrient changes, the fate of faecal coliforms, and biopesticide property assessment. *Bioresource Technology* 311: 123523.

Rajiv, P., Narendhran, S., Subhash, K.M., Sankar, A., Sivaraj, R. and Venckatesh, R. (2013). *Parthenium hysterophorus* L. compost: Assessment of its physical properties and allelopathic effect on germination and growth of *Arachis hypogaea* L. *International Research Journal of Environmental Sciences* 2: 1–5.

Saini, A., Aggarwal, N.K., Sharma, A., Kaur, M. and Yadav, A. (2014). Utility Potential of *Parthenium hysterophorus* for its strategic management. *Advances in Agriculture 2014*: 381859.

Saravanane, P., Nanjappa, H.V. and Ramachandrappa, B.K. (2008). Effect of weeds utilization as nutrient source on soil fertility and tuber yield of potato. *Mysore Journal of Agricultural Sciences* 42: 464–467.

Saravanane, P., Poonguzhalan, R. and Chellamuthu, V. (2012). *Parthenium* (*Parthenium hysterophorus* L.) distribution and its bioresource potential for rice production in Puducherry, India. *Pakistan Journal of Weed Science Research* 18: 551–555.

Singh, M.N. and Beck, M.H. (2006). *Parthenium* contact sensitivity travels to the UK. *British Journal of Dermatology* 155: 847–848.

Suryawanshi, D.S. (2011). Utilization of weed biomass as an organic source in maize. *Life Science Bulletin* 8: 10–12.

Tanveer, A., Khaliq, A., Ali, H.H, Mahajan, G. and Chauhan, B.S. (2015). Interference and management of Parthenium: The world's most important invasive weed. *Crop Protection* 68: 49–59.

Tudor, G.D., Ford, A.L., Armstrong, T.R. and Bromage, E.K. (1982). Taints in meat from sheep grazing *Parthenium hysterophorus*. *Australian Journal of Experimental Agriculture and Animal Husbandry* 22: 43–46.

Vyankatrao, N.P. (2017). Conversion of *Parthenium hysterophorus* L. Weed to compost and vermicompost. *Bioscience Discovery* 8: 619–627.

Wang, Y., Gong, J., Li, J., Xin, Y., Hao, Z., Chen, C., Li, H., Wang, B., Ding, M., Li, W., Zhang, Z., Xu, P., Xu, T., Ding, G. and Li, J. (2020). Insights into bacterial diversity in compost: Core microbiome and prevalence of potential pathogenic bacteria. *Science of the Total Environment* 718: 137304.

Wubneh, W.Y. (2019). *Parthenium hystrophorus* in ethiopia: Distribution, impact and management – A review. *World Scientific News* 130: 127–136.

15 Microbial Fermentation of Waste Oils for Production of Added-Value Products

Naganandhini Srinivasan,[a] *Kiruthika Thangavelu,*[bc] *and Sivakumar Uthandi*[a]

[a]Biocatalysts Laboratory, Department of Agricultural Microbiology, Tamil Nadu Agricultural University, Coimbatore, Tamil Nadu, India, 641 003

[b]Department of Renewable Energy Engineering, Agricultural Engineering College and Research Institute, Tamil Nadu Agricultural University, Coimbatore, Tamil Nadu, India, 641 003

[c]Department of Agriculture Engineering, Mahendra Engineering College, Namakkal, Tamil Nadu, India, 637 503

CONTENTS

DOI: 10.1201/9781003191247-15

15.1 INTRODUCTION

Edible oils are indispensable in the Indian kitchen and are a major source of dietary fat in Indian diets, and it provides energy, fat-soluble vitamins, and vital fatty acids. After the United States, China, and Brazil, India's vegetable oil economy ranked fourth place worldwide. India is the biggest importer and the third-largest consumer in the world. According to the most recent data, each Indian consumed an average of 19.7 kg of cooking oil per annum in the year 2019–2020 (*Times of India*, 2021). In 2021, over 36 million metric tons of oilseeds were produced with soybean, the highest produced oilseed with over 13 million metric tons in India (Statistica, 2021). The country's unique agro-ecological conditions are ideal for cultivating nine annual oilseed plants, including seven palatable oilseeds (sunflower, peanut, rapeseed, niger, soybean, sesame, safflower, and mustard) and two nonpalatable oilseeds (castor and linseed) (Jha et al. 2012).

15.2 PHYSICOCHEMICAL PROPERTIES OF WCO

Oil is frequently utilized in the most famous gourmet preparation called frying, in salad dressing, and food emulsions both in industrial and domestic food preparations. Cooking food at 150 to 190°C (higher) temperature in a preheated deep oil/fat is known as deep-fat frying (Farkas, Singh, and Rumsey 1996; Debnath et al. 2009). It magnifies the sensory qualities of the meal, such as the distinct fried flavor, golden-brown color, and crunchy texture, despite keeping it moist and delicate. Deep-fat frying is among the most widely used food preparation procedures due to the pleasant flavor and the significant economic benefit. The approach is based on the interaction of high-temperature oil with food, which roasts and dehydrates the food, causing physico-biochemical changes that lead to the creation of noxious compounds through chemical reactions (Tsoutsos et al. 2016; Nayak et al. 2016). When the water in the food attacks the ester link of triacylglycerols (TAGs) in the oil, it produces monoacylglycerols (MAG), free fatty acids (FFA), diacylglycerols (DAG), and glycerol together at once. The food properties, oil type and surface/capacity ratio, degree of air amalgamation into the oil, heating temperature and process, immersion duration, and the material used to build the boiling container all have an impact on food and oil modifications. Several chemical reactions in the deep-fat frying processes include oxidation, polymerization, and hydrolysis (Choe and Min 2007; Velasco, Marmesat, and Dobarganes 2008). The oxidation of deep-fried oil is the most significant deteriorative effect. In deep-fried oil, three types of oxidation occur: autoxidation, photosensitized oxidation, and thermal oxidation (Nayak et al. 2016). Another key reaction that happens in fried food is browning, called Maillard.

The primary molecules impacting sugars during frying are free amino groups from amino acids, peptides, and proteins, carbonyl groups or other aldehydes, as well as ketones of sugars. At frying temperatures, numerous intermediary products known as Amadori products or premelanoidins quickly polymerize, generating dark-colored molecules (melanoidins). Oxidized monomer, nonpolar dimer alcohols, aldehydes, alkanes, ketones, acids, furfurals, pyridines, lactones, and other chemicals are produced due to this reaction (Silvagni et al. 2012; Panadare 2015). The edible oil after the frying process may contain mutagenic, carcinogenic, neurotoxic, and hepatotoxic compounds, among other things (Bastida and Sánchez-Muniz 2002; Paul, Mittal, and Chinnan 1997; Tsoutsos et al. 2016).

15.3 WASTE COOKING OIL (WCO)

Commercial use of cooking oil on a large scale by the snack manufacturing industry and F&B (food and beverage) enterprises, *viz.*, hotels, canteens, restaurants, roadside dhabas, bistros, and road sellers, leads to the generation of gallons of WCO with no planned and technical way of disposal (Greenea 2018). It is ironic that while our mouth waters on the view of those pooris, potato finger fries, jalebis, chips, samosas, or chicken lollipops sizzling in the fryer, we rarely contemplate where the excess fat and oil will be discarded at the end of the day. It will either be reused or discarded in the sewer system or drain. In both cases, it is going to damage somebody or something.

WCO is an oil-based substance composed of palatable vegetable matter that has been used in food preparation but is no longer fit to eat (Kalam et al. 2011). This wide variation is due to regional differences in culinary habits and consumption trends, which also influence the biochemical properties, nature, and contaminants levels in the cooking oils used (Rincón, Cadavid, and Orjuela 2019).

15.4 STATISTICS OF WCO GENERATION

Data on the quantity of WCO generated worldwide is extremely limited and hard to discover, as cooking oil production and consumption vary widely between different regions and countries. Vegetable oil consumption is estimated to be approximately 200 million tons worldwide. According to estimates, 32% of edible oil consumed is wasted, implying that 64 million tons of WCO will be available globally (Teixeira, Nogueira, and Nunes 2018). WCO production in the European Union is projected to be around 0.9 million tons per year. Huge amounts of oil waste are produced in densely populated countries, such as China, the United States, Japan, India, Germany, Spain, the Republic of Korea, and Canada, producing 5.6 Mt (million tons), 1.2 Mt, 570 kt (thousand tons), 1.1 Mt, 493 kt, 300 kt, 411 kt, and 148 kt, respectively (Teixeira, Nogueira, and Nunes 2018).

In India, 2,467 crore liters of edible oil are consumed annually (speech of Sandeep Chaturwedi, BDAI President, 2019) (*The Hindu*, 2019). Households and food business operators (FBOs) consume approximately two-thirds of these, while soap and oleochemical firms use one-third. In India, the total amount of WCO produced by FBOs is expected to be 11.45 lakh tons.

15.5 MANAGEMENT OF WCO

Due to WCO's environmental concerns, recycling and valorizing these polluting wastes have become critical. Many industrialized nations have previously established guidelines for the deployment of discriminatory routes for WCO assortment, transportation, treatment, and retrieval by appropriate operators who have a license. At present, a small volume of WCO is directed to use as an animal feed component (Tres et al. 2013) or as a biodiesel feedstock, or as an oleochemical feedstock (Lam et al. 2016; Tres et al. 2013; Chrysikou et al. 2019) by licensed waste haulers. Another way to reuse WCO is to use it as a component of fermentation media to cultivate various microbes for the synthesis of high-value chemicals. The accessibility, large volume, and low price of WCO are appealing for different bioprocess-based product development, stability, and economic viability, while at the same time, reducing environmental issues that arise from it.

We get rid of WCO through drains and syphons at households and small food vendors. In big food, operators do some processes such as filtration, decoloring, and refining before discarding or reusing. After this process, they resell WCO again to small food companies that do not give importance to the quality or redirect it as feedstock for oleochemical production. As of comatose behaviors, lack of guidelines, or dearth of law implementation, WCO mismanagement causes a diversity of cascading issues that include sewage blockages, wastewater overflows, infrastructure damage, vectors and pests, nauseating odors, and central wastewater treatment plants running up expenses, in addition to ecosystem pollution and public health consequences (Orjuela and Clark 2020). WCO has chemicals that can stay in the atmosphere for several years. It raises the organic pollutants on feedwater sources and generates a thin film on the water's surface, which diminishes the amount of dissolved oxygen (DO) required by subaquatic species, causing ecological change (Contreras-Andrade, Sierra-Vargas, and Guerrero-Fajardo 2013). Despite the low volume ratio, a 1 L of oil could potentially pollute the water of 1 M liters (million liters) (Kingston 2002; Saadoun 2015). These lipid-rich wastes can delay wastewater treatment systems by adsorption of long-chain fatty acids (LCFA) of oil into the biomass, causing mass transfer issues, sludge flotation, and foam formation, and finally, microbial communities that live in anaerobic environments are suppressed (Alves et al. 2009; Appels et al. 2011). When WCO is improperly disposed in kitchen faucets, it solidifies and clogs sewer systems. Metal and concrete parts in pipes may corrode if WCO decomposition persists (Refaat 2010).

It is critical to manage, recycle, and valorize WCO to limit the harmful effects on ecosystems. Several developed countries implement rules that punish the illegal act of releasing WCO into the public sewer system. In 2018, the Food Safety and Standards Authority of India (FSSAI) gave a direction note on the standard operating procedure (SOP) for treatment and disposal of WCO. The FSSAI SOPs restrict improper disposal of WCO into public sewers or drains. Instead, it had better be disposed into the environment safely and securely, ideally through delivery to authorized WCO aggregators/collectors who have been officially registered with authorities, viz., the Biodiesel Association of India and the State Biodiesel Board.

The FSSAI also endorses mixing WCO with different available absorbent materials such as saw filling/sawdust, sand, used cotton, waste cloth, towel, wastepaper, and so on at the household level and then disposing of it in dustbins. The discarding of any dangerous material into water bodies in the environment is a punishable offense under the Water Act 1974.

15.6 EFFECTS ON THE ENVIRONMENT AND HUMAN HEALTH

In general, the dumping activity of untreated WCO into landfills or rivers causes an adverse impact on the ecosystem. One of the primary environmental problems that develop is eutrophication, which occurs when a thin layer of oil blocks sunlight from penetrating a river's surface, eventually disturbing the oxygen supply of aquatic life. The imbalance equilibrium of aquatic ecosystems in lakes, rivers, and marine systems has affected water quality as well (Kingston 2002; Saadoun 2015). Reusing cooking oil can lead to an increase in free radicals in the body, which can lead to inflammation, which is the basis of many disorders, such as obesity, heart disease, and diabetes. Inflammation in the body can lower immunity and make you more susceptible to infections (FSSAI 2019).

15.7 MICROBIAL CONVERSION OF WCO INTO VALUE-ADDED PRODUCTS

WCO as a direct substrate for microbial practices which is a perfect way to lower the cost of producing valuable chemicals while simultaneously increasing the economic worth of these wastes. As WCO is considered hazardous to the surroundings and has greater energy-demanding breakdown processes, using it simply as a microbial process feedstock is a wonderful chance to minimize the production costs of derived biochemical products while also increasing the economic worth of these wastes (El Bialy, Gomaa, and Azab 2011). In the metabolic reaction, WCO can be utilized as a carbon and energy source by some species of yeast, fungi, and bacteria, which convert them into many platform biochemicals, as shown in Figure 15.1. The following products can be produced from WCO.

15.7.1 Biodiesel

It is a combination of fatty acid methyl/ethyl esters made from lipid-based feedstocks (animal fats/vegetable oils) by the transesterification process. It has the potential to replace fossil fuels, which helps alleviate the overuse of nonrenewable fuels. The majority of commercially produced biodiesel is now made from vegetable oils using an alkali-catalyst method. The use of WCO is a key component in reducing biodiesel manufacturing costs by 60–90% (del Pilar Rodriguez et al. 2016) and frequent ways of its reuse and recycling. It allows the saving of 21% and 96% of crude oil and fossil energy (Corral Bobadilla et al. 2017). The tons of produced WCOs can be turned into biodiesel with maximum yields, ensuring a steady supply of sustainable and green fuel.

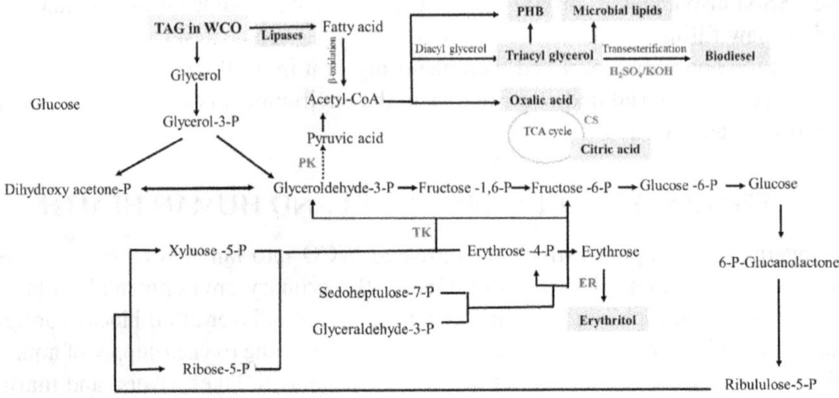

FIGURE 15.1 Pathway involved in value-added compounds production from WCO. PK, phosphokinase; TK, transketolase; ER, erythrose reductase; CS, citrate synthase; TCA, tricarboxylic acid; and CoA, coenzyme.

At present, 850 million liters of diesel are used up each month in India. By 2030, the country wants to mix 5% of biodiesel into diesel. As per this estimate, in a year, 5 billion liters of biodiesel are in need. In India, 2,700 billion liters of cooking oil are consumed annually, of which 140 billion WCO can be collected from big consumers, such as restaurants, hotels, and caterers, to transformation into biodiesel, which would yield roughly 110 billion liters of biodiesel (*The Economic Times*, 2019). State-owned oil marketing firms like Indian Oil and Bharat Petroleum started a program to acquire biodiesel made from WCO in one hundred cities across the nation. The idea of substituting WCO for fresh vegetable oils in this process proposed three-pronged answers: financial, ecological, and waste managerial way. As a result, numerous process explorations and feasibility evaluations of WCO-based biodiesel production have been carried out and reported (Sirisomboonchai et al. 2015; Zhang et al. 2003).

Katre et al. (2012) grew five strains of *Yarrowia lipolytica* on glucose, WCO, and waste motor oil (WMO) for biodiesel production. The neutral lipid made up the majority of lipids accumulated by all strains, which totaled more than 20% (w/w, dry) of their cell mass. Among them, *Y. lipolytica* NCIM 3589 possesses to have the highest lipid-to-biomass coefficient ($Y_{L/X}$) of 0.29 g/g on the glucose (30 g/L) and 0.43 g/g on WCO (100 g/L), with rich content of saturated fatty acids (SFA) and monounsaturated fatty acids (MUFA), comparable to typical vegetable oils used in biodiesel synthesis. Strain 3589's empirically determined and anticipated biodiesel qualities were found to be highly correlated with industry standards.

Vastano et al. (2019) valorized WCO into high-content FFAs. Both recombinant *Escherichia coli* and native *Pseudomonas resinovorans* polyhydroxyalkanoates (PHAs) generating cell factories were used to validate the bioprocess. WCO was microbially fermented to minimize the level of FFAs and produce a polymer. With appropriate strain design and process optimization, up to 1.5 g/L of the medium-chain–length polyhydroxyalkanoates (mcl-PHAs) were achieved, as well as WFO efficient conversion (Waste Frying Oil) into biodiesel (80% yield).

Sarno and Iuliano (2019) developed a nanoparticle-based catalyst for biodiesel synthesis from WCO. The lipase from the *Thermomyces lanuginosus* (TL) was fixed on Fe_3O_4/Au-based nanoparticles via physical interactions. Using it as a catalyst, a ~90% yield was gained without any pretreatment and at a 20% lipase unit, 45°C temperature reaction, and 1:6 oil-to-methanol ratio after twenty-four hours. After the first three cycles of use, 74% of the immobilized lipase activity stays. In particular, the produced biodiesel has a 97.8% ester content and a 0.53% α-linolenic methyl ester content, which meets EN14214 standards.

Binhayeeding et al. (2020) produced biodiesel by exploring WCO as a substrate through an enzymatic transesterification reaction. For this, lipases of *Rhizomucor miehei* and *Candida rugosa* were immobilized on polyhydroxybutyrate, acting as an environment-protection catalyst. This enzyme mixture produced 96.5% of biodiesel under the following temperature of 45°C, 1% mixed lipase (50% of every lipase), 5% water content, 6:1 methanol-to-oil proportion, and 250 rpm shaking speed. This mixture catalyst could be recycled for up to six cycles before the activity of lipase starts to decline. The chemical and fuel properties have been fulfilled with EN 14214 and ASTM D 6751 specifications.

Bhatia et al. (2021) investigated the impact of several surfactants on *Rhodococcus* growth and lipid synthesis using WCO as a carbon source. It produced 3.42 g/L of biomass and 2.39 g/L of lipids, with 70% lipid accumulation from 1.15% input of WCO. The fatty acid methyl esters (FAMEs) contains 61.68% palmitoleic acid, > 21.48% palmitic acid, > 12.95% myristic acid, > 2.35% stearic acid, > 0.74% pentadecanoic acid, > 0.72% heptadecanoic acid, and > 2.35% oleic acid. The qualities of the biodiesel generated from WCO comply with EN14214 and IS15607 standards.

15.7.2 Biosurfactants

Biosurfactants are surface-active substances with emulsifying properties. These compounds are classified as amphiphilic molecules because they have both the hydrophilic and hydrophobic ends; they can interact at the aqueous-nonaqueous interface (Marchant and Banat 2012). The global biosurfactant market is expected to reach USD 2.6–5.5 billion by the year 2023, increasing at a 5.6% annual pace from the year 2017 to 2023 (Singh, Patil, and Rale 2019). Despite the increased market requirements, the cost of biosurfactant production is still higher than that of synthetic alternatives. Its manufacturing in the industrial sector continues to be hampered by high costs and low yields (Marchant and Banat 2012). Hence, exploring WCO or any low-price substrates for biosurfactant synthesis is a cost-cutting strategy, as it reduces total production costs by up to 10–30% (Zenati et al. 2018).

WCO-based biosurfactant products are sturdily connected with the microbes like *Pseudomonas aeruginosa* (Chen et al. 2018), *Bacillus* sp. (Durval et al. 2019), *Streptomyces* sp. (Santos et al. 2019), and *Candida tropicalis* (da Rocha Junior et al. 2019). A broad range of WCO concentration levels of 9–100 g/L and biosurfactant production of 0.6–67 g/L has been recorded, based on the microbial species and cultivation environments of culture (Csutak, Corbu, and Vassu 2017; Kourmentza et al. 2018; Oliveira and Garcia-Cruz 2013; Venkatesh and Vedaraman 2012; Dzięgielewska and Adamczak 2013). When cultivated on the yeast extract and peptone added with 1% petroleum, n-hexadecane, or fried sunflower oil, the yeast strain *Candida tropicalis* CMGB114 produced substantial levels of biosurfactants. In all the examined growth mediums, the cells have multiplying lipid bodies, demonstrating active lipogenesis and a link between the biosurfactant and lipid production (Csutak, Corbu, and Vassu 2017).

Hasanizadeh, Moghimi, and Hamedi (2017) used waste frying oil to grow *Mucor circinelloides*, which produced biosurfactants. After three days of incubation, it lowers culture medium surface tension (26.6 mN/m) and forms a clear zone diameter of 12.9 cm in an oil-spreading test. *M. circinelloides* consumed 8% of WFO and 87.6% of crude oil in five and twelve days of the incubation period, respectively.

De Souza et al. (2018) produced surfactant by *Cunninghamella echinulata*, exploiting soybean oil waste (SOW) and corn steep liquor (CSL), and obtained 2% and 8.82% of biosurfactant in SOW and CSL, respectively. A 5.18 g/L of biosurfactant yield was obtained in 120 hours of growth. The low surface tension of 31.7 mN/m and an anionic nature were among the appealing qualities of the biosurfactant generated. Apart from that, it excels at removing spilled oil like diesel of 98.7% and kerosene of 92.3% from maritime sand.

Hisham (2019) used *Bacillus* sp. HIP3 to produce biosurfactant from WCO for heavy metal removal. The culture gave a high supplanting area with 38 mN/m of surface tension in seven days of incubation period when grown on a mineral salt medium with WCO of 2% (v/v) at a temperature of 30°C, and 200 rpm agitation. The extraction procedure yielded 99.5 g/L of biosurfactant when chloroform and methanol (2:1) were used as solvents. The produced biosurfactant is lipopeptide in nature, like standard surfactin. In addition, it is capable of eliminating 13.57% copper, 12.71% lead, 2.91% zinc, 1.68% chromium, and 0.7% cadmium from the artificially contaminated water, demonstrating its bioremediation potential as similar to standard surfactin. Shi et al. (2021) isolated oil-degrading and rhamnolipids-producing unique halotolerant *Pseudomonas aeruginosa* M4. For optimizing the oil degradation conditions, different sources of carbon and nitrogen, incubation temperature, inoculum load, and medium pH were evaluated. The degradation rate of oil was 85.20%, and 23.86 U/mL of lipase activity was attained under the ideal conditions of pH 8, 5% inoculum load, and 35°C incubation temperature. The predominant degradation intermediates were acetic and n-butyric acids. Salt tolerance of *P. aeruginosa* M4 was good up to a level of 70 g/L. When *P. aeruginosa* M4 was given WCO as a single carbon source, the highest rhamnolipid content of 1119.87 mg/L was obtained.

Kim et al. (2021) used *Starmerella bombicola* to metabolize and transform WCO into a high-value biosurfactant sophorolipid (SLs) through fed-batch fermentation. When compared to flask culture, the SLs content was enhanced 3.7-fold (315.6 g/L vs. 84.8 g/L), which was WCO's highest ever value obtained. SLs were also used to make methyl hydroxy–branched fatty acids (MHBFAs), which are vital compound used in different kinds of industries but are difficult to generate via conventional procedures. This is the first time MHBFAs created from SLs have been applied to plastic building blocks.

15.7.3 Polyhydroxyalkanoates (PHAs)

PHAs are microbial polyesters synthesized by a range of microbes under imbalanced feeding environments as carbon and energy storage compounds. PHAs are elastomers, thermoplastic polyesters, and sticky resins that consist of numerous R-OH-alkanoic acids based on their chain length. Cosmetics, food industry (packing, molding, and coating), pharmaceuticals, agriculture, tissue engineering, and water and wastewater treatment, and denitrification are just a few sectors where it's used (Kourmentza et al. 2018; Muhammadi et al. 2015). In 2021, the market for PHAs was around 84.4 million (Kourmentza et al. 2018). However, the price of the substrate (mostly carbon source), which accounts for 50% of the total manufacturing cost, continues to be a barrier to commercial-scale PHA synthesis (Song et al. 2008).

PHAs made from WCO were shown to be high quality, with molecular weight, chemical and thermal qualities comparable to those made from purified glucose or oils (Kamilah et al. 2018; Sharma et al. 2017; Verlinden et al. 2011). However, as the number of single bonds between carbon atoms and free fatty acids appear to be key determinants for biopolymer formation, WCO's origin and composition may alter PHA yields.

WCO is the most explored food waste for the production of PHA at the lab scale (Cruz et al. 2016; Rodriguez-Perez et al. 2018). The bacterial species capable of accumulating PHAs from WCO are *Cupriavidus necator*, *Burkholderia thailandensis*, *Klebsiella pneumonia*, *Pseudomonas* sp., and *Bacillus* sp. (Benesova et al. 2017; Kamilah et al. 2018; Kourmentza et al. 2018). WCO concentrations ranging from 10 to 40 g/L were utilized for the PHA production reliant on bacterial strains. Many bacteria have been shown to produce biosurfactants at a statistically significant concentration of WCO (De Souza et al. 2018; Niu et al. 2019; Ozdal, Gurkok, and Ozdal 2017; Venkatesh and Vedaraman 2012). Oliveira and Garcia-Cruz (2013), on the other hand, found no link between WCO concentration and biosurfactant generated from *Bacillus pumilus*. Furthermore, bacterial species produce significantly more PHAs from oily substrates (0.6–0.8 g/L) than produce from sugars (0.3–0.4 g/L). Apart from bacterial species, all these differences might be attributed to culture environments, as various operational and nutritive parameters influence WCO-based PHA synthesis.

Cupriavidus necator is one of the most-studied candidates for PHB production. Different forms of rapeseed oil, like waste frying, heated, and pure oil, were studied by Verlinden et al. (2011) as components of culture media to produce PHB. The WCO, heated oil, and pure oil were used to produce 1.2, 0.9, and 0.62 g/L of PHB, respectively, after seventy-two hours. In comparison to USFA, SFA contributes to the formation of energy-rich PHB. When using WCO and urea as a carbon and nitrogen source, respectively, the identical strain of *C. necator* H16 and its transformed mutant, *C. necator* PHB4, produced excellent PHA (Ka, Sa, and Aya 2013).

Gatea, Haider, and Khudair (2017) described that the culture *Pseudomonas aeruginosa* D7 produced a maximum of 1.74 and 0.62 g/L of dry biomass and PHB, respectively, after forty-eight hours of growing on WCO, while using the waste corn oil with a 2% initial concentration obtained 1.99 g/L of dry wt. with 0.75 of PHB producing 37% (PHB/biomass dry wt.).

Kongpeng, Iewkittayakorn, and Chotigeat (2017) produced PHA from WCO and compared it with PHAs from palm oil. To determine the optimum conditions, the two WCO storage duration (four and ten weeks) and three WCO concentration inputs (10, 20, and 30 g/L) were employed. The medium containing 30 g/L of WCO yielded a peak of 5.26 g/L PHA content after four weeks of storage, with 27.36 wt.% of cell dry weight (CDW). The results were similar to those obtained with cell growth employing 20 g/L of palm oil and 1% of fructose (CDW 5.93 g/L, and PHA content 26.96 wt. %). Furthermore, the content of PHA was higher when the cultures grown in 10 g/L WCO were preserved for ten weeks compared to 20 and 30 g/L of WCO. As a result, it is possible that WCO could be utilized to make PHA.

Sangkharak et al. (2020) used WCO to make PHA from *Bacillus thermoamylovorans*. This was also utilized as a biofuel feedstock, particularly 3-hydroxyalkanoate methyl ester (3HAME). While providing the optimal concentration of 4%, WCO yielded a 4 g/L of biomass and 3.5 g/L of PHA under batch cultivation at 45°C, 150 rpm, and forty-eight hours. The isolated polymer was found as poly 3-hydroxybutyrate-co-3-hydroxyvalerate [P(3HB-co-3HV)] with 85% 3-hydroxybutyrate (3HB) and 15% 3-hydroxyvalerate (3HV). The copolymer's characteristics were very comparable to commercial P (3HB-co-3HV).

Pan et al. (2021) acquired 7.6 g/L PHB yield using *C. necator* under the conditions of 30°C, 150 r/min, and pH 7.0 by utilizing 3% waste oil. Furthermore, in a 5 L bioreactor, the yield of PHB reached 8.25 g/L after fermentation for ninety-six hours by batch culture. The synthesized PHB upon characterization revealed an average molecular mass of 30 kD and a polydispersity index (PDI) of 1.44. The melting temperature and decomposition temperature of PHB were 175.7°C and 285.5°C, respectively, suggesting that it meets the demand for thermoplastic materials.

15.7.4 Lipases

Lipases (triacylglycerol acyl-hydrolases) are ubiquitous hydrolytic enzymes that catalyze the breakdown of fats and oils into free fatty acids, monoacylglycerols, diacylglycerols, and glycerol. It has a wide range of applications in different industries like detergent, pharmacy, cosmetic, food technology, agrochemical, textile, and biodiesel (Darvishi et al. 2017; Salihu and Alam 2015). It is produced in the interface of oil and water by several fungi, like *Aspergillus* sp., *Rhizopus* sp., *Rhizomucor* sp., *Penicillium* sp., *Mucor* sp., and *Geotrichum* sp., and yeasts (*Yarrovia lypolitica*, *Candida* spp., *Pichia* spp., *Rhodotorula* spp., etc.)., *Burkholderia* spp., *Pseudomonas* spp., and *Bacillus* spp. are reported as bacterial lipase producers (Treichel et al. 2010). The long-chain acylglycerols synthesis and hydrolysis are catalyzed by microbial lipases.

By 2023, the lipase market is anticipated to touch $590.233 million (www.marketsandmarkets.com/PressReleases/lipase.asp). Several bacteria utilize vegetable oils as carbon sources, and extracellular lipases are released to break down triglycerides into FFA (free fatty acids).

In both shake flasks and a bench-scale bioreactor, Domínguez et al. (2010) investigated the biodegradation of WCO and applied it as a lipase production inducer for *Y. lipolytica* CECT 1240 cultures. The strain CECT 1240 degrades spent oil and decreases oil COD to 90% within three days in bioreactor. Compared to oil-free medium, WCO-added medium increased extracellular lipase production in the yeast.

Sunil Kumar and Sangeeta Negi (2014) explored the extracellular lipase of *Penicillium chrysogenum* for WCO hydrolysis and observed 17% and 5% of oleic acid and stearic acid release, respectively, with a 1:1 proportion of lipase-to-WCO incorporation. Suci, Arbianti, and Hermansyah (2018) cultured *B. subtilis* in WCO through the submerged fermentation (SmF) to produce lipase. The highest lipase activity of 4.96 U/mL was achieved while manipulating the concentration of inducer, inoculum, Ca^{2+} ion, nitrogen source, and substrate. The maximum activity was observed under the following conditions: 4% WCO, 5% inoculum, 0.25% olive oil, 0.5% yeast extract, and 10 mM Ca^{2+} at 30°C for eighty-four-hour fermentation.

Lopes et al. (2019) assessed *Y. lipolytica* W29 to degrade WCO and produce lipase simultaneously. They employed the Taguchi method to evaluate the upshot of concentration of initial WCO, Arabic gum and pH on lipase production. It was observed that the initial medium pH and interaction between the WCO and concentration of Arabic gum were found to be the most noteworthy factors and obtained maximum lipase activity of 12,000 U/L.

15.7.5 Biogas

Biogas (methane-rich gas) is produced from the anaerobic digestion of various kinds of wastes that could be exploited in different ways, including as vehicle fuel, heating, and electricity generation. Free fatty acid, high SFA, and toxic components in the WCO affect the methanogenesis process. Pretreatments and other substrate-based co-digestion helps to mitigate methanogenesis failure and improve CH_4 production rates.

Recently, Marchetti et al. (2020) have reviewed the importance of recycling WCO for biogas production. Microbial populations, process variables (pH, temperature, lipid concentration, and agitation), raw material pretreatment, and fat co-digestion with other feedstocks are all major concerns. The designs of the lipid digestion reactor were also examined. All this makes it easier to apply technology to find methods to tackle the biological and engineering barriers to its spread, and researchers have studied the importance of recycling WCO for biogas production. Microbial populations, process variables (temperature, lipid concentration, agitation, and pH), raw material pretreatment, and fat codigestion with other feedstocks are all major concerns. The state-of-art designs in the lipid-digesting reactors were also investigated. All this makes it easier to apply technology to find the solution to overcome biological and engineering barriers.

In stirring conditions, Meng et al. (2015) evaluated the methane generation capacity of floating oil skim from food waste and achieved 900 mL of CH_4/mL fat, over a considerable lag period. In semicontinuous anaerobic reactors, He et al. (2018) employed WCO as the only carbon source and achieved the most remarkable results of 1.5 g/volatile solids (VS) L day at organic loading rates (OLRs), with a specific yield of methane at 0.78 L per g VS day.

Affes et al. (2017) tripled the solubilization of COD using lipases from the *Staphylococcus xylosus* at the temperature of 37°C with a pH of 7 for six days and obtained 0.6 L/g COD of biogas yield by batch-fermenting hydrolyzed grease with a slaughterhouse waste at 25:75% v/v in mesophilic environments. The amount of hydrolyzed grease in biogas production decreased as the proportion of hydrolyzed grease increased.

WCO was co-digested with the remaining *Y. lipolytica* biomass utilized for citric acid synthesis by Moeller et al. (2018). Under mesophilic settings and a yeast-to-WCO ratio of 1:20 w/w, they obtained 0.75 Nm^3 of CH_4/kg VS, boosting the methane output from yeast and WCO in a monodigestion by 31% and 6%, correspondingly. In addition, methane production was raised even more in semicontinuous operations (0.884 Nm^3/kg VS).

Damasceno et al. (2018) used a mixture containing enzymes from *Penicillium brevicompactum* and a rhamnolipid biosurfactant from *Pseudomonas aeruginosa* to pretreat poultry slaughterhouse wastewater, resulting in an 88% increase in CH_4 and the removal of 95.8% of fats related to control.

Marchetti et al. (2020) co-digested WCO and pig slurry at a 76:24 ratio based on total VS and obtained 811 mL CH_4/g VS. Pig slurry increased WCO digestion by shortening the lag period and allowing more CH_4 to be produced per reactor.

He et al. (2018) examined the effect of OLR and microbial community dynamics on methane production using skimmed oil waste. At an OLR of 1.0, the specific biogas/methane yields were 1.44/0.98 L/g VS. During the anaerobic digestion (AD) process, the most frequent archaea were *Methanosaeta*, while *Syntrophomona*, *Anaerovibrio*, and *Synergistaceae* were commonly found bacteria. Besides, redundancy analysis revealed that WCO's OLR had a more significant impact on bacterial populations than archaeal communities.

15.7.6 Microbial Lipids

Oils formed by microorganisms, also recognized as single-cell oils (SCO), are mostly made up of triacylglycerols (TAGs) and neutral lipids that accumulate in the cytoplasm in a construct called the lipid body. It could be used as food supplements or feedstock for biodiesel synthesis (Béligon et al. 2016; Carsanba, Papanikolaou, and Erten 2018).

Despite numerous works described, usage of WCO as the substrate for the accumulation of intracellular lipids and the use of bacterial species were limited. The biodegradation of butter and olive oil cooking fats was investigated utilizing *Y. lipolytica* LFMB 20, as well as *Pseudomonas putida* CP1 and *Bacillus* spp. based bioaugmented product. The microbes were cultivated under aerobic conditions in a shake flask containing a medium supplied with approximately 0.85% of waste fat by weight. At the close of the incubation, analysis of the leftover substance revealed that the yeast had removed almost 90% fat, while bacteria had eliminated both fats by approximately 95% (Tzirita et al. 2018). The bacterial consortium gave 42 and 63% lipid yields for olive oil and butter, respectively. *Y. lipolytica* yielded 22% for both lipids.

In contrast, oleaginous yeast *Y. lipolytica* is a prominent biolipids producer and has been utilized as a typical eukaryotic organism for *de novo* lipid synthesis and fatty acid metabolism. For certain *Y. lipolytica* strains, the influence of WCO input on *ex novo* lipid synthesis was investigated, and the optimum level was determined by yeast strain. *Y. lipolytica* 3589 and SWJ-1b showed no evidence of WCO substrate inhibition on the lipid content up to 100 g/L and 140 g/L (Katre et al. 2012, Liu et al. 2015), respectively; however, W29 and 3472 strains accumulated the most lipids with WCO of 30 g/L (Lopes et al. 2019, Katre et al. 2012).

In order to boost lipid buildup by microbial species, supplementing WCO-based media with other surfactants, emulsifiers, or nutrients had been tried. When magnesium and phosphate were added to the *Y. lipolytica* cells, the level of lipid content yield was increased (Liu et al. 2015). Chemical surfactants or emulsifiers like Tween 80 were added to the medium to improve the WCO's absorption and stability in the microbial cells (Arous et al. 2017).

Papanikolaou et al. (2011) cultivated *Aspergillus* sp. (five strains) on the waste cooking olive oil to produce lipid-rich biomass. A 15 g/L waste oil was supplemented to the growth medium. All *Aspergillus* accumulated cellular lipids in prominent quantities, and the strain ATHUM 3482 synthesized the highest lipids of 64% (w/w) on a dry-weight basis.

15.7.7 Platform Chemicals

The aforementioned microbial-mediated products from WCO have received the most attention in lab-scale investigations, and bioconversion of WCO into other significant platform chemicals has also received much attention.

15.7.7.1 Bioemulsifier

Liu et al. (2011) proposed to synthesize bioemulsifiers from WFOs using a high-efficiency demulsifying strain, *Alcaligenes* sp. S-XJ-1. It is a cost-effective approach to lower synthesis costs and broadens its use in oil field applications. The yield of the bioemulsifier rose by 4.06 g/L when the pH of the growth medium increased to 4. Furthermore, using WCO as an additional carbon source in the fed-batch culture of *Alcaligenes* sp. proved to be a promising technique for enhancing bioemulsifier output (5.96 g/L).

15.7.7.2 Organic Acid

Papanikolaou et al. (2011) used *Aspergillus* sp. ATHUM 3482 and *P. expansum* NRRL 973 to produce oxalic acid and citric acid as by-products from waste cooking olive oil. Liu et al. (2015) used the yeast strain *Y. lipolytica* and successfully produced citric acid (CA) of 31.7 g/L, 6.5 g/L isocitric acids, and 5.9 g/L biomass from 80 g/L WCO in 336 hours of fermentation, which suggests that WCO appears to be a promising carbon source for citric acid production. It was also proposed that additional magnesium and nitrogen, instead of phosphate and vitamin B1, were required for the CA production.

15.7.7.3 Riboflavin

Riboflavin is a water-soluble vitamin that acts as a precursor for flavin adenine dinucleotide and flavin mononucleotide coenzymes, which are engaged in oxidation-reduction chemical reactions in all organisms (Revuelta et al. 2016). Globally, it is used as a nutritional supplement for animals and humans. Fermentation produces vitamin B2 in a single reaction, making it more cost-effective and environmentally friendly than chemical synthesis. Riboflavin is derived from industrial strains of *Bacillus subtilis*, *Ashbya gossypii*, and *Candida famata* var. *flareri*, with titers up to 14, 15, and 20 g/L, respectively, by global manufacturers like DSM (earlier Roche; Netherlands), Hubei Guangji Pharmaceuticals, Badische Anilin-und SodaFabrik (BASF) (Germany), and Shanghai Acebright (Lim, Lee, and Heo 2001; Revuelta et al. 2016).

Wei et al. (2013) bioconverted WCO into riboflavin by *A. gossypii* and attained a 4.78 g/L riboflavin yield with initial pH of 6.5 and input of 40 g/L WCO in the medium. No residual WCO was found at the end of fermentation. In addition, they monitored reduced sugar, biomass, pH, and free amino nitrogen for the efficient utilization of WCO for riboflavin production. During the fermentation process, when pH ranged from 6.5 to 6.8, the free amino nitrogen levels and decreased sugar were used very effectively, and riboflavin yield was enhanced to 6.76 g/L.

15.7.7.4 Vitamin B_{12}

Propionibacterium freudenreichii PTCC 1674 synthesized vitamin B_{12}, acetic acid, and propionic acid concurrently in a frying medium based on waste sunflower oil. In addition to WCO concentration, dimethyl benzimidazolyl, $CoCl_2$, $FeSO_4$, and $CaCl_2$ were found to have a considerable impact on vitamin B_{12} synthesis. The optimized medium provides 170% higher vitamins (2.60 mg/L) than the original medium. WCO as a low-cost carbon source for industrial manufacture of Vit B_{12} was explored in-depth by Hajfarajollah et al. (2015).

15.7.7.5 Carotene

Nanou and Roukas (2016) synthesized carotene by *Blakeslea trispora* using WCO. The maximum carotene yield of 2,021 mg/L was obtained with WCO of 50 g/L as a carbon source, added with butylated hydroxytoluene (BHT) and corn steep liquor (CSL). The oxidative stress caused by hydroperoxides and BHT significantly improved the carotene yield, but the external addition of glucose, Span 80, yeast extract, and others did not improve yield. The carotenes produced in this process contain 74.2% beta-carotene, 23.2% gamma-carotene, and 2.6% lycopene. Once the process was scaled up to a 1.4-L bubble column bioreactor, carotene synthesis was not sustained and yielded a lower yield (980.0 mg/L) than flask trials (Nanou et al. 2017).

15.7.7.6 Erythritol

Erythritol, 4C sugar alcohol, is currently utilized as a sucrose substitute with 70–80% of sucrose's sweetness. Xiaoyan et al. (2017) used *Y. lipolytica* M53 for coproducing lipase and erythritol from WCO, demonstrating that WCO was better for erythritol synthesis than pure vegetable oils, including olive, soybean, and rapeseed oil. The most influential factors for the efficient biosynthesis of erythritol include WCO concentration, the quantity of an osmotic agent, NaCl, C/N ratio, and nitrogen source. For the coproduction of erythritol and lipase, ammonium oxalate proved to be the optimal nitrogen source. The process was upscaled from a shake flask (250 mL) to a fermenter (5L) and registered a modest increase in yield (20.5 g/L to 22.1 g/L) after seventy-two hours, while the greatest lipase activity of 12.7 U/mL was reached in just twenty-four hours.

Later, Liu et al. (2018) produced erythritol and citric acid from WCO under varied media conditions using *Y. lipolytica* M53. The maximum erythritol yield of 21.8 g/L and high lipase activity was observed under higher osmotic pressure (OP) of 2.76 osmol/L and a low pH of 3.0, but the maximum CA synthesis (12.6 g/L) was found at a high pH (6.0) and low OP (0.75 osmol/L). The activities of glycerol kinase, transketolase, citrate synthase, and erythrose reductase involved in the biochemical pathway were identified and confirmed that erythritol and citric acid metabolic flux modulation occurred mainly at the posttranscriptional stage.

15.8 CONCLUSION AND PERSPECTIVE

WCOs are valuable waste discarded after the cooking process, generated in abundant quantities. Despite the presence of some contaminants in WCO, the predominant constituents of these oily wastes are TAGs, glycerides, and fatty acids, making these viable carbon bases for growth and metabolite synthesis in the microorganism. At present, WCO valorization using microorganisms to create value-added products, including enzymes, microbial lipids, biosurfactants, bioplastics, and other platform chemicals, is a promising biotechnological strategy, and significant market potential is being addressed. The inclusion of WCO in the medium composition might improve its production economics. Many firms have started to use WCO as feedstock for biodiesel production, enabling a circular economy. Despite the fact that a circular economy model centered on WCOs appears appealing, fundamental obstacles must be addressed. The future prospects might be centered on dealing with WCO's heterogeneous nature and harmful by-products, revamping processes, setting up WCO collection centers, and improving traceability from household, small factories, and large industries, enforcing the law and policies to avoid improper disposal and WCO management.

TABLE 15.1
Added-Value Products Derived from WCO through Microbial Fermentation Process

S. No.	Process & Type of Fermentation	Added-Value Products	Substrates	Yield	Reference
1.	Aerobic & submerged fermentation	Biodiesel	WCO	~90%	Sarno and Iuliano (2019)
2.	Anaerobic & surface fermentation	Biogas	Codigestion WCO + *Yarrowia lipolytica* biomass	0.884 Nm^3/kg VS	Moeller et al. (2018)
3.	Aerobic & submerged fermentation	Microbial lipids	WCO	63% (butter) & 42% (olive oil)	Tzirita et al. (2018)

(*Continued*)

TABLE 15.1 ***(Continued)***
Added-Value Products Derived from WCO through Microbial Fermentation Process

S. No.	Process & Type of Fermentation	Added-Value Products	Substrates	Yield	Reference
4.	Aerobic & submerged fermentation	Microbial lipids	WCO	24–75%	Katre et al. 2012; Liu et al. 2019; Lopes et al. (2019)
5.	Aerobic & submerged fermentation	Thermostable lipase	WCO	340 U/mL	Awad et al. (2015)
6.	Aerobic & submerged fermentation	Surfactant	Soybean oil waste (SOW)	5.18 g/L	De Souza et al. (2018)
7.	Aerobic & solid state fermentation	Oleic acid and stearic acid	Emulsified cooking oil	17% and 5%	Kumar and Negi (2015)
8.	Aerobic & submerged fermentation	Citric acid	WCO (80 g/L)	31.7 g/L of citric acid	Liu et al. (2015)
9.	Aerobic & submerged fermentation	Polyhydroxy-alkonate (PHA) & biodiesel	WCO	1.5 g/L & 86%	Vastano et al. (2019)
10.	-	Glycerol	Waste olive oil	-	Maegala et al. (2020)
11.	Aerobic & submerged fermentation	Riboflavin	WCO (40 g/L)	4.78 g/L	Wei et al. (2013)
12.	Aerobic & submerged fermentation	Oxalic acid	Waste olive oil (15 g/L)	5 g/L	Papanikolaou et al. (2011)
13.	Aerobic & submerged fermentation	Citric acid	Waste olive oil (15 g/L)	3.5 g/L	Papanikolaou et al. (2011)
14.	Aerobic & submerged fermentation (336 h)	Citric acid	WCO (80 g/L)	31.7 g/L	Liu et al. (2015)
15.	Aerobic & submerged fermentation (72 h)	Erythritol	WCO (30 g/L)	22.1 g/L	Xiaoyan et al. (2017)
16.	Aerobic & submerged fermentation (72 h)	Erythritol	WCO (30 g/L)	21.8 g/L	Liu et al. (2018)

TABLE 15.1 *(Continued)*
Added-Value Products Derived from WCO through Microbial Fermentation Process

S. No.	Process & Type of Fermentation	Added-Value Products	Substrates	Yield	Reference
17.	Aerobic & submerged fermentation (120 h)	Vit B_{12}	Waste frying sunflower oil (4% w/v)	2.60 mg/L	Hajfarajollah et al. (2015)
18.	Aerobic & submerged fermentation (7 d)	Bioemulsifier	Used vegetable oil & lard (3%)	5.96 g/L	Liu et al. (2011)
19.	Aerobic, submerged fermentation and pellet formed (6 d)	Carotene	WCO (50 g/L)	2021 mg/L	Nanou and Roukas (2016)

REFERENCES

Affes, Maha, Fathi Aloui, Fatma Hadrich, Slim Loukil, and Sami Sayadi. 2017. "Effect of bacterial lipase on anaerobic co-digestion of slaughterhouse wastewater and grease in batch condition and continuous fixed-bed reactor." *Lipids in Health and Disease* 16 (1):1–8.

Alves, M. Madalena, M. Alcina Pereira, Diana Z. Sousa, Ana J. Cavaleiro, Merijn Picavet, Hauke Smidt, and Alfons J. M. Stams. 2009. "Waste lipids to energy: How to optimize methane production from long-chain fatty acids (LCFA)." *Microbial Biotechnology* 2 (5):538–550.

Appels, Lise, Joost Lauwers, Jan Degrève, Lieve Helsen, Bart Lievens, Kris Willems, Jan Van Impe, and Raf Dewil. 2011. "Anaerobic digestion in global bio-energy production: Potential and research challenges." *Renewable and Sustainable Energy Reviews* 15 (9):4295–4301.

Arous, Fatma, Imen Ben Atitallah, Moncef Nasri, and Tahar Mechichi. 2017. "A sustainable use of low-cost raw substrates for biodiesel production by the oleaginous yeast *Wickerhamomyces anomalus*." *3 Biotech* 7 (4):1–10.

Awad, Ghada E. A., Hanan Mostafa, Enas N. Danial, Nayera A. M. Abdelwahed, and Hassan M. Awad. 2015. "Enhanced production of thermostable lipase from *Bacillus cereus* ASSCRC-P1 in waste frying oil based medium using statistical experimental design." *Journal of Applied Pharmaceutical Science* 5 (9):007–015.

Bastida, S., and F. J. Sánchez-Muniz. 2002. "Polar content vs. TAG oligomer content in the frying-life assessment of monounsaturated and polyunsaturated oils used in deep-frying." *Journal of the American Oil Chemists' Society* 79 (5):447–451.

Béligon, Vanessa, Gwendoline Christophe, Pierre Fontanille, and Christian Larroche. 2016. "Microbial lipids as potential source to food supplements." *Current Opinion in Food Science* 7:35–42.

Benesova, P., D. Kucera, I. Marova, and S. Obruca. 2017. "Chicken feather hydrolysate as an inexpensive complex nitrogen source for PHA production by *Cupriavidus necator* on waste frying oils." *Letters in Applied Microbiology* 65 (2):182–188.

Bhatia, Shashi Kant, Ranjit Gurav, Yong-Keun Choi, Hong-Ju Lee, Sang Hyun Kim, Min Ju Suh, Jang Yeon Cho, Sion Ham, Sang Ho Lee, and Kwon-Young Choi. 2021. "*Rhodococcus* sp. YHY01 a microbial cell factory for the valorization of waste cooking oil into lipids a feedstock for biodiesel production." *Fuel* 301:121070.

Binhayeeding, Narisa, Sappasith Klomklao, Poonsuk Prasertsan, and Kanokphorn Sangkharak. 2020. "Improvement of biodiesel production using waste cooking oil and applying single and mixed immobilised lipases on polyhydroxyalkanoate." *Renewable Energy* 162:1819–1827.

Carsanba, E., S. Papanikolaou, and HÜSeyİN Erten. 2018. "Production of oils and fats by oleaginous microorganisms with an emphasis given to the potential of the nonconventional yeast *Yarrowia lipolytica*." *Critical Reviews in Biotechnology* 38 (8):1230–1243.

Chen, Chunyan, Ni Sun, Dongsheng Li, Sihua Long, Xiaoyu Tang, Guoqing Xiao, and Linyuan Wang. 2018. "Optimization and characterization of biosurfactant production from kitchen waste oil using *Pseudomonas aeruginosa*." *Environmental Science and Pollution Research* 25 (15):14934–14943.

Choe, E., and D. B. Min. 2007. "Chemistry of deep-fat frying oils." *Journal of Food Science* 72 (5):R77–R86.

Chrysikou, Loukia P., Vasiliki Dagonikou, Athanasios Dimitriadis, and Stella Bezergianni. 2019. "Waste cooking oils exploitation targeting EU 2020 diesel fuel production: Environmental and economic benefits." *Journal of Cleaner Production* 219:566–575.

Contreras-Andrade, Ignacio, F. E. Sierra-Vargas, and C. A. Guerrero-Fajardo. 2013. "Biodiesel production from waste cooking oil by enzymatic catalysis process." *Journal of Chemistry and Chemical Engineering* 7:993–1000.

Corral Bobadilla, Marina, Rubén Lostado Lorza, Rubén Escribano García, Fátima Somovilla Gómez, and Eliseo P. Vergara González. 2017. "An improvement in biodiesel production from waste cooking oil by applying thought multi-response surface methodology using desirability functions." *Energies* 10 (1):130.

Cruz, Madalena V., Filomena Freitas, Alexandre Paiva, Francisca Mano, Madalena Dionísio, Ana Maria Ramos, and Maria A. M. Reis. 2016. "Valorization of fatty acids-containing wastes and byproducts into short-and medium-chain length polyhydroxyalkanoates." *New Biotechnology* 33 (1):206–215.

Csutak, Ortansa, Viorica Corbu, and Tatiana Vassu. 2017. "Studies on the correlation between biosurfactant and lipid synthesis in *Candida tropicalis* CMGB114 using hydrocarbons and vegetable oil wastes." *Revista de Chimie (Bucharest)* 68 (2):255–259.

da Rocha Junior, Rivaldo B., Hugo M. Meira, Darne G. Almeida, Raquel D. Rufino, Juliana M. Luna, Valdemir A. Santos, and Leonie A. Sarubbo. 2019. "Application of a low-cost biosurfactant in heavy metal remediation processes." *Biodegradation* 30 (4):215–233.

Damasceno, Fernanda R. C., Elisa D. Cavalcanti-Oliveira, Ioannis K. Kookos, Apostolis A. Koutinas, Magali C. Cammarota, and Denise M. G. Freire. 2018. "Treatment of wastewater with high fat content employing an enzyme pool and biosurfactant: Technical and economic feasibility." *Brazilian Journal of Chemical Engineering* 35:531–542.

Darvishi, Farshad, Zahra Fathi, Mehdi Ariana, and Hamideh Moradi. 2017. "*Yarrowia lipolytica* as a workhorse for biofuel production." *Biochemical Engineering Journal* 127: 87–96.

De Souza, Patrícia Mendes, Nadielly R. Andrade Silva, Daniele G. Souza, Thayse A. Lima e Silva, Marta C. Freitas-Silva, Rosileide F. S. Andrade, Grayce K. B. Silva, Clarissa D. C. Albuquerque, Arminda Saconi Messias, and Galba M. Campos-Takaki. 2018. "Production of a biosurfactant by *Cunninghamella echinulata* using renewable substrates and its applications in enhanced oil spill recovery." *Colloids and Interfaces* 2 (4):63.

Debnath, Sukumar, N. K. Rastogi, A. G. Gopala Krishna, and B. R. Lokesh. 2009. "Oil partitioning between surface and structure of deep-fat fried potato slices: A kinetic study." *LWT-Food Science and Technology* 42 (6):1054–1058.

del Pilar Rodriguez, Maria, Ryszard Brzezinski, Nathalie Faucheux, and Michèle Heitz. 2016. "Enzymatic transesterification of lipids from microalgae into biodiesel: A review." *AIMS Energy* 4 (6):817–855.

Domínguez, Alberto, Francisco J. Deive, M. Angeles Sanromán, and María A. Longo. 2010. "Biodegradation and utilization of waste cooking oil by *Yarrowia lipolytica* CECT 1240." *European Journal of Lipid Science and Technology* 112 (11):1200–1208.

Durval, Italo José B., Ana Helena M. Resende, Mariana A. Figueiredo, Juliana M. Luna, Raquel D. Rufino, and Leonie A. Sarubbo. 2019. "Studies on biosurfactants produced using *Bacillus cereus* isolated from seawater with biotechnological potential for marine oil-spill bioremediation." *Journal of Surfactants and Detergents* 22 (2):349–363.

Dzięgielewska, Ewelina, and Marek Adamczak. 2013. "Evaluation of waste products in the synthesis of surfactants by yeasts." *Chemical Papers* 67 (9):1113–1122.

El Bialy, Heba, Ola M. Gomaa, and Khaled Shaaban Azab. 2011. "Conversion of oil waste to valuable fatty acids using oleaginous yeast." *World Journal of Microbiology and Biotechnology* 27 (12):2791–2798.

Farkas, B. E., R. P. Singh, and T. R. Rumsey. 1996. "Modeling heat and mass transfer in immersion frying. I, model development." *Journal of Food Engineering* 29 (2):211–226.

FSSAI. 2019. The Use and Reuse of Cooking Oil - Disadvantages, Regulations and all you need to know. https://fssai.gov.in/upload/media/FSSAI_NEws_Oil_Insider_30_09_2019.pdf

Gatea, Iman H., Nadhim H. Haider, and Saad H. Khudair. 2017. "Bioplastic (Poly-3-Hydroxybutyrate) production by local *Pseudomonas aeruginosa* isolates utilizing waste cooking oil." *World Journal of Pharmacology Research* 6 (8).

Greenea. 2018. And do you recycle your used cooking oil at home? https://www.greenea.com/wp-content/uploads/2017/03/Greenea-article-UCO-household collection-309 2017.pdf

Hajfarajollah, Hamidreza, Babak Mokhtarani, Hamidreza Mortaheb, and Ali Afaghi. 2015. "Vitamin B 12 biosynthesis over waste frying sunflower oil as a cost effective and renewable substrate." *Journal of Food Science and Technology* 52 (6):3273–3282.

Hasanizadeh, Parvin, Hamid Moghimi, and Javad Hamedi. 2017. "Biosurfactant production by *Mucor circinelloides* on waste frying oil and possible uses in crude oil remediation." *Water Science and Technology* 76 (7):1706–1714.

He, Jing, Xing Wang, Xiao-bo Yin, Qiang Li, Xia Li, Yun-fei Zhang, and Yu Deng. 2018. "Insights into biomethane production and microbial community succession during semi-continuous anaerobic digestion of waste cooking oil under different organic loading rates." *AMB Express* 8 (1):1–11.

Hisham, Nurul Hanisah Md Badrul. 2019. "Production of biosurfactant from used cooking oil by local bacterial isolates for heavy metals removal." *Molecules* 24 (14):1–16.

Jha, Girish Kumar, Suresh Pal, V. C. Mathur, Geeta Bisaria, P. Anbukkani, R. R. Burman, and S. K. Dubey. 2012. *Edible Oilseeds Supply and Demand Scenario in India: Implications for Policy*, New Delhi, Division of Agricultural Economics, Indian Agricultural Research Institute.

Ka, Hanisah, Kumar Sa, and Tajul Aya. 2013. "The management of waste cooking oil: A preliminary survey." *The Journal of Environmental Health* 4:76–81.

Kalam, M. A., H. H. Masjuki, M. H. Jayed, and A. M. Liaquat. 2011. "Emission and performance characteristics of an indirect ignition diesel engine fuelled with waste cooking oil." *Energy* 36 (1):397–402.

Kamilah, Hanisah, Adel Al-Gheethi, Tajul Aris Yang, and Kumar Sudesh. 2018. "The use of palm oil-based waste cooking oil to enhance the production of Polyhydroxybutyrate [P (3HB)] by *Cupriavidus necator* H16 strain." *Arabian Journal for Science & Engineering (Springer Science & Business Media BV)* 43 (7).

Katre, Gouri, Chirantan Joshi, Mahesh Khot, Smita Zinjarde, and Ameeta RaviKumar. 2012. "Evaluation of single cell oil (SCO) from a tropical marine yeast *Yarrowia lipolytica* NCIM 3589 as a potential feedstock for biodiesel." *AMB Express* 2 (1):1–14.

Kim, Jeong-Hun, Yu-Ri Oh, Juyoung Hwang, Jaeryeon Kang, Hyeri Kim, Young-Ah Jang, Seung-Soo Lee, Sung Yeon Hwang, Jeyoung Park, and Gyeong Tae Eom. 2021. "Valorization of waste-cooking oil into sophorolipids and application of their methyl hydroxyl branched fatty acid derivatives to produce engineering bioplastics." *Waste Management* 124:195–202.

Kingston, Paul F. 2002. "Long-term environmental impact of oil spills." *Spill Science & Technology Bulletin* 7 (1–2):53–61.

Kongpeng, Chatsuda, Jutarut Iewkittayakorn, and Wilaiwan Chotigeat. 2017. "Effect of storage time and concentration of used cooking oil on polyhydroxyalkanoates (phas) production by *Cupriavidus necator* H16." *Sains Malaysiana* 46 (9):1465–1469.

Kourmentza, C., J. Costa, Z. Azevedo, C. Servin, Christian Grandfils, V. De Freitas, and M. A. M. Reis. 2018. "*Burkholderia thailandensis* as a microbial cell factory for the bioconversion of used cooking oil to polyhydroxyalkanoates and rhamnolipids." *Bioresource Technology* 247:829–837.

Kumar, Sunil, and Sangeeta Negi. 2015. "Transformation of waste cooking oil into C-18 fatty acids using a novel lipase produced by *Penicillium chrysogenum* through solid state fermentation." *3 Biotech* 5 (5):847–851.

Lam, Su Shiung, Rock Keey Liew, Ahmad Jusoh, Cheng Tung Chong, Farid Nasir Ani, and Howard A. Chase. 2016. "Progress in waste oil to sustainable energy, with emphasis on pyrolysis techniques." *Renewable and Sustainable Energy Reviews* 53:741–753.

Lim, HyeonSook, K. S. N. Lee, and YoungRan Heo. 2001. "The relationships of health-related lifestyles with homocysteine, folate, and vitamin B12 status in Korean adults." *Korean Journal of Community Nutrition* 6 (Suppl. 3):507–515.

Liu, Jia, Kaiming Peng, Xiangfeng Huang, Lijun Lu, Hang Cheng, Dianhai Yang, Qi Zhou, and Huiping Deng. 2011. "Application of waste frying oils in the biosynthesis of biodemulsifier by a demulsifying strain *Alcaligenes* sp. S-XJ-1." *Journal of Environmental Sciences* 23 (6):1020–1026.

Liu, Xiaoyan, Jinshun Lv, Jiaxing Xu, Jun Xia, Aiyong He, Tong Zhang, Xiangqian Li, and Jiming Xu. 2018. "Effects of osmotic pressure and pH on citric acid and erythritol production from waste cooking oil by *Yarrowia lipolytica*." *Engineering in Life Sciences* 18 (6):344–352.

Liu, Xiaoyan, Jinshun Lv, Jiaxing Xu, Tong Zhang, Yuanfang Deng, and Jianlong He. 2015. "Citric acid production in *Yarrowia lipolytica* SWJ-1b yeast when grown on waste cooking oil." *Applied Biochemistry and Biotechnology* 175 (5):2347–2356.

Liu, Xiaoyan, Yubo Yan, Pusu Zhao, Jie Song, Xinjun Yu, Zhipeng Wang, Jun Xia, and Xiaoyu Wang. 2019. "Oil crop wastes as substrate candidates for enhancing erythritol production by modified *Yarrowia lipolytica* via one-step solid state fermentation." *Bioresource Technology* 294:122194.

Lopes, Marlene, Sílvia M. Miranda, Joana M. Alves, Ana S. Pereira, and Isabel Belo. 2019. "Waste cooking oils as feedstock for lipase and lipid-rich biomass production." *European Journal of Lipid Science and Technology* 121 (1):1800188.

Maegala, N. M., S. Anupriya, A. Hazeeq Hazwan, Y. Nor Suhaila, and A. Hasdianty. 2020. "Conversion of waste cooking oil to glycerol by halal microbial lipase." *IOP Conference Series: Earth and Environmental Science. 6th International Conference on Environment and Renewable Energy,* 23–25th February, Vietnam 505: 1–7.

Marchant, Roger, and Ibrahim M. Banat. 2012. "Microbial biosurfactants: Challenges and opportunities for future exploitation." *Trends in Biotechnology* 30 (11):558–565.

Marchetti, Rosa, Ciro Vasmara, Lorenzo Bertin, and Francesca Fiume. 2020. "Conversion of waste cooking oil into biogas: Perspectives and limits." *Applied Microbiology and Biotechnology* 104 (7):2833–2856.

Meng, Ying, Sang Li, Hairong Yuan, Dexun Zou, Yanping Liu, Baoning Zhu, Akiber Chufo, Muhammad Jaffar, and Xiujin Li. 2015. "Evaluating biomethane production from anaerobic mono-and co-digestion of food waste and floatable oil (FO) skimmed from food waste." *Bioresource Technology* 185:7–13.

Moeller, Lucie, Aline Bauer, Andreas Zehnsdorf, Mi-Yong Lee, and Roland Arno Müller. 2018. "Anaerobic co-digestion of waste yeast biomass from citric acid production and waste frying fat." *Engineering in Life Sciences* 18 (7):425–433.

Muhammadi, Shabina, Muhammad Afzal, and Shafqat Hameed. 2015. "Bacterial polyhydroxyalkanoates-eco-friendly next generation plastic: Production, biocompatibility, biodegradation, physical properties and applications." *Green Chemistry Letters and Reviews* 8 (3–4):56–77.

Nanou, Konstantina, and Triantafyllos Roukas. 2016. "Waste cooking oil: A new substrate for carotene production by *Blakeslea trispora* in submerged fermentation." *Bioresource Technology* 203:198–203.

Nanou, Konstantina, Triantafyllos Roukas, Emmanuel Papadakis, and Parthena Kotzekidou. 2017. "Carotene production from waste cooking oil by *Blakeslea trispora* in a bubble column reactor: The role of oxidative stress." *Engineering in Life Sciences* 17 (7):775–780.

Nayak, Prakash Kumar, U. M. A. Dash, Kalpana Rayaguru, and Keasvan Radha Krishnan. 2016. "Physio-chemical changes during repeated frying of cooked oil: A Review." *Journal of Food Biochemistry* 40 (3):371–390.

Niu, Yongwu, Jianan Wu, Wei Wang, and Qihe Chen. 2019. "Production and characterization of a new glycolipid, mannosylerythritol lipid, from waste cooking oil biotransformation by *Pseudozyma aphidis* ZJUDM34." *Food Science & Nutrition* 7 (3):937–948.

Oliveira, Juliana Guerra de, and Crispin Humberto Garcia-Cruz. 2013. "Properties of a biosurfactant produced by *Bacillus pumilus* using vinasse and waste frying oil as alternative carbon sources." *Brazilian Archives of Biology and Technology* 56:155–160.

Orjuela, Alvaro, and James Clark. 2020. "Green chemicals from used cooking oils: Trends, challenges, and opportunities." *Current Opinion in Green and Sustainable Chemistry* 26:100369.

Ozdal, Murat, Sumeyra Gurkok, and Ozlem Gur Ozdal. 2017. "Optimization of rhamnolipid production by *Pseudomonas aeruginosa* OG1 using waste frying oil and chicken feather peptone." *3 Biotech* 7 (2):1–8.

Pan, Lan-jia, Jie Li, Qing-huai Lin, and Yin Wang. 2021. "Polyhydroxybutyrate production from mixed waste cooking oil by *Cupriavidus necator*." *Biotechnology Bulletin* 37 (4):127.

Panadare, D. C. 2015. "Applications of waste cooking oil other than biodiesel: A review." *Iranian Journal of Chemical Engineering (IJChE)* 12 (3):55–76.

Papanikolaou, Seraphim, Adreas Dimou, S. Fakas, P. Diamantopoulou, A. Philippoussis, M. Galiotou-Panayotou, and G. Aggelis. 2011. "Biotechnological conversion of waste cooking olive oil into lipid-rich biomass using *Aspergillus* and *Penicillium* strains." *Journal of Applied Microbiology* 110 (5):1138–1150.

Paul, S., G. S. Mittal, and M. S. Chinnan. 1997. "Regulating the use of degraded oil/fat in deep-fat/oil food frying." *Critical Reviews in Food Science and Nutrition* 37 (7):635–662.

Refaat, A. A. 2010. "Different techniques for the production of biodiesel from waste vegetable oil." *International Journal of Environmental Science & Technology* 7 (1):183–213.

Revuelta, José L., Rodrigo Ledesma-Amaro, Alberto Jiménez. 2016. "Industrial production of vitamin B2 by microbial fermentation." *Industrial Biotechnology of Vitamins Biopigments, and Antioxidants*:15–40.

Rincón, Luz A., Juan G. Cadavid, and J. Alvaro. 2019. "Used cooking oils as potential oleochemical feedstock for urban biorefineries—Study case in Bogota, Colombia." *Waste Management Orjuela* 88:200–210.

Rodriguez-Perez, Santiago, Antonio Serrano, Alba A. Pantión, and Bernabé Alonso-Fariñas. 2018. "Challenges of scaling-up PHA production from waste streams. A review." *Journal of Environmental Management* 205:215–230.

Saadoun, Ismail M. K. 2015. "Impact of oil spills on marine life." *(Emerging Pollutants in the Environment-Current Further Implications)*:75–104.

Salihu, Aliyu, and Md Zahangir Alam. 2015. "Solvent tolerant lipases: A review." *Process Biochemistry* 50 (1):86–96.

Sangkharak, Kanokphorn, Pimchanok Khaithongkaeo, Teeraphorn Chuaikhunupakarn, Aopas Choonut, and Poonsuk Prasertsan. 2020. "The production of polyhydroxyalkanoate from waste cooking oil and its application in biofuel production." *Biomass Conversion and Biorefinery*:1–14.

Santos, E. F., M. F. S. Teixeira, A. Converti, A. L. F. Porto, and L. A. Sarubbo. 2019. "Production of a new lipoprotein biosurfactant by *Streptomyces* sp. DPUA1566 isolated from lichens collected in the Brazilian Amazon using agroindustry wastes." *Biocatalysis and Agricultural Biotechnology* 17:142–150.

Sarno, Maria, and Mariagrazia Iuliano. 2019. "Biodiesel production from waste cooking oil." *Green Processing and Synthesis* 8 (1):828–836.

Sharma, Parveen K., Riffat I. Munir, Teresa de Kievit, and David B. Levin. 2017. "Synthesis of polyhydroxyalkanoates (PHAs) from vegetable oils and free fatty acids by wild-type and mutant strains of *Pseudomonas chlororaphis*." *Canadian Journal of Microbiology* 63 (12):1009–1024.

Shi, Juan, Yichao Chen, Xiaofeng Liu, and Dong Li. 2021. "Rhamnolipid production from waste cooking oil using newly isolated halotolerant *Pseudomonas aeruginosa* M4." *Journal of Cleaner Production* 278:123879.

Silvagni, Adriano, Lorenzo Franco, Alessandro Bagno, and Federico Rastrelli. 2012. "Thermo-induced lipid oxidation of a culinary oil: The effect of materials used in common food processing on the evolution of oxidised species." *Food Chemistry* 133 (3):754–759.

Singh, Pooja, Yogesh Patil, and Vinaykumar Rale. 2019. "Biosurfactant production: Emerging trends and promising strategies." *Journal of Applied Microbiology* 126 (1):2–13.

Sirisomboonchai, Suchada, Maidinamu Abuduwayiti, Guoqing Guan, Chanatip Samart, Shawket Abliz, Xiaogang Hao, Katsuki Kusakabe, and Abuliti Abudula. 2015. "Biodiesel production from waste cooking oil using calcined scallop shell as catalyst." *Energy Conversion and Management* 95:242–247.

Song, Jin-Hwan, Che-Ok Jeon, Mun-Hwan Choi, Sung-Chul Yoon, and Woo-Jun Park. 2008. "Polyhydroxyalkanoate (PHA) production using waste vegetable oil by *Pseudomonas* sp. strain DR2." *Journal of Microbiology and Biotechnology* 18 (8):1408–1415.

Statistica. 2021. Used cooking oil as feedstock for fuel biodiesel production in India from 2011 to 2020. https://www.statista.com/statistics/1053190/india-used-cooking-oil-feedstock-fuel-biodiesel-production/.

Suci, M., Rita Arbianti, and Heri Hermansyah. 2018. "Lipase production from *Bacillus subtilis* with submerged fermentation using waste cooking oil." *IOP Conference Series Earth Environment Science*. 2nd International Tropical Renewable Energy Conference (i-TREC), Bali, Indonesia 105, 1–6.

Teixeira, Margarida Ribau, Ricardo Nogueira, and Luís Miguel Nunes. 2018. "Quantitative assessment of the valorisation of used cooking oils in 23 countries." *Waste Management* 78:611–620.

The Economic Times. 2019. Government launches programme for converting used cooking oil into biodiesel in 100 cities. https://m.economictimes.com/industry/energy/oil-gas/government-launches-programme-for-converting-used-cooking-oil-into-biodiesel-in-100-cities/articleshow/70617703

Treichel, Helen, Débora de Oliveira, Marcio A. Mazutti, Marco Di Luccio, and J. Vladimir Oliveira. 2010. "A review on microbial lipases production." *Food and Bioprocess Technology* 3 (2):182–196.

Tres, A., R. Bou, F. Guardiola, C. D. Nuchi, N. Magrinyà, and R. Codony. 2013. "Use of recovered frying oils in chicken and rabbit feeds: Effect on the fatty acid and tocol composition and on the oxidation levels of meat, liver and plasma." *Animal* 7 (3):505–517.

Tsoutsos, T. D., S. Tournaki, O. Paraíba, and S. D. Kaminaris. 2016. "The used cooking oil-to-biodiesel chain in Europe assessment of best practices and environmental performance." *Renewable and Sustainable Energy Reviews* 54:74–83.

Tzirita, Markella, Seraphim Papanikolaou, Afroditi Chatzifragkou, and Bríd Quilty. 2018. "Waste fat biodegradation and biomodification by *Yarrowia lipolytica* and a bacterial consortium composed of *Bacillus* spp. and *Pseudomonas putida*." *Engineering in Life Sciences* 18 (12):932–942.

Vastano, Marco, Iolanda Corrado, Giovanni Sannia, Daniel K. Y. Solaiman, and Cinzia Pezzella. 2019. "Conversion of no/low value waste frying oils into biodiesel and polyhydroxyalkanoates." *Scientific Reports* 9 (1):1–8.

Velasco, Joaquín, Susana Marmesat, and M. Carmen Dobarganes. 2008. "Chemistry of frying." *Advances in Deepfat Frying of Foods*, London, New York, Boca Raton, CRC Press, Taylor & Francis Group:33–56.

Venkatesh, Narayana Murthy, and Nagarajan Vedaraman. 2012. "Remediation of soil contaminated with copper using rhamnolipids produced from *Pseudomonas aeruginosa* MTCC 2297 using waste frying rice bran oil." *Annals of Microbiology* 62 (1):85–91.

Verlinden, Rob A. J., David J. Hill, Melvin A. Kenward, Craig D. Williams, Zofia Piotrowska-Seget, and Iza K. Radecka. 2011. "Production of polyhydroxyalkanoates from waste frying oil by *Cupriavidus necator*." *AMB Express* 1 (1):1–8.

Wei, Shiping, Fang Zhao, Zhenglong Jiang, and Dongsheng Zhou. 2013. "Microbial conversion of waste cooking oil into riboflavin by *Ashbya gossypii*." *Bioscience Journal* 29 (4).

Xiaoyan, Liu, Xinjun Yu, Jinshun Lv, Jiaxing Xu, Jun Xia, Zhen Wu, Tong Zhang, and Yuanfang Deng. 2017. "A cost-effective process for the coproduction of erythritol and lipase with *Yarrowia lipolytica* M53 from waste cooking oil." *Food and Bioproducts Processing* 103:86–94.

Zenati, Billal, Alif Chebbi, Abdelmalek Badis, Kamel Eddouaouda, Hocine Boutoumi, Mohamed El Hattab, Dorra Hentati, Manel Chelbi, Sami Sayadi, and Mohamed Chamkha. 2018. "A non-toxic microbial surfactant from *Marinobacter hydrocarbonoclasticus* SdK644 for crude oil solubilization enhancement." *Ecotoxicology and Environmental Safety* 154:100–107.

Zhang, Yen, M. A. Dube, D. D. L. McLean, and M. Kates. 2003. "Biodiesel production from waste cooking oil: 1. Process design and technological assessment." *Bioresource Technology* 89 (1):1–16.

16 Concepts and Recent Trends in Life Cycle Analyses in Waste Valorization

Valeria Caltzontzin-Rabell, Sergio Iván Martínez-Guido, Claudia Gutiérrez-Antonio, Juan Fernando García-Trejo, and Ana Angélica Feregrino-Pérez
Facultad de Ingeniería, Universidad Autónoma de Querétaro, Campus Amazcala, Carretera a Chichimequillas s/n km. 1, El Marqués, Querétaro, 76225, México

CONTENTS

16.1 INTRODUCTION

The globalization process is driven by for-profit motives, facilitated by a combination of neoliberal politics, privatization, and information-technological developments (Elzinga, 2001). In addition, globalization improves the optimized disposition of productive elements and resources (Xiaoxuan et al., 2009). Therefore, globalization

DOI: 10.1201/9781003191247-16

has influenced the growth of the economic sectors; this development of the economic sectors can be observed through the gross domestic product indicator. According to the World Bank (2021a), the growth domestic product was 87,735 billion USD; this value represents an increment of 16.64% with respect to the growth domestic product of 2015. Moreover, the development of the economic sectors has contributed to satisfy the increase in the demand of goods and services for the world population. At the same time, energy consumption and waste generation have augmented. Nevertheless, this scenario has changed significantly in 2020 due to the appearance of the SARS-CoV-2 virus. The COVID-19 pandemic has caused profound influences for many industries, including agriculture, manufacturing, finance, education, health care, sports, tourism, and food (Nicola et al., 2020; Jiang et al., 2021). Regarding energy consumption in 2020, there was a 5% decrease in the worldwide energy demand, and it is expected that the energy demand levels observed in 2019 can be reached between 2022 and 2025 (IEA, 2021a). On the other hand, waste generation has also been impacted by COVID-19 pandemic. According to Fan et al. (2021), the main changes include waste amount, composition, timing/frequency (temporal), distribution (spatial), and risk. Therefore, it is necessary to develop advanced engineering and management tools for the generated waste (Klemeš et al., 2020).

In this context, the International Energy Agency has proposed, in collaboration with the International Monetary Fund, a sustainable recovery plan focused on boosting economic growth, creating jobs, and building more resilient and cleaner energy systems (IEA, 2021b). The plan includes policies, investments, and measures to accelerate the deployment of six key areas, presented in Figure 16.1, which are

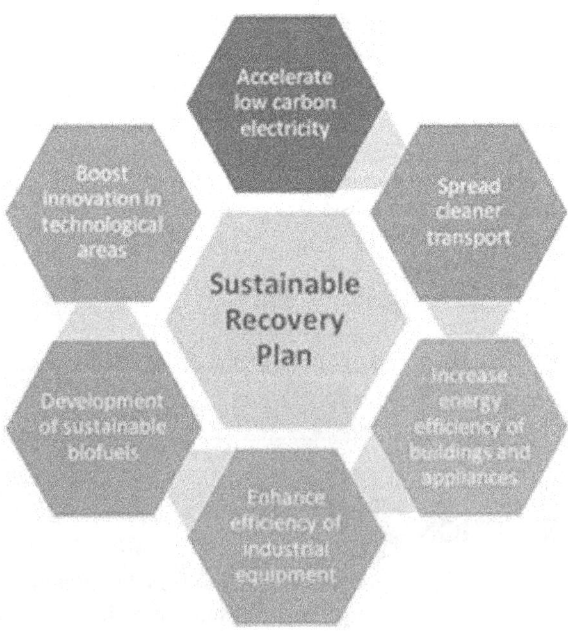

FIGURE 16.1 Key areas established in the sustainable recovery plan.

planned to be implemented during the period 2021–2023 (IEA, 2021b). Together with this plan, the vaccination of the entire world population will allow the economic sectors to recover, thereby increasing waste generation again.

According to Thürer et al. (2017), *waste* is defined as any system input (transformed resources, transforming resources) that is not transformed into a system output that is valued by customers (fulfilled customer demand; this is neither unfulfilled nor exceeded) just in time. Based on this definition, the global waste composition includes rubber and leather (2%), wood (2%), metal (4%), glass (5%), plastic (12%), paper and cardboard (17%), food and green (44%), and other residues integrating the 14% remaining (World Bank, 2021b). The forecast indicates that the global waste will increase to 3.40 billion tons, which represent more than double the population growth over the same period of 2021–2050 (World Bank, 2021b). An important fact is that waste generation depends on income level; the World Bank (2021b) reports that 16% of the world's population, which live in high-income countries, generate about 34% of the global waste. Nowadays, waste management comprises disposal in open dump (33%), landfill (25.2%), incineration (11.1%), landfill with gas collection (7.7%), controlled landfill (3.7%), and other options (0.3%); on the other hand, 13.5% is recycled, while 5.5% is composted. The actual management strategies contribute with 5% of the total carbon dioxide emissions, from which 50% are associated with food waste (World Bank, 2021b). In addition, the operating costs for integrated waste management, including collection, transport, treatment, and disposal, generally range from 35 USD to 100 USD per ton for low- and high-income countries, respectively (World Bank, 2021b). In brief, waste generation represents a big problem for society due to its high volume, constant growth, pollution, and health problems, as well as high management costs. Thus, it is necessary to develop alternatives for solving this multidimensional problem.

In particular, the greater percentage of global waste corresponds to food and green (44%), which are part of the organic waste. Organic waste is produced in the food and beverage industry, paper industry, agriculture, forestry and gardening activities, households, as well as livestock, poultry, pig, and fish industry (Bijmans et al., 2011). Organic waste has a complicated management due to its elevated amount of water, which facilitates its decomposition, but not necessarily its degradation. Moreover, its accumulation represents a pollution problem, which affects the environment as well as the health of society. Based on the data reported by World Bank (2021b), composting is the main treatment alternative for organic waste. Nevertheless, there are other alternatives that allow the revaluation of organic waste, since these materials contain value-added components that can be extracted, or they can be processed to generate new products. Some revaluation processes include the use of microbial electrosynthesis and electrofermentation (Jiang et al., 2019), reforming (Kurniawan et al., 2021), bioprocessing through bacteria, fungi, and crude enzyme (Usmani et al., 2021), anaerobic digestion and composting (Lin et al., 2018; Panigrahi and Dubey, 2019; Kumar et al., 2021), fermentation (Karthick and Nanthagopal, 2021), biological and thermochemical processes (Munir et al., 2018; Montalvo et al., 2020), biorefineries (Awasthi, Ferreira et al., 2021; Liu et al., 2021). Organic waste revalorization can solve residue-accumulation problem at the same time that allows it to obtain an economic benefit of new product insertion at market. In spite of these positive effects, it is important to ensure that this revaluation process does not cause a bigger problem than the one that it is trying to solve, especially in terms of the environmental impact related to energy, land use, and even the generation of additional waste. In this context, life cycle analysis is a valuable tool for the revaluation of the process's impacts. Life cycle analysis (LCA), also known as life cycle assessment, is a methodology used to evaluate the environmental impact of a product through its life cycle, encompassing extraction and processing of the raw materials, manufacturing, distribution, use, recycling, and final disposal (Ilgin and Gupta, 2010). This tool has been applied to evaluate modifications to existing processes or to compare different processes that obtain the same product from different raw materials. In addition, this tool can be applied to the revaluation process of organic waste. Indeed, Melikoglu (2020) affirmed that full life cycle assessments coupled with carbon, land, and water footprint analyses should be made for any potential waste reutilization option before realizing any large-scale investments.

In the literature, reported reviews focused on the use of life cycle analysis to evaluate the revaluation of municipal solid waste mainly (Pujara et al., 2019; Mukherjee et al., 2020; Awasthi, Sarsaiya et al., 2021). However, another type of organic waste must also be considered. Therefore, this chapter is focused on the application of life cycle analysis for waste valorization. Section 2 presents basic concepts related to waste biomass, while its valorization methods are presented in Section 3. Later, a revision of the literature where life cycle analysis is applied to waste valorization is analyzed (Section 4) in order to identify the future trends in this area (Section 5).

16.2 WASTE BIOMASS

As mentioned before, *waste* is defined as any system input (transformed resources, transforming resources) that is not transformed into a system output that is valued by customers (fulfilled customer demand; this is neither unfulfilled nor exceeded) just in time (Thürer et al., 2017). Thus, waste generation involves both environmental and economic problems. When discarded, they reach landfills or are incinerated, which generates almost 3.6 billion metric tons of greenhouse gases, like carbon dioxide and methane, that pollute the environment (Kudakasseril et al., 2013; Kaur et al., 2020); also, due to their composition, they emit odors and attract potentially pathogenic organisms that can spread diseases (Rehman et al., 2017). Economically, their management costs around $1.2 trillion per year (Surendra et al., 2020).

Wastes are part of the second-generation biomass, and their use has gained relevance since it does not compete with food cultivation, eliminating the need for land (Esteban and Ladero, 2018). Also, they tend to have low cost as the needed resources comprise their handling and management for their treatment, discarding production costs. Several residues coming from agricultural/agro-industrial, forestry, municipal, and food industries can be found among these biomass waste (Chandel et al., 2018; Osman et al., 2019). It is important to mention that energy crops are also found in the second-generation biomass; however, they will not be described since they are not residues. Next, we will provide information about the main types of organic waste.

16.2.1 Agro-Residues

Agro-residues include agricultural and agro-industrial wastes, which are composed of by-products from agricultural production and product management. Some examples of these wastes can include crop waste (stalks or leaves, rice straw, rice husk, wheat straw, corn stover, husks, and cobs), sugarcane bagasse, olive stones, peel from fruits, pomaces, and animal manure (Kudakasseril et al., 2013; Maity, 2015; Alalwan et al., 2019); Figure 16.2 shows residues from the cultivation of tomato. Approximately 2.2 billion tons are generated (Millati et al., 2019), and they can be used as animal feed, domestic fuels, source for protein recovery, or they are treated in order to obtain biofuels, such as bioethanol, biogas, biodiesel, and even polymers (Alalwan et al., 2019; Contreras et al., 2019; Hassan et al., 2019). Although large quantities of these types of waste are generated, they cannot be used entirely since part of them must be placed on the field to prevent erosion. The percentage of wastes available would be of about 15 to 40% of the total residues produced (Kudakasseril et al., 2013).

Agro-residue wastes are mainly composed of approximately 40% cellulose, 30% hemicellulose, 20% lignin, and 10% of protein, ash, or extractive compounds (Liew et al., 2014; Maity, 2015; Carrillo-Nieves et al., 2019).

16.2.2 Forestry Residues

Forestry wastes include dead trees, chips, sawdust, particles, or wood that was not collected after wood is harvested or processed (Kudakasseril et al., 2013; Maity, 2015). Almost 0.2 billion tons of forestry residue are generated (Millati et al., 2019); part of them are pelletized and used as heating source or are treated to produce biofuels, mainly in solid state (Chandel et al., 2018; Tauro et al., 2018). Figure 16.3 shows

FIGURE 16.2 Residues generated from the cultivation of tomato crop.

FIGURE 16.3 Residues generated from gardening activities.

gardening residues. This biomass is mainly composed of 40–45% cellulose, 20–30% hemicellulose, 20–32% lignin, and the rest of pectin, ash, proteins, and other components, such as sugars (Maity, 2015; Millati et al., 2019).

16.2.3 Municipal Waste

Municipal waste includes used paper, cardboard, sludge, as well as garden and kitchen waste (Kudakasseril et al., 2013; Maity, 2015), as shown in Figure 16.4. The World Bank estimates that each year 1.3 billion tons of this type of wastes are generated (Hoornweg and Bhada-Tata, 2012). Some of these residues, such as paper, glass, or metal, can be recovered for recycling, while others are treated with thermal or biological processes to produce energy or compost, respectively. Therefore, their composition results vary since products such as paper, plastic, glass, metal, and textiles are found mixed with other organic residues. The chemical composition of municipal waste will also vary between each country, depending on if they are low, intermediate, or high income, as well as their consumption habits. Overall, it is reported that this type of wastes can have a moisture content of 29–55% and an ash content of 16–28% (Millati et al., 2019).

16.2.4 Food Waste

From the global production of food for human consumption, almost 1,400 and 1,700 million tons are discarded each year. This enormous quantity requires activities for its elimination, which include its use as swine feed, in incineration, in anaerobic digestion to produce ethanol or methane, in extraction of components like phenolic

FIGURE 16.4 Municipal solid wastes.

FIGURE 16.5 Food domestic wastes.

compounds, and even in composting in order to generate fertilizers (Esteban and Ladero, 2018). Among food wastes can be found shells, bagasse, seeds, as well as food that does not meet the standards for sale or that is in the putrefaction process. Figure 16.5 shows domestic food residues as an example. These wastes are mainly composed of starch, carbohydrates, fats, proteins, cellulose, sugars, and minerals (Esteban and Ladero, 2018).

16.3 REVALUATION PROCESSES OF WASTE BIOMASS

As mentioned in the previous section, organic wastes represent an environmental, social, and economic problem. In this context, biomass waste use, or their

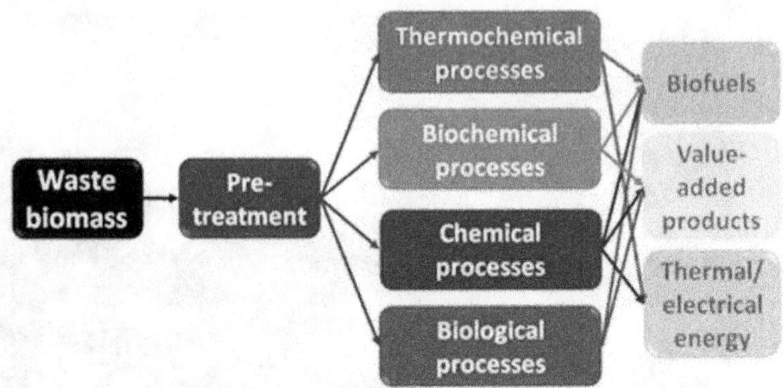

FIGURE 16.6 Revaluation pathways for the conversion of waste biomass.

valorization, offers an opportunity to obtain different products and reduce their volume at the same time. This valorization depends on the waste composition and, therefore, can go through different processes to obtain energy, biofuels, or value-added products. These processes include thermochemical (combustion, pyrolysis, and gasification), biochemical (fermentation, anaerobic digestion) (Cherubini, 2010), chemical (transesterification, hydrotreating, oligomerization), and biological (microalgae cultivation, composting, and insect treatment) processes (Liew et al., 2014). These revaluation pathways can convert the biomass into different value-added products, biofuels, and thermal or electrical energy (Figure 16.6). Next, we will provide the main information about each revaluation pathway.

16.3.1 Thermochemical Processes

Thermochemical processes use heating to decompose and degrade biomass in order to obtain products like energy, fuels, and chemicals; these processes require low residence times and have lower costs than other processes (Alalwan et al., 2019). Among thermochemical processes, combustion, pyrolysis, and gasification are found (Ng et al., 2017; Giwa et al., 2018).

Combustion requires oxygen and temperatures between 900°C and 1,100°C. It produces heat or electricity and ashes, and helps in reducing wastes/volume by 80–90%. However, it also releases gases, such as carbon monoxide and carbon dioxide, that contain harmful chemicals and heavy metals, which require special management (Osman et al., 2019; Kaur et al., 2020). On the other hand, pyrolysis also heats biomass for short periods of time, but in the absence of oxygen and temperatures between 400 and 600°C. Products such as bio-oil, char, and gaseous products are obtained after this process; the operating conditions of the pyrolysis can be modified in order to maximize the production of one product. Bio-oil viscosity is affected by feedstock composition, and due to the existence of chemical compounds such as aldehydes, bio-oil has short storage time (Ghatak, 2011; Ng et al., 2017; Osman et al., 2019; Kaur et al., 2020). Finally, gasification requires a controlled oxygen environment and temperatures of 800–900°C to be carried out. Products like biofuels, electricity, and heat are generated. The main product is syngas, which is a mixture of hydrogen, carbon dioxide, methane, and carbon monoxide; syngas can be used as fuel or as feedstock to obtain biofuels or chemicals. Although it has more energy conversion efficiency than combustion, moisture content, density, and ash level seem to be limiting factors for gas production (Ghatak, 2011; Liew et al., 2014; Kaur et al., 2020).

Thermochemical processes imply the conversion of waste biomass at elevated temperatures; therefore, value-added products cannot be obtained through this type of revaluation pathway. Wastes with elevated amount of lignin and low moisture are the most preferable for this type of conversion process.

16.3.2 Biochemical Processes

Biochemical processes seek biomass conversion into smaller molecules using bacteria or enzymes. Anaerobic digestion and fermentation are part of these processes, and both of them require a biomass pretreatment in order to facilitate access of microorganisms to carbohydrates that will be digested or fermented (Maity et al., 2015; Osman et al., 2019). Some of their limitations include low reaction rates and affections by phenol compounds, which are currently being researched to develop better and stronger microorganisms (Maity, 2015; Negro et al., 2017; Alalwan et al., 2019).

Anaerobic digestion uses bacteria for biomass degradation; it requires a free-oxygen environment and a temperature between 30 and 65°C in order to produce biogas (Cherubini, 2010; Awasthi et al., 2019). This biochemical process has an efficiency between 65 and 81%, and the methane contained can be used to produce methanol or electricity. However, some limitations include slow conversion rates and having optimum conditions for the microorganisms (Ng et al., 2017). Also, a solid residue called digestate is also obtained, which can be used as fertilizer.

The fermentation process can produce alcohol or organic acids by using microorganisms like yeasts, bacteria, mold, or enzymes (Cherubini, 2010; Alalwan et al., 2019). Usually, the production of alcohol is low, since at concentrations greater than 15% the microorganisms become inactivated. Therefore, the separation of ethanol and water is highly expensive, since the solution is diluted (Ng et al., 2017). In addition, value-added products such as levulinic acid and lactic acid can be obtained from fermentation. Finally, another product generated in the fermentation process is carbon dioxide.

Biochemical processes imply the conversion of biomass waste at moderate temperatures; therefore, thermal or electrical energy cannot be obtained through this type of revaluation pathway. Wastes with elevated amount of cellulose and hemicellulose are the most preferable for this type of conversion process.

16.3.3 Chemical Processes

The chemical process allows the conversion of waste biomass into chemicals, biofuels, value-added products, as well as thermal or electrical energy. The chemical process usually requires catalysts as well as moderate to high temperature and pressure conditions to perform the conversion of the biomass. A great amount of chemical processes can be found in literature, such as transesterification, hydrotreating, cracking, and reforming.

Transesterification is defined as the conversion of triglyceride feedstock with an alcohol in the presence of a basic or acid catalyst to produce biodiesel and glycerol. Biodiesel can be used in mixtures with fossil diesel, allowing a reduction of carbon dioxide emissions. Moreover, glycerol can be used for the production of cosmetics, soaps, as well as paints.

On the other hand, the hydroprocessing of triglyceride feedstock allows the production of renewable hydrocarbons; in this conversion pathway, the reactions are carried out at high temperature and pressure in the presence of hydrogen. In addition, water and carbon dioxide are generated in the deoxygenation reaction. The renewable hydrocarbons can be used to generate heat, electricity or fuels (light gases, naphtha, biojet fuel, green diesel).

Chemical processes imply the conversion of waste biomass at moderate to high temperatures and pressures; in spite of this, it is possible to generate thermal or electrical energy, biofuels, chemicals, as well as some value-added products. Wastes with elevated amount of triglycerides are the most preferable for this type of conversion process.

16.3.4 Biological Processes

Biological processes use organisms to degrade biomass and, therefore, revalue it. Microalgae, composting, and insects are part of these (Rehman et al., 2017; Khoo et al., 2019; Kaur et al., 2020).

Microalgae are unicellular plants that generate, through photosynthesis, various molecules of interest. Their cultivation requires sunlight and a nutrient-rich effluent, and depending on the type of strain and substrate for their growing, different products such as lipids, protein, chemicals, pharmaceuticals, or biomaterials can be obtained (Khoo et al., 2019). Some advantages regarding their use are a high growth rate and lipid accumulation (Alalwan et al., 2019); however, there is still more research required in order to select promising strains that are stronger, absorb more nutrients, or accumulate more lipids (Pleissner et al., 2018). Composting is another process that uses microorganisms to reduce almost 40% of the waste, obtaining as the main product compost that benefits the soil. These process requires controlling conditions such as temperature, moisture, pH, and carbon/nitrogen ratio. Although its use is old and popular, there are still some limitations since it requires space and uses waste, which affects its composition (Kaur et al., 2020). On the other hand, insect cultivation is another treatment for biomass conversion; it uses organisms like insects to degrade organic waste. This degradation is called biotransformation, since waste biomass is transformed into insect biomass. Their potential to biotransform large amounts of waste and use little water and space has been reported, and more research could improve their performance and escalation process. Insects are mainly composed of fat and protein, which can be used to produce fuels or animal feed, and the leftovers can be used as fertilizer (Rehman et al., 2017; Girotto and Cossu, 2019).

Biological processes imply the conversion of waste biomass at low temperatures and pressures; therefore, the production of thermal or electrical energy is not feasible through this conversion pathway. Waste with elevated amount of triglycerides, sugar, and starch is the most preferable for this type of conversion process.

On the other hand, the waste biomass can be processed in order to extract value-added compounds that already are present in it; this strategy must be considered as a priority for all areas of the industrial sector since it satisfies environmental policy by establishing the foundations of the circular economy (Romaní et al., 2018). Moreover, waste biomass is a good source of compounds that are of interest to various industries, such as pharmaceuticals, cosmetics, food, agriculture, energy, etc. Despite the fact that the extraction of value-added compounds is a good alternative for the reuse of waste biomass, there is still a certain degree of limitations to the use of this type of material, among which the next ones can be mentioned:

- Biomass waste complexity in terms of the presence of various types of biomass is a limitation for their separation, classification, and use.
- The volume of waste biomass is a limitation for extraction times and processes.
- There is no standardization for the extraction of the compounds derived from the nature of the biomass and the compounds of interest to be extracted.
- There is no regulation that ensures their quality and safety.

Thus, in order to carry out an extraction process, it is necessary to consider some aspects. The first one is that waste biomass is composed of various fractions of the raw material. In the case of plants, for instance, the wastes can contain different parts of the plant, and therefore different concentrations or content of compounds derived from the matrix itself (SaraJliJa et al., 2012; Rehman et al., 2020). The second aspect is the selection of the compound of interest to be extracted based on the characteristics of the waste biomass, since they can contain a high amount of moisture, in fresh form, as well as low amount of moisture, in dry form, or with low or almost no-humidity concentration. The dry form is more recommended; nevertheless, it is necessary to establish the drying technique, since it can influence the quality of the compounds of interest to be extracted from the biomass. Furthermore, the low moisture content reduces the probability of contamination, fermentation, and/or decomposition of the waste biomass; these situations are highly probable to occur with fresh waste biomass or with a high moisture content. In this context, it is suggested that fresh samples be processed quickly, recommending handling no longer than three hours after harvesting or processing (Azwanida, 2015). The third aspect is the particle size of the residual biomass, since it is a factor that helps establish the degree of penetration of the solvent and obtaining the compound of interest. According to literature, an efficient extraction can be performed with particle sizes ranging from 0.5 mm to 125 mm (Azwanida, 2015; Yeop et al., 2017). The fourth aspect is the temperature used during the extraction process, since the extraction efficiency of thermolabile compounds is directly affected (Herrero et al., 2006).

In addition, it is essential to detect the most adequate route for the recovery of the compounds of interest by means of a suitable extraction technique. With regard to extraction processes, there are various techniques; however, in general they can be divided into two categories: classical and modern techniques.

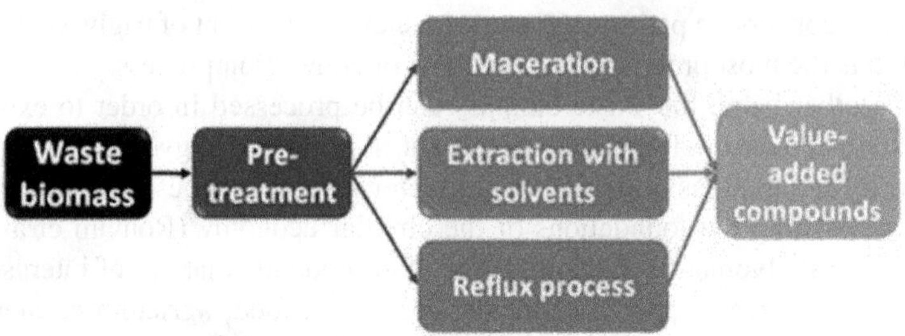

FIGURE 16.7 Classical techniques for the extraction of value-added compounds.

The classical techniques have several limitations, such as the indiscriminate use of solvents, long extraction times, and in some cases, heating requirements that can affect the integrity of the bioactive compounds. Moreover, the classic techniques also cause toxic effects for human health and the environment, due to prolonged exposure and the release of greenhouse gases by the use of organic solvents as well as the generation of additional waste (Add Aziz et al., 2021). The advantage of these techniques lies in their simplicity, easy handling, and standardization. The classical techniques are composed of simple operations, such as maceration, extraction with solvents (Soxhlet), and reflux process (Latiff, 2015), Figure 16.7.

Maceration is a solid-liquid extraction process where the value-added compounds contained in the solid interact with a liquid solvent, which can be polar or nonpolar. The choice of solvent will depend on the type of compound to be extracted. This interaction process requires long exposure times and occasional agitation. It is established that the minimum interaction time is three days (Handa et al., 2008). The advantages of this process are that it is simple, it does not require advanced equipment to be realized, and its energy consumption is minimal. In addition, maceration provides good yields with the appropriate solvent, and it does not affect the characteristics of the compounds of interest. The main disadvantage of this technique is the long exposure time required for the interaction between the waste biomass and the solvent, as well as the generation of residual by-products.

Soxhlet extraction, or extraction with solvents, consists of the interaction of the sample embedded in a filter paper or cellulose cartridge with the extraction solvent, which is heated through an equipment called Soxhlet extractor; this extractor carries out the heating process of the solvent, followed by its evaporation and condensation. Soxhlet extraction allows the volatile compounds of the solvent to interact with the compounds to be extracted and to be carried by the solvent in the condensation step (Soxhlet, 1879; Azwanida, 2015). The advantage of this process is the speed of obtaining the compounds with respect to the maceration. The disadvantages of Soxhlet extraction include the generation of large amount of volatile compounds that are toxic to the environment as well as human health; in addition, this process uses dangerous and flammable materials, and it is necessary to use dry residual biomass with a small particle size.

The reflux process consists of a constant evaporation of the solvent, followed by its condensation; this process is controlled for a set time without generating loss of the solvent. The advantages of the reflux process include that it is a widely used technique in the industry, it is profitable, and also, it is easy to operate (Wang et al., 2013). Among the disadvantages, it can be listed that it is not friendly to the environment, and the extraction times are prolonged, since approximately two days are required to obtain good yields (Easmin et al., 2015).

As can be seen, the classic techniques are not friendly to the environment, and also, they can cause damage to the health of human beings. Moreover, these techniques can generate other residual by-products and require long times for extraction, which makes them unsuitable techniques for obtaining compounds of interest at industrial scale.

On the other hand, modern techniques have great advantages, such as being friendly to the environment, reduction in energy consumption due to the use of alternative solvents, and renewable natural sources that positively impact the quality and safety of the compounds; moreover, these techniques allow reduction in exposure times, use of solvents, and degradation of compounds (Easmin et al., 2015). The main limitations of these techniques are the costs, access, and management of the equipment necessary for its implementation. Examples of this type of processing are pressurized liquid extraction (PLE), ultrasound-assisted extraction (EAU), microwave-assisted extraction (MAE), and supercritical fluid extraction (SFE) (Add Aziz et al., 2021) (Figure 16.8).

Pressurized liquid extraction is a process that uses solvents at high pressure and temperature but maintaining the liquid state of the solvent, that is, without exceeding the critical points, thereby reducing the viscosity of the solvent and the surface tension. Due to the low surface tension, the interaction with the waste biomass is facilitated, allowing a greater penetration of the solvent in more distal areas and increasing the surface contact (Herrero et al., 2006; Rodrigues et al., 2020). The technique can be used with organic solvents and also with water; in the last case, the technique is called subcritical water extraction (SWE), and it uses liquid water at higher temperature and pressure, which allows the extraction of less polar compounds, given the modified properties of water (Plaza and Turner, 2017). The main disadvantage of this type of process is the degradation of some compounds due to high temperatures and the decrease in the selectivity of the compound to be obtained.

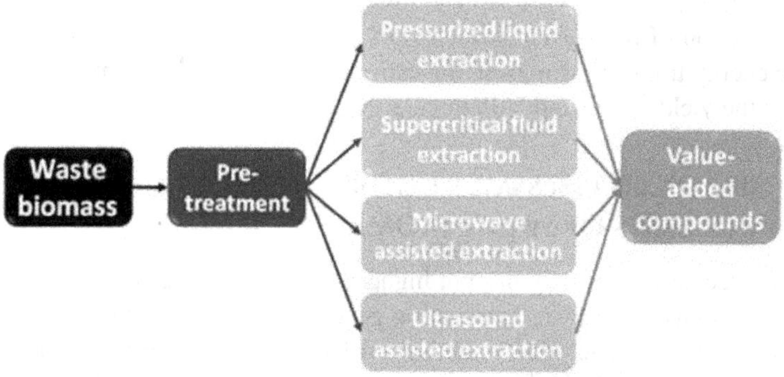

FIGURE 16.8 Modern techniques for the extraction of value-added compounds.

The advantages are the reduction of the use of solvents, extraction time, as well as being friendly with the environment. This type of extraction process is increasingly accepted for application at industrial level (del Pilar Snchez-Camargo et al., 2017; Plaza and Turner, 2017).

Supercritical fluid extraction (SFE) is a technique that uses carbon dioxide (CO_2) as a dense fluid solvent, allowing the variation of the CO_2 density over wide ranges with the modification of temperature and pressure without mixing with other solvents (Brunner, 2005). This characteristic facilitates the extraction of low-volatility compounds, such as essential oils, lipophilic compounds, and slightly polar compounds (Ravetti Duran et al., 2018). This process has several advantages, such as its low thermal degradation of thermosensitive compounds, and the use of CO_2 provides several attributes, such as its being nontoxic, nonflammable, environmentally safe, and having greater acceptance in the industrial sector. The main disadvantage is its low extraction capacity for high-molecular-weight compounds, given the low polarity of CO_2. However, this obstacle can be corrected by mixing the CO_2 with a solvent or cosolvent, which will increase the solubility of the more polar compounds (Mazzutti et al., 2020).

Microwave-assisted extraction (MAE) consists of the extraction of compounds of interest using microwave energy (Paré et al., 1994). Microwave radiation generates heat, which interacts with compounds (water and organic components), allowing the transfer of solutes from a solid matrix to a solvent phase (Sun et al., 2016). Its advantages are the significant reduction of solvent, reduction of process time and energy consumption, as well as high performance and less thermal degradation (Llompart et al., 2017; Vila Verde et al., 2018). The main disadvantage is the cost of access to microwave equipment and establishment of conditions based on the compound and the extraction matrix.

Ultrasound-assisted extraction is a technique that uses ultrasonic waves to cause cavitation in the extraction matrix, which in turn induces macroturbulence between particles at high speed and the microporous particles of the extraction matrix; this allows obtaining a rapid diffusion of the compounds towards the solvent (Mason et al., 1996; Azmir et al., 2013). This process allows the extraction of thermolabile and unstable compounds. Compounds such as carotenoids, protein polysaccharides, phenolic compounds, aromatics, and sterols have been successfully extracted (Vinatoru, 2001; Jadhav et al., 2009). The main advantages are the low use of solvents, energy, temperature, and extraction time. The disadvantages include the required training for the selection of the solvent, which is of great importance due to cavitation and acoustic energy transfer. Moreover, the extraction matrix and solvent interaction also influence the yield (Soria and Villamiel, 2010).

16.4 LIFE CYCLE ASSESSMENT OF REVALUATION PROCESSES OF WASTE BIOMASS

In the last decades, humanity has put higher attention in terms of natural resources depletion and environmental degradation generated by anthropogenic activities. In this way, many industries and scientists around the world have responded to this aggravating situation, designing greener products and processes. Hence, the use of pollution-prevention strategies and environmental management systems has resulted in attractive alternatives to achieve environmental improvement. Particularly, life cycle assessment (LCA) was described by Curran (2012) as a tool used to give sustainability support, based in technical changes in production process, as the selection of one material over another. According to the Environmental Protection Agency of the United States (EPA) (2017), LCA allows to quantify potential environmental impacts generated by a product, material, process, or activity; in this tool are included all environmental affections across the life cycle product system, from material acquisition to manufacturing, use, and final disposition.

Therefore, the International Organization for Standardization (ISO), through ISO 14000, proposed the standards which include environmental labels and declarations, life cycle assessment, and eco-design. In addition, ISO 14000 manifests that LCA is constituted by four main phases, as shown in Figure 16.9 (Lee and Inaba, 2004):

1) Goal and scope definition, in which systems boundaries and goal and scope of the studies are defined.
2) The life cycle inventory, in which obtained are all the flows quantities of all the substances from and to natural environment.
3) Life cycle impact assessment, in which all identified substances are characterized in terms of their impacts to the environment.
4) Interpretation phase, where results and uncertainties are displayed and recommendations are proposed (Wowra et al., 2020).

Under this perspective, Guerrero and Muñoz (2017) integrated the LCA (environmental impact and energy balances) into ethanol production of the second generation from banana agro-residues. The proposed approach included, firstly, system boundaries delimitation under a well-to-wheels consideration; this means that environmental

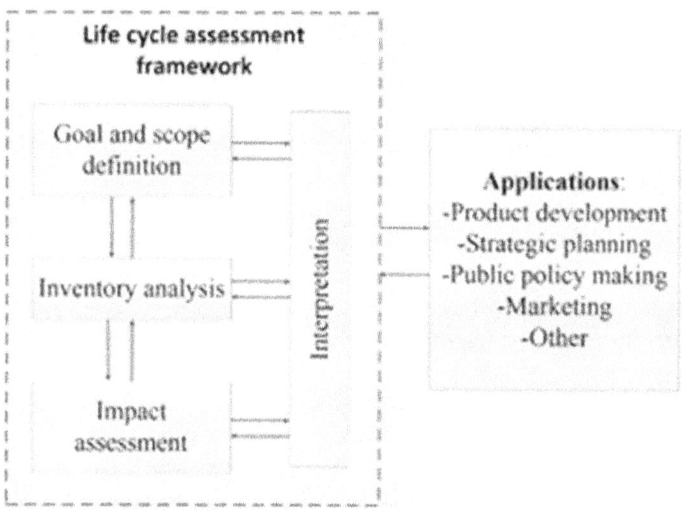

FIGURE 16.9 Life cycle assessment process diagram.

impacts were evaluated from raw material extraction (from the well) to its transformation, distribution, and final use (to the wheels). Afterwards, an environmental database was mainly obtained from ECoinvent v.3 (Wernet et al., 2016) and SimaPro 8.0.4.30 library (PRé-Sustainability, 2020); in addition, ReCiPe midpoint (Huijbregts et al., 2017) was used to calculate environmental impact assessment. In this way, results show a reduction of 64.77% in CO_2 emissions in comparison with the report for regular gasoline, highlighting that 52.3% of generated emissions are given by wastewater production in bioethanol process. Similarly, Gullón et al. (2018) proposed LCA use to evaluate the revalorization of vine shoots from a biorefinery perspective. The goal and scope considered in their work included the production of biopolymers and energy generation as revalorization routes; system delimitation was considered from vine shoots harvesting to the generation of revalorized product. From these delimited systems were obtained all the inventory parameters (raw materials, chemical compounds, energy). Environmental impact assessment was calculated through CML 2001 method v2.05 (Guinée, 2002). Obtained results were grouped in midpoint categories; in addition, electricity requirements were identified as the main environmental hotspot. Hence, the authors proposed that it is necessary to consider other energy sources different to the fossil fuel sources. LCA also has been proposed as a measuring tool in revalorization problems, with the goal of including waste in circularity economy, as performed by Monsiváis-Alonso et al., 2020, in which fish oil waste was used to obtain omega-3 as a revalorization product. ReCiPe was used as the LCA methodology estimation, quantifying the impact over human health repercussions, resources depletion, and ecosystem damage generated by omega-3 production. In this study, LCA was integrated with social and economic impacts, giving a multiobjective optimization problem, with the goal to improve and achieve circular economy in the seafood industry. The integration of waste fish oil into omega-3 production process was compared with the current use of this residue (as heat energy source at the fish industry); the results showed that revalorization has an increase of 98% in terms of environmental impact. However, this integration improves social and economic aspects with USD$334,363/y of new income. Afterwards, Ita-Nagy et al. (2020) evaluated environmentally bagasse fiber reintegration into biocomposite production, with the goal of obtaining bio-based material substitutes of fossil fuel. In this way, the Intergovernmental Panel on Climate Change (IPCC) methodology (Stocker, 2014) was used as an environmental tool, quantifying the GHG emissions over a one-hundred-year period. Mainly, results show that it is possible to achieve GHG emission reduction of 39–40% when biocomposites replace fossil-based polyethylene pellets, under a relation 1:1. In addition, almost 40% of GHG emission generated by biocomposite production is given by the sugarcane production; therefore, harvesting and cultivation need to be carried out under greener techniques.

Under this premise, Christensen et al. (2020) presented the use of LCA into waste management systems, given the huge current and increasing problem due to system complexity and the necessity of sustainable solutions, which can include circular economy. Authors identified some points to consider in the LCA application in residue management; first, it is necessary to understand how existing waste management systems works to identify ways to improve current systems. Between the alternatives to achieve improvements, there are development and prospective technologies; particularly in this point, LCA application has higher impact. Thus, it can be defined if a waste management alternative process at lab scale has the opportunity to be scaled into a pilot process or not from an environmental point of view; in addition, economic and social impacts can be added in this kind of sustainability analysis, with the goal of proposing green optimal solutions. On the other hand, new national and international regulations and strategic plans for waste management systems are required; in this way, future policy development can be formulated from LCA mathematical modeling strategies, in which is included is the environmental performance of the entire system management.

Afterwards, Mendieta et al. (2021) integrated the LCA in the revalorization of non-centrifugal cane sugar sector, with the goal of achieving a circular bioeconomy. The revaluation process produces biofertilizer and biogas through low-cost biodigestion technology, which can be integrated into sugarcane cultivation and sugar production process. Following the framework described by the ISO 14,040 (ISO/TC 207, 2006a) and 14,044 (ISO/TC 207, 2006b) standards, LCA was included. SimaPro 8 software and ReCipe midpoint method were used to measure the environmental impact, in which were considered were climate change, ozone depletion, terrestrial acidification, freshwater eutrophication, marine eutrophication, photochemical oxidant formation, particulate matter formation, metal depletion, and fossil depletion categories. Results show that it is possible to reduce environmental impact into 5% if biodigesters are used instead of synthetic fertilizers and firewood cookstove. Particularly, marine and freshwater eutrophication categories show higher decreases in environmental impact terms, being 99.36% and 87.61% less when biodigesters are used as a revalorization alternative.

As it is possible to notice, LCA implementation as an improvement tool in residue revalorization has become a versatile methodology, which can complement economic and social impacts, with the goal of achieving process sustainability. In this way, several methods and eco-indicators that help quantify environmental damage have been developed, each one of these with a variable number of category damage considered; nevertheless, even when LCA represents real-life process modeling, still exist some limitations linked with the methodologies. Some of the disadvantages or limitations are given by the huge amount of needed data, which in some cases are not available; hence, conclusions lack solidity. Moreover, LCA relies on assumptions, scenarios, and scope from one study to other; therefore, results obtained from one study are hard to compare with those obtained from other studies. However, nowadays LCA is still an alternative which represents the real-life world in a simplified way and gives us a global idea about the nearest environmental benefit of a residue revalorization pathway process.

16.5 CONCLUSIONS AND PERSPECTIVES

Without any doubt, the revaluation of waste is necessary from social, environmental, and economic points of view. Considering the high volume, low density, differences in composition, and easy decomposition rate, the technical challenge to manage them is big. Nevertheless, there are several alternatives for the revaluation of waste, which allow the extraction and/or generation of value-added products, as well as biofuels and bioenergy (thermal and electrical). Each alternative has different energy requirements, efficiency, and also products and by-products; thus, it is important to consider these aspects as well as the environmental impact of the revaluation process. In other words, the revaluation process must be sustainable. In this context, life cycle analysis is a powerful tool to compare different revaluation processes.

As described previously, there are different advantages and limitations when LCA is used as a decision variable in the waste revalorization process. Most of them have resulted from the difficulties that researchers and decision makers face when they try to compare obtained results from different analyses. Hence, higher efforts in terms of policy generation related to guidelines and appropriate results comparison remain as opportunity fields. In addition, teamwork on the part of the different metrics and used methodologies is necessary to obtain equivalences from their results, increasing the compatibility between them.

Kaur and Amor (2017) suggested that a simplified assumptions in LCA performances also has a negative effect on obtained results, adding a greater uncertainty degree. In some cases, simplifications or even a huge amount of assumptions in environmental impact assessment evaluations result from low data availability in particular processes. Social development also plays a crucial role in data availability limitations because LCA methodologies and inventory creation are widely performed in developed countries in comparison with developing ones. In this way, it is necessary to add robustness in database systems, with the goal of avoiding problems of this precedence. Particularly, in new alternatives of waste revalorization, lack of data availability is directly linked to the novelty of the process's self-essence; due to this, new pathways have been analyzed only at laboratory level, in which considered are the total control of all the variables, such as temperature, pressure, pH, humidity, and concentration, among others.

On the other hand, potential change in emissions over a long period of tracking is another topic of interest in the future trends in LCA methodologies. In this way, dynamic mathematical models with environmental objectives have become a new field to develop in the last decade. New formulations and programming tools are part of the current research fields; however, there is still an uncertainty in how to couple emission changes without adding complexity in LCA methods, without compromising robustness of results and excessive computational expense.

LCA results in an attractive methodology to be applied into circular economy works, studies, developments, and investigation, considering that this model refers to a production and consumption model that involves sharing, renting, reusing, repairing, renewing, and recycling existing materials and products as many times as possible to create added value, resulting to this integration offering a more environmentally friendly panorama.

16.6 ACKNOWLEDGMENTS

Financial support provided by SEP, through grant PRODEP-UAQ/332/19-for the postdoctoral stay of S. I. Martínez-Guido, and from CONACyT, through the grant 320583 as well as the scholarship of V. Caltzontzin-Rabell for the realization of her postgraduate studies, is gratefully acknowledged. Additionally, the authors are grateful for the financial support provided by the Universidad Autónoma de Querétaro through the research fund (FIN202107) Investigación vinculada a la atención de problemas nacionales 2021 of the Faculty of Engineering for the development of this project.

REFERENCES

Aziz, Nur Amanina, Hasham Rosnani, Roji Sarmidi Moahamad, Hasyimah Suhaimi Siti, and Hafiz Idris Moahamad Khairul. 2021. A review on extraction techniques and therapeutic value of polar bioactives from Asian medical herbs: Case study on Orthosiphon aristatus, Eurycoma longifolia and Andrographis paniculata. *Saudi Pharmaceutical Journal*, 29: 143–165. https://doi.org/10.1016/j.jsps.2020.12.016.

Azmir, J., I.S.M. Zaidul, M.M. Rahman, K.M. Sharif, A. Mohamed, F. Sahena, M.H.A. Jahurul, K. Ghafoor, N.A.N. Norulaini, A.K.M. Omar. 2013. Techniques for extraction of bioactive compounds from plant materials: A review. *Journal of Food Engineering*, 117(4): 426–436, ISSN 0260-8774, https://doi.org/10.1016/j.jfoodeng.2013.01.014.

Alalwan, Hayder A., Alaa H. Alminshid, and Haydar A. S. Aljaafari. 2019. Promising evolution of biofuel generations. Subject review. *Renewable Energy Focus*, 28 (March): 127–139. https://doi.org/10.1016/j.ref.2018.12.006.

Romaní, Aloia, Michele Michelin, Lucília Domingues, José A. Teixeira. 2016. Chapter 16 – Valorization of Wastes From Agrofood and Pulp and Paper Industries Within the Biorefinery Concept: Southwestern Europe Scenario. Editor(s): Thallada Bhaskar, Ashok Pandey, S. Venkata Mohan, Duu-Jong Lee, Samir Kumar Khanal, Waste Biorefinery, Elsevier, 487–504. ISBN 9780444639929.

Awasthi, Mukesh Kumar, Jorge A. Ferreira, Ranjna Sirohi, Surendra Sarsaiya, Benyamin Khoshnevisan, Samin Baladi, Raveendran Sindhu, Parameswaran Binod, Ashok Pandey, Ankita Juneja, Deepak Kumar, Zengqiang Zhang, and Mohammad J. Taherzadeh. 2021. A critical review on the development stage of biorefinery systems towards the management of apple processing-derived waste. *Renewable and Sustainable Energy Reviews*, 143: 110972. ISSN 1364–0321. https://doi.org/10.1016/j.rser.2021.110972.

Awasthi, Mukesh Kumar, Surendra Sarsaiya, Steven Wainaina, Karthik Rajendran, Sanjeev Kumar Awasthi, Tao Liu, Yumin Duan, Archana Jain, Raveendran Sindhu, Parameswaran Binod, Ashok Pandey, Zengqiang Zhang, and Mohammad J. Taherzadeh. 2021. Techno-economics and life-cycle assessment of biological and thermochemical treatment of bio-waste. *Renewable and Sustainable Energy Reviews*, 144: 110837. ISSN 1364–0321. https://doi.org/10.1016/j.rser.2021.110837.

Awasthi, Mukesh Kumar, Surendra Sarsaiya, Steven Wainaina, Karthik Rajendran, Sumit Kumar, Wang Quan, Yumin Duan, et al. 2019. A critical review of organic manure biorefinery models toward sustainable circular bioeconomy: Technological challenges, advancements, innovations, and future perspectives. *Renewable and Sustainable Energy Reviews*, 111 (November 2018): 115–131. https://doi.org/10.1016/j.rser.2019.05.017.

Azwanida, N. N. 2015. A review on the extraction methods use in medicinal plants, principle strength and limitation. *Medicinal & Aromatic Plants*, 4(196): 2167–10412. https://doi.org/10.4172/2167-0412.1000196.

Brunner, Gerd. 2005. Supercritical fluids: Technology and application to food processing. *Journal of Food Engineering*, 67 (1–2): 21–33, ISSN 0260-8774, https://doi.org/10.1016/j.jfoodeng.2004.05.060.

Bijmans, M. F. M., C. J. N. Buisman, R. J. W. Meulepas, and P. N. L. Lens. 2011. Sulfate reduction for inorganic waste and process water treatment. Editor(s): Murray Moo-Young, *Comprehensive Biotechnology* (Third Edition), Pergamon, 384–395. ISBN 9780444640475. https://doi.org/10.1016/B978-0-444-64046-8.00367-0.

Carrillo-Nieves, Danay, Magdalena J. Rostro Alanís, Reynaldo de la Cruz Quiroz, Héctor A. Ruiz, Hafiz M. N. Iqbal, and Roberto Parra-Saldívar. 2019. Current status and future trends of bioethanol production from agro-industrial wastes in Mexico. *Renewable and Sustainable Energy Reviews*, 102 (November 2018): 63–74. https://doi.org/10.1016/j.rser.2018.11.031.

Chandel, Anuj Kumar, Vijay Kumar Garlapati, Akhilesh Kumar Singh, Felipe Antonio Fernandes Antunes, and Silvio Silvério da Silva. 2018. The path forward for lignocellulose biorefineries: Bottlenecks, solutions, and perspective on commercialization. *Bioresource Technology*, 264 (June): 370–381. https://doi.org/10.1016/j.biortech.2018.06.004.

Cherubini, Francesco. 2010. The biorefinery concept: Using biomass instead of oil for producing energy and chemicals. *Energy Conversion and Management*, 51(7): 1412–1421. https://doi.org/10.1016/j.enconman.2010.01.015.

Christensen, T. H., A. Damgaard, J. Levis, Y. Zhao, A. Björklund, U. Arena, M. A. Barlaz, V. Starostina, A. Boldrin, T. F. Astrup, and V. Bisinella. 2020. Application of LCA modelling in integrated waste management. *Waste Management*, 118: 313–322. doi:10.1016/j.wasman.2020.08.034.

Contreras, María del Mar, Antonio Lama-Muñoz, José Manuel Gutiérrez-Pérez, Francisco Espínola, Manuel Moya, and Eulogio Castro. 2019. Protein extraction from agri-food residues for integration in biorefinery: Potential techniques and current status. *Bioresource Technology*, 280 (February): 459–477. https://doi.org/10.1016/j.biortech.2019.02.040.

Curran, M. A. 2012. *Life Cycle Assessment Handbook: A Guide for Environmentally Sustainable Products*, John Wiley & Sons, Inc.

del Pilar Sánchez-Camargo, A., N. Pleite, M. Herrero, A. Cifuentes, E. Ibáñez, and B. Gilbert-López. 2017. New approaches for the selective extraction of bioactive compounds employing bio-based solvents and pressurized green processes. *The Journal of Supercritical Fluids*, 128: 112–120. https://doi.org/10.1016/j.supflu.2017.05.016.

Easmin, M. S., M. Z. I. Sarker, S. Ferdosh, S. H. Shamsudin, K. B. Yunus, M. S. Uddin, and H. A. Khalil. 2015. Bioactive compounds and advanced processing technology: Phaleria macrocarpa (sheff.) Boerl, a review. *Journal of Chemical Technology & Biotechnology*, 90(6): 981–991. https://doi.org/10.1002/jctb.4603.

Elzinga, science and technology: Internationalization. 2001. Editor(s): Neil J. Smelser and Paul B. Baltes, *International Encyclopedia of the Social & Behavioral Sciences*, Pergamon, 13633–13638. ISBN 9780080430768. https://doi.org/10.1016/B0-08-043076-7/03166-1.

EPA "United States Environmental Protection Agency". 2017. Design for the environment life-cycle assessment. https://archive.epa.gov/epa/saferchoice/design-environment-life-cycle-assessments.html (Accessed on 03.2021).

Esteban, Jesus, and Miguel Ladero. 2018. Food waste as a source of value-added chemicals and materials: A biorefinery perspective. *International Journal of Food Science and Technology*, 53(5): 1095–1108. https://doi.org/10.1111/ijfs.13726.

Fan, Yee Van, Peng Jiang, Milan Hemzal, and Jiří Jaromír Klemeš. 2021. An update of COVID-19 influence on waste management. *Science of the Total Environment*, 754: 142014. ISSN 0048–9697. https://doi.org/10.1016/j.scitotenv.2020.142014.

Ghatak, Himadri Roy. 2011. Biorefineries from the perspective of sustainability: Feedstocks, products, and processes. *Renewable and Sustainable Energy Reviews*, 15(8): 4042–4052. https://doi.org/10.1016/j.rser.2011.07.034.

Girotto, Francesca, and Raffaello Cossu. 2019. Role of animals in waste management with a focus on invertebrates' biorefinery: An overview. *Environmental Development*, 32 (July 2018): 0–1. https://doi.org/10.1016/j.envdev.2019.08.001.

Giwa, Adewale, Idowu Adeyemi, Abdallah Dindi, Celia García Baños Lopez, Catia Giovanna Lopresto, Stefano Curcio, and Sudip Chakraborty. 2018. Techno-economic assessment of the sustainability of an integrated biorefinery from microalgae and Jatropha: A review and case study. *Renewable and Sustainable Energy Reviews*, 88 (February 2017): 239–257. https://doi.org/10.1016/j.rser.2018.02.032.

Guerrero, B., and E. Muñoz. 2017. Life cycle assessment of second generation ethanol derived from banana agricultural waste: Environmental impacts and energy balance. *The Journal of Cleaner Production*, 174: 710–717. doi:10.1016/j.jclepro.2017.10.298.

Guinée, J. B. 2002. Handbook on life cycle assessment operational guide to the ISO standards. *The International Journal of Life Cycle Assessment*, 7: 311–313. doi:10.1007/BF02978897.

Gullón, P., B. Gullón, I. Dávila, J. Labidi, and S. González-Garcia. 2018. Comparative environmental life cycle assessment of integral revalorization of vine shoots from a biorefinery perspective. *Science of the Total Environment*, 624: 225–240. doi:10.1016/j.scitotenv.2017.12.036.

Handa, S. S., S. P. S. Khanuja, G. Longo, and D. D. Rakesh. 2008. Extraction technologies for medicinal and aromatic plants. *Earth, Environmental and Marine Sciences and Technologies, International Centre for Science and High Technology* ICS-UNIDO, AREA Science Park. https://www.unido.org/sites/default/files/2009-10/Extraction_technologies_for_medicinal_and_aromatic_plants_0.pdf#page=25

Hassan, Shady S., Gwilym A. Williams, and Amit K. Jaiswal. 2019. Moving towards the second generation of lignocellulosic biorefineries in the EU: Drivers, challenges, and opportunities. *Renewable and Sustainable Energy Reviews*, 101 (November 2018): 590–599. https://doi.org/10.1016/j.rser.2018.11.041.

Herrero, M., A. Cifuntes, and E. Ibanez. 2006. Sub- and supercritical fluid extraction of functional ingredients from different natural sources: Plants, food-by-products, algae and microalgae: A review. *Food Chemistry*, 98: 136–148. https://doi.org/10.1016/j.foodchem.2005.05.058.

Hoornweg, Daniel, and Perinaz Bhada-Tata. 2012. *What a Waste: A Global Review of Solid Waste Management*. Urban Development Series; Knowledge, Papers No. 15, World Bank, Washington, DC. https://openknowledge.worldbank.org/handle/10986/17388.

Huijbregts, M. A. J., Z. J. N. Steinmann, P. M. F. Elshout, G. Stam, F. Verones, M. Vieira, M. Zijp, A. Hollander, and R. Zelm. 2017. ReCiPe2016: A harmonised life cycle impact assessment method at midpoint and endpoint level. *The International Journal of Life Cycle Assessment*, 22: 138–147. doi:10.1007/s11367-016-1246-y.

Ilgin, Mehmet Ali, and Surendra M. Gupta. 2010. Environmentally conscious manufacturing and product recovery (ECMPRO): A review of the state of the art. *Journal of Environmental Management*, 91(3): 563–591. ISSN 0301–4797. https://doi.org/10.1016/j.jenvman.2009.09.037.

International Energy Agency, Sustainable Recovery. 2021a. Global primary energy demand growth by scenario 2019–2030. www.iea.org/data-and-statistics/charts/global-primary-energy-demand-growth-by-scenario-2019-2030 (Accessed on 10.05.2021).

International Energy Agency, Sustainable Recovery. 2021b. www.iea.org/reports/sustainable-recovery (Accessed on 10.05.2021).

ISO/TC 207/SC 5. 2006a. ISO 14040:2006 Environmental management Life cycle assessment principles and framework. *International Organization for Standardization*. www.iso.org/obp/ui/#iso:std:iso:14040:ed-2:v1:en (Accessed on 03.2021).

ISO/TC 207/SC 5. 2006b. ISO 14044:2006 Environmental management Life cycle assessment requirements and guidelines. *International Organization for Standardization*. www.iso.org/standard/38498.html (Accessed on 03.2021).

Ita-Nagy, D., I. Vázquez-Rowe, R. Kahhat, I. Quispe, G. Chinga-Carrasco, N. M. Clauser, and M. C. Area. 2020. Life cycle assessment of bagasse fiber reinforced biocomposites. *Science of the Total Environment*, 720: 137586. doi:10.1016/j.scitotenv.2020.137586.

Jadhav, D., B. N. Rekha, P. R. Gogate, and V. K. Rathod. 2009. Extraction of vanillin from vanilla pods: A comparison study of conventional soxhlet and ultrasound assisted extraction. *Journal of Food Engineering*, 93: 421–426. https://doi.org/10.1016/j.jfoodeng.2009.02.007.

Jiang, Peng, Yee Van Fan, and Jiří Jaromír Klemeš. 2021. Impacts of COVID-19 on energy demand and consumption: Challenges, lessons and emerging opportunities. *Applied Energy*, 285: 116441. ISSN 0306–2619. https://doi.org/10.1016/j.apenergy.2021.116441.

Jiang, Yong, Harold D. May, Lu Lu, Peng Liang, Xia Huang, and Zhiyong Jason Ren. 2019. Carbon dioxide and organic waste valorization by microbial electrosynthesis and electro-fermentation. *Water Research*, 149: 42–55. ISSN 0043–1354. https://doi.org/10.1016/j.watres.2018.10.092.

J.R.J. Paré, J.M.R. Belanger, S.S. Stafford, A new tool for the analytical laboratory, *Trends in Analytical Chemistry*, 13, pp. 176–184, 1994.

Karthick, C., and K. Nanthagopal. 2021. A comprehensive review on ecological approaches of waste to wealth strategies for production of sustainable biobutanol and its suitability in automotive applications. *Energy Conversion and Management*, 239: 114219. ISSN 0196–8904. https://doi.org/10.1016/j.enconman.2021.114219.

Kaur, C., and B. Amor. 2017. Recent developments, future challenges and new research directions in LCA of buildings: Acritical review. *Renewable and Sustainable Energy Reviews*, 67: 408–416. doi:10.1016/j.rser.2016.09.058.

Kaur, Jaskiran, Gini Rani, and K. N. Yogalakshmi. 2020. Problems and issues of food waste-based biorefineries. *Food Waste to Valuable Resources*: 343–357. INC. https://doi.org/10.1016/b978-0-12-818353-3.00016-x.

Khoo, Choon Gek, Yaleeni Kanna Dasan, Man Kee Lam, and Keat Teong Lee. 2019. Algae biorefinery: Review on a broad spectrum of downstream processes and products. *Bioresource Technology*, 292 (June): 121964. https://doi.org/10.1016/j.biortech.2019.121964.

Klemeš, Jiří Jaromír, Yee Van Fan, Raymond R. Tan, and Peng Jiang. 2020. Minimising the present and future plastic waste, energy and environmental footprints related to COVID-19. *Renewable and Sustainable Energy Reviews*, 127: 109883. ISSN 1364–0321. https://doi.org/10.1016/j.rser.2020.109883.

Kudakasseril Kurian, Jiby, Gopu Raveendran Nair, Abid Hussain, and G. S. Vijaya Raghavan. 2013. Feedstocks, logistics and pre-treatment processes for sustainable lignocellulosic biorefineries: A comprehensive review. *Renewable and Sustainable Energy Reviews*, 25: 205–219. https://doi.org/10.1016/j.rser.2013.04.019.

Kumar, Manish, Shanta Dutta, Siming You, Gang Luo, Shicheng Zhang, Pau Loke Show, Ankush D. Sawarkar, Lal Singh, and Daniel C. W. Tsang. 2021. A critical review on biochar for enhancing biogas production from anaerobic digestion of food waste and sludge. *Journal of Cleaner Production*, 305: 127143. ISSN 0959–6526. https://doi.org/10.1016/j.jclepro.2021.127143.

Kurniawan, Tonni Agustiono, Ram Avtar, Deepak Singh, Wenchao Xue, Mohd Hafiz Dzarfan Othman, Goh Hui Hwang, Iswanto Iswanto, Ahmad B. Albadarin, and Axel Olaf Kern. 2021. Reforming MSWM in Sukunan (Yogjakarta, Indonesia): A case-study of applying a zero-waste approach based on circular economy paradigm. *Journal of Cleaner Production*, 284: 124775. ISSN 0959–6526. https://doi.org/10.1016/j.jclepro.2020.124775.

Latiff, N. A. 2015. Fractionation and characterization of polyphenols rich extract from Labisia pumila Leaves (Doctoral dissertation). Universiti Teknologi Malaysia.

Lee, K. M., and A. Inaba. 2004. Life cycle assessment: Best practices of ISO 14040 series. www.apec.org › APEC › 04_cti_scsc_lca_rev (Accessed on 03.2021).

Liew, Weng Hui, Mimi H. Hassim, and Denny K. S. Ng. 2014. Review of evolution, technology and sustainability assessments of biofuel production. *Journal of Cleaner Production*, 71: 11–29. https://doi.org/10.1016/j.jclepro.2014.01.006.

Lin, Long, Fuqing Xu, Xumeng Ge, Yebo Li. 2018. Improving the sustainability of organic waste management practices in the food-energy-water nexus: A comparative review of anaerobic digestion and composting. *Renewable and Sustainable Energy Reviews*, 89, 151–167. ISSN 1364–0321. https://doi.org/10.1016/j.rser.2018.03.025.

Liu, Huimin, Shiyi Qin, Ranjna Sirohi, Vivek Ahluwalia, Yuwen Zhou, Raveendran Sindhu, Parameswaran Binod, Reeta Rani Singhnia, Anil Kumar Patel, Ankita Juneja, Deepak Kumar, Zengqiang Zhang, Jitendra Kumar, Mohammad J. Taherzadeh, and Mukesh Kumar Awasthi. 2021. Sustainable blueberry waste recycling towards biorefinery strategy and circular bioeconomy: A review. *Bioresource Technology*, 332: 125181. ISSN 0960–8524. https://doi.org/10.1016/j.biortech.2021.125181.

Llompart, M., M. Celeiro, C. Garcia-Jares, and T. Dagnac. 2017. Microwave-assisted extraction of pesticides and emerging pollutants in the environment. *Comprehensive Analytical Chemistry*: 131–201. https://doi.org/10.1016/bs.coac.2017.01.004.

Mason, T.J., L. Paniwnyk, and J.P. Lorimer. 1996. The uses of ultrasound in food technology. *Ultrasonics Sonochemistry*, 3(3): S253–S260, ISSN 1350-4177, https://doi.org/10.1016/S1350-4177(96)00034-X.

Maity, Sunil K. 2015. Opportunities, recent trends and challenges of integrated biorefinery: Part I. *Renewable and Sustainable Energy Reviews*, 43: 1427–1445. https://doi.org/10.1016/j.rser.2014.11.092.

Melikoglu, Mehmet. 2020. Reutilisation of food wastes for generating fuels and value added products: A global review. *Environmental Technology & Innovation*, 19: 101040. ISSN 2352–1864. https://doi.org/10.1016/j.eti.2020.101040.

Mendieta, O., L. Castro, H. Escalante, and M. Garfí. 2021. Low-cost anaerobic digester to promote the circular bioeconomy in the non-centrifugal cane sugar sector: A life cycle assessment. *Bioresource Technology*, 326: 124783. doi:10.1016/j.biortech.2021.124783.

Millati, Ria, Rochim Bakti Cahyono, Teguh Ariyanto, Istna Nafi Azzahrani, Rininta Utami Putri, and Mohammad J. Taherzadeh. 2019. Agricultural, Industrial, municipal, and forest wastes: An overview. *Sustainable Resource Recovery and Zero Waste Approaches*. Elsevier B. V. https://doi.org/10.1016/B978-0-444-64200-4.00001-3.

Monsiváis-Alonso, R., S. Soheil-Mansouri, and A. Román-Martínez. 2020. Life cycle assessment of intensified processes towards circular economy: Omega-3 production from waste fish oil. *Chemical Engineering and Processing*, 158: 108171. doi:10.1016/j.cep.2020.108171.

Montalvo, S., C. Huiliñir, R. Borja, E. Sánchez, and C. Herrmann. 2020. Application of zeolites for biological treatment processes of solid wastes and wastewaters—A review. *Bioresource Technology*, 301: 122808. ISSN 0960–8524. https://doi.org/10.1016/j.biortech.2020.122808.

Mukherjee, C., J. Denney, E. G. Mbonimpa, J. Slagley, and R. Bhowmik. 2020. A review on municipal solid waste-to-energy trends in the USA. *Renewable and Sustainable Energy Reviews*, 119: 109512. ISSN 1364–0321. https://doi.org/10.1016/j.rser.2019.109512.

Munir, M. Tajammal, Seyed Soheil Mansouri, Isuru A. Udugama, Saeid Baroutian, Krist V. Gernaey, and Brent R. Young. 2018. Resource recovery from organic solid waste using hydrothermal processing: Opportunities and challenges. *Renewable and Sustainable Energy Reviews*, 96: 64–75. ISSN 1364–0321. https://doi.org/10.1016/j.rser.2018.07.039.

Negro, María José, Paloma Manzanares, Encarnación Ruiz, Eulogio Castro, and Mercedes Ballesteros. 2017. The biorefinery concept for the industrial valorization of residues from olive oil industry. Olive mill waste: Recent advances for sustainable management. Elsevier Inc. https://doi.org/10.1016/B978-0-12-805314-0.00003-0.

Ng, Denny K. S., Kok Siew Ng, and Rex T. L. Ng. 2017. Integrated biorefineries. *Encyclopedia of Sustainable Technologies*, 4. Elsevier. https://doi.org/10.1016/B978-0-12-409548-9.10138-1.

Nicola, Maria, Zaid Alsafi, Catrin Sohrabi, Ahmed Kerwan, Ahmed Al-Jabir, Christos Iosifidis, Maliha Agha, and Riaz Agha. 2020. The socio-economic implications of the coronavirus pandemic (COVID-19): A review. *International Journal of Surgery*, 78: 185–193. ISSN 1743–9191. https://doi.org/10.1016/j.ijsu.2020.04.018.

Osman, Ahmed I., Adel Abdelkader, Charlie Farrell, David Rooney, and Kevin Morgan. 2019. Reusing, recycling and up-cycling of biomass: A review of practical and kinetic modelling approaches. *Fuel Processing Technology*, 192 (April): 179–202. https://doi.org/10.1016/j.fuproc.2019.04.026.

Panigrahi, Sagarika, and Brajesh K. Dubey. 2019. A critical review on operating parameters and strategies to improve the biogas yield from anaerobic digestion of organic fraction of municipal solid waste. *Renewable Energy*, 143: 779–797. ISSN 0960–1481. https://doi.org/10.1016/j.renene.2019.05.040.

Plaza, M., and C. Turner. 2017. Pressurized hot water extraction of bioactives. *TrAC Trends in Analytical Chemistry*: 53–82. https://doi.org/10.1016/bs.coac.2016.12.005.

Pleissner, Daniel, and Birgit A. Rumpold. 2018. Utilization of organic residues using heterotrophic microalgae and insects. *Waste Management*, 72: 227–239. https://doi.org/10.1016/j.wasman.2017.11.020.

PRé-Sustainability. 2020. SimaPro database manual, Methods library. https://simapro.com/wp-content/uploads/2020/10/DatabaseManualMethods.pdf (Accessed on 03.2021).

Pujara, Yash, Pankaj Pathak, Archana Sharma, and Janki Govani. 2019. Review on Indian municipal solid waste management practices for reduction of environmental impacts to achieve sustainable development goals. *Journal of Environmental Management*, 248: 109238. ISSN 0301–4797. https://doi.org/10.1016/j.jenvman.2019.07.009.

Ravetti Duran, R., P. Escudero Falsetti, L. Muhr, R. Privat, and D. Barth. 2018. Phase equilibrium study of the ternary system CO2. H2O. ethanol at elevated pressure: Thermodynamic model selection. Application to supercritical extraction of polar compounds. *The Journal of Supercritical Fluids*, 138: 17–28. https://doi.org/10.1016/j.supflu.2018.03.016.

Rehman, Kashif ur, Minmin Cai, Xiaopeng Xiao, Longyu Zheng, Hui Wang, Abdul Aziz Soomro, Yusha Zhou, Wu Li, Ziniu Yu, and Jibin Zhang. 2017. Cellulose decomposition and larval biomass production from the co-digestion of dairy manure and chicken manure by Mini-Livestock (Hermetia Illucens L.). *Journal of Environmental Management*, 196: 458–465. https://doi.org/10.1016/j.jenvman.2017.03.047.

Rehman, M. U., F. Khan, and K. Niaz. 2020. Introduction to natural products analysis. *Recent Advances in Natural Products Analysis*: 3–15. Elsevier. https://doi.org/10.1016/B978-0-12-816455-6.00001-9.

Rodrigues, Luiz Gustavo G., Simone Mazzutti, Ilyas Siddique, Mayara da Silva, Luciano Vitali, Sandra Regina Salvador Ferreira. 2020. Subcritical water extraction and microwave-assisted extraction applied for the recovery of bioactive components from Chaya (Cnidoscolus aconitifolius Mill.). *The Journal of Supercritical Fluids*, 165: 104976.

SaraJliJa, H., N. Cˇ ukelJ, G. N. D. Mršic´, M. Brncˇic´, and D. C´ uric´ D. 2012. Preparation of flaxseed for lignan determination by gas chromatography-mass spectrometry method. *Czech Journal of Food Sciences*, 30(1): 45–52. https://doi.org/10.17221/107/2010-CJFS.

Soria, A. C., and M. Villamiel. 2010. Effect of ultrasound on the technological properties and bioactivity of food: A review. *Trends in Food Science and Technology*, 21: 323–331. https://doi.org/10.1016/j.tifs.2010.04.003.

Soxhlet, F. 1879. Die gewichtsanalytische Bestimmung des Milchfettes. *Dinglers Polytechnisches Journal*, 232: 461–465.

Stocker, T. 2014. *Climate Change 2013: The Physical Science Basis: Working Group I Contribution to the Fifth Assessment Report of the Intergovernmental Panel on Climate Change*, Cambridge University Press.

Sun, J., W. Wang, and Q. Yue. 2016. Review on microwave-matter interaction fundamentals and efficient microwave-associated heating strategies. *Materials*, 9: 231. https://doi.org/10.3390/ma9040231.

Surendra, K. C., Jeffery K. Tomberlin, Arnold van Huis, Jonathan A. Cammack, Lars Henrik L. Heckmann, and Samir Kumar Khanal. 2020. Rethinking organic wastes bioconversion: Evaluating the potential of the black soldier fly (Hermetia Illucens (L.)) (Diptera: Stratiomyidae) (BSF). *Waste Management*, 117: 58–80. https://doi.org/10.1016/j.wasman.2020.07.050.

Tauro, Raúl, Carlos A. García, Margaret Skutsch, and Omar Masera. 2018. The potential for sustainable biomass pellets in Mexico: An analysis of energy potential, logistic costs and market demand. *Renewable and Sustainable Energy Reviews*, 82 (September 2017): 380–389. https://doi.org/10.1016/j.rser.2017.09.036.

Thürer, Matthias, Ivan Tomašević, and Mark Stevenson. 2017. On the meaning of 'waste': Review and definition. *Production Planning & Control*, 28(3): 244–255. doi:10.1080/09537287.2016.1264640.

Usmani, Zeba, Minaxi Sharma, Abhishek Kumar Awasthi, Nallusamy Sivakumar, Tiit Lukk, Lorenzo Pecoraro, Vijay Kumar Thakur, Dave Roberts, John Newbold, and Vijai Kumar Gupta. 2021. Bioprocessing of waste biomass for sustainable product development and minimizing environmental impact. *Bioresource Technology*, 322: 124548. ISSN 0960–8524. https://doi.org/10.1016/j.biortech.2020.124548.

Vila Verde, G. M., D. A. Barros, M. Oliveira, G. Aquino, D. M. Santos, J. de Paula, L. Dias, M. Piñeiro, and M. M. Pereira. 2018. A green protocol for microwave-assisted extraction of volatile oil terpenes from pterodon emarginatus vogel. (Fabaceae). *Molecules*, 23: 651. https://doi.org/10.3390/molecules23030651.

Vinatoru, M. 2001. An overview of the ultrasonically assisted extraction of bioactive principles from herbs. *Ultrasonics Sonochemistry*, 8: 303–313. https://doi.org/10.1016/S1350-4177(01)00071-2.

Wang, Dong-Geng, Wen-Ying Liu, Guang-Tong Chen. 2013. A simple method for the isolation and purification of resveratrol from Polygonum cuspidatum. *Journal of Pharmaceutical Analysis*, 3(4): 241–247, ISSN 2095-1779, https://doi.org/10.1016/j.jpha.2012.12.001.

Wernet, G., C. Bauer, B. Steubing, J. Reinhard, E. Moreno-Ruiz, and B. Weidema. 2016. The ecoinvent database version 3 (part I): Overview and methodology. *The International Journal of Life Cycle Assessment*, 21: 1218–1230. doi:10.1007/s11367-016-1087-8.

World Bank. 2021a. Growth Domestic Product at current values, DataBank. https://data.worldbank.org/indicator/NY.GDP.MKTP.CD (Accessed on 09.05.2021).

World Bank. 2021b. Trends in solid waste management, DataBank. https://datatopics.worldbank.org/what-a-waste/trends_in_solid_waste_management.html#:~:text=The%20world%20generates%202.01%20billion,from%200.11%20to%204.54%20kilograms. (Accessed on 09.05.2021).

Wowra, K., V. Zeller, and L. Schebek. 2020. Nitrogen in life cycle assessment (LCA) of agricultural crop production systems: Comparative analysis of regionalization approaches. *Science of the Total Environment*, 763: 143009. doi:10.1016/j.scitotenv.2020.143009.

Xiaoxuan, Kou, Zhu Hong, Liu Lijun, and Yue Jibo. 2009. Chapter 3 — The relationship between economic globalization and higher education internationalization in China**-This chapter is part of a project supported by Tianjin education science research programming project (No. ZZG100). Editor(s): Parikshit K. Basu and Yapa M. W. Y. Bandara, *Chandos Asian Studies Series, WTO Accession and Socio-Economic Development in China*, Chandos Publishing, 43–58. ISBN 9781843345473. https://doi.org/10.1016/B978-1-84334-547-3.50003-5.

Yeop, A., J. Sandanasam, S. F. Pan, S. Abdulla, M. M. Yusoff, and J. Gimbun. 2017. The effect of particle size and solvent type on the gallic acid yield obtained from Labisia pumila by ultrasonic extraction. *MATEC Web of Conferences*, 111: 02008. EDP Sciences. https://doi.org/10.1051/matecconf/201711102008.

Index

For Product Safety Concerns and Information please contact our EU
representative GPSR@taylorandfrancis.com
Taylor & Francis Verlag GmbH, Kaufingerstraße 24, 80331 München, Germany

www.ingramcontent.com/pod-product-compliance
Lightning Source LLC
Chambersburg PA
CBHW060758220326
41598CB00022B/2474

* 9 7 8 1 0 3 2 0 4 2 7 4 9 *